Intuitive Calculus

WITH COLLEGE ALGEBRA

Intuitive Calculus

WITH COLLEGE ALGEBRA

for Biological, Management, and Social Sciences

LOHMAN

POWELL

SHOOP

WALKER
KENT STATE UNIVERSITY

 D. VAN NOSTRAND COMPANY

New York   Cincinnati   Toronto   London   Melbourne

D. Van Nostrand Company Regional Offices:
New York   Cincinnati   Milbrae

D. Van Nostrand Company International Offices:
London   Toronto   Melbourne

Library of Congress Catalog Card Number 72-11661
ISBN: 0-442-26622-7

Published by D. Van Nostrand Company
450 West 33rd Street, New York, N.Y. 10001

Published simultaneously in Canada by
Van Nostrand Reinhold Ltd.

10 9 8 7 6 5 4 3 2 1

Preface

This text is intended to provide the necessary background in differential and integral calculus of functions of one variable as a basic tool in such fields as the social sciences, management, biology, and the behavioral sciences.

The material is presented in a purposely informal manner, with proofs of theorems frequently omitted. Instead, the concepts appearing in definitions and in the statements of theorems are illustrated through both mathematical and nonmathematical examples. Students who wish to continue their study of calculus beyond this text will have little difficulty in making the transition to other standard works. The text is self-contained and is organized in a continuous manner with few extraneous sections.

Each section contains many completely solved examples for both motivational and explanatory purposes. The practice of stating a problem and following the statement by a detailed pursuit of its solution is followed. The exercise sets at the end of each section constitute a vital part of the text. Each set consists of two parts separated by three stars (***). Problems in the first part are designed to lead the student through the important methods of the preceding section and as such should be considered an essential portion of each assignment. Problems in the second part are designed to further solidify the skills just learned and occasionally to anticipate material to be presented in later sections. Problems preceded by a star (*) usually call for a greater degree of understanding and intuition on the student's part. Answers to all exercises in the first part and to selected exercises in the second part are provided in an appendix.

Throughout the text the propositional statement of the form "If . . . , then . . ." which appears in definitions and theorems is emphasized by capitalizing the words "IF" and "THEN." In this way the hypothesis and the conclusion of a formal statement can readily be discerned.

The organization of the book involves three principal segments: Chapters 1-5 constitute basic college algebra; Chapters 6-8 deal with the calculus of functions of one variable; and Chapter 9 provides an introduction to linear algebra and matrices. Consistent with the continuous exposition of topics mentioned earlier, it is suggested that Chapters 1-8 be presented in their natural ordering. If the instructor wishes, he may present Chapter 9 immediately after Chapter 2, provided he introduces the Σ-notation found in section 8.1. The authors have learned, however, that the additional mathematical maturity obtained from the study of Chapters 3-8 serves to facilitate the coverage of the slightly more difficult material in Chapter 9.

The opinion of the authors is that verbal applications of mathematical constructs "to the real world" should be nontrivial (not solvable by methods more elementary than those being illustrated) and eclectic. Indeed, since the text is designed to appeal to students in a variety of fields, an attempt has been made to include verbal problems from as many disciplinary areas as possible. It is hoped that the motivational factor achieved by the verbal problems will not only convince the student that mathematics is useful in his own field but also appeal to his imagination and challenge him to apply the material spontaneously to problems he will encounter later in his studies.

We would like to thank the students and instructors at Kent State University who, during the academic years of 1970-1971 and 1971-1972, provided us with invaluable criticism, encouragement, and feedback regarding the original and the revised manuscripts. To Mrs. Darlene May we extend our deep appreciation for her patience, skill, and endurance in typing both the original and revised manuscripts. To our departmental chairman, Professor Richard K. Brown, we express our sincerest appreciation for his encouragement and assistance in implementing this project.

Contents

1 | *Sets and the Real Number System*

Practically all fields of human endeavor have a language all their own. The nuclear physicist speaks of neutrons and fission and the biologist speaks of cells and myosis. The economist speaks of gross national product and inflation and the chemist speaks of hydrocarbons and reaction. In other words, special names are given to the particular objects, operations, and processes encountered in these fields. Mathematics, too, has a language of its own which describes the objects (integers, rational numbers, sets, etc.) and operations (addition, multiplication, union, etc.) encountered in this field.

Many fields also have written languages of their own. Consider the symbolic statement $Na + Cl \longrightarrow NaCl$. When a chemist writes this he actually means: "one molecule of sodium combined with one molecule of chlorine yield one molecule of sodium chloride." The symbolic statement is just a shorter, more convenient way of writing out the longer, more cumbersome statement. The same is true of mathematical notation. It is used to denote the objects we study in mathematics and to allow us to write statements concisely.

In this chapter we begin the study of some of the written and spoken language of mathematics. At the same time, we examine those properties of the real number system which will be used in the remainder of the text.

1.1. SETS

We will frequently find it necessary to talk about collections of objects such as certain collections of numbers. A collection of objects of any type will be called a *set*. Some examples of sets are the following:

(1) The set of students at Kent State University,

(2) The set consisting of the letters d, T, and a,

(3) The set of women over twenty and under forty,

(4) The set of men who are accountants or lawyers,

(5) The set consisting of the numbers 4, 1, 7, and 16,

(6) The set of children who do not attend kindergarten, and

(7) The set consisting of Mary, America, 17, apple, and pizza.

If an object belongs to a set then the object is an *element* or *member* of the set. Thus,

(1) the letter a is an element of the set consisting of the letters d, T, and a.

(2) Mary is an element of the set consisting of Mary, America, and 17, apple, and pizza, and

(3) the number 1 is an element of the set consisting of 4, 1, 7, and 16.

So far, our methods of describing sets are rather long-winded. There are two acceptable methods of describing sets in a more concise manner. These two methods are the *listing method* and the *rule method*, which we now describe.

The listing method is exactly what its name describes. We list, between braces, the elements of the set and agree that the resulting symbol represents the set whose elements are precisely those between the braces. For example,

(1) $\{d, T, a\}$ represents the set consisting of the letters d, T, and a,

(2) $\{4, 1, 7, 16\}$ represents the set consisting of the numbers 4, 1, 7, and 16, and,

(3) $\{Mary, America, 17, apple, pizza\}$ represents the set consisting of Mary, America, 17, apple, and pizza.

The listing method, however, has a serious drawback. Consider the set consisting of the names of all living human beings. In order to describe this set using the listing method, we would have to list the name of each living human being between braces. This, of course, is very impractical because of the extremely large number of objects appearing in such a list. Therefore, it appears that, in order to describe this kind of set with much success, we need another method of description.

This brings us to the rule method. It is a second method used to describe sets. We place a certain statement between braces. We then agree that the resulting symbol represents the set of objects for which the statement is true. For example, the symbol

$$\{x: \ x \text{ is a living human being}\},$$

read as "the set of all x such that x is a living human being," represents the set of all living human beings. The letter x appearing in the symbol

$$\{x: \ x \text{ is a living human being}\}$$

need not be confusing. It should be thought of as a placeholder. Thus, given an object o, we determine whether or not the statement "*o* is a living human being"

is true. If it is, then *o* belongs to the set so described. For example, if *o* is the living object John Jones, then "John Jones is a living human being" is a true statement. Thus John Jones belongs to the set. On the other hand, if *o* is the object guitar, then "guitar is a living human being" is not a true statement. Consequently, guitar is not an element of the set

$$\{x: \ x \text{ is a living human being}\}.$$

Since the letter *x* is only a placeholder we could have used any other letter in its place. This means, for instance, that

$$\{y: \ y \text{ is a living human being}\}$$

also represents the set of all living human beings. The following are other examples of sets described by the rule method:

(1) The set of all students at Kent State University can be represented by
$$\{x: \ x \text{ is a student at Kent State University}\},$$
(2) The set of all accountants or lawyers can be represented by
$$\{y: \ y \text{ is an accountant or } y \text{ is a lawyer}\},$$
(3) The set of all women over twenty and under forty can be represented by
$$\{z: \ z \text{ is a woman, } z \text{ is over twenty, and } z \text{ is under forty}\},$$
(4) The set of all adults who did not attend kindergarten can be represented by
$$\{n: \ n \text{ is an adult and } n \text{ did not attend kindergarten}\}.$$

In using the rule method to describe a set, there may be no object which satisfies the statement which qualifies objects for membership in the set. For instance, no object *x* satisfies the statement

"*x* is a man who is 19 feet tall."

This means that the set

$$\{x: \ x \text{ is a man and } x \text{ is 19 feet tall}\}$$

contains no elements. We call this set the *empty set*. In describing such a set by the listing method, we would place all its elements (there are none!) between braces. Hence we use the symbol { } to represent the empty set.

Since we are interested in the elements of a set, we want to have notation which indicates that an object is an element of a set or that it is not an element of a set. If A is a set, we write $x \in A$ to indicate that the object x is an element of A. If x is not an element of A we write $x \notin A$. The following examples illustrate the use of this notation.

(1) $d \in \{d, T, a\}, 4 \notin \{d, T, a\}, a \in \{d, T, a\}$,
$h \notin \{d, T, a\}, T \in \{d, T, a\}$
(2) John Doe $\notin \{w: \ w \text{ is a woman}\}$,

(3) John Doe $\in$ {x: x is a human being and x is not a woman}, and

(4) Mrs. Jones $\notin$ {y: y is a human being and y is not a woman}.

Since a set can have, for its elements, other sets we can have {a} $\in$ {{a}, d, T}. Here, the set on the right-hand side has as its elements {a}, d, and T. Thus

$$a \notin \{\{a\}, d, T\}, \quad \{d\} \notin \{\{a\}, d, T\}, \quad d \in \{\{a\}, d, T\}.$$

Exercise Set 1.1

1. Determine whether each of the following statements is true or false:
   a. The letter q is an element of the set consisting of p, q, and r.
   b. The number one is an element of the set consisting of the numbers eleven, twelve, thirteen, and fourteen.
   c. The number 0 is an element of the empty set.
   d. Multiplication is an element of the set consisting of addition, subtraction, multiplication, and division.
   e. $s \in \{p, q, r\}$
   f. $1 \notin \{11, 12, 13, 14\}$
   g. $0 \in \{\ \}$
   h. $+ \in \{+, -, \cdot\}$

2. Determine whether each of the following statements is true or false:
   a. Ohio is an element of the set consisting of all the states in the United States.
   b. San Francisco is an element of the set consisting of all the states in the United States.
   c. The number 2 is an element of the set consisting of all integers which are not multiples of 5.
   d. Wheat is an element of the set consisting of all grains.
   e. Alaska $\in$ {x: x is a state in the United States}.
   f. San Francisco $\in$ {y: y is a city in the United States}.

3. Describe the following sets using the listing method:
   a. The set consisting of Bob, Carol, Ted, and Alice.
   b. The set consisting of the days in a week.
   c. The set consisting of the number of days in a week.
   d. The set consisting of those states of which Philadelphia is the capital city.
   e. The set consisting of 3, cat, ?, and T.
   f. The set consisting of the pairs of numbers which can appear on a throw of one green die and one red die.

4. Describe the following sets using the rule method:
   a. The set of all living men.
   b. The set of all doctors who are surgeons.
   c. The set {a, e, i, o, u}.
   d. The set of all even integers.

e. The set of all students who do not hold a scholarship.
f. The empty set.

* * *

5. Determine whether each of the following statements is true or false:
   a. $a \in \{a\}$
   b. $\{1, 6, g\} \in \{1, 6, g\}$
   c. $\{a\} \in \{a\}$
   d. $\{1, 6, g\} \in \{\{1, 6, g\}\}$
   e. $\{ \ \} \in \{a, b, c, d\}$
   f. $5 \notin \{x: \ x$ is a number and $x \neq 14\}$

6. Use set notation to write the following statements as concisely as possible:
   a. 5 is an element of the set consisting of 4, 5, and John Doe.
   b. France is an element of the set of all countries.
   c. 3 is not an element of the set consisting of $\$$, a, and Hollywood.
   d. Lake Erie is not an element of the set of all unpolluted bodies of water.
   e. The number twelve is an element of the set which consists of the number twelve.
   f. Putter is an element of a set consisting of golf clubs.
   g. The set consisting of a, g, h is not an element of the set of all letters of the alphabet.
   h. The United States Senate is not an element of the set consisting of all United States senators.

1.2 SUBSETS

Let L denote the set of all living human beings and let M denote the set of all living male human beings. It is then clear that L and M are comparable in the following sense: every living male human being is a living human being. Likewise, if C denotes the set of all civil service employees and P denotes the set of all mail carriers, then C and P are comparable in a similar manner. Namely, every mail carrier is a civil service employee. Since this will occur frequently in our work, we would like a shorthand notation to indicate when sets are comparable, as in the preceding examples.

1.1 Definition. *Let A and B be sets. The set B is a **subset** of the set A (or B is contained in A) provided each element of B is an element of A. In this case write $B \subseteq A$.*

Using the sets L, M, C, and P, we have that M is a subset of L. Therefore, we write $M \subseteq L$. Similarly, $P \subseteq C$. Other examples are
   (1) $\{2, 4\} \subseteq \{1, 2, 3, 4\}$,
   (2) $\{h, a, d, j, b\} \subseteq \{j, l, k, p, 4, \text{Mary}, a, h, 16, b, j, d\}$,

(3) $\{x\colon x$ is a student$\} \subseteq \{y\colon y$ is a person$\}$,
(4) $\{$New York$\} \subseteq \{3,$ Jack, $c,$ New York, $1, 10, z, b\}$,
(5) $\{a\colon a$ is either a plumber or a carpenter$\} \subseteq \{b\colon b$ is a skilled worker$\}$,
(6) $\{y\colon y$ is not a college student$\}$
  $\subseteq \{z\colon z$ is not a freshman college student$\}$,
(7) $\{z\colon z$ is a plumber and z is a homeowner$\} \subseteq \{w\colon w$ is a plumber$\}$, and
(8) $\{n\colon n$ is a doctor$\} \subseteq \{m\colon m$ is a doctor or m is a lawyer$\}$.

We can visualize the notion of one set being a subset of another set quite eas-ily. In Figure 1.1, let the points in the darkly shaded region represent the ele-ments of the set B and let the points in the shaded region represent the elements of the set A. The statement $B \subseteq A$ means that every element of B is an element of A. Pictorially, this means that every point in the darkly shaded region must also lie in the shaded region.

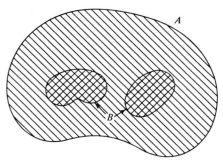

Figure 1.1

Given two sets A and B, it is not necessarily true that B is a subset of A or vice versa. According to the definition, B is not a subset of A if there is at least one element of B which is not an element of A. Thus

(1) $\{b, 3, a, k\}$ is not a subset of $\{6, c, a,$ Jack, $3\}$ since $b \in \{b, 3, a, k\}$ but $b \notin \{6, c, a,$ Jack, $3\}$ (note that k has this property also) and
(2) $\{x\colon x$ is an "A" student$\}$ is not a subset of $\{y\colon y$ is a student on a schol-arship$\}$ since there are "A" students who do not hold scholarships.

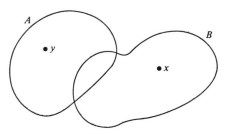

Figure 1.2

Figure 1.2 illustrates a case where B is not a subset of A and A is not a subset of B. The element x is an element of B but not of A and the element y is an element of A but not of B.

To show that B is *not* a subset of A it is sufficient to exhibit one element in B which is not an element of A. However, to show that $B \subseteq A$ it is necessary to demonstrate that *every* element of B is also an element of A.

The following statements are readily shown to be true. We omit their verifications. If A, B, and C are sets then

(1) $A \subseteq A$ $(B \subseteq B$, etc.).

Moreover,

(2) if $A \subseteq B$ and $B \subseteq C$, then $A \subseteq C$.

Notice that statements (1) and (2) are in the form

$$\text{"IF} \ldots, \quad \text{THEN} \ldots \text{."}$$

This is a very common form of every day communication. For example, one might hear "IF the rain stops, THEN we will go on a picnic." That is, the statement "we will go on a picnic" depends directly upon the statement "the rain stops" being satisfied. We call the statement "the rain stops" the *hypothesis* and the statement "we will go on a picnic" the *conclusion*. Thus, when the hypothesis is satisfied the conclusion will follow. Many statements in mathematics take this form also. We will highlight these by capitalizing the words "if" and "then" in the formal statements of theorems and definitions throughout the text.

Exercise Set 1.2

1. For each of the given sets A, use set notation to describe a set B such that $B \subseteq A$ and a set C such that C is not a subset of B.
   a. $A = \{1, 2, 3\}$
   b. $A = \{l, a, t, q, m\}$
   c. $A = \{9\}$
   d. $A = \{x: \ x \text{ is a student}\}$
   e. $\{y: \ y \text{ is a state in the United States}\}$
   f. $\{\triangle, \square, \bigcirc\}$
   g. $\{t: \ t \text{ is an ocean}\}$
   h. $\{p: \ p \text{ is a selling price greater than } \$10\}$

2. For each of the given sets A, use set notation to describe three different subsets of A.
   a. $A = \{1, 2, 3\}$
   b. $A = \{l, a, t, q, m\}$
   c. $A = \{9, 4\}$
   d. $A = \{y: \ y \text{ is a city}\}$
   e. $\{z: \ z \text{ is a person under 18 years age}\}$
   f. $\{\text{Jane, Jean, Joan, Jill}\}$
   g. $\{q: \ q \text{ is a pilot}\}$
   h. $\{b: \ b \text{ is not a professional football player}\}$

* * *

3. Determine whether each of the following statements is true or false:
   a. $\{a\} \subseteq \{a, 6, g\}$
   b. $\{a, 2, 6, g\} \subseteq \{a, 6, g\}$
   c. $\{a, 6, g\} \subseteq \{a, 6, g\}$
   d. $a \subseteq \{1, 9, 7\}$
   e. $a \subseteq \{1, a, 7\}$
   f. $\{ \ \} \subseteq \{a, 6, g\}$
   g. $\{a, 6, g\} \subseteq \{1, a, 7\}$
   h. $\left\{\dfrac{2}{3}, \dfrac{3}{4}, \dfrac{4}{5}\right\} \subseteq \left\{\dfrac{2}{3} + \dfrac{3}{4} + \dfrac{4}{5}\right\}$

4. Determine whether each of the following statements is true or false:
   a. $\{x: \ x \text{ is a college student}\} \subseteq \{y: \ y \text{ is a student}\}$
   b. $\{p: \ p \text{ is an apple pie}\} \subseteq \{p: \ p \text{ is a pecan pie}\}$
   c. $\{x: \ x \text{ is a noun}\} \subseteq \{n: \ n \text{ is a noun}\}$
   d. $\{\text{Ace of Clubs}\} \subseteq \{a: \ a \text{ is a playing card}\}$
   e. $\{\text{United States}\} \subseteq \{x: \ x \text{ is a state in the United States}\}$
   f. $\left\{\pi, \dfrac{\pi}{2}, \dfrac{\pi}{4}\right\} \subseteq \{4\pi, 2\pi, \pi\}$

5. Complete each of the following statements with one of the symbols $\in, \notin, \subseteq$, or "is not a subset of."
   a. $\{3, 6, 10\}$ _____ $\{1, 3, 9, 6, 10, 11\}$
   b. $\{6\}$ _____ $\{1, 3, 9, 6, 10, 11\}$
   c. 6 _____ $\{1, 3, 9, 6, 10, 11\}$
   d. -6 _____ $\{1, 3, 9, 6, 10, 11\}$
   e. $\{ \ \}$ _____ $\{1, 3, 9, 6, 10, 11\}$
   f. $\{10, 3, 90, 6, 100, 11\}$ _____ $\{1, 3, 9, 6, 10, 11\}$
   g. $\{\text{good, bad, small}\}$ _____ $\{t: \ t \text{ is an adjective}\}$
   h. $\{x: \ x \text{ is a ball}\}$ _____ $\{y: \ y \text{ is a football}\}$
   i. $\{A\}$ _____ $\{\{A\}, \{B\}, \{C\}\}$

6. Let A, B, and C be sets. Show that:
   a. $A \subseteq A$
   b. $\{ \ \} \subseteq B$
   c. if $A \subseteq B$ and $B \subseteq C$, then $A \subseteq C$

1.3 EQUALITY OF SETS

In this section we wish to formulate reasonable conditions under which we will say that two sets are equal. Consider the sets

$$\{b, 1, 2, c, \text{Bob}\} \text{ and } \{\text{Bob}, 2, b, 1, c\}.$$

In both cases, the elements belonging to the sets are the same. Only the orders in which the elements are listed are different. But that only means that we specified (i.e., listed) the "same" collection of elements in two different ways. For our purposes, however, the ways in which the membership of the two preceding sets is specified is not important. What is important is the fact that both of these sets consist of the same elements. This means that

$$\{\text{Bob}, 2, b, 1, c\} \subseteq \{b, 1, 2, c, \text{Bob}\}$$

and

$$\{b, 1, 2, c, \text{Bob}\} \subseteq \{\text{Bob}, 2, b, 1, c\}.$$

We use this observation as the criterion for defining equality.

1.2 Definition. *Let A and B be sets. IF $A \subseteq B$ and $B \subseteq A$, THEN A equals B (written $A = B$). IF A does not equal B, THEN write $A \neq B$.*

Using the definition of equality, we see that

(1) $\{b, 1, 2, c, \text{Bob}\} = \{\text{Bob}, 2, b, 1, c\}$

(2) $\{a, 3, \text{Joe}\} = \{a, \text{Joe}, 3\} = \{3, a, \text{Joe}\} = \{3, \text{Joe}, a\}$
$\qquad = \{\text{Joe}, a, 3\} = \{\text{Joe}, 3, a\}$

(3) $\{\#, \$, 2, \text{Mary}\} \neq \{2, \$, \text{Jack}, \text{Mary}, \#\}$.

Note that, in the last example, we have

$$\{\#, \$, 2, \text{Mary}\} \subseteq \{2, \$, \text{Jack}, \text{Mary}, \#\}.$$

The reason the sets are not equal is because $\{2, \$, \text{Jack}, \text{Mary}, \#\}$ is not a subset of $\{\#, \$, 2, \text{Mary}\}$. Thus the condition in the definition, that each set must be a subset of the other in order to say they are equal, is not satisfied. Therefore, according to the definition, the sets are not equal.

The definition of equality of sets illustrates a sometimes confusing property of the listing method for describing sets. Consider the sets $A = \{a, 3, \text{Joe}\}$ and $B = \{3, a, \text{Joe}, a, 3\}$. Are these sets equal? In order to answer this question we must revert back to the definition of equality for sets. In other words, are the statements $A \subseteq B$ and $B \subseteq A$ true statements? Certainly every element of A is an element of B. Therefore $A \subseteq B$. On the other hand, we also see that every element of B is an element of A even though some of the elements of B, namely a and 3, are listed twice. For instance, the fact that the element 3 is listed twice does not mean that we are listing two different elements. Hence $B \subseteq A$ so that $A = B$. In other words, if an element is repeated in describing a set using the listing method, this repetition does not change the set. Naturally, since this is the case, there is no point in listing elements more than once and this practice should be avoided.

Exercise Set 1.3

Given sets A and B, determine if $A \subseteq B$, $B \subseteq A$, $A = B$, or that neither set is a subset of the other.

1. $A = \{\text{December}, \text{Ruth}, 6\}$ $B = \{\text{Ruth}, 6, \text{December}\}$

2. $A = \{a, b, 3, 7, \$\}$ $B = \{3, b, \$, 7\}$

3. $A = \{x : x$ is a letter in the word buffalo$\}$ $B = \{u, o, a, l, f, b\}$

4. $A = \{2, 1, 0, -1, -2\}$ $B = \left\{\dfrac{4}{2}, \dfrac{6}{6}, \dfrac{-6}{3}, \dfrac{3}{-3}, 0\right\}$

5. $A = \{W: \ W \text{ is a bird}\}$ $B = \{z: \ z \text{ is an insect}\}$

6. $A = \{t, r, s\}$ $B = \{\alpha, \beta, \gamma\}$

* * *

7. $A = \{\text{quarter, dime}\}$ $B = \{c: \ c \text{ is a U.S. coin with face value of } 10\text{¢ or } 25\text{¢}\}$

8. $A = \{\theta, \Delta, \square, /\}$ $B = \{\square, \theta, \Delta, X\}$

9. $A = \{p: \ p \text{ is a 50-legged cow}\}$ $B = \{r: \ r \text{ is a triangle with four sides}\}$

10. $A = \{t: \ t \text{ is a 5-sided figure}\}$ $B = \{t: \ t \text{ is a pentagon}\}$

11. $A = \{\text{north, south, east, west}\}$ $B = \{m: \ m \text{ is a direction}\}$

12. $A = \{y: \ y \text{ is a number between 1 and 2}\}$ $B = \{ \ \}$

Determine whether each of the following statements is true or false:

13. $\{\text{December, Ruth, 6}\} = \{\text{Ruth, December, 10, 6}\}$

14. $\{ \ \} = 0$

15. $\{ \ \} = \{0\}$

16. $\{t\} = t$

17. $\{\#, 1, c, 7, \$\} = \{1, \$, \{7\}, \#, c\}$

18. $\{10, 4, \text{Bob}, d\} = \{d, 4, \text{Bob}, \{d\}, 10\}$

19. $\{\text{time}\} = \{\text{clock}\}$

20. $\{\text{Joe}, \{k, 2\}, 10, \text{Chicago}\} = \{10, \text{Joe}, \text{Chicago}, \{2, k\}\}$

21. $\{x: \ x \text{ is a football player and a basketball player}\} = \{y: \ y \text{ is a football player}\}$

22. $\{x: \ x \text{ is a doctor}\} = \{y: \ y \text{ is a doctor}\}$

23. $\{p: \ p \text{ is a professor and an accountant}\} = \{p: \ p \text{ is a professor or an accountant}\}$

24. $\{S: \ S \text{ is not a student or } S \text{ is not a woman}\} = \{u: \ u \text{ is not a student and a woman}\}$

1.4 THE INTEGERS

We recall that the simplest type of counting one learns in grade school is done using the set of positive integers, also called the set of natural numbers. The set of *positive integers* (*natural numbers*) is the set whose elements are numbers 1,

2, 3, etc. This set can be described using the listing method by the symbol $\{1, 2, 3, \ldots\}$, where the three dots after the number 3 indicate that the list continues according to the pattern indicated at the beginning of the list. After our initial exposure to the set of positive integers, we learned how to find the sum $m + n$ and the product $m \cdot n$ of two positive integers m and n, noting that $m + n$ and $m \cdot n$ are again positive integers. For example, $5 + 3 = 8$ and $5 \cdot 3 = 15$.

A second set of numbers to which we were exposed is the set of *integers*. Before describing this set, let us review a few facts. There is a number called zero, denoted by 0, having the property that

$$n + 0 = 0 + n = n$$

for every positive integer n. Also, given a positive integer n, there is a number called the *additive inverse* of n, and denoted by $-n$, such that

$$n + (-n) = (-n) + n = 0.$$

The set

$$\{-n: \; n \text{ is a positive integer}\}$$

is called the set of *negative integers*. Thus the set of negative integers is the set $\{-1, -2, -3, \ldots\}$. The set of integers is the set $\{0, 1, -1, 2, -2, 3, -3, \ldots\}$. Naturally, the sum $m + n$ and the product $m \cdot n$ of two integers m and n are integers.

There is a way of visualizing the set of integers geometrically. The procedure we are about to describe is one which enables us to associate with each integer n a unique point on a given line. We then can think of the points so described as representing the integers matched to them. To begin with, we let L be an arbitrary line. Select any point on the line and associate this point with the integer zero. Call this point 0. We now fix a certain length u and call it unit length. There are only two rays we can form which have 0 as an endpoint. We select one of these and call it the positive ray (or axis). The other is called the negative ray (or axis) as is indicated in Figure 1.3.

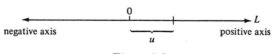

Figure 1.3

Now consider the point on the positive axis a unit length u from 0: call this point 1. Consider a point lying a unit length u in the positive direction from 1: call this point 2. Continuing in this manner, associate with each positive integer n a unique point on L (see Figure 1.4).

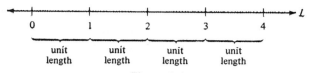

Figure 1.4

Moreover, if n is a positive integer, then the point on L corresponding to n is n units from 0 along the positive axis. For example, the point corresponding to 6 is 6 units from 0 along the positive axis.

Consider now the point on the negative axis a unit length u from 0: call this point -1. Similarly, the point lying a unit length u in the negative direction from -1 is called -2 and so on. We associate with each negative integer m a unique point on L by continuing in this manner (see Figure 1.5).

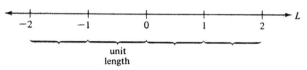

Figure 1.5

If $-m$ is a negative integer, then the point on L corresponding to $-m$ is m units from 0 along the negative axis. For example, the point on L corresponding to -4 is 4 units from 0 along the negative axis.

By this construction, we associate with each integer a unique point on L. Moreover, distinct integers are associated with distinct points on L. As noted before, the integer n associated with a point P on the line L gives the location of P on L. The integer n associated with P is called the *coordinate* of P.

Note that the line L does not necessarily have to be a horizontal line as pictured in Figures 1.3-1.5. Other orientations for L (see Figure 1.6) are allowable.

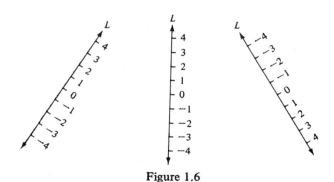

Figure 1.6

In fact, in Chap. 2, we will see that both vertical and horizontal lines of this type are used to construct the rectangular coordinate system.

Let m and n be integers. We say that n *divides* m and call n a *factor* of m in case there is an integer k such that $m = k \cdot n$. Thus

(1) 2 divides 6 and 2 is a factor of 6 because $6 = 3 \cdot 2$,

(2) 3 divides 63 and 3 is a factor of 63 because $63 = 21 \cdot 3$,

(3) 1 divides -7 and 1 is a factor of -7 because $-7 = (-7) \cdot 1$,

(4) 1 divides any integer m and 1 is a factor of m because $m = m \cdot 1$, and

(5) -12 divides 132 and -12 is a factor of 132 because $132 = (-11) \cdot (-12)$.

We see, however, that 3 does not divide 1. This means that there is no integer k such that $1 = k \cdot 3$. Also, 4 does not divide -22 because there is no integer k such that $-22 = k \cdot 4$. Hence, if m and n are integers with $n \neq 0$, there does not necessarily exist a number k in the set of integers such that $m = k \cdot n$. This situation is remedied in the next section by considering the set of rational numbers.

Exercise Set 1.4

1. Which of the following numbers are elements of the set

$$\{x: \ x \text{ is an integer}\}?$$

$$3, \ -3, \ (2)(3), \ 2 + 3, \ 2 - 3, \ \frac{2}{3}, \ (3) \cdot (-3),$$

$$1{,}067{,}582, \ 0, \ -5280, \ 1.4.$$

2. Draw a line L. On L, locate a point and call it 0. Choose a unit length u. On the line that you have drawn, locate the point that is associated with each of the following integers:

$1, \ 5, \ -2, \ 0, \ -1, \ 1 + 3, \ 1 - 3, \ -2 + 5, \ -2 - 5, \text{ and} -2 + 2.$

3. Complete each of the following statements:

a. 3 divides 12 because $12 = $ _____ .

b. -3 divides 12 because $12 = $ _____ .

c. -1 divides 10 because $10 = $ _____ .

d. 4 divides -12 because $-12 = $ _____ .

* * *

4. Given sets A and B, determine if $A \subseteq B$, $B \subseteq A$, $A = B$ or that neither set is a subset of the other.

a. $A = \{x: \ x \text{ is a positive integer}\}$ $B = \{1, 2, 3, \ldots\}$

b. $A = \{0\}$ $B = \{x: \ x \text{ is a positive integer}\}$

c. $A = \{x: \ x \text{ is a positive integer}\}$ $B = \{y: \ y \text{ is a negative integer}\}$

d. $A = \{0\}$ $B = \{z: \ z \text{ is a negative integer}\}$

e. $A = \{1, 2, 3, \ldots\}$ $B = \{y: \ y \text{ is an integer}\}$

f. $A = \{-1, -2, -3, \ldots\}$ $B = \{x: \ x \text{ is an integer}\}$

5. a. Find all the positive factors of: 2, 12, 60, −21, 7, 1.
   b. Find all the factors of: 2, 12, −60, −21, 7, −1.
   c. Find all the factors of 0.

6. The set $\{3n: \ n$ is a positive integer$\}$ describes the set of all positive multiples of 3 by the rule method. This set can also be described by the listing method as $\{3, 6, 9, \ldots\}$. Describe the following sets using both the rule and listing methods:
   a. The set of all positive multiples of 5.
   b. The set of all negative multiples of 5.
   c. The set of all multiples of 5.
   d. The set of all positive multiples of −5.

The following problems concern the arithmetic of integers. The formal properties which must be used are stated, as a review, in Sec. 1.6.

7. Find $1001 + 99$, $1001 - 99$, $99 + 1001$, $99 - 1001$.

8. Find $(5) \cdot (8)$, $(-5) \cdot (8)$, $(5)(-8)$, $(-5)(-8)$.

9. Find $(-4) \cdot (7)$, $(7)(-4)$, $-(4 \cdot 7)$, $-[(-4)(7)]$.

10. Find $(-257)(1)$, $(-1)(-1)$, $(-257)(0)$, $(-257)(1 - 1)$.

11. Complete the following statements:
   $15 + \underline{\hspace{1.5cm}} = 0$, $-15 + \underline{\hspace{1.5cm}} = 0$,
   $8 + \underline{\hspace{1.5cm}} = 0$, $-3000 + \underline{\hspace{1.5cm}} = 0$.

12. Find $(2 + 9) + 17$, $2 + (9 + 17)$, $(2 - 9) + 17$, $2 - (9 + 17)$.

13. Find $(2 \cdot 9) \cdot 5$, $2 \cdot (9 \cdot 5)$, $2 \cdot [(9) \cdot (-5)]$, $(2 \cdot 9) \cdot (-5)$.

14. Find $(2 + 5)9$, $(2)(9) + (5)(9)$, $(2 \cdot 9) + 5$, $2 \cdot (9 + 5)$.

15. If m and n are positive integers, then $(-m)(-n) = \underline{\hspace{2cm}}$.
   If m and n are positive integers, then $(-m) \cdot n = m \cdot (-n) = \underline{\hspace{2cm}}$.

16. Determine which of the following belongs to the set

$$\{n: \ n \text{ is a positive integer}\}:$$

34, -34, $-(-34)$, a, $-a$, $(a) \cdot (a)$.

17. Let L be a horizontal line. Suppose that the integers have been associated with a subset of the set of points of L as described in Sec. 1.4, where the positive axis is to the right. Let P and Q be points on L.
   (a) If the coordinate of P is 4 and Q lies 3 units to the right of P, what is the coordinate of Q?
   (b) If the coordinate of P is 4 and Q lies 3 units to the left of P, what is the coordinate of Q?
   (c) If the coordinate of P is 5 and Q lies 5 units to the left of P, what is the coordinate of Q?
   (d) If the coordinate of P is −10 and Q lies 7 units to the right of P, what is the coordinate of Q?
   (e) If the coordinate of P is −6 and Q lies 11 units to the left of P, what is the coordinate of Q?

1.5 THE RATIONAL NUMBERS

The next set of numbers we consider is the set of *rational numbers* (i.e., commonly called fractions). The set of rational numbers can be described by the rule method as

$$\left\{ \frac{m}{n} : \quad m \text{ and } n \text{ are integers, } n \neq 0 \right\}.$$

If $n = 1$, then we write $\frac{m}{1} = m$. Consequently, the set of integers is a subset of the set of rational numbers.

Two rational numbers $\frac{m}{n}$ and $\frac{m'}{n'}$ are equal if and only if $m \cdot n' = n \cdot m'$. Thus $\frac{6}{8} = \frac{3}{4}$ because $6 \cdot 4 = 24 = 8 \cdot 3$. In general, we can remove any *common divisor* (i.e., *factor*) of both the *numerator* m and the *denominator* n in the fraction $\frac{m}{n}$ and not change the number. Hence if $m = k \cdot m'$ and $n = k \cdot n'$, where k, m', and n' are integers, then $\frac{m}{n} = \frac{m'}{n'}$. Using this definition of equality, we see that

(1) $\dfrac{-16}{6} = \dfrac{2(-8)}{2(3)} = \dfrac{-8}{3}$,

(2) $\dfrac{12}{3} = \dfrac{3(4)}{3(1)} = \dfrac{4}{1} = 4$, and

(3) $\dfrac{364}{42} = \dfrac{(14)(26)}{(14)(3)} = \dfrac{26}{3}$.

This process is frequently referred to as "reducing the fraction to lowest form."

Given two rational numbers we can find their sum and product. Let m, n, m', and n' be integers with n and n' nonzero. The sum of $\frac{m}{n}$ and $\frac{m'}{n'}$, denoted by $\frac{m}{n} + \frac{m'}{n'}$ is obtained by

$$\frac{m}{n} + \frac{m'}{n'} = \frac{m \cdot n' + n \cdot m'}{n \cdot n'}.$$

For example,

$$\frac{3}{4} + \frac{10}{7} = \frac{3 \cdot 7 + 4 \cdot 10}{4 \cdot 7} = \frac{21 + 40}{28} = \frac{61}{28} \text{ and}$$

$$\frac{2}{3} + \frac{4}{5} = \frac{2 \cdot 5 + 4 \cdot 3}{3 \cdot 5} = \frac{10 + 12}{15} = \frac{22}{15}.$$

The product of $\dfrac{m}{n}$ and $\dfrac{m'}{n'}$, denoted by $\dfrac{m}{n} \cdot \dfrac{m'}{n'}$, is obtained by multiplying numerators and denominators. Hence,

$$\frac{m}{n} \cdot \frac{m'}{n'} = \frac{m \cdot m'}{n \cdot n'}.$$

For example,

$$\frac{-12}{35} \cdot \frac{5}{-3} = \frac{(-12) \cdot 5}{35 \cdot (-3)} = \frac{-60}{-105} = \frac{4}{7} \text{ and } \frac{7}{9} \cdot \frac{-2}{5} = \frac{7 \cdot (-2)}{9 \cdot 5} = \frac{-14}{45}.$$

Since the sum and product of integers are integers, the numbers $m \cdot n' + n \cdot m'$, $n \cdot n'$, and $m \cdot m'$ are integers. Hence the sum and product of rational numbers are rational numbers.

Suppose that m and n are nonzero integers. Then $m \cdot n = n \cdot m$ (properties concerning multiplication and addition of numbers are stated more fully in Sec. 1.6). Thus, it follows that $\dfrac{n}{m}$ and $\dfrac{m}{n}$ are rational numbers,

$$\frac{m}{n} \cdot \frac{n}{m} = \frac{m \cdot n}{n \cdot m} = \frac{m \cdot n}{m \cdot n} = 1,$$

and

$$\frac{n}{m} \cdot \frac{m}{n} = \frac{n \cdot m}{m \cdot n} = \frac{n \cdot m}{n \cdot m} = 1.$$

Hence if r is a nonzero rational number, there is a rational number called the *multiplicative inverse* of r, denoted by r^{-1}, such that

$$r \cdot r^{-1} = r^{-1} \cdot r = 1.$$

In fact, we have shown that if $r = \dfrac{m}{n}$, then $r^{-1} = \dfrac{n}{m}$. In particular, if $r = m = \dfrac{m}{1}$, then $r^{-1} = m^{-1} = \dfrac{1}{m}$, provided $m \neq 0$. For example, $\left(\dfrac{3}{4}\right)^{-1} = \dfrac{4}{3}$ and $\left(\dfrac{-7}{2}\right)^{-1} = \dfrac{2}{-7}$.

Let r and s be rational numbers with $s \neq 0$. We write $\dfrac{r}{s}$ to mean $r \cdot s^{-1}$. Thus if $r = \dfrac{m}{n}$ and $s = \dfrac{m'}{n'}$, then

$$\frac{r}{s} = \frac{\dfrac{m}{n}}{\dfrac{m'}{n'}} = \frac{m}{n} \cdot \left(\frac{m'}{n'}\right)^{-1} = \frac{m}{n} \cdot \frac{n'}{m'} = \frac{m \cdot n'}{n \cdot m'}.$$

This, of course, is the familiar "invert and multiply" rule commonly used to divide one fraction by another. Thus,

$$\frac{\frac{3}{4}}{\frac{5}{-7}} = \frac{3}{4} \cdot \frac{-7}{5} = \frac{3 \cdot (-7)}{4 \cdot 5} = \frac{-21}{20}.$$

As in the case of integers there is a simple way of visualizing the set of rational numbers. We proceed as before. Let the line L, the point 0 corresponding to the integer zero, and the positive and negative axes be chosen. Assume, too, that unit length has been chosen and that the points on L corresponding to the integers have been assigned as before. Now let r be any nonzero rational number. Then we can write $r = \frac{m}{n}$, where m and n are integers. Since we also have $r = \frac{-m}{-n}$ and since either n or $-n$ is a positive integer, we can always write $r = \frac{m}{n}$ where n is a positive integer. We now subdivide each line segment between consecutive integers into n segments of equal length. This divides L (*partitions L*) into smaller line segments.

If m is a positive integer, the point on L corresponding to $\frac{m}{n}$ is the endpoint, farthest from 0, of the m-th segment $\left(\text{of length } \frac{1}{n}\right)$ in the positive direction from 0. Therefore, to find the point on L corresponding to the rational number $\frac{8}{3}$, the line segments between consecutive integers are each subdivided into 3 segments of equal length (see Figure 1.7). We let $\frac{8}{3}$ correspond to the endpoint, farthest from 0, of the eighth segment in the positive direction from 0. Naturally the other points indicated in Figure 1.7 correspond to the rational numbers $\frac{1}{3}, \frac{2}{3}, \frac{3}{3} = 1, \frac{4}{3}, \frac{5}{3}, \frac{6}{3} = 2, \frac{7}{3}$, and $\frac{9}{3} = 3$.

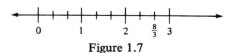

Figure 1.7

Note that to find the point corresponding to $\frac{3}{3}$, each line segment between consecutive integers is subdivided into 3 segments of equal length. The point corresponding to $\frac{3}{3}$ is the endpoint, farthest from 0, of the 3rd segment of length $\frac{1}{3}$ in the positive direction. This is precisely the point corresponding to

1. A similar argument shows, for example, that the points corresponding to $\frac{6}{3}$ and 2, $\frac{4}{6}$ and $\frac{2}{3}$, $\frac{20}{8}$ and $\frac{5}{2}$ are the same.

If m is a positive integer, the point on L corresponding to $\frac{-m}{n}$ is the endpoint, farthest from 0, of the m-th segment $\left(\text{of length } \frac{1}{n}\right)$ in the negative direction from 0. Hence to find the point L corresponding to the rational number $\frac{-7}{6}$, the line segments between consecutive integers are each subdivided into 6 segments of equal length (see Figure 1.8). Thus, $\frac{-7}{6}$ corresponds to the endpoint, farthest from 0, of the seventh segment in the negative direction from 0.

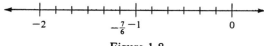

Figure 1.8

The other points indicated in Figure 1.8 correspond to the rational numbers

$$\frac{-1}{6}, \frac{-2}{6} = \frac{-1}{3}, \frac{-3}{6} = \frac{-1}{2}, \frac{-4}{6} = \frac{-2}{3}, \frac{-5}{6}, \frac{-6}{6} = -1,$$

$$\frac{-8}{6} = \frac{-4}{3}, \frac{-9}{6} = \frac{-3}{2}, \frac{-10}{6} = \frac{-5}{3}, \frac{-11}{6}, \frac{-12}{6} = -2.$$

If the rational number r is associated with the point P on the line L, we call r the coordinate of P. The coordinate of the point gives the location of the point on L. This correspondence between the set of rational numbers and a subset of the set of points of L has the following properties:

Each rational number corresponds to a unique point on L.
Distinct rational numbers are associated with distinct points on L.

Exercise Set 1.5

1. Determine why each of the following numbers is an element of the set $\{r:$ r is a rational number$\}$:

$$15, \; -2, \; \frac{15}{-2}, \; 1 \cdot 5, \; \frac{3}{7+9}, \; \frac{3}{7-9}, \; 4\text{-}15, \; 0,$$

$$\frac{251}{75} + \frac{681}{796}, \; \frac{251}{75} \cdot \frac{681}{796}$$

2. On a coordinate line locate the points which correspond to the following rational numbers:

$$\frac{1}{2}, \frac{6}{3}, \frac{4}{9}, \frac{-2}{3}, \frac{-9}{9}, \frac{4}{8}, \quad 2, \quad \frac{-7}{3}, \frac{7}{-3}, \frac{-7}{-3}, \frac{12}{9}, \frac{4}{3}$$

3. Determine which of the following pairs of rational numbers are equal:

$$\frac{3}{12} \text{ and } \frac{7}{28}, \quad \frac{-3}{8} \text{ and } \frac{3}{-8}, \quad \frac{19}{21} \text{ and } \frac{23}{25}, \quad \frac{-12}{51} \text{ and } \frac{-16}{68}$$

4. Find $r + s$, $r \cdot s$, and $\frac{r}{s}$ where

$$r = \frac{2}{5}, \ s = \frac{5}{3}; \quad r = \frac{-15}{8}, \ s = \frac{4}{9}; \quad r = 4, \ s = \frac{9}{10}; \quad \text{and } r = \frac{3}{-8}, \ s = \frac{-6}{7}.$$

* * *

5. Given the sets A and B, determine if $A \subseteq B$, $B \subseteq A$, $A = B$, or that neither set is a subset of the other.

   a. $A = \{x: \ x \text{ is an integer}\}$ $B = \{r: \ r \text{ is a rational number}\}$

   b. $A = \{n: \ n \text{ is a positive integer}\}$ $B = \{r: \ r \text{ is a rational number}\}$

   c. $A = \left\{\dfrac{3}{4}, \dfrac{4}{5}, \dfrac{5}{6}, \dfrac{6}{7}, \dfrac{7}{8}\right\}$ $B = \left\{\dfrac{-7}{-8}, \dfrac{-5}{6}, \dfrac{4}{-5}, \dfrac{-6}{-7}, \dfrac{3}{-4}\right\}$

   d. $A = \left\{\dfrac{1}{m}: \ m \text{ is an integer and } m \neq 0\right\}$ $B = \left\{1, \dfrac{1}{2}, \dfrac{1}{3}\right\}$

   e. $A = \{-1, 0, 1\}$ $B = \left\{\dfrac{-3}{3}, \dfrac{0}{3}, \dfrac{3}{3}\right\}$

   f. $A = \left\{\dfrac{n}{m}: \ n \text{ and } m \text{ are integers and } m \neq 0\right\}$ $B = \{r: \ r \text{ is a rational number}\}$

6. Reduce each of the following rational numbers to lowest form:

$$\frac{-24}{12}, \quad \frac{0}{295}, \quad \frac{-72}{-54}, \quad \frac{4 \cdot 7 \cdot 5 \cdot 9}{8 \cdot 21 \cdot 15}, \quad \frac{-60}{108}, \quad \frac{3}{111}$$

7. Find the multiplicative inverse of each of the given rational numbers:

$$\frac{2}{3}, \quad \frac{-2}{3}, \quad \frac{.3}{-2}, \quad \frac{2158}{7943}, \quad -22, \quad \frac{1}{45}, \quad \frac{-12}{-13}, \quad \frac{5}{4}$$

8. In each of the following find $r + s, r \cdot s,$ and $\frac{r}{s}$.

a. $r = \frac{4}{3}, \quad s = \frac{1}{6}$

b. $r = \frac{-6}{7}, \quad s = \frac{3}{4}$

c. $r = \frac{-3}{5}, \quad s = \frac{4}{-11}$

d. $r = \frac{-3}{5}, \quad s = -4$

e. $r = \frac{7}{2}, \quad s = \frac{-7}{2}$

f. $r = \frac{13}{5}, \quad s = \frac{5}{13}$

g. $r = 0.9, \quad s = .01$

h. $r = 6, \quad s = -\frac{5}{3}$

9. Compute:

a. $\frac{2 + 5}{4}, \qquad \frac{2}{4} + \frac{5}{4}$

b. $\frac{4}{2 + 5}, \qquad \frac{4}{2} + \frac{4}{5}$

c. $\frac{7 + 2}{7 \cdot 2}, \qquad \frac{7}{7} + \frac{2}{2}$

d. $\frac{15 + 16}{5 + 4}, \qquad \frac{15}{5} + \frac{16}{4}$

e. $\frac{15 \cdot 16}{5 \cdot 4}, \qquad \frac{15}{5} \cdot \frac{16}{4}$

f. $\frac{3}{7} + \frac{-3}{7}, \qquad \frac{3}{7} + \frac{3}{-7}$

10. Determine whether each of the following is true or false. If the statement is false give an example to demonstrate this. In each case assume that the denominators are not zero:

a. $\frac{a \cdot b}{c \cdot b} = \frac{a}{c}$

b. $\frac{a + b}{c + b} = \frac{a}{c}$

c. $\frac{m + 1}{n + 1} = \frac{m}{n}$

d. $\frac{m \cdot \frac{1}{a}}{n \cdot \frac{1}{a}} = \frac{m}{n}$

e. $\frac{m + n}{a + b} = \frac{m}{a} + \frac{n}{b}$

f. $\frac{m}{n} + \frac{(-m)}{n} = \frac{m}{n} + \frac{m}{(-n)}$

1.6 THE REAL NUMBER SYSTEM

Up to this point, the numbers which we have considered are relatively uncomplicated. That is, they are either integers or rational numbers which can be expressed simply as quotients of integers. If we were to proceed, operating strictly within the confines of the rational number system, would this be adequate for our work? The answer is no. One of the topics of study in this text is the solving of equations. It can be shown that there is no rational number x such that $x^2 = 2$ (where $x^2 = x \cdot x$). Consequently, a very simple equation such as $x^2 = 2$ will not have a solution in the rational number system. Thus, since we wish to solve equations of this type later in the text, we will find it necessary to work with a number system that is "large enough" to meet our future needs.

Fortunately, there is an extremely important and useful set of numbers larger than the set of rational numbers. This set is called the set of *real numbers* and is denoted by $\Re$. Since we will work with the real numbers throughout the rest of the text, this section is devoted to a discussion of the properties of $\Re$.

The elements of $\Re$ are called real numbers. The set of rational numbers is a subset of $\Re$. This means that every rational number is a real number. If a and b are real numbers, there correspond unique real numbers called the sum and product of a and b. The sum and product of a and b are denoted by $a + b$ and $a \cdot b$, respectively. In other words, there are two ways of combining pairs of real numbers to obtain other real numbers. The method of combining real numbers using the symbol "+" is called the operation of addition. The method of combining real numbers using the symbol "·" is called the operation of multiplication. Moreover, if a and b are rational numbers, hence real numbers, then $a + b$ and $a \cdot b$ are still found just as the sum and product of rational numbers were found in the previous section.

The operations of addition and multiplication individually satisfy some very important properties. Many of these properties are used while doing ordinary arithmetic. We list these properties for the sake of completeness. However, the formal use of these properties in this text is of short duration. Instead, we will assume familiarity with these properties in later sections.

Properties of Addition

1. $a + 0 = 0 + a = a$ for every real number a (0 is the *additive identity*).
2. $a + b = b + a$ for all real numbers a and b (*commutative property* for addition).
3. For each real number a, there is a real number $-a$ such that $a + (-a) = 0$ ($-a$ is the *additive inverse* of a).
4. $a + (b + c) = (a + b) + c$ for all real numbers a, b, and c (*associative property* for addition).

5. IF a, b, c, d are real numbers, $a = b$ and $c = d$, THEN $a + c = b + d$. In particular, IF $a = b$, THEN $a + c = b + c$.

Properties of Multiplication

1. $1 \cdot a = a \cdot 1 = a$ for every real number a (1 is the *multiplicative identity*).
2. $a \cdot b = b \cdot a$ for all real numbers a and b (*commutative property* for multiplication).
3. IF a is a real number and $a \neq 0$, THEN there is a real number a^{-1} such that $a \cdot a^{-1} = a^{-1} \cdot a = 1$ (a^{-1} is the *multiplicative inverse* of a).
4. $a \cdot (b \cdot c) = (a \cdot b) \cdot c$ for all real numbers a, b, and c (*associative property* for multiplication).
5. IF a, b, c, d are real numbers, $a = b$ and $c = d$, THEN $ac = bd$. In particular, IF $a = b$, THEN $a \cdot c = b \cdot c$.

Properties Relating Addition and Multiplication

1. $a \cdot (b + c) = a \cdot b + a \cdot c$ for all real numbers a, b, and c ⎫ (*distributive*

2. $(b + c) \cdot a = b \cdot a + c \cdot a$ for all real numbers a, b, and c ⎭ *properties*)

When no confusion can arise we will write $a \cdot b$ as ab. It should be noted that because of the associative property for both addition and multiplication we can write $a + b + c$ to represent $(a + b) + c$ or $a + (b + c)$ (i.e., they are the same number) and abc to represent $a(bc)$ or $(ab)c$ (again, they are equal).

The following examples illustrate some of the properties:
1. $-6 + 0 = 0 + (-6) = -6$ because $a + 0 = 0 + a = a$.
2. $(73)(21) = (21)(73)$ because $ab = ba$,
3. $\dfrac{2}{3} + \left(\dfrac{7}{3} + \dfrac{3}{2}\right) = \left(\dfrac{2}{3} + \dfrac{7}{3}\right) + \dfrac{3}{2} = \dfrac{9}{3} + \dfrac{3}{2} = 3 + \dfrac{3}{2} = \dfrac{6}{2} + \dfrac{3}{2} = \dfrac{9}{2}$ because $a + (b + c) = (a + b) + c$, and
4. $4(3 + 6) = (4 \cdot 3) + (4 \cdot 6) = 12 + 24 = 36$ because $a(b + c) = ab + ac$.

The preceding list of properties of the real numbers is usually referred to as a set of *axioms*. This means that we accept these properties without proof. In certain courses in mathematics, we would be interested in starting with a set of axioms for the real numbers and using these axioms to prove additional properties of the real numbers. It is not our intent to proceed in this direction. We merely state some important facts about the set of real numbers, noting only that all of the following statements can be deduced from the axioms for the real numbers.

Additional Facts

1. IF a, b, c are real numbers and $a + c = b + c$, THEN $a = b$.
2. IF a and b are real numbers and $a + b = 0$, THEN $b = -a$.

3. $-(-a) = a$ for every real number a.
4. IF a and b are real numbers THEN $-(a + b) = (-a) + (-b)$.
5. IF a, b, c are real numbers, $c \neq 0$ and $ac = bc$, THEN $a = b$.
6. IF a, b are real numbers, $a \neq 0$ and $ab = 1$, THEN $b = a^{-1}$.
7. IF $a \neq 0$, THEN $a^{-1} \neq 0$ and $(a^{-1})^{-1} = a$.
8. $a^{-1} b^{-1} = (ba)^{-1}$ for all nonzero real numbers a and b.
9. IF a and b are real numbers, THEN $-(ab) = (-a)(b) = a(-b)$.
10. $(-a)(-b) = ab$ for all real numbers a and b.
11. $a0 = 0a = 0$ for every real number a.
12. IF a and b are real numbers and $ab = 0$, THEN $a = 0$ or $b = 0$.

The following examples illustrate some of the additional facts of the real numbers that we have stated:

(1) $-(-4) = 4$ because $-(-a) = a$,
(2) $(-3)(-5) = 3 \cdot 5 = 15$ because $(-a)(-b) = a \cdot b$,
(3) $-(5 + (-16)) = -5 + (-(-16)) = -5 + 16 = 11$ because $-(a + b) = (-a) + (-b)$, and
(4) $\left(\left(\frac{7}{6} \right)^{-1} \right)^{-1} = \frac{7}{6}$ because $(a^{-1})^{-1} = a$.

The operation of subtraction of real numbers has not been mentioned up to this point. We discuss subtraction in terms of the operation of addition.

1.3 Definition. *IF a and b are real numbers, THEN the real number $a - b$, read "a minus b," is defined by $a - b = a + (-b)$.*

Thus, according to the definition,

(1) $5 - 2 = 5 + (-2) = 3$,
(2) $-3 - 6 = -3 + (-6) = -9$, and
(3) $9 - 9 = 9 + (-9) = 0$.

There are various properties of subtraction with which we are already familiar. We list these properties for reference.

Properties of Subtraction

1. $a(b - c) = ab - ac$ for all real numbers a, b, and c.
2. $-a + b = b - a$ for all real numbers a and b.
3. $-(a - b) = b - a$ for all real numbers a and b.

Therefore, we have the following:

(1) $4(10 - 8) = 4 \cdot 10 - 4 \cdot 8 = 40 - 32 = 8$ because $a(b - c) = ab - ac$,
(2) $-5 + 2 = 2 - 5 = -3$ because $-a + b = b - a$, and
(3) $-(7 - 1) = 1 - 7 = -6$ because $-(a - b) = b - a$.

There is one final operation that we wish to define for pairs of real numbers. This is the operation of division, which was mentioned briefly in Sec. 1.5.

1.4 Definition. *IF a and b are real numbers with $b \neq 0$, THEN the real number $\frac{a}{b}$, called the quotient of a by b, is defined by $\frac{a}{b} = ab^{-1}$.*

We also call $\frac{a}{b}$ the fraction a over b and call a and b the numerator and denominator, respectively, of the fraction a over b. If $b \neq 0$, it is immediate from the definition that $\frac{1}{b} = 1 \cdot b^{-1} = b^{-1}$. That is, $\frac{1}{b}$ is a number such that $b \cdot \frac{1}{b} = \frac{1}{b} \cdot b = 1$. Thus,

$$\frac{5}{2} = 5 \cdot \frac{1}{2}, \frac{\frac{5}{4}}{\frac{2}{3}} = \frac{5}{4} \cdot \frac{3}{2}, \text{ and } 5^{-1} = \frac{1}{5}.$$

According to the definition, $\frac{a}{b}$ is only defined when $b \neq 0$. This is the case because 0 does not have a multiplicative inverse. Consequently, the symbol $\frac{a}{0}$ has no meaning.

As is the case with the operations of addition, multiplication, and subtraction, there are basic properties possessed by the operation of division. We close this section by listing some of these properties.

Properties of Division

IF a, b, c, d are real numbers with $b \neq 0, d \neq 0$, THEN

1. $\frac{a}{b} = \frac{c}{d}$ if and only if $ad = bc$.

2. $\frac{ac}{bc} = \frac{a}{b}$,

3. $\frac{a}{b} \cdot \frac{c}{d} = \frac{ac}{bd}$,

4. $\frac{a}{b} + \frac{c}{d} = \frac{ad + bc}{bd}$, and

5. $-\left(\frac{a}{b}\right) = \frac{-a}{b} = \frac{a}{-b}$.

Examples of these properties are:

(1) $\frac{4}{7} = \frac{12}{21}$ because $4(21) = 7(12)$ and

(2) $\frac{3}{5} + \frac{4}{15} = \frac{3(15) + 4(5)}{5(15)} = \frac{45 + 20}{75} = \frac{65}{75}$ because $\frac{a}{b} + \frac{c}{d} = \frac{ad + bc}{bd}$

Also, $\frac{65}{75} = \frac{13(5)}{15(5)} = \frac{13}{15}$ because $\frac{ac}{bc} = \frac{a}{b}$.

Exercise Set 1.6

1. Indicate those numbers which are integers. Indicate those numbers which are rational numbers but not integers:

$$7, \quad \frac{6}{3}, \quad \frac{1}{3}, \quad -4, \quad 0, \quad -\frac{3}{2}, \quad \left(\frac{1}{2}\right)^{-1}, \quad \frac{8+6}{4+2}, \quad \frac{8 \cdot 6}{4 \cdot 2}$$

2. Which of the properties of the real number system are illustrated by each of the following statements?

a. $t \cdot t^{-1} = 1$

b. $17\left(\frac{5280}{5280}\right) = 17$

c. $3(2 + 5) = 3(5 + 2)$

d. $3(2 + 5) = 3 \cdot 2 + 3 \cdot 5$

e. $\frac{3}{12} \cdot \frac{15}{10} = \frac{1}{4} \cdot \frac{3}{2}$

f. $(-2 + 7)(601 + 1)$
$= (601 + 1)(-2 + 7)$

g. $\left(\frac{376}{51} + \frac{-32}{109}\right) + \frac{14}{3} = \frac{376}{51} +$
$\left(\frac{-32}{109} + \frac{14}{3}\right)$

h. $5, 682, 951 + 0 = 5, 682, 951$

i. $2(x + 4y) = 2x + 8y$

j. $\left(4 \cdot \frac{3}{4}\right)\pi = 4 \cdot \left(\frac{3\pi}{4}\right)$

k. $-\sqrt{2} + \sqrt{2} = 0$

3. Which of the additional facts about the real numbers are illustrated by each of the following statements?

a. $-\left(\frac{-\sqrt{3}}{6}\right) = \frac{\sqrt{3}}{6}$

b. $(-2)(\sqrt{5}) = -2\sqrt{5}$

c. $-(2 + \sqrt{5}) = -2 - \sqrt{5}$

d. If $x + \frac{2}{5} = 0$, then $x = -\frac{2}{5}$.

e. If $\frac{4}{3}x = 1$, then $x = \frac{3}{4}$.

f. $(622^{-1})^{-1} = 622$

g. $(-369)(-7895) = (369 \cdot 7895)$

h. $\left(\frac{2}{3}\right)^{-1}\left(\frac{9}{2}\right)^{-1} = \left(\frac{2}{3} \cdot \frac{9}{2}\right)^{-1}$

i. $27 \cdot 0 = 15 \cdot 0$

j. If $t + 24 = 6 + 24$, then $t = 6$.

k. If $24t = 24 \cdot (24)^{-1}$, then $t = (24)^{-1}$.

l. If $(x - 1)(y + 8) = 0$, then $x - 1 = 0$ or $y + 8 = 0$.

* * *

Simplify each of the following expressions:

4. $(-3 + 8) + (12 + (-14))$

5. $(-3 + 8)(12 + (-14))$

6. $-3(8 + (12 + (-14)))$

7. $(-3)(8) + (12)(-14)$

8. $-[((3 + 8) - 12) + 14]$

9. $(3)(8)^{-1} + (12)(14)^{-1}$

10. $3(8 + 12)^{-1} + (-14)$

11. $\left(\dfrac{15}{7}\right)\left[\left(\dfrac{-3}{5}\right)\left(\dfrac{10}{21}\right)^{-1}\right]$

12. $\dfrac{3 \cdot 6 \cdot 5 \cdot 6 \cdot 2 \cdot 1}{5 \cdot 6 \cdot 3 \cdot 12 \cdot 3}$

13. $\dfrac{\sqrt{2}}{2} + \dfrac{\sqrt{2}}{5}$

14. $\dfrac{\sqrt{3}}{8} - \dfrac{2}{5}$

15. $\dfrac{2}{\sqrt{3}} + \dfrac{2}{3}$

16. $\dfrac{4}{3}(7 + (\sqrt{2} - x))$

17. $[4^{-1}(10)](4)$

18. $5[(a + 1) + (-a)]$

19. $\left(4 \cdot \dfrac{5}{9}\right)16$

20. $\dfrac{259}{584} \cdot \dfrac{6488}{-789} \cdot \left(\dfrac{259}{584}\right)^{-1}$

21. $\dfrac{\dfrac{x + y}{2}}{\dfrac{x + y}{3}}$

22. $\dfrac{\dfrac{4}{3} + \left(\dfrac{8}{9}\right)^{-1}}{\left(\dfrac{4}{3} + \dfrac{8}{9}\right)^{-1}}$

23. $\dfrac{\pi + 2\pi}{2 + 3\pi}$

24. $\dfrac{\pi}{2} - \dfrac{\pi}{6}$

25. $\dfrac{\dfrac{54}{81} - \dfrac{30}{45}}{\dfrac{16}{27} \cdot \dfrac{54}{81}}$

*As mentioned in the text, axioms for the real numbers are used while doing ordinary arithmetic. We show now, for example, why $3^{-1}(5 \cdot 3) = 5$, listing the reasons to the right of each step:

$$\begin{aligned}
3^{-1}(5 \cdot 3) &= 3^{-1}(3 \cdot 5) & &\text{Multiplication is commutative} \\
&= (3^{-1} \cdot 3)5 & &\text{Multiplication is associative} \\
&= 1 \cdot 5 & &a^{-1}a = 1 \\
&= 5 & &1 \cdot a = a
\end{aligned}$$

Show why the following statements are correct, listing the axioms and definitions which are used to the right of each step:

26. $\left(4 \cdot \dfrac{5}{9}\right)16 = 16\left(\dfrac{5}{9} \cdot 4\right)$

27. $(-6)\left[1 - \left(\dfrac{1}{-2}\right)\right] = -9$

28. $rs + s = (r + 1)s$

29. $\dfrac{1}{12}\left[\dfrac{1}{5} - \dfrac{\dfrac{2}{3}}{\dfrac{2}{7}}\right] = -\dfrac{8}{45}$

30. $\left(\dfrac{r}{s}\right)\left(\dfrac{s - ts}{\dfrac{r}{t}}\right) = t(1 - t)$

1.7 ORDER PROPERTIES OF ℜ

From past experience, we are aware of the fact that some numbers are called positive while others are called negative. The concepts of positive and negative numbers follow from the *order properties* for ℜ. These are now stated.

Properties of Order

1. There is a nonempty subset P of ℜ called the set of positive real numbers. Given a real number a, *exactly one* of the following is true:
   $a = 0$, a is positive (i.e., $a \in P$), or $-a$ is positive (i.e., $-a \in P$).
2. IF a and b are positive numbers, THEN $a + b$ and ab are positive.

Naturally, we want to ask the question: What are some typical positive numbers? It can be shown that the number 1 is positive. Moreover, this fact agrees with our previous dealings with the number 1. By the second order axiom, $1 + 1 = 2$ is positive. Similarly, $2 + 1 = 3$ is positive and so on. In other words, $\{1, 2, 3, \ldots\} \subseteq P$. That is, every positive integer is a positive real number.

If a is a nonzero real number and $-a \in P$, we say a is a *negative* real number. Consequently, -5 is negative because $-(-5) = 5$ is positive. One should not mistakingly think that $-a$ is negative for any real number a. As we have just seen, $-a$ is positive when $a = -5$. It is easily seen that a is positive if and only if $-a$ is negative.

A common way of comparing unequal numbers is by using inequalities. Note that if a and b are not equal then $a - b \neq 0$. Therefore, by the first order axiom, either $a - b$ or $-(a - b) = b - a$ is positive. We use this observation in the following definition of inequality.

1.5 Definition. *IF a and b are real numbers, THEN a is less than b (written $a < b$), provided $b - a$ is positive.*

For example,
$3 < 5$ because $5 - 3 = 2$ is positive,
$\dfrac{1}{3} < \dfrac{10}{3}$ because $\dfrac{10}{3} - \dfrac{1}{3} = \dfrac{9}{3} = 3$ is positive,
$-4 < -\dfrac{1}{2}$ because $-\dfrac{1}{2} - (-4) = -\dfrac{1}{2} + 4 = \dfrac{7}{2}$ is positive, and
$-6 < 0$ because $0 - (-6) = 0 + 6 = 6$ is positive.

If $a \neq b$, then either $a < b$ or $b < a$ since either $b - a$ or $a - b$ is positive. This means that for any two real numbers a and b, exactly one of the following holds:

$$a = b, \quad a < b, \quad \text{or} \quad b < a.$$

We also write $b > a$ and say b is greater than a to mean $a < b$, i.e., a is less than b. Thus,

$$5 > 3, \quad \frac{10}{3} > \frac{1}{3}, \quad -\frac{1}{2} > -4, \quad \text{and} \quad 0 > -6.$$

Another consequence of the definition of inequality is that a is positive if and only if $0 < a$ (equivalently, $a > 0$). Also, a is negative if and only if $a < 0$ (equivalently, $0 > a$).

Given the order axioms for the real numbers and the definition of inequality, it is now possible to prove many other facts about inequalities. We list these facts below:

Properties of Inequalities

1. $a < b$ if and only if there is a positive number c such that $a + c = b$.
2. IF $a < b$ and $b < c$, THEN $a < c$.
3. IF $a < b$, THEN $a + c < b + c$.
4. IF $a < b$ and $c < d$, THEN $a + c < b + d$.
5. IF $a < b$ and $c > 0$, THEN $ac < bc$.
6. IF $a < b$ and $c < 0$, THEN $ac > bc$.
7. $ab > 0$ if and only if $a > 0$ and $b > 0$ or $a < 0$ and $b < 0$.
8. $ab < 0$ if and only if $a > 0$ and $b < 0$ or $a < 0$ and $b > 0$.
9. IF $a \neq 0$, THEN $a^2 > 0$.
10. IF $a > 0$, THEN $a^{-1} > 0$.
11. IF $b > 0$ and $d > 0$, THEN $\frac{a}{b} < \frac{c}{d}$ if and only if $ad < bc$.

Properties 5 and 6 simply state that multiplication by a positive number does not change the "sense" of an inequality, while multiplication by a negative number changes the "sense" of an inequality. It follows, from properties 7 and 10, that $\frac{m}{n} > 0$ if m and n are positive integers. This is because $\frac{m}{n} = m \cdot \frac{1}{n} = mn^{-1}$ and both m and n^{-1} are positive. As illustrations of some of the preceding facts about inequalities, we have:

1. $-4 + 7 < -1 + 7$ because $-4 < -1$ (property 3),
2. $\frac{-10}{3} + 6 < -1 + 8$ because $\frac{-10}{3} < -1$ and $6 < 8$ (property 8),
3. $12 \cdot 2 < 13 \cdot 2$ because $12 < 13$ and $2 > 0$ (property 5),
4. $12(-2) > 13(-2)$ because $12 < 13$ and $-2 < 0$ (property 6),
5. $(-5)(-3) > 0$ because $-5 < 0$ and $-3 < 0$ (property 7),

6. $5(-3) < 0$ because $5 > 0$ and $-3 < 0$ (property 8), and

7. $\dfrac{3}{2} < \dfrac{49}{32}$ since $3(32) = 96$, $2(49) = 98$ and $96 < 98$ (property 11).

Let a and b be real numbers. We say a is *less than or equal to* b and write $a \leqslant b$ if and only if $a = b$ or $a < b$. We also write $b \geqslant a$ and say b is *greater than or equal to* a to mean $a \leqslant b$. Thus, while the symbol "$<$" denotes strict inequality, the symbol "$\leqslant$" allows for possible equality. Therefore

$$2 \leqslant \frac{7}{2}, \quad 2 \leqslant 2, -6 \leqslant -1, \quad -6 \leqslant -6, \quad 0 \leqslant 0, \quad 0 \leqslant 1, \quad \text{and} \quad -1 \leqslant 0.$$

Finally, suppose two inequalities, both involving the same real number, are given. For example, $-1 < 3$ and $3 < 4$. Instead of writing both inequalities separately, we combine the statements and write $-1 < 3 < 4$. Similarly, $-2 > -7$ and $-7 > -10$ can be combined by writing $-2 > -7 > -10$ or, equivalently, $-10 < -7 < -2$.

Exercise Set 1.7

1. Between each of the following pairs of numbers place either $<$, $>$, or $=$, whichever will make a true statement:

   a. $\dfrac{-5}{7}$ 0

   b. -16 -17

   c. $\dfrac{-5}{3}$ $\dfrac{-4}{3}$

   d. $\dfrac{1}{3}$ $\dfrac{1}{2}$

   e. $\dfrac{11}{-6}$ $\dfrac{-22}{12}$

   f. $\dfrac{13}{-14}$ $\dfrac{-27}{28}$

   g. $\dfrac{11}{-6}$ -2

   h. $\dfrac{375}{742}$ $\dfrac{231}{457}$

   i. 10 $\dfrac{28}{3}$

   j. $\dfrac{-15}{-8}$ $\dfrac{15}{-8}$

2. Which of the following statements are always true?
   a. x is a positive number where $x \in \mathcal{R}$.
   b. $-x$ is a negative number where $x \in \mathcal{R}$.
   c. $-(-x)$ is a positive number where $x \in \mathcal{R}$.
   d. If $p < q$ and $q < r$, then $p < r$.
   e. If $p < q$ and $p < r$, then $q < r$.
   f. If $a < 0$ and $b > 0$, then $a < b$.
   g. If $a + 19 = b$, then $a < b$.
   h. If $a > c$ and $a < d$, then $c < a < d$.

* * *

Use the properties of inequalities to complete the following statements:

3. If $x < 3$, then $x + 2 < \underline{\hspace{2cm}}$.

4. If $x - 2 < 0$, then $x + 2 < $ _____.

5. If $x - 7 < - 12$, then $x - 3 < $ _____.

6. If $- 3 < x < \frac{1}{2}$, then _____ $< x + \frac{1}{4} < $ _____.

7. If $t > 3$, then $12t > $ _____.

8. If $t > 3$, then $-2t < $ _____.

9. If $- 1 < t + 2 < 1$, then _____ $< t < $ _____.

10. If $t < S$, then $t - 5 < S + $ _____.

11. If $xy > 0$ and $x > \sqrt{2}$, then $y > $ _____.

12. If $xy < 0$ and $x < -\frac{3}{4}$, then y _____.

13. If $2r - 6 < - 3$, then $2r < $ _____.

14. If $r \leqslant 5$, then $\frac{7}{3} r \leqslant $ _____.

15. If $\frac{-3}{2} \leqslant r \leqslant \frac{-1}{2}$, then _____ $\leqslant r + 1 \leqslant $ _____.

16. If $-4r < - 6$, then $r > $ _____.

17. If $\sqrt{5} \, r < \sqrt{5}$, then $r < $ _____.

18. If $x + 3 > y + 3$, then $x > $ _____.

19. If $- 1 < x < 4$, then _____ $< x + 2 < $ _____.

20. If $x > 10$, then $x + 2 > $ _____ and $3(x + 2) > $ _____.

*Using the order axioms, the definition of inequality, and the properties of the real number system, prove the following properties of inequalities:

21. If $a < b$ and $b < c$, then $a < c$.

22. If $a < b$, then $a + c < b + c$.

23. If $a < b$ and $c > 0$, then $ac < bc$.

24. If $a < b$ and $c < 0$, then $ac > bc$.

25. If $a \neq 0$, then $a^2 > 0$.

1.8 THE REAL NUMBER LINE

In Sec. 1.5 we showed how to establish a one-one correspondence between the set of rational numbers and a subset of the set of points on a line L. We noted in Sec. 1.6 that the rational numbers are real numbers and that there is no rational number x such that $x^2 = 2$. There is, however, a positive real number x such that $x^2 = 2$. This number is denoted by $\sqrt{2}$, read "the square root of 2."

Hence $\sqrt{2}$ is not a rational number. Such numbers are called *irrational numbers.* Therefore, the set of real numbers contains numbers other than rational numbers. These numbers are not yet associated with points on L. In order to have a reasonable geometric representation for the set of real numbers, we want every real number to correspond to a point on the line L in such a way that no points of L are omitted. We, therefore, make the following assumptions:

1. For each real number a there corresponds a unique point on L. Moreover, distinct real numbers correspond to distinct points on L and each point on L is the correspondent of some real number.

2. If r is a rational number, then the point on L corresponding to r is determined as in Sec. 1.5.

3. Let a, b be real numbers and let A, B be the points on L corresponding to a and b, respectively. If $a < b$ then B lies on L in the positive direction from A.

The last statement says, for example, that if L is horizontal with positive direction to the right, then $a < b$ implies B lies to the right of A. Similarly, if L is vertical with positive direction upward, then $a < b$ implies B lies above A. If A is the point on L corresponding to the real number a, we say a is the coordinate of A. Thus every point on L has a unique coordinate. We call L a *coordinate line* or a *real number line.*

Given two points A and B on a coordinate line L, it is natural to want to give measure to the segment from A to B. This segment is denoted by AB. Before doing this, it is necessary to discuss the notion of absolute value.

1.7 Definition. *Let a be a real number. The absolute value of a, denoted by $|a|$, is defined by the rule*

$$|a| = \begin{cases} a, \text{ if } a \geqslant 0 \\ -a, \text{ if } a < 0. \end{cases}$$

According to the definition,

$$\left|\frac{7}{6}\right| = \frac{7}{6}, |-1| = -(-1) = 1, |0| = 0, |-4| = -(-4) = 4.$$

It is apparent from the definition that $|a| \geqslant 0$ for every real number a. We are now ready to define length.

1.8 Definition. *Let A and B be two points on the coordinate line L and assume A and B have coordinates a and b, respectively. The length of AB, also called the distance between A and B, is denoted by $|AB|$ and defined by the formula* $|AB| = |b - a|$.

Lengths of several line segments are shown in Figure 1.9.

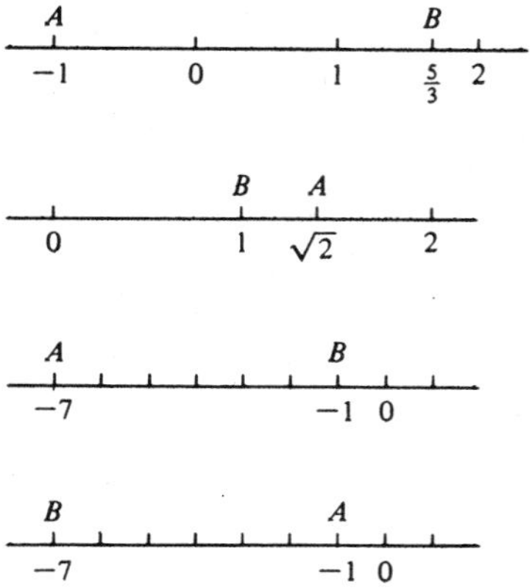

$$|AB| = |\frac{5}{3} - (-1)| = |\frac{5}{3} + 1| = |\frac{8}{3}| = \frac{8}{3}$$

$$|AB| = |1 - \sqrt{2}| = -(1 - \sqrt{2}) = \sqrt{2} - 1$$

$$|AB| = |-1 - (-7)| = |-1 + 7| = |6| = 6$$

$$|AB| = |-7 - (-1)| = |-7 + 1| = |-6| = 6$$

Figure 1.9

If A and B are points on the coordinate line L, then the distance $|AB|$ between A and B does not tell us the relative positions of A and B on L. For this additional information we introduce the idea of directed distance.

1.9 Definition. *Let A and B be points on the coordinate line L and assume A and B have coordinates a and b, respectively. The **directed distance** from A to B, denoted by $\overline{AB}$, is defined by the formula*

$$\overline{AB} = b - a.$$

Figure 1.10 shows directed distances corresponding to the distance between the points A and B in Figure 1.9.

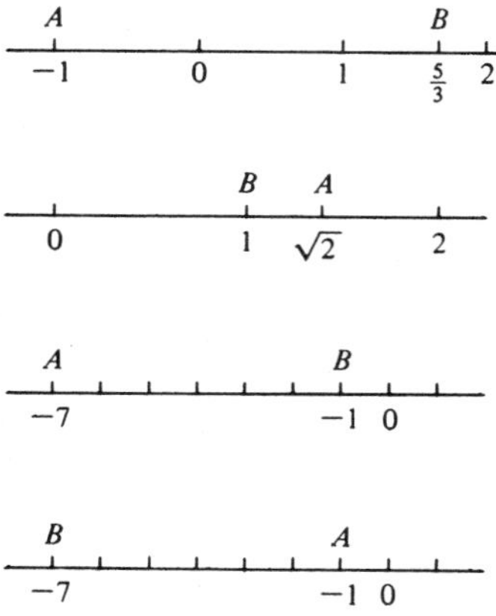

$$\overline{AB} = \frac{5}{3} - (-1) = \frac{5}{3} + 1 = \frac{8}{3}$$

$$\overline{AB} = 1 - \sqrt{2}$$

$$\overline{AB} = -1 - (-7) = -1 + 7 = 6$$

$$\overline{AB} = -7 - (-1) = -7 + 1 = -6$$

Figure 1.10

We see that $|\overline{AB}| = |AB|$. Therefore the directed distance $\overline{AB}$ from A to B tells us not only the distance between A and B but it also tells us the relative positions of A and B on L. Namely, if $\overline{AB} > 0$, then B lies on L in the positive direction from A. If $\overline{AB} < 0$, then B lies on L in the negative direction from A.

Before closing this chapter, we wish to introduce notation for some special sets of real numbers which will appear in later chapters. Let a and b be real numbers. The set $[a, b]$ of real numbers is called the *closed interval* from a to b and is defined by

$$[a, b] = \{x: \ a \leqslant x \leqslant b\}.$$

The set (a, b) of real numbers is called the *open interval* from a to b and is defined by

$$(a, b) = \{x: \ a < x < b\}.$$

Thus,
1. $[3, 7] = \{x: \ 3 \leqslant x \leqslant 7\}$,
2. $[-1, 0] = \{y: \ -1 \leqslant y \leqslant 0\}$,
3. $(3, 7) = \{z: \ 3 < z < 7\}$, and
4. $(-1, 0) = \{w: \ -1 < w < 0\}$.

The sets $[a, b]$ and (a, b) are easily visualized as the darkened segments in Figure 1.11 when $a < b$.

$$[a, b] \qquad\qquad (a, b)$$

Figure 1.11

Note that $a \notin (a, b)$ and $b \notin (a, b)$.

Exercise Set 1.8

1. Let $A = \{x: \ x$ is a point on a line $L\}$. Is there a one-to-one correspondence between the sets A and B if:
   a. $B = \{j: \ j$ is an integer$\}$?
   b. $B = \{t: \ t$ is a rational number$\}$?
   c. $B = \{r: \ r$ is a real number$\}$?

2. Let C and D be points on a horizontal line L corresponding to the real numbers c and d. For the given c and d, determine whether C lies to the right or left of D:

   a. $c = \sqrt{2}, \ d = \dfrac{7}{5}$

   b. $c = -5, \ d = -\dfrac{300}{61}$

   c. $c = \dfrac{\sqrt{4}}{3}, \ d = \dfrac{-2}{3}$

   d. $c = \pi, \ d = \dfrac{22}{7}$

3. Find $|a|$ if:
   a. $a = 38$ b. $a = -38$ c. $a = \sqrt{7}$ d. $a = -\sqrt{7}$
   e. $a = 38 - \sqrt{7}$ f. $a = -38 + \sqrt{7}$ g. $a = 0$ h. $a = |\pi - 3|$

4. Find $|AB|$ and $\overline{AB}$ if:
   a. $A = \dfrac{1}{4}$, $B = 1$ b. $A = -\dfrac{1}{2}$, $B = \dfrac{1}{3}$ c. $A = -2$, $B = -1$
   d. $A = \sqrt{7}$, $B = -\sqrt{7}$ e. $A = -3$, $B = -10$ f. $A = -\sqrt{3}$, $B = -2$

5. Which of the following intervals are closed intervals and which are open intervals?
   a. $[-2, 0]$ b. $[-3, -2]$ c. $(-2, 3)$ d. $(835, 1259)$
   e. $\left\{ x: \ 5 \leqslant x \leqslant \dfrac{21}{2} \right\}$ f. $\{y: \ \pi < y < 2\pi\}$ g. $\{t: \ -1 \leqslant t \leqslant 1\}$

                         * * *

6. Determine which of the following are true statements:
   a. $|-2| = |2|$ b. $|5| = |5^{-1}|$
   c. $|-2| + |3| = |-2 + 3|$ d. $|-12| \cdot |6| = |(-12) \cdot 6|$
   e. $|a - b| = |b - a|$ f. $|a^2| = a^2$
   g. $|a| + |b| = |a + b|$ h. $|a - b| = -|b - a|$

7. Determine which of the following are true statements:
   a. $2 \in [2, 8]$ b. $\dfrac{1}{2} \in (-1, 1)$
   c. $\sqrt{2} \in (1, 1.4)$ d. $\sqrt{2} \in [1, 1.4]$
   e. $-30.003 \in (-30.003, -30.001)$ f. $\dfrac{39}{5} \in [0, 10]$
   g. $|1 - \sqrt{3}| \in [-1, 0]$ h. $|1 - \sqrt{3}| \in (-1, 0)$

8. Change each of the following to set notation:
   a. $\left(\dfrac{-2}{3}, \dfrac{1}{3} \right)$ b. $\left[\dfrac{-2}{3}, \dfrac{1}{3} \right]$ c. $\left[-\sqrt{5}, -\dfrac{\sqrt{2}}{2} \right]$
   d. $[1.001, 1.01]$ e. $(1.2, 2.1)$ f. $\left(\dfrac{7}{5}, \dfrac{8}{5} \right)$

9. Change each of the following to interval notation:
   a. $\{ x: \ 1 \leqslant x \leqslant 75 \}$ b. $\left\{ t: \ \dfrac{-1}{7} < t < \dfrac{35}{4} \right\}$
   c. $\{y: \ \sqrt{21} < y < 22\}$ d. $\{z: \ 0 < t + 1 < 2\}$
   e. $\left\{ p: \ -\dfrac{19}{5} \leqslant p \leqslant -2 \right\}$ f. $\{q: \ -8 < q < 3\}$

10. Show on a line L the set of points which corresponds to the given set of real numbers:
   a. $[4, 5]$ b. $[4, 6]$ c. $(4, 5)$ d. $(4, 6)$
   e. $\left[-6, \dfrac{-2}{3} \right]$ f. $(0, 1)$ g. $\left[-1, \dfrac{4}{5} \right]$ h. $\left(\dfrac{3}{8}, \dfrac{1}{2} \right)$

11. Show on a line L the set of points which corresponds to the given set of real numbers:

a. $\{y: \ -1 \leqslant y \leqslant 6\}$

b. $\left\{x: \ \sqrt{2} < x < \dfrac{7}{3}\right\}$

c. $\{t: \ 0 \leqslant t \leqslant 50\}$

d. $\{t: \ -50 \leqslant t \leqslant 0\}$

e. $\{p: \ 0 < p < 1\}$

f. $\{p: \ p = 1\}$

12. Indicate on a line L the set of points which corresponds to the given set of real numbers:

a. $[2, 5) = \{x: \ 2 \leqslant x < 5\}$

b. $(-1, 3] = \{x: \ -1 < x \leqslant 3\}$

c. $\left[-\dfrac{1}{2}, \infty\right) = \left\{x: \ x \geqslant -\dfrac{1}{2}\right\}$

d. $\left(-\dfrac{1}{2}, \infty\right) = \left\{x: \ x > -\dfrac{1}{2}\right\}$

e. $(-\infty, 0) = \{x: \ x < 0)$

f. $(-\infty, 0] = \{x: \ x \leqslant 0\}$

g. $(-\infty, \infty) = \{x: \ x \in \mathcal{R}\}$

2 / Linear Equations and Inequalities

In Chap. 1 we examined equalities and inequalities and discussed some of their properties. For example, given the numbers $\frac{3}{2}$ and $\frac{49}{32}$, we were interested in determining whether $\frac{3}{2} > \frac{49}{32}$, $\frac{3}{2} = \frac{49}{32}$, or $\frac{3}{2} < \frac{49}{32}$. By using the methods available to us, we ascertained that $\frac{3}{2} < \frac{49}{32}$. In this chapter we are going to investigate problems involving equalities and inequalities, but these problems will be of a slightly different nature. One such problem might be to determine all real numbers x such that $3x - 2 < 5$. This type of inequality is called a *conditional inequality* since the order is preassigned and we must find conditions on x such that the inequality holds. We see that if $x = 4$, then $3x - 2 > 5$ since $3(4) - 2 = 10$ and $10 > 5$. However, if $x = 2$, then $3x - 2 < 5$ since $3(2) - 2 = 4$ and $4 < 5$. We say that $x = 2$ is a *solution* of the inequality $3x - 2 < 5$, while $x = 4$ is not a solution. Note, however, that $x = 4$ is a solution of the inequality $3x - 2 > 5$. We will show how to determine the entire collection of solutions for such conditional inequalities. In addition, we will determine all solutions of *conditional equalities*, also called *equations*, such as $3x - 2 = 5$. Thus, all solutions of each of the conditional statements,

$$3x - 2 > 5, \quad 3x - 2 = 5, \quad 3x - 2 < 5,$$

will be completely determined.

The rectangular coordinate system is introduced in this chapter. We will use this coordinate system to locate points in the plane in much the same manner that latitude and longitude locate points on a globe. Once this is done, we can then use our coordinate system as an aid in sketching "pictures" related to the mathematical objects and problems which we will examine.

2.1 LINEAR EQUATIONS

A *linear equation* (in one variable) is a (conditional) statement of the form

$$ax + b = 0$$

where $a, b \in \mathfrak{R}$. For example,

$$7x - 2 = 0$$

is a linear equation (in the variable x), where $a = 7$ and $b = -2$. We will discuss the reason why we use the term "linear" in this expression in Sec. 2.5.

A *solution* of the linear equation $ax + b = 0$ is a number x which makes the statement $ax + b = 0$ a true statement. The process of finding all solutions is known as solving the equation. Our method for solving a linear equation will be to either add (or subtract) equal numbers from each side of the equality or to multiply (or divide) each side of the equality by equal numbers. In other words, we will be using the properties

$$\text{(i)} \quad \text{if } a = b \text{ then } a + c = b + c$$

and

$$\text{(ii)} \quad \text{if } a = b \text{ then } a \cdot c = b \cdot c.$$

We apply (i) and (ii) in such a way that *equivalent equations* (i.e., equations with the same solutions) are obtained. In fact, to solve an equation, we obtain equations which are equivalent to the original equation but whose solutions are easily found. For example, if:

$$
\begin{aligned}
7x - 2 &= 0, \\
\text{then} \quad (7x - 2) + 2 &= 0 + 2 \quad \text{(adding 2 to both sides of the equality)} \\
7x + (-2 + 2) &= 0 + 2 \quad \text{(associative law for addition)} \\
7x + 0 &= 0 + 2 \quad \text{(property of additive inverse)} \\
7x &= 2 \quad \text{(property of 0)}
\end{aligned}
$$

Similarly, if we multiply one side of an equality by a number, then we must multiply the other side of the equality by the same number so that the resulting equation is equivalent. Continuing our example, we have that

$$7x = 2$$

(remember, this is equivalent to the equation $7x - 2 = 0$). So

$$\frac{1}{7}(7x) = \frac{1}{7}(2) \quad \left(\text{multiplying both sides of the equality by } \frac{1}{7}\right)$$

$$\left(\frac{1}{7} \cdot 7\right)x = \frac{2}{7} \quad \text{(associative law for multiplication and property of multiplication of fractions)}$$

$$1 \cdot x = \frac{2}{7} \quad \text{(property of multiplicative inverse)}$$

$$x = \frac{2}{7} \quad \text{(property of 1)}.$$

Thus, $x = \frac{2}{7}$ is equivalent to saying $7x = 2$, which, in turn, is equivalent to saying $7x - 2 = 0$. We have shown that $7x - 2 = 0$ implies that $x = \frac{2}{7}$. Moreover, if $x = \frac{2}{7}$ then $7x - 2 = 0$ since

$$7\left(\frac{2}{7}\right) - 2 = 2 - 2 = 0.$$

Therefore, $7x - 2 = 0$ if and only if $x = \frac{2}{7}$. We have determined the exact conditions on x such that $7x - 2 = 0$. Namely, $x = \frac{2}{7}$.

In the preceding example we indicated every step which was used in solving the equation. Also included were the properties and axioms which justified these steps. In the future, inessential steps and reasons will be omitted since, in practice, we usually take several arithmetic steps at once and are not interested in their axiomatic justification.

Problem. Find all values of x such that

$$-2x + 3 = 1.$$

Solution. Note that this is equivalent to the linear equation $-2x + 2 = 0$ since we can add -1 to each side of the equation and obtain

$$(-2x + 3) - 1 = 1 - 1,$$
or
$$-2x + 2 = 0.$$

Now to solve the equation (i.e., to find all of the values of x such that $-2x + 3 = 1$) we isolate the terms involving the variable on one side of the equation and the constant terms on the other side of the equation. Proceeding with this aim in mind, if

$$-2x + 3 = 1,$$

then $\quad(-2x+3)-3=1-3$ (add -3 to each side of the equation),
$$-2x=-2,$$
$$\left(-\frac{1}{2}\right)(-2x)=\left(-\frac{1}{2}\right)(-2)\left(\text{multiplying both sides of the equality by } -\frac{1}{2}\right),$$
$$x=1.$$

Therefore $-2x+3=1$ and $x=1$ are equivalent equations. It follows that the solution of the equation $-2x+3=1$ is $x=1$.

Problem. Find all values of x such that

$$2x+7=5x-1.$$

Solution. First, we want to get all of the variable terms on one side of the equation and all of the constant terms on the other side of the equation. This can be accomplished by subtracting $2x$ and adding 1 to each side of the equation. If

then
$$2x+7=5x-1,$$
$$(2x+7)+(-2x+1)=(5x-1)+(-2x+1),$$
$$8=3x,$$
$$\frac{1}{3}\cdot 8=\frac{1}{3}(3x)$$
$$\frac{8}{3}=x.$$

Thus, if $2x+7=5x-1$, then $x=\frac{8}{3}$. Conversely, if $x=\frac{8}{3}$, then $2x+7=5x-1$. In other words, $2x+7=5x-1$ and $x=\frac{8}{3}$ are equivalent equations so that the solution of $2x+7=5x-1$ is $x=\frac{8}{3}$.

We see that, in each of our examples, there is exactly one solution. Also, there is a definite method of finding that solution, namely, placing all of the variable terms on one side of the equation and all of the constant terms on the other side of the equation.

The total collection of solutions is called the *solution set.* For example, the solution set of the equation $7x-2=0$ is the set $\left\{\frac{2}{7}\right\}$, because $7x-2=0$ if and only if $x=\frac{2}{7}$. In terms of equality of sets this means

$$\{x:\ 7x-2=0\}=\left\{\frac{2}{7}\right\}.$$

From our other examples we have

$$\{x: \ -2x + 3 = 1\} = \{1\}$$

and

$$\{x: \ 2x + 7 = 5x - 1\} = \left\{\frac{8}{3}\right\}.$$

We should note here that the letter x carries no importance. For example, we could use any other symbol or letter in its place such as u, t, or Δ. Thus $\{1\} = \{x: \ -2x + 3 = 1\} = \{u: \ -2u + 3 = 1\} = \{\Delta: \ -2\Delta + 3 = 1\} = \{t: \ -2t + 3 = 1\}$ as long as it is understood that x, u, Δ or t represents a real number.

Problem. Find $\{s: \ 3 = 9s + 5\}$.

Solution. We want to find all values s such that $3 = 9s + 5$. Following the steps in our previous examples, we have

$$3 = 9s + 5,$$
$$3 - 5 = (9s + 5) - 5,$$
$$-2 = 9s,$$
$$\frac{1}{9}(-2) = \frac{1}{9}(9s),$$
$$-\frac{2}{9} = s.$$

Therefore $3 = 9s + 5$ is equivalent to $-\frac{2}{9} = s$ so that $\{s: \ 3 = 9s + 5\} = \left\{-\frac{2}{9}\right\}$.

Exercise Set 2.1

1. Which of the following equations are linear equations in one variable?

   a. $3x + 2 = 0$ b. $-3x + 2 = 0$ c. $0 = 3x + 2$

   d. $3x = 0$ e. $x^2 + 3 = 0$ f. $\dfrac{3x + 2}{2} = 0$

   g. $3x - 2 = 0$ h. $3x + 2 = y$ i. $y = -3x + 2$

2. Rewrite each of the following linear equations in the form $ax + b = 0$:

   a. $3x = 8$ b. $x = -12$ c. $7 - 5 = 4x + 2x$

   d. $-x + \dfrac{1}{2} = x - \dfrac{1}{2}$ e. $4x + 7 - x = 11x + 7$

   f. $\dfrac{1}{3}x - \dfrac{5}{8} + \dfrac{15}{8} = \dfrac{2}{5}x$ g. $cx + p = dx - r$

3. Determine a and b for each of the equations in problem 2.

4. Complete each of the following, by choosing a number which will make the resulting statement true:
   a. If $2x = 7$, then $2x + 4 = $ _____.
   b. If $2x = 7$, then $2x - 12 = $ _____.
   c. If $2x = 7$, then $3(2x) = $ _____.
   d. If $2x = 7$, then $-\dfrac{4}{3}(2x) = $ _____.
   e. If $2x - 6 = 0$, then $2x = $ _____.
   f. If $2x - 6 = 3x$, then $-x = $ _____.
   g. If $2x = -2$, then $x = $ _____.
   h. If $\dfrac{1}{2}x = -2$, then $x = $ _____.

* * *

In problems 5–16, find the solution set of the given equation.

5. $12x + 4 = 0$

6. $8x - 2 = 0$

7. $3t - 6 = 8t + 4$

8. $-5t + 1 = 4t - 8$

9. $\dfrac{1}{2}s - \dfrac{5}{4} = \dfrac{3}{4}s + 1$

10. $0.2s + .07 = 0.5s - .01$

11. $(3y + 1) - (y - 1) = (3y - 1) + (4 - y)$

12. $(3y + 1) - (y + 4) = (y - 5) + (y + 2)$

13. $\left(\dfrac{6}{5} - \dfrac{1}{3}\right) - x = \dfrac{8}{7}x + 3$

14. $1 - \left(\dfrac{6}{5} - \dfrac{1}{3}\right)x = x + 4$

15. $W = 3\left(\dfrac{4}{9}W + \dfrac{4}{3}\right)$

16. $0 = -2\left(\dfrac{1}{3}W + \dfrac{1}{4}W + \dfrac{1}{6}W - \dfrac{1}{4} + \dfrac{1}{12}\right)$

17. Find $\{z: \quad 13z + 7 = 12 - z\}$.

18. Find $\left\{z: \dfrac{z}{8} + 8 = z + \dfrac{1}{8}\right\}$.

19. Find $\{x: \quad ax + b = 0\}$ if $a \neq 0$.

20. Find $\left\{y: \dfrac{1}{c}y + \dfrac{1}{d} = 0\right\}$.

2.2. LINEAR INEQUALITIES

A *linear inequality* (in one variable) is a (conditional) statement of the form

$$ax + b < 0 \text{ (or } ax + b > 0)$$

where $a, b \in \mathcal{R}$. As usual, we read $ax + b < 0$ as "$ax + b$ is less than 0." We read $ax + b > 0$ as "$ax + b$ is greater than 0." For example, $3x + 2 < 0$ is a linear inequality in the variable x.

To solve a linear inequality $ax + b < 0$ means to find all numbers x such that $ax + b < 0$. Such values of x are called *solutions*. Writing this in the notation from the last section we want to find $\{x: \ ax + b < 0\}$. This is our *solution set*.

To do this we will employ the following properties of inequalities:

(i) if $a < b$ then $a + c < b + c$,
(ii) if $a < b$ and $c > 0$ then $ac < bc$,
(iii) if $a < b$ and $c < 0$ then $ac > bc$.

Our method of attack is similar to the one used in solving linear equations. That is, we will apply (i), (ii), and (iii) in such a way that simpler, *equivalent inequalities* (i.e., inequalities with same solutions) are obtained. In particular, we collect all constant and variable terms as before.

Problem. Find all numbers x such that

$$7x - 2 < 0.$$

Solution. This is a linear inequality and to solve it we isolate the variable terms on one side of the inequality and the constant terms on the other side. To do this, we add 2 to both sides of $7x - 2 < 0$. This means that if

$$7x - 2 < 0,$$
then $\qquad 7x - 2 + 2 < 0 + 2,$
or $\qquad 7x < 2.$

In order to solve for x, we must multiply by $\frac{1}{7}$. Hence, if

$$7x < 2,$$
then $\qquad \frac{1}{7}(7x) < \frac{1}{7}(2),$

or $\qquad x < \frac{2}{7}.$

By reversing the steps, we see that if $x < \frac{2}{7}$, then $7x - 2 < 0$. This means that $7x - 2 < 0$ and $x < \frac{2}{7}$ are equivalent inequalities. Therefore $\{x: \ 7x - 2 < 0\} =$

$\{x: \ x < \frac{2}{7}\}$. For example, -1 is a solution of the inequality, since $-1 < \frac{2}{7}$. However, $\frac{19}{3}$ is not a solution since $\frac{2}{7} < \frac{19}{3}$.

Problem. Find all numbers x such that

$$-2x + 3 > 1.$$

Solution. Note that this inequality is equivalent to a linear inequality since it is equivalent to $-2x + 3 - 1 > 0$. Therefore, once again we wish to isolate the variable and the constant terms. This can be accomplished by adding -3 to both sides of the inequality. If

then

or

$$-2x + 3 > 1,$$
$$-2x + 3 - 3 > 1 - 3,$$
$$-2x > -2.$$

In order to solve for x, we will have to multiply by $-\frac{1}{2}$, a negative number. This will reverse the sense of the inequality. Keeping this in mind, if

then

or

$$- 2x > - 2,$$
$$-\frac{1}{2}(-2x) < -\frac{1}{2}(-2),$$
$$x < 1.$$

If the steps are reversed, then $x < 1$ implies $-2x + 3 > 1$. In other words, $-2x + 3 > 1$ and $x < 1$ are equivalent inequalities so that $\{x: \ -2x + 3 > 1\} = \{x: \ x < 1\}$.

Problem. Find all numbers t such that

$$4t - 7 < 2t + 4.$$

Solution. We see that this inequality is equivalent to the inequality

$$(4t - 7) - (2t + 4) < (2t + 4) - (2t + 4),$$

(i.e., $2t - 11 < 0$). Thus it is equivalent to a linear inequality.

To solve the inequality we attempt to isolate the variable terms on one side of the inequality and the constant terms on the other side. To do this we add $- 2t + 7$ to both sides of $4t - 7 < 2t + 4$. That means that if

then

or

$$4t - 7 < 2t + 4,$$
$$(4t - 7) + (- 2t + 7) < (2t + 4) + (- 2t + 7),$$
$$2t < 11.$$

In order to solve for t, we must multiply by $\frac{1}{2}$. Hence if

$$2t < 11,$$

then

$$\frac{1}{2}(2t) < \frac{1}{2}(11),$$

or

$$t < \frac{11}{2}.$$

By reversing the steps, we see that if $t < \frac{11}{2}$, then $4t - 7 < 2t + 4$. This means that $4t - 7 < 2t + 4$ and $t < \frac{11}{2}$ are equivalent inequalities. Therefore $\{t: \ 4t - 7 < 2t + 4\} = \left\{t: \ t < \frac{11}{2}\right\}$. For example, 4 is a solution of the inequality $\left(\text{since } 4 < \frac{11}{2}\right)$ however $\frac{17}{3}$ is not a solution $\left(\text{since } \frac{11}{2} < \frac{17}{3}\right)$.

Problem. Find all numbers x such that

$$-\frac{12}{5}x - \frac{7}{3} > \frac{2}{9}.$$

Solution. Note that this is equivalent to a linear inequality since it is equivalent to $-\frac{12}{5}x - \frac{7}{3} - \frac{2}{9} > 0$.

Once again, we wish to isolate the variable and constant terms. This can be accomplished by adding $\frac{7}{3}$ to both sides of the inequality. If

$$-\frac{12}{5}x - \frac{7}{3} > \frac{2}{9},$$

then

$$-\frac{12}{5}x - \frac{7}{3} + \frac{7}{3} > \frac{2}{9} + \frac{7}{3},$$

or

$$-\frac{12}{5}x > \frac{23}{9}.$$

In order to solve for x, we will have to multiply by $-\frac{5}{12}$, a negative number. This will reverse the sense of inequality. Keeping this in mind, if

$$-\frac{12}{5}x > \frac{23}{9},$$

then

$$-\frac{5}{12}\left(-\frac{12}{5}x\right) < -\frac{5}{12}\left(\frac{23}{9}\right),$$

or
$$x < -\frac{115}{108}.$$

If the steps are reversed, then $x < -\frac{115}{108}$ implies $-\frac{12}{5}x - \frac{7}{3} > \frac{2}{9}$. In other words, $-\frac{12}{5}x - \frac{7}{3} > \frac{2}{9}$ and $x < -\frac{115}{108}$ are equivalent inequalities so that

$$\left\{x: \quad -\frac{12}{5}x - \frac{7}{3} > \frac{2}{9}\right\} = \left\{x: \quad x < -\frac{115}{108}\right\}.$$

We combine the statements $ax + b = 0$ or $ax + b < 0$ to read $ax + b \leqslant 0$, read "$ax + b$ is less than or equal to 0." Similarly $ax + b \geqslant 0$ means either $ax + b = 0$ or $ax + b > 0$. It now makes sense to talk about finding solutions of $ax + b \geqslant 0$, $ax + b \leqslant 0$, or equivalent inequalities.

Problem. Solve the inequality $3x - 7 \geqslant 2$.

Solution. To do this we solve $3x - 7 = 2$ and $3x - 7 > 2$. The solution set of $3x - 7 = 2$ is $\{3\}$, while the solution set of $3x - 7 > 2$ is $\{x: \ x > 3\}$. Combining the two solution sets, we see that

$$\{x: \ 3x - 7 \geqslant 2\} = \{x: \ x \geqslant 3\}.$$

Exercise Set 2.2

1. Translate each of the following notational statements into a verbal statement:
   a. $3x + 5 > 1$ b. $3x + 5 \leqslant 1$ c. $-5x - 1 \geqslant 20$

2. For which of the given inequalities is -2 a solution?
   a. $x + \frac{1}{2} < 325$ b. $x + \frac{1}{2} < -1$ c. $x + 2 < 0$
   d. $-x - 2 \geqslant 0$ e. $-5x + 4 > 15$ f. $-5x - 30 > -15$

3. Given the set of symbols, $\{<, >, \leqslant, \geqslant, =\}$, fill in each blank with the symbol which will make the statement true:
   a. If $2x < 7$, then $2x + 4$ _____ $7 + 4$.
   b. If $2x > 7$, then $2x - 12$ _____ -5.
   c. If $2x \geqslant 7$, then $3(2x)$ _____ 21.
   d. If $2x < 7$, then $-\frac{4}{3}(2x)$ _____ $-\frac{28}{3}$.
   e. If $\frac{x}{2} - 6 \geqslant 0$, then $2x$ _____ 24.
   f. If $2x - 6 < 3x$, then x _____ -6.
   g. If $-2x \leqslant -2$, then x _____ 1.
   h. If $\frac{1}{2x} > 0$, then $2x$ _____ 0.

4. Fill in each blank with a real number which will make the statement true.
   a. If $x < -2$, then $x + 2 < $ _____ .
   b. If $x \geqslant -7$, then $x - 4 \geqslant $ _____ .
   c. If $2x + 1 > -5$, then $2x + 6 > $ _____ .
   d. If $2x + 1 > -5$, then $2x - 3 > $ _____ .
   e. If $2x + 1 > 8$, then $2x > $ _____ .
   f. If $-4 < x < -1$, then _____ $< x + 1 < $ _____ .
   g. If $-1 < x < \dfrac{3}{2}$, then _____ $< x + 1 < $ _____ .
   h. If $\dfrac{3}{2} < x$, then _____ $< x + 1$.

* * *

In problems 5-16, find the solution set for each of the given linear inequalities or equalities:

5. $-3x + 5 > 0$

6. $5t + 2 \geqslant 9t - 4$

7. $-4t - 1 < 4t + 7$

8. $-16y - 12 = 0$

9. $-16y - 12 < 0$

10. $-16y - 12 > 0$

11. $\dfrac{3}{5} W + \dfrac{6}{5} W = 3$

12. $\dfrac{3}{5} W + \dfrac{6}{5} W \geqslant 3$

13. $2.2x - 3.5x < .4(.6x - 1.1)$

14. $18x - 2 \leqslant 18x + 3$

15. $\dfrac{12}{11} W - 5 < \dfrac{13}{12} W$

16. $\dfrac{4}{5}(x + 1) - \left(\dfrac{3}{8} - \dfrac{1}{3}\right)x > \dfrac{11}{12} - 4x$

17. Find $\left\{z: \ z + \dfrac{1}{2} < z - \dfrac{1}{2}\right\}$

18. Find $\left\{z: \ \dfrac{z}{2} - \dfrac{-z}{4} \geqslant \dfrac{z}{-3} + \dfrac{-z}{5}\right\}$

19. Find $\{x: \ ax + b < 0 \text{ for } a > 0\}$

20. Find $\left\{y: \ \dfrac{1}{c}y + \dfrac{1}{d} < 0 \text{ for } c < 0\right\}$

2.3 CARTESIAN (RECTANGULAR) COORDINATE SYSTEM

Take two real number lines and intersect them at right angles, the common point of intersection being the number 0 on each line (see Figure 2.1). We usually assume that the same unit length is used on each number line. The horizontal

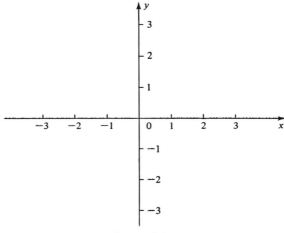

Figure 2.1

line is called the *x-axis* and has a positive direction from left to right. The vertical line is called the *y-axis* and has its positive direction upward. These two lines are called the *coordinate axes*. The point of intersection (0 on each axis) is called the *origin*. We shall now see how these two lines can be used to give the location of any point in the plane containing these two axes (think of the page as an unending sheet in any direction and of each axis as an unending line in either direction). Let P be any point in the plane (see Figure 2.2). Construct lines through P parallel to the x-axis and the y-axis. Let Q and T be the

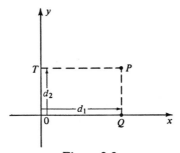

Figure 2.2

points of intersection on the x-axis and y-axis, respectively. Let d_1 denote the directed distance on the x-axis from 0 to Q, and let d_2 denote the directed distance on the y-axis from 0 to T. We assign to the point P the ordered pair (d_1, d_2) of real numbers. Actually, the ordered pair (d_1, d_2) serves as a unique set of directions used to locate P. For if we know d_1 and d_2, then we can determine the position of P in the plane. Thus for each point in the plane there corresponds one and only one ordered pair (d_1, d_2) of real numbers and vice versa. Note that while the standard notation (d_1, d_2) for the ordered pair is the same as the standard notation for the open interval from d_1 to d_2, the meaning of the symbol (d_1, d_2) will be clear from the context in which it is used.

Let us stop for a moment and examine the term "ordered pair (d_1, d_2)." The word "pair" means, of course, two objects (in this case d_1 and d_2). The word "ordered" means that it is important to consider which object occurs in the first position (i.e., d_1) and which one occurs in the second position (i.e., d_2). For example, (3, 1) and (1, 3) are different ordered pairs and represent different points in the plane (see Figure 2.3).

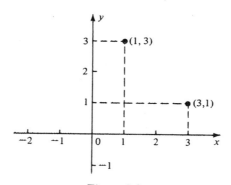

Figure 2.3

Returning to our discussion of points (d_1, d_2) we call d_1 the *x-coordinate of the point P* and d_2 the *y-coordinate of the point P*. For this reason a point is frequently written (x, y). The set of all ordered pairs (x, y), where x and y are real numbers, is called the *Cartesian* (rectangular) *coordinate system* of the plane. Plotting points in the plane is the process of locating the point P given the coordinates (x, y) of P.

Problem. Plot the points $(-2, -1)$, $\left(4, \dfrac{1}{2}\right)$, $\left(\dfrac{5}{2}, -3\right)$, and $\left(0, \dfrac{5}{4}\right)$.

Solution. These points are indicated in Figure 2.4.

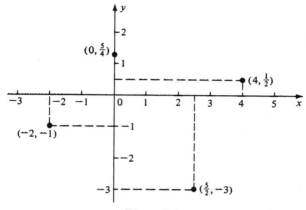

Figure 2.4

Problem. Plot the points $(x, 4x - 1)$, where $x = -1, 0, 1, 2$.

Solution. The following table lists the points we wish to plot. They are shown in Figure 2.5.

x	$(x, 4x - 1)$
-1	$(-1, 4(-1)-1) = (-1, -5)$
0	$(0, 4(0)-1) = (0, -1)$
1	$(1, 4(1)-1) = (1, 3)$
2	$(2, 4(2)-1) = (2, 7)$

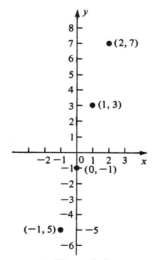

Figure 2.5

Exercise Set 2.3

1. Consider the following figure:

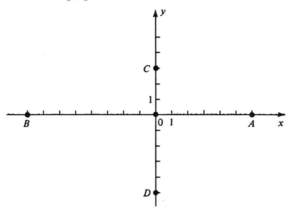

 a. Which line is the x-axis?
 b. Which line is the y-axis?
 c. Locate the origin.
 d. What is the directed distance from 0 to A?
 e. What is the directed distance from 0 to B?
 f. What is the directed distance from 0 to C?
 g. What is the directed distance from 0 to D?
 h. What is the directed distance from 0 to 0?

2. Consider the following figure:

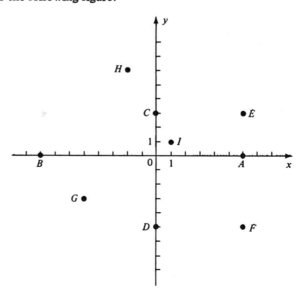

What is the ordered pair that is assigned to:

a. Point E?   b. Point F?   c. Point I?   d. Point G?

e. Point A?   f. Point D?   g. Point H?   h. Point B?

3. Plot the given points in the Cartesian coordinate system:

a. $(4, -1)$ b. $\left(\dfrac{2}{3}, \pi\right)$ c. $(0, 0)$ d. $(-3, -2)$

e. $(3, 2)$ f. $\left(1, -\dfrac{1}{2}\right)$ g. $(0, 4)$ h. $\left(-\dfrac{5}{2}, 0\right)$

* * *

In exercises 4–7, plot the points in the Cartesian coordinate system:

4. Points $(x, 3x - 2)$ for $x = -1, 0, 1, 2$

5. Points $(x, -2x + 1)$ for $x = 0, 1, -2, 3$

6. Points $(x, 3)$ for $x = -2, -1, 0, 1$

7. Points $(x, 4x + 3)$ for $x = -1, -\dfrac{1}{2}, 0, \dfrac{1}{2}, 1, \dfrac{3}{2}$

In exercises 8–15, plot at least four points belonging to each set:

8. $\{(x, x) : x \in \mathcal{R}\}$

9. $\{(x, y) : y = x\}$

10. $\{(x, -2x) : x \in \mathcal{R}\}$

11. $\{(x, y) : y = -2x\}$

12. $\{(x, ax) : a < 0 \text{ and } x \in \mathcal{R}\}$

13. $\{(x, y) : 3y = 5x\}$

14. $\{(x, \sqrt{3}\, x) : x \text{ is an integer}\}$

15. $\{(x, y) : y = 3.2\}$

2.4 THE DISTANCE FORMULA

Suppose we have two points in the plane. What is the straight line distance between these two points? The answer to this question will extend the notion of distance between two points on a coordinatized line and will prove useful in later work.

Before answering this question we must first briefly investigate the meaning of the square root of a number and the square of a number.

If $r \in \mathcal{R}$ then the *square* of r, written r^2, is the number $r \cdot r$. The 2 (in r^2) is the power (or exponent) of r. We will study this further in Chap. 3.

Example.   (i) $\left(-\dfrac{7}{4}\right)^2 = \left(-\dfrac{7}{4}\right)\left(-\dfrac{7}{4}\right) = \dfrac{49}{16}$,

(ii) $(14)^2 = (14)(14) = 196$,

(iii) $(-14)^2 = (-14)(-14) = 196$,

(iv) $\left(\dfrac{1}{3}\right)^2 = \left(\dfrac{1}{3}\right)\left(\dfrac{1}{3}\right) = \dfrac{1}{9}$.

If r is a nonnegative real number, then the principal *square root* of r, written $\sqrt{r}$, is that nonnegative number t such that $t^2 = r$.

Example.   (i) $\sqrt{16} = 4$ since $4^2 = 16$,

(ii) $\sqrt{6.25} = 2.5$ since $(2.5)^2 = 6.25$,

(iii) $\sqrt{\dfrac{1}{4}} = \dfrac{1}{2}$ since $\left(\dfrac{1}{2}\right)^2 = \dfrac{1}{4}$.

If a and b are positive, then $\sqrt{ab} = \sqrt{a}\,\sqrt{b}$ and $\sqrt{\dfrac{a}{b}} = \dfrac{\sqrt{a}}{\sqrt{b}}$.

Example.   (i) $\sqrt{15} = \sqrt{3 \cdot 5} = \sqrt{3}\,\sqrt{5}$,

(ii) $\sqrt{28} = \sqrt{4 \cdot 7} = \sqrt{4}\,\sqrt{7} = 2\sqrt{7}$,

(iii) $\sqrt{\dfrac{5}{7}} = \dfrac{\sqrt{5}}{\sqrt{7}}$,

(iv) $\sqrt{\dfrac{16}{75}} = \dfrac{\sqrt{16}}{\sqrt{75}} = \dfrac{4}{\sqrt{3(25)}} = \dfrac{4}{5\sqrt{3}}$.

There is one pitfall that should carefully be avoided when using square roots. Namely, if a and b are positive, then, in general, $\sqrt{a+b}$ and $\sqrt{a} + \sqrt{b}$ are unequal. For example, $\sqrt{16+9} = \sqrt{25} = 5$ while $\sqrt{16} + \sqrt{9} = 4 + 3 = 7$. Therefore $\sqrt{16+9} \neq \sqrt{16} + \sqrt{9}$.

Finally, before we examine distance, we should note the use of subscripts. We have used d_1 and d_2 (read "d sub 1" and "d sub 2") in Sec. 2.3. The numerals 1 and 2 are called subscripts. This is a very handy method of notation and we shall make frequent use of it.

Now, in order to find the distance between two points (x_1, y_1) and (x_2, y_2) in the plane, we make use of Pythagorean theorem which states that given a right triangle, the sum of the squares of the two sides is equal to the square of the hypotenuse. Let d denote the straight line distance between (x_1, y_1) and (x_2, y_2) (see Figure 2.6). Construct lines parallel to the x-axis through the points (x_1, y_1) and (x_2, y_2) and construct lines parallel to the y-axis through the points (x_1, y_1) and (x_2, y_2). These four lines form a rectangle having vertices (x_1, y_1), P, (x_2, y_2), Q. Consider the right triangle (x_1, y_1), P, (x_2, y_2). The distance, s, from (x_1, y_1) to P is given by $|x_2 - x_1|$ and the distance, t, from P to (x_2, y_2) is given by $|y_2 - y_1|$. So, applying the Pythagorean theorem,

$$d^2 = s^2 + t^2$$
$$= |x_2 - x_1|^2 + |y_2 - y_1|^2.$$

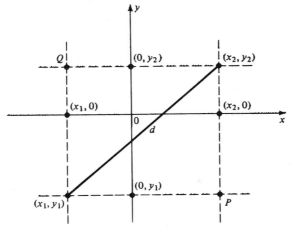

Figure 2.6

Since $|x_2 - x_1|^2 = (x_2 - x_1)^2$ and $|y_2 - y_1|^2 = (y_2 - y_1)^2$ (see Exercise Set 2.4, problem 17), we have that

$$d = \sqrt{(x_2 - x_1)^2 + (y_2 - y_1)^2}.$$

2.1 Theorem. *The distance d between two points (x_1, y_1) and (x_2, y_2) in the plane is given by the formula*

$$d = \sqrt{(x_2 - x_1)^2 + (y_2 - y_1)^2}.$$

Problem. Find the distance between the points $(3, -1)$ and $(-2, -4)$.

Solution. Let $(x_1, y_1) = (3, -1)$ and $(x_2, y_2) = (-2, -4)$; then $x_1 = 3, y_1 = -1$, $x_2 = -2$, and $y_2 = -4$. Therefore

$$d = \sqrt{(-2 - 3)^2 + (-4 - (-1))^2} = \sqrt{(-5)^2 + (-3)^2}$$
$$= \sqrt{25 + 9} = \sqrt{34}.$$

If we wanted to estimate $\sqrt{34}$, we see that

$$5 < \sqrt{34} < 6$$

since $5^2 = 25$ and $6^2 = 36$.

Problem. Show that the points $(3, 1)$, $(-2, -3)$, and $(2, 2)$ are vertices of an isosceles triangle (see Figure 2.7).

Solution. Let us determine the distances between the points. We will see that two of the distances are equal. Let d_1 be the distance from $(3, 1)$ to $(-2, -3)$.

Then

$$d_1 = \sqrt{(3 - (-2))^2 + (1 - (-3))^2} = \sqrt{5^2 + 4^2}$$
$$= \sqrt{25 + 16} = \sqrt{41}.$$

Let d_2 be the distance from $(3, 1)$ to $(2, 2)$. Then

$$d_2 = \sqrt{(3 - 2)^2 + (1 - 2)^2} = \sqrt{1^2 + (-1)^2} = \sqrt{2}.$$

Finally, letting d_3 be the distance from $(2, 2)$ to $(-2, -3)$, we have

$$d_3 = \sqrt{(2 - (-2))^2 + (2 - (-3))^2} = \sqrt{4^2 + 5^2}$$
$$= \sqrt{16 + 25} = \sqrt{41}.$$

We see that $d_1 = d_3$, which implies that the triangle is isosceles.

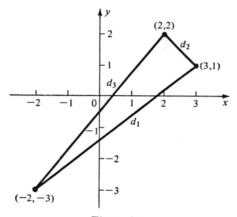

Figure 2.7

The following problem is somewhat more difficult. The solution is given, however, since this interesting fact is the converse of the Pythagorean theorem.

Problem. Let a triangle PQR have sides of positive lengths d_1, d_2, and d_3 and assume that $d_1^2 = d_2^2 + d_3^2$. Does this imply the triangle is a right triangle?

Solution. We introduce a coordinate system in the plane so that P and Q lie on the nonnegative x-axis with P at the origin (see Figure 2.8). Thus P has coordinates $(0, 0)$, Q has coordinates $(x_2, 0)$, and R has coordinates (x_3, y_3). By the distance formula,

$$d_1 = \sqrt{(x_3 - 0)^2 + (y_3 - 0)^2},$$
$$d_2 = \sqrt{(x_2 - 0)^2 + (0 - 0)^2},$$
$$d_3 = \sqrt{(x_3 - x_2)^2 + (y_3 - 0)^2}.$$

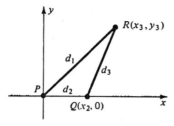

Figure 2.8

Since $d_1^2 = d_2^2 + d_3^2$, it follows that

$$x_3^2 + y_3^2 = x_2^2 + (x_3 - x_2)^2 + y_3^2,$$
$$x_3^2 + y_3^2 = x_2^2 + x_3^2 - 2x_2x_3 + x_2^2 + y_3^2,$$
$$0 = 2x_2^2 - 2x_2x_3,$$
$$0 = 2x_2(x_2 - x_3).$$

Because d_2 is positive, $x_2 \neq 0$ hence $x_2 - x_3 = 0$. Since $x_2 = x_3$, the line through P and Q is parallel to the y-axis so that PQR is a right triangle.

Problem. Determine if the points (7, 0), (4, 1), (6, 7) are vertices of a right triangle.

Solution. We find the lengths of the three sides and see if the converse of the Pythagorean theorem holds. First, the distance from (7, 0) to (4, 1) is

$$d_1 = \sqrt{(7 - 4)^2 + (0 - 1)^2} = \sqrt{3^2 + (-1)^2} = \sqrt{9 + 1} = \sqrt{10},$$

the distance from (4, 1) to (6, 7) is

$$d_2 = \sqrt{(4 - 6)^2 + (1 - 7)^2} = \sqrt{(-2)^2 + (-6)^2}$$
$$= \sqrt{4 + 36} = \sqrt{40},$$

and the distance from (6, 7) to (7, 0) is

$$d_3 = \sqrt{(6 - 7)^2 + (7 - 0)^2} = \sqrt{(-1)^2 + 7^2}$$
$$= \sqrt{1 + 49} = \sqrt{50}.$$

Since $d_3^2 = 50, d_1^2 + d_2^2 = 10 + 40$, and $d_3^2 = d_1^2 + d_2^2$, we have a right triangle.

Example. Let P and Q be distinct points in the plane (see Figure 2.9). The *midpoint*, M, of the segment of the line from P to Q is that point on the line from P to Q where the distance between P and M is equal to the distance between M and Q.

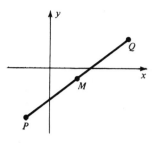

Figure 2.9

If P has coordinates (x_1, y_1) and Q has coordinates (x_2, y_2), then we claim M has coordinates $\left(\dfrac{x_1 + x_2}{2}, \dfrac{y_1 + y_2}{2}\right)$. To check this let $\overline{M}$ have coordinates $\left(\dfrac{x_1 + x_2}{2}, \dfrac{y_1 + y_2}{2}\right)$, let d_1 be the distance from P to $\overline{M}$, let d_2 be the distance from $\overline{M}$ to Q, and let d_3 be the distance between P and Q. First, note that we have $d_1 = d_2$ because

$$d_1 = \sqrt{\left(x_1 - \frac{x_1 + x_2}{2}\right)^2 + \left(y_1 - \frac{y_1 + y_2}{2}\right)^2}$$

$$= \sqrt{\left(\frac{x_1 - x_2}{2}\right)^2 + \left(\frac{y_1 - y_2}{2}\right)^2}$$

and

$$d_2 = \sqrt{\left(x_2 - \frac{x_1 + x_2}{2}\right)^2 + \left(y_2 - \frac{y_1 + y_2}{2}\right)^2}$$

$$= \sqrt{\left(\frac{x_2 - x_1}{2}\right)^2 + \left(\frac{y_2 - y_1}{2}\right)^2}$$

$$= \sqrt{\left(\frac{x_1 - x_2}{2}\right)^2 + \left(\frac{y_1 - y_2}{2}\right)^2}.$$

This places $\overline{M}$ on the perpendicular line which bisects the segment from P to Q.

Now, if $d_1 + d_2 = d_3$, then $\overline{M}$ is on the line segment (why?) and is, indeed, the midpoint. Since $d_1 = d_2$ we have that $d_1 + d_2 = 2d_2$. So we want to check to determine whether $2d_2 = d_3$. We have

$$2d_2 = 2\sqrt{\left(\frac{x_2 - x_1}{2}\right)^2 + \left(\frac{y_2 - y_1}{2}\right)^2}$$

$$= 2\sqrt{\frac{1}{4}[(x_2 - x_1)^2 + (y_2 - y_1)^2]} = 2\sqrt{\left(\frac{1}{2}\right)^2 [(x_2 - x_1)^2 + (y_2 - y_1)^2]}$$

$$= 2 \cdot \frac{1}{2}\sqrt{(x_2 - x_1)^2 + (y_2 - y_1)^2} = \sqrt{(x_2 - x_1)^2 + (y_2 - y_1)^2}$$

and

$$d_3 = \sqrt{(x_2 - x_1)^2 + (y_2 - y_1)^2}.$$

So $2d_2 = d_3$. Thus $\overline{M}$ and M are the same point, which proves our assertion.

Example. The midpoint of the line segment from $\left(3, \dfrac{1}{2}\right)$ to $(-7, 4)$ is the point

$$\left(\frac{3 + (-7)}{2}, \frac{\dfrac{1}{2} + 4}{2}\right) = \left(-2, \frac{9}{4}\right).$$

Exercise Set 2.4

1. For the given r, determine r^2:

   a. $r = 7$ b. $r = -\dfrac{2}{3}$ c. $r = \dfrac{11}{4}$ d. $r = 1.732$

   e. $r = (-2)^2$ f. $r = (3^2)(2^2)$ g. $r = [(3)(2)]^2$ h. $r = 0$

   i. $r = \sqrt{5}$ j. $r = -\sqrt{\dfrac{5}{22}}$ k. $r = \sqrt{a}$ l. $r = -\sqrt{c}$

2. For the given r, determine $\sqrt{r}$:

   a. $r = 9$ b. $r = \dfrac{4}{9}$ c. $r = 4^2$ d. $r = (-3)^2$

   e. $r = (2^2)(4^2)$ f. $r = [(2)(4)]^2$ g. $r = 49$ h. $r = 0$

   i. $r = \pi^2$ j. $r = 1.21$ k. $r = .64$ l. $r = .064$

3. Evaluate each of the following:

   a. $\sqrt{9 + 16}$

   b. $\sqrt{9} + \sqrt{16}$

   c. $\sqrt{30 + 6}$

   d. $\sqrt{30} + \sqrt{6}$

   e. $(5 + 2)^2$

   f. $5^2 + 2^2$

   g. $(\sqrt{2} - 3)^2$

   h. $(\sqrt{2})^2 - 3^2$

4. Given the points $(2, 5)$ and $(3, -4)$:

   a. Let $P = (2, 5) = (x_1, y_1)$. What is x_1? y_1?

   b. Let $Q = (3, -4) = (x_2, y_2)$. What is x_2? y_2?

   c. Find the distance between the points $(2, 5)$ and $(3, -4)$ by substituting these values for x_1, y_1, x_2, y_2 in the distance formula.

   d. Now let $P = (3, -4) = (x_1, y_1)$. What is x_1? y_1?

   e. Let $Q = (2, 5) = (x_2, y_2)$. What is x_2? y_2?

   f. Find the distance between the points P and Q by substituting these values for x_1, y_1, x_2, and y_2 in the distance formula.

* * *

In exercises 5–12, use the distance formula to determine the distance between P and Q:

5. $P = (4, 9)$ $Q = (1, 6)$

6. $P = (2, 5)$ $Q = (2, 9)$

7. $P = (-1. 5)$ $Q = (1, -4)$

8. $P = (-8, 3)$ $Q = (2, -1)$

9. $P = \left(\dfrac{1}{2}, -\dfrac{2}{3}\right)$ $Q = \left(-\dfrac{3}{2}, -\dfrac{4}{3}\right)$

10. $P = \left(\dfrac{2}{5}, -\dfrac{2}{3}\right)$ $Q = \left(-\dfrac{8}{5}, -\dfrac{3}{2}\right)$

11. $P = (2\sqrt{2}, -\sqrt{5})$ $Q = (3\sqrt{2}, 2\sqrt{5})$

12. $P = \left(-\sqrt{3}, -\dfrac{1}{2}\right)$ $Q = (4\sqrt{3}, -8)$

13. Find the midpoint of the line segment from P to Q for:

    a. $P = (4, 9)$ $Q = (1, 6)$

    b. $P = (-1, 5)$ $Q = (1, -4)$

    c. $P = \left(\dfrac{1}{2}, -\dfrac{2}{3}\right)$ $Q = \left(-\dfrac{3}{2}, -\dfrac{4}{3}\right)$

    d. $P = \left(\dfrac{2}{5}, -\dfrac{2}{3}\right)$ $Q = \left(-\dfrac{8}{5}, -\dfrac{3}{2}\right)$

14. Which of the following points are a distance of 13 from the origin?

    a. $(-5, 12)$ b. $(\sqrt{133}, -6)$ c. $(1, 13)$ d. $(9, -9)$

15. If a triangle has vertices $(0, 2)$, $(\sqrt{3}, -1)$, and $(0, -2)$, prove that it is a right triangle.

16. If a triangle has vertices $(-1, -1)$, $(-2, 5)$, and $(1, 1)$, determine whether or not it is a right triangle.

17. Show that $|x - y|^2 = (x - y)^2$. (Use the definition of absolute value.)

*18. Find a real number M such that $(15, 4)$, $(-7, 8)$, and $(M, -4)$ are vertices of a right triangle with the right angle at the vertex $(M, 1)$.

*19. Given the right triangle with vertices $(7, 0)$, $(4, 1)$, and $(6, 7)$, find the midpoints of the three sides. Determine the area of the original triangle and compare that to the area of the triangle having the midpoints as vertices.

2.5 THE GRAPH OF A LINEAR EQUATION

Let us consider what we mean by a straight line. We can construct a straight line by taking a straight edge, laying it on a piece of paper, and marking along one edge with a pencil. We can roughly determine whether or not a point is on the line visually. However, if we want to completely characterize all the points on the line, this procedure is not sufficient. By introducing a coordinate system in the plane, we can consider just the coordinates of the points on the line. Can we characterize, algebraically, the coordinates of points on the line? In other words, can we write an algebraic expression which identifies this line? If so, how can we distinguish this line algebraically from any other line? Do two distinct points determine a line uniquely as we were taught in plane geometry? We will answer these questions and discover other facts concerning lines in this section.

Certainly, one of the distinguishing features of a straight line is its inclination to the x-axis. For example, the lines l_1 and l_2 in Figure 2.10 have different inclinations to the x-axis.

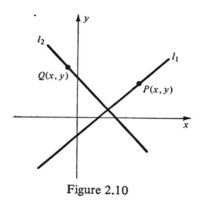

Figure 2.10

This is indicated by the fact that the point $P(x, y)$ on l_1 rises as x increases, while $Q(x, y)$ on l_2 falls as x increases. The following definition gives us a way of measuring this inclination.

2.2 Definition. *Let l be a nonvertical line and let (a, b), (c, d) be distinct points on l. The slope of l, denoted by m, is defined by the formula*

$$m = \frac{d - b}{c - a}.$$

Consider, for example, the line shown in Figure 2.11. Since $a < c$ and $b < d$, the numbers $d - b$ and $c - a$ are positive. Hence the slope m is positive so that lines inclined in this manner have positive slope.

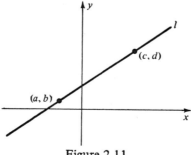

Figure 2.11

On the other hand, consider the line shown in Figure 2.12. Since $a < c$ and $d < b$, the number $c - a$ is positive but $d - b$ is negative. Consequently, m is negative so that lines inclined in this manner have negative slope.

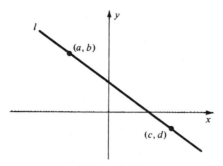

Figure 2.12

Note that we do not define the slope of a vertical line l, because when l is parallel to the y-axis the x-coordinates of all points on l are equal. Thus $c - a = 0$ and division by 0 is not permitted. If l is a horizontal line, then the y-coordinates of all points on l are equal. Thus $d - b = 0$, implying $m = 0$. Therefore horizontal lines have slope zero. We might also note here that it does not make any difference as to the order in which we take (a, b) and (c, d). However, once we have decided which point we are going to treat as the first point we must subtract coordinates in the denominator in the same order as we subtract them in the numerator. Notice that

$$\frac{b - d}{a - c} = \frac{(-1)(b - d)}{(-1)(a - c)} = \frac{d - b}{c - a}.$$

Hence $\dfrac{b - d}{a - c}$ also gives us the slope.

Problem. Find the slope of the line l if the points $(7, 1)$ and $(-1, 4)$ are on the line.

Solution. Using the formula $\dfrac{d - b}{c - a}$ we have that the slope is given by $\dfrac{4 - 1}{-1 - 7} = \dfrac{3}{-8} = -\dfrac{3}{8}.$

According to the definition of slope, the number $m = \dfrac{d - b}{c - a}$ appears to depend on the pair of points (a, b) and (c, d) of l which were used to form the quotient.

Fortunately, this is not the case. The verification of this statement is a fairly straightforward geometric argument using similar triangles. We omit the proof but record this important fact.

2.3 Theorem. *Let l be a nonvertical line. IF (a, b), (c, d) are distinct points on l and (a', b'), (c', d') are distinct points on l,*

THEN

$$\frac{d - b}{c - a} = \frac{d' - b'}{c' - a'}.$$

The preceding theorem now enables us to characterize a line algebraically. Let us investigate a specific example before considering the general case. Suppose that a line passes through the two points (4, 1) and (−2, 2). Thus, the line has slope

$$m = \frac{1 - 2}{4 - (-2)} = \frac{-1}{6} = -\frac{1}{6}.$$

Let (x, y) be any other point on the line. We know that the slope of the line can be determined by any two points on the line. For example, using the points (x, y) and (4, 1), we have

$$\frac{y - 1}{x - 4} = -\frac{1}{6}.$$

Multiplying both sides by $6(x - 4)$ yields $-(x - 4) = 6(y - 1)$. Equivalently, this means

$$x + 6y = 10.$$

We have just seen that every point on the line must satisfy this equation. This is called the equation of the line.

In general, if a line has slope m and passes through the point (a, b), then the equation of the line is

$$m = \frac{y - b}{x - a}.$$

The latter equation is called the *point-slope form* of the equation of the line through the point (a, b) with slope m. The equation can be rewritten as

$$m(x - a) = y - b$$

or
$$y = mx + (b - am).$$

Problem. Find the equation of the line passing through the points $(-1, 6)$ and $(2, 3)$.

Solution. This line has slope $m = \dfrac{6 - 3}{-1 - 2} = \dfrac{3}{-3} = -1$. Now, let (x, y) be any other point on the line. Thus

$$-1 = \frac{y - 6}{x - (-1)} \quad \text{(using the points } (x, y) \text{ and } (-1, 6))$$

or $-(x + 1) = y - 6$.

Therefore the equation of the line is

$$y = -x + 5.$$

Let us convince ourselves that using $(-1, 6)$ instead of $(2, 3)$ made no difference in the resulting equation of the line. Again, let (x, y) be any other point on the line. Thus

$$-1 = \frac{y - 3}{x - 2} \quad \text{(using } (x, y) \text{ and } (2, 3))$$

or $(-1)(x - 2) = y - 3$.

Therefore

$$-x + 2 = y - 3$$

or $y = -x + 5$.

Notice that we can determine the equation of a line if we have the slope and a point on the line (and to determine the slope we need two points on the line). This means that two points determine a line uniquely.

We have shown that, given a nonvertical line l, there exist real numbers A, B, and C, not all zero, such that every point (x, y) on l satisfies an equation of the form

$$Ax + By = C.$$

Such an equation is called a *linear equation.* It is also a fact that any point (x, y), whose coordinates satisfy the preceding linear equation, is a point on l (see Exercise Set 2.5, problem 23).

Thus every line is represented by a linear equation. Let us now show that every linear equation represents a straight line. Suppose A, B, and C are not all zero. If $B = 0$ and $Ax + By = Ax + 0 \cdot y = C$, then $x = \dfrac{C}{A}$. Hence (x, y) lies on the vertical line which intersects the x-axis at $\left(\dfrac{C}{A}, 0\right)$. If $B \neq 0$, then $Ax + By = C$ if and only if $y = -\dfrac{A}{B}x + \dfrac{C}{B}$. But all points satisfying the latter equation lie on the line with slope $-\dfrac{A}{B}$ which passes through $\left(0, \dfrac{C}{A}\right)$. Hence $Ax + By = C$ represents a straight line.

2.4 Theorem. *Let A, B, C be real numbers, not all zero. The set of all points (x, y) satisfying the linear equation*

$$Ax + By = C$$

is a straight line. Moreover, every straight line is the set òf points (x, y) satisfying a linear equation.

Therefore all lines are represented by linear equations and all linear equations represent lines. Given a linear equation

$$Ax + By = C$$

the set $\{(x, y): Ax + By = C\}$ is called the *graph* of the equation. Moreover, this graph is a straight line.

Problem. Sketch the graphs of the equations

(i) $3x + 2y = 5$

and (ii) $5x - 3y = 4$.

Solution. As we have seen, it is only necessary to find two points on the line. We can do this by letting x (or y) be any arbitrary value and find the corresponding value of y (or x). Letting $y = 0$ in (i), for example, yields $3x + 2(0) = 5$ or

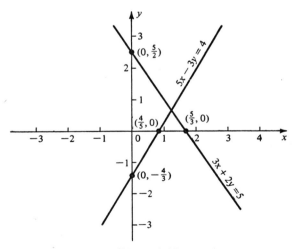

Figure 2.13

$x = \frac{5}{3}$. Thus the point $\left(\frac{5}{3}, 0\right)$ is on this line and is the point at which the line crosses the x-axis. Appropriately, such a point is called the *x-intercept* of the line. If $x = 0$ in (i), then $3(0) + 2y = 5$ so that $\left(0, \frac{5}{2}\right)$ is on the line. This is the point at which the line crosses the y-axis, and it is called the *y-intercept* of the line. In case (ii) the x-intercept is $\left(\frac{4}{5}, 0\right)$ and the y-intercept is $\left(0, -\frac{4}{3}\right)$. The graphs of these lines are sketched in Figure 2.13.

Exercise Set 2.5

1. On a set of coordinate axes, sketch a line such that:
   a. its slope is positive
   b. its slope is negative
   c. its slope is zero
   d. its slope is undefined

2. Find the slope of the line passing through the points A and B if:
   a. $A = (2, 3)$ $B = (4, 9)$
   b. $A = (2, 9)$ $B = (4, 3)$
   c. $A = (-4, 6)$ $B = (3, -8)$
   d. $A = (-5, -7)$ $B = (4, -12)$
   e. $A = \left(\frac{1}{2}, \frac{3}{4}\right)$ $B = \left(-\frac{5}{8}, \frac{1}{4}\right)$
   f. $A = (-3, -7)$ $B = (3, 7)$
   g. $A = (-3, 7)$ $B = (3, 7)$
   h. $A = \left(\frac{25}{21}, -\frac{6}{7}\right)$ $B = \left(-\frac{2}{3}, -\frac{5}{14}\right)$

3. Find an equation of the line passing through the point A with a slope equal to m:
   a. $A = (-3, 5)$, $m = 1$
   b. $A = (-3, 5)$, $m = 4$
   c. $A = (-3, 5)$, $m = -2$
   d. $A = (-3, 5)$, $m = 0.$

4. Given the linear equation $2x + y = 6$:
   a. Complete the following table:

x	0	1	-1	4	-4	$\frac{1}{2}$	$-\frac{1}{2}$	$\frac{7}{2}$	$-\frac{7}{2}$
y									

   b. On a set of coordinate axes, plot the points whose coordinates are the pairs (x, y) in the table.

c. Sketch the line which passes through the points.

d. What is the equation of this line?

* * *

In problems 5–8, find an equation of the line having slope m and passing through the given point P:

5. a. $m = -\dfrac{3}{5}$ $P = (3, -2)$ b. $m = -\dfrac{3}{5}$ $P = (3, 2)$

6. a. $m = 0$ $P = (4, 1)$ b. $m = 0$ $P = (-12, 2)$

7. a. $m = \dfrac{3}{11}$ $P = \left(\dfrac{1}{2}, -\dfrac{2}{3}\right)$ b. $m = \dfrac{3}{11}$ $P = \left(2, -\dfrac{3}{2}\right)$

8. a. $m = 1003$ $P = (2, 1)$ b. $m = 1003$ $P = (0, 0)$

In problems 9–16, find an equation of the line passing through the two given points:

9. a. $(1, 7) \ (4, -2)$ b. $(1, 7) \ (-4, 2)$

10. a. $\left(\dfrac{2}{3}, -\dfrac{3}{5}\right)\left(\dfrac{12}{7}, 2\right)$ b. $\left(\dfrac{2}{3}, -\dfrac{3}{5}\right)\left(-\dfrac{3}{5}, \dfrac{2}{3}\right)$

11. a. $(-6, 1) \ (12, -2)$ b. $(-6, 1) \ (0, 0)$

12. a. $\left(-\dfrac{2}{5}, \dfrac{1}{2}\right) (0, 0)$ b. $\left(-\dfrac{2}{5}, \dfrac{1}{2}\right) (12, -2)$

13. a. $(4, -18) \ (29, -18)$ b. $(4, -18) \ (4, 18)$

14. a. $(-15, 0) \ (-15, 25)$ b. $(-15, 0) \ (0, -15)$

15. a. $(3, \sqrt{2}) \ (3, -\sqrt{2})$ b. $(3, \sqrt{2}) \ (-3, \sqrt{2})$

16. a. $(\sqrt{2}, \sqrt{5}) \left(4, \dfrac{1}{2}\right)$ b. $\left(-\sqrt{2}, \dfrac{1}{2}\right) (\sqrt{5}, 4)$

17. Determine the x-intercept and the y-intercept of the line given by the equation:

a. $x + y = 1$ b. $-x + y = -1$

c. $3x + 6y = 2$ d. $\dfrac{4}{5}x - \dfrac{3}{4}y = \dfrac{7}{60}$

In problems 18–22, sketch the graph of the line determined by the given equation:

18. a. $3x - 2y = 5$ b. $-3x + 2y = 5$

19. a. $-6x + 7 = 2y - 4x$ b. $-6x - 7 = 2y - 4x$

20. a. $-y = -x + 3$ b. $y = x - 3$

21. a. $\dfrac{2}{7}x + \dfrac{3}{4} = 2y + 9$ b. $-\dfrac{2}{7}x + \dfrac{3}{4} = 2y + 9$

22. a. $\sqrt{2}x - 1 = y + 1$ b. $2x - 1 = y + \sqrt{2}$

*23. Fill in all of the algebraic steps and give the indicated reasons in the following argument:

Show that $\{(x, y): Ax + By = C, A \neq 0, B \neq 0\}$ represents a line (geometrically).

Let (x_1, y_1), (x_2, y_2), (x_3, y_3) be three distinct points in the given set and assume that $x_1 < x_2 < x_3$. Let d_1 be the distance between (x_1, y_1) and (x_2, y_2), let d_2 be the distance between (x_2, y_2) and (x_3, y_3), and let d_3 be the distance between (x_3, y_3) and (x_1, y_1). If we prove that $d_3 = d_1 + d_2$, then $\{(x, y): Ax + By = C, A \neq 0, B \neq 0\}$ represents a line geometrically.

Reason: ?_____

We have that $y = -\frac{A}{B}x + \frac{C}{B}$; thus $y_1 = -\frac{A}{B}x_1 + \frac{C}{B}$, $y_2 = -\frac{A}{B}x_2 + \frac{C}{B}$, and $y_3 = -\frac{A}{B}x_3 + \frac{C}{B}$.

Reason: ?_____

So

$$d_1 = \sqrt{(x_2 - x_1)^2 + (y_2 - y_1)^2} = \sqrt{1 + \frac{A^2}{B^2}}\, (x_2 - x_1),$$

$$d_2 = \sqrt{(x_3 - x_2)^2 + (y_3 - y_2)^2} = \sqrt{1 + \frac{A^2}{B^2}}\, (x_3 - x_2),$$

and $$d_3 = \sqrt{(x_1 - x_3)^2 + (y_1 - y_3)^2} = \sqrt{1 + \frac{A^2}{B^2}}\, (x_3 - x_1).$$

Reason: ?_____

Therefore $d_3 = d_1 + d_2$, i.e., (x_1, y_1), (x_2, y_2), (x_3, y_3) are on the same line.

2.6 THE GRAPHS OF LINEAR INEQUALITIES

A *linear inequality* (in the two variables x and y) is a (conditional) statement of the form

$$y < Ax + B \text{ or } y > Ax + B,$$

where $A, B \in \mathfrak{R}$. The set $\{(x, y): y < Ax + B\}$ is called the *graph* of $y < Ax + B$. Similarly, $\{(x, y): y > Ax + B\}$ is called the graph of $y > Ax + B$.

How do we sketch the graphs

$$\{(x, y): \ y < Ax + B\} \text{ and } \{(x, y): \ y > Ax + B\}?$$

Suppose we choose a value of x, say $x = x_0$. Consider the number $y_0 = Ax_0 + B$. The point $(x_0, y) \in \{(x, y): \ y < Ax + B\}$ provided $y < y_0$ (see Figure 2.14) and $(x_0, y) \in \{(x, y): \ y > Ax + B\}$ provided $y > y_0$.

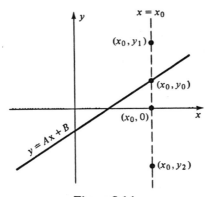

Figure 2.14

That is, $y < Ax_0 + B$ if and only if (x_0, y) lies below the line $y = Ax + B$ and $y > Ax_0 + B$ if and only if (x_0, y) lies above the line $y = Ax + B$. For example, in Figure 2.14, we see that $(x_0, y_1) \in \{(x, y): y > Ax + B\}$ and $(x_0, y_2) \in \{(x, y): y < Ax + B\}$. Extending this idea (by letting x_0 be any value on the x-axis) we have the graph in Figure 2.15.

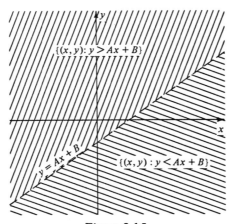

Figure 2.15

The line $y = Ax + B$ is dotted in Figure 2.15 in order to indicate that its points belong neither to $\{(x, y): y > Ax + B\}$ nor to $\{(x, y): y < Ax + B\}$.

Problem. Sketch the graphs

    (i) $\{(x, y): y < -3x + 7\}$ and (ii) $\{(x, y): 2x - y \leqslant -3\}$.

Solution. In case (i) we graph the line $y = -3x + 7$. Since the points $(0, 7)$ and $\left(\frac{7}{3}, 0\right)$ lie on the line, a dotted line is drawn through these points to designate the graph of the line (see Figure 2.16). Now, let $x = x_0$. The point (x_0, y_0) is on the line, where $y_0 = -3x_0 + 7$. So $(x_0, y) \in \{(x, y): \ y < -3x + 7\}$ provided $y < y_0$. Thus we have our graph, the shaded region below the line. The line is *not* included (that is why we made it a dotted line).

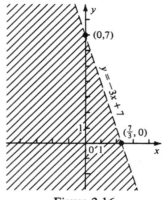

Figure 2.16

In case (ii) we write $\{(x, y): \ 2x - y \leqslant -3\} = \{(x, y): \ y \geqslant 2x + 3\}$ and follow the same type of analysis (as in case (i)). However, in this case, we obtain the region above the line and, also, we include the line (see Figure 2.17). The line $y = 2x + 3$ is left solid in this case in order to indicate that points on the line belong to the graph.

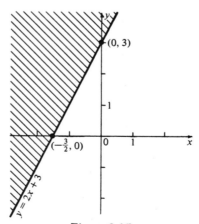

Figure 2.17

Exercise Set 2.6

1. Which of the following are linear inequalities in two variables?
   a. $y < 4x - 8$
   b. $2y > 4x - 8$
   c. $t < -6 + 5s$
   d. $t < 5s + 3r$
   e. $y^2 > 1 - x^2$
   f. $y = 3x - 10$

2. Given the expression $2x - 7$, choose y such that:
   a. $y = 2x - 7$ for $x = 1$
   b. $y = 2x - 7$ for $x = -2$
   c. $y < 2x - 7$ for $x = 1$
   d. $y < 2x - 7$ for $x = -2$
   e. $y > 2x - 7$ for $x = 1$
   f. $y > 2x - 7$ for $x = -2$

3. a. Sketch the graph of the line $y = 2x - 7$.
   b. Label the set of points in the plane for which $y = 2x - 7$.
   c. Label the set of points in the plane for which $y < 2x - 7$.
   d. Label the set of points in the plane for which $y > 2x - 7$.
   e. Does every point in the plane lie in one and only one of the sets described in parts b, c, and d?

4. How does the graph of the inequality $3x + 2y > 4$ differ from the graph of the inequality $3x + 2y \geqslant 4$?

5. Which of the following sets of points contain the origin?
   a. $\{(x, y): \ y < x + 1\}$
   b. $\{(x, y): \ y > x + 1\}$
   c. $\{(x, y): \ y < -2x + 5\}$
   d. $\{(x, y): \ y > -2x + 5\}$
   e. $\{(x, y): \ y < -x - 8\}$
   f. $\{(x, y): \ y > -x - 8\}$
   g. $\{(x, y): \ y < x\}$
   h. $\{(x, y): \ y > x\}$

* * *

In problems 6–11, graph the given linear inequality. If the line is to be included then make it a solid line, otherwise make it a dotted line:

6. a. $y < x + 1$
   b. $y > x + 1$

7. a. $y < -2x + 5$
   b. $y > -2x + 5$

8. a. $y < -x - 8$
   b. $y > -x - 8$

9. a. $y < x$
   b. $y > x$

10. $3x + 2y > 4$

11. $2x + 5y < x - 2$

In problems 12–19, sketch the indicated graph:

12. $\left\{(x, y): \ \frac{2}{3}x - y \geqslant 7\right\}$

13. $\{(x, y): \ 4y - 7x \leqslant 3y + 2\}$

14. $\{(x, y): \ -3y + 7 \leqslant 5x + 3\}$

15. $\{(x, y): \ -x - y > 0\}$

16. $\{(x, y): \ y \leqslant \sqrt{2}\ x\}$

17. $\left\{(x, y): \ \dfrac{5}{2}x - \dfrac{2}{3}y > -\dfrac{4}{9}\right\}$

18. $\left\{(x, y): \ \dfrac{x}{5} - \dfrac{y}{2} \geqslant \sqrt{3}\ \right\}$

19. $\left\{(x, y): \ -12y < \dfrac{1}{3}x - 3\right\}$

20. Graphically, find all points common to the sets:

$$\{(x, y): \ y < 2x - 7\} \text{ and } \{(x, y): \ y > -3x + 4\}.$$

2.7 UNION AND INTERSECTION OF SETS

In Sec. 2.6 we saw that $\{(x, y): \ y \geqslant 2x + 3\}$ is obtained by combining $\{(x, y): \ y > 2x + 3\}$ with the set of points on the line $y = 2x + 3$. There are several ways of combining sets to form new sets. Two such ways of combining sets are discussed in this section.

The *union of the sets A and B*, written $A \cup B$, is the set of all elements belonging to *either A or B* (or both). We can write this, in set notation, as

$$A \cup B = \{x: \ x \in A \text{ or } x \in B\}.$$

The *intersection of the sets A and B*, written $A \cap B$, is the set of all elements belonging to *both A and B*. Writing this in set notation, we have

$$A \cap B = \{x: \ x \in A \text{ and } x \in B\}.$$

In the following drawing the set A is shaded by \\\\\ and the set B is shaded by /////. The resulting picture is called a *Venn diagram* (see Figure 2.18).

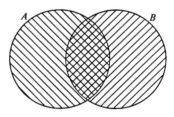

Figure 2.18

$A \cup B$ is the entire shaded area while $A \cap B$ is the double shaded area.

Example. Let A be the set of people who drive automobiles and let B be the set of people who fly airplanes. Then $A \cup B$ is the set of people who drive an automobile or fly an airplane and $A \cap B$ is the set of people who drive an automobile and also fly an airplane.

Example. Let A be the set of nearsighted people and let B be the set of bird-watchers. Then $A \cup B$ is the set of people who are nearsighted or are bird-watchers and $A \cap B$ is the set of nearsighted birdwatchers.

Example. If $A = \{1, 0, 4, 2, 3, 5\}$ and $B = \left\{-1, \frac{3}{2}, 2, 0, 1, -2\right\}$, then

$$A \cup B = \left\{1, 0, 4, 2, 3, 5, -1, \frac{3}{2}, -2\right\}$$

and

$$A \cap B = \{1, 0, 2\}.$$

Example. If $A = \left\{x: \ x > \frac{4}{3}\right\}$ and $B = \{x: \ x \text{ is an integer less than } 10\}$, then

$$A \cup B = \left\{x: \ x > \frac{4}{3} \text{ or } x \text{ is an integer less than } 10\right\}$$

and

$$A \cap B = \{2, 3, 4, 5, 6, 7, 8, 9\}.$$

Example. If $A = \left\{x: \ x > \frac{13}{4}\right\}$ and $B = \left\{x: \ x < \frac{2}{9}\right\}$, then

$$A \cup B = \left\{x: \ x > \frac{13}{4}\right\} \cup \left\{x: \ x < \frac{2}{9}\right\} = \left\{x: \ x > \frac{13}{4} \text{ or } x < \frac{2}{9}\right\}$$

and $A \cap B = \{\}$ $\Big($since there do not exist real numbers which are *both* less than $\frac{2}{9}$ and greater than $\frac{13}{4}\Big)$.

Example. We found, in sketching graphs of linear inequalities, that

$$\{(x,y): \ y \leqslant Ax + B\} = \{(x,y): \ y = Ax + B\} \cup \{(x,y): \ y < Ax + B\}.$$

That is, we graph the line $\{(x, y): \ y = Ax + B\}$ and include it with the graph $\{(x,y): \ y < Ax + B\}$ in order to obtain $\{(x,y): \ y \leqslant Ax + B\}$.

Problem. Use Venn diagrams to show that $A \cap (B \cup C) = (A \cap B) \cup (A \cap C)$.

Solution. This can be accomplished by showing that the region representing $A \cap (B \cup C)$ is the same as the region representing $(A \cap B) \cup (A \cap C)$. In Figure 2.19 we have shaded in $B \cup C$ and $A \cap (B \cup C)$.

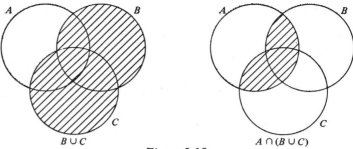

Figure 2.19

The sets $A \cap B$, $A \cap C$, and $(A \cap B) \cup (A \cap C)$ are shaded in Figure 2.20.

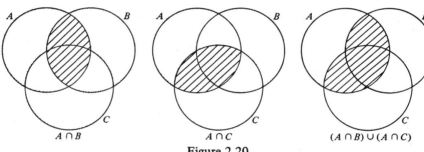

Figure 2.20

A comparison of the shaded regions representing $A \cap (B \cup C)$ and $(A \cap B) \cup (A \cap C)$ shows they are the same. Consequently, $A \cap (B \cup C) = (A \cap B) \cup (A \cap C)$.

Exercise Set 2.7

1. For the given sets A and B, find $A \cup B$ and $A \cap B$:
   a. $A = \{-1, 0, 1\}$ $B = \{0, 1, 2\}$
   b. $A = \{1, 2, 3, 4\}$ $B = \{7, 6, 5, 4, 3, 2, 1\}$
   c. $A = \{x: \ x \text{ is a positive integer}\}$ $B = \{x: \ x > 0\}$
   d. $A = \{x: \ x^2 = 1\}$ $B = \{x: \ x \neq 1\}$
   e. $A = \{x: \ x \leq 2\}$ $B = \{x: \ x > -5\}$
   f. $A = \{x: \ x \geq 2\}$ $B = \{x: \ x < 2\}$
   g. $A = \{x: \ (x - 2)(x + 3) = 0\}$ $B = \{x: \ x > -3\}$
   h. $A = \{x: \ (x - 2)(x + 3) > 0\}$ $B = \{x: \ (x - 2)(x + 3) \leq 0\}$

2. Express each of the following sets as the union of two sets:
   a. $\{t: \ t \leq 2\}$
   b. $\{y: \ 2y \leq x + 1\}$
   c. $\{s: \ -s \geq u + 1\}$
   d. $\{x: \ x < -1 \text{ or } x > 1\}$
   e. $\{W: \ W^2 = 1\}$
   f. $\{x: \ (x - 2)(x + 3) = 0\}$

3. Express each of the following sets as the intersection of two sets:
   a. $\{t: \ t<2 \text{ and } t>0\}$
   b. $\{y: \ 2y<x \text{ and } y>x+1\}$
   c. $\{M: \ M \text{ is a multiple of } 10\}$
   d. $\{x: \ 2<x<4\}$
   e. $\{i: \ i \text{ is a positive integer}\}$
   f. $\{y: \ y \text{ is an integer and } y \text{ is an even number}\}$

4. Use a Venn diagram to illustrate each of the following sets:
   a. $A \cup B$
   b. $A \cap B$
   c. $(A \cup B) \cup C$
   d. $(A \cap B) \cap C$
   e. $(A \cup C) \cup B$
   f. $(A \cup C) \cap B$

* * *

In problems 5–14, let $A = \{-1, 0, 1\}$, $B = \left\{x: \ x>\dfrac{1}{2}\right\}$, $C = \{x: \ -1<x<4\}$, and $D = \{x: \ -1 \leqslant x \leqslant 1\}$.

5. Find $A \cap B$.

6. Find $B \cap C$.

7. Find $B \cup C$.

8. Find $A \cap (B \cup D)$.

9. Find $(A \cap B) \cup D$. How does this result compare with problem 8?

10. Find $(A \cup C) \cap D$.

11. Find $(A \cap B) \cup (A \cap D)$. How does this result compare with problem 8?

12. Find $B \cap C \cap D$.

13. Find $(B \cup C) \cap (C \cup D)$.

14. Find $A \cup B \cup C \cup D$.

In problems 15 and 16 use a diagram of the following type:

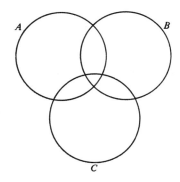

*15. Draw Venn diagrams to show that

$$\text{(i) } A \cup (B \cap C) = (A \cup B) \cap (A \cup C)$$
$$\text{(ii) } A \cap (B \cup C) = (A \cap B) \cup (A \cap C).$$

*16. Draw Venn diagrams to show that

and
(i) $A \cup (B \cup C) = (A \cup B) \cup C$
(ii) $A \cap (B \cap C) = (A \cap B) \cap C.$

This is why the symbols $A \cup B \cup C$ and $A \cap B \cap C$ are not ambiguous. Is the symbol $A \cap B \cup C$ ambiguous (see problems 8 and 9)?

2.8 SYSTEMS OF LINEAR EQUATIONS

Let us now return to the concept of the slope of a line. We found in Sec. 2.6 that if m is the slope of the line l and (a, b) is a point on the line l, then the equation of l is $y = mx + (b - am)$. Let us write this as $y = mx + B$, where $B = b - am$. This representation, $y = mx + B$, is called the *slope-intercept* form of l. If $m = 0$ then the line is parallel to the x-axis. Any other line parallel to l will also be parallel to the x-axis (thus its slope will be 0). Therefore, if two lines are parallel and are each parallel to the x-axis, then their slopes are the same.

Now, assume that l is not parallel to the x-axis, that is, $m \neq 0$. Suppose that the line l_1 is parallel to l and that l intersects the x-axis at $(a, 0)$ and l_1 intersects the x-axis at $(b, 0)$ (see Figure 2.21).

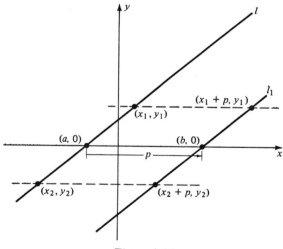

Figure 2.21

Let (x_1, y_1) and (x_2, y_2) be distinct points on l. Construct lines parallel to the x-axis through (x_1, y_1) and (x_2, y_2). These lines will meet the line l_1 at the points $(x_1 + p, y_1)$ and $(x_2 + p, y_2)$ respectively (where p is the directed distance from a to b on the x-axis). So, the slope of l_1 (say m_1) is given by

$$m_1 = \frac{y_2 - y_1}{(x_2 + p) - (x_1 + p)}$$

$$= \frac{y_2 - y_1}{x_2 + p - x_1 - p}$$

$$= \frac{y_2 - y_1}{x_2 - x_1} = m.$$

Therefore, parallel lines have equal slopes. Moreover, by the same type of argument, nonparallel lines have unequal slopes.

2.5 Theorem. *Two nonvertical lines are parallel if and only if they have the same slope.*

Recall that, given a linear equation, in order to determine the slope of the line we merely write the equation in the slope-intercept form

$$y = mx + B$$

and examine the coefficient m of x. This number is the slope of the line.

Problem. What is the slope of the line $3x - 2y + 4 = 2x + 6$?

Solution. First we subtract $(3x + 4)$ from each side of the equation. We obtain $3x - 2y + 4 - (3x + 4) = 2x + 6 - (3x + 4)$, or

$$-2y = -x + 2.$$

Thus

$$y = \frac{1}{2}x - 1.$$

Therefore the slope of the line is $\frac{1}{2}$ (the coefficient of x when we put the equation in the form $y = mx + B$).

Problem. Under what conditions are the lines

$$l_1: \quad a_1x + b_1y = c_1 \ (b_1 \neq 0)$$

and

$$l_2: \quad a_2x + b_2y = c_2 \ (b_2 \neq 0)$$

parallel.

Solution. Solving for y in the equation of the line l_1, we have

$$b_1y = -a_1x + c_1$$

or
$$y = -\frac{a_1}{b_1} x + \frac{c_1}{b_1}.$$

It follows that $-\frac{a_1}{b_1}$ is the slope of l_1. Similarly, we can write the equation of

the line l_2 as $y = -\frac{a_2}{b_2} x + \frac{c_2}{b_2}$. Thus the slope of l_2 is $-\frac{a_2}{b_2}$. Therefore l_1 and l_2

are parallel if and only if $-\frac{a_1}{b_1} = -\frac{a_2}{b_2}$ or $\frac{a_1}{b_1} = \frac{a_2}{b_2}$. We have just proved the fol-

lowing handy result.

2.6 Theorem. *IF $b_1 \neq 0$ and $b_2 \neq 0$, THEN the lines represented by the equations*

$$a_1 x + b_1 y = c_1$$

$$a_2 x + b_2 y = c_2$$

are parallel if and only if $\frac{a_1}{b_1} = \frac{a_2}{b_2}$.

Now, suppose we have two lines l_1 and l_2 given by

$$l_1: \quad a_1 x + b_1 y = c_1$$

$$l_2: \quad a_2 x + b_2 y = c_2.$$

To *solve this system of two linear equations in two unknowns (x and y) simultaneously* means to find all ordered pairs (x_0, y_0) such that if $x = x_0$ and $y = y_0$ then the equations of both lines are satisfied. Such pairs of real numbers are called *solutions* of the system. Geometrically, we want to find all points (x_0, y_0) which lie on both lines. Therefore, there exist three distinct possibilities:

(i) The lines are not parallel. Thus they meet at exactly one point, and there is exactly one solution (x_0, y_0) of our system of equations. Notice that this is the point of intersection of the two lines. That is,

$$\{(x_0, y_0)\} = l_1 \cap l_2$$

or

$$\{(x_0, y_0)\} = \{(x, y): \ a_1 x + b_1 y = c_1\} \cap \{(x, y): \ a_2 x + b_2 y = c_2\}.$$

(ii) The lines are parallel and are distinct. In this case l_1 and l_2 have no points in common so that there is no solution to the system of equations. It follows that $l_1 \cap l_2 = \{ \ \}$.

(iii) The lines are identical. In this case they are parallel and, in fact, the same line. Thus they meet at every point on the line, so that there are an infinite number of solutions to the system of equations.

In case (i), when the lines are not parallel, the slopes of the lines are unequal. Such a system is said to be *consistent*. In cases (ii) and (iii) the slopes of the lines are equal. In case (ii) the system is said to be *inconsistent*, while in case (iii) the system is said to be *dependent*.

If we have the system of equations

$$l_1: \quad a_1 x + b_1 y = c_1$$
$$l_2: \quad a_2 x + b_2 y = c_2,$$

each of the possible cases is illustrated in Figure 2.22.

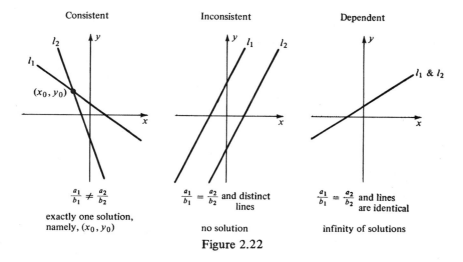

Consistent	Inconsistent	Dependent
$\frac{a_1}{b_1} \neq \frac{a_2}{b_2}$	$\frac{a_1}{b_1} = \frac{a_2}{b_2}$ and distinct lines	$\frac{a_1}{b_1} = \frac{a_2}{b_2}$ and lines are identical
exactly one solution, namely, (x_0, y_0)	no solution	infinity of solutions

Figure 2.22

Problem. Determine if the following system of linear equations is consistent, inconsistent, or dependent. If consistent, find the solution:

$$l_1: \quad 3x - 5y = 2$$
$$l_2: \quad 2x - 3y = 4.$$

Solution. We can write the equation of l_1 as $y = \frac{3}{5}x - \frac{2}{5}$ and the equation of l_2 as $y = \frac{2}{3}x - \frac{4}{3}$. Hence l_1 has slope $\frac{3}{5}$ and l_2 has slope $\frac{2}{3}$. Therefore the system is consistent (the slopes are unequal). To find the solution we can use one of two methods (we will demonstrate both methods). If (x, y) is the solution, then x and y satisfy both equations. That is,

$$l_1: \quad 3x - 5y = 2$$
$$l_2: \quad y = \frac{2}{3}x - \frac{4}{3}.$$

Using $y = \frac{2}{3}x - \frac{4}{3}$ in the equation for l_1, we have

$$3x - 5\left(\frac{2}{3}x - \frac{4}{3}\right) = 2$$

or

$$-\frac{1}{3}x = -\frac{14}{3}.$$

Hence $x = 14$ and, substituting this value into one of the equations, say the equation of l_2, we have

$$y = \frac{2}{3}(14) - \frac{4}{3} = \frac{28}{3} - \frac{4}{3} = \frac{24}{3} = 8.$$

Therefore $(14, 8)$ is the solution to our system.

The second method proceeds as follows: We begin with the system

$$l_1: \quad 3x - 5y = 2$$
$$l_2: \quad 2x - 3y = 4.$$

We want to multiply the equation of l_1 by some constant and the equation of l_2 by another constant so that when we add them, one of the unknowns (either x or y) is eliminated. This will allow us to solve for the other unknown. In this case if we multiply the equation for l_1 by 2 and the equation for l_2 by -3 we have

$$l_1: \quad 2(3x - 5y) = 2(2)$$
$$l_2: \quad -3(2x - 3y) = -3(4)$$

or

$$l_1: \quad 6x - 10y = 4$$
$$l_2: \quad -6x + 9y = -12.$$

Now, if we add the two equations we have

$$(6x - 10y) + (-6x + 9y) = 4 + (-12)$$

or

$$-y = -8.$$

Therefore $y = 8$. Now, by substituting $y = 8$ into one of the original equations, say the equation of l_2, we have

$$3x - 3(8) = 4.$$

So $2x = 4 + 24$ or $x = 14$. Therefore, the solution is $(14, 8)$.

In general, we select the variable to be eliminated. The equations of l_1 and l_2 are then multiplied by the appropriate numbers so that, in the resulting equations, the coefficients of the variable which is to be eliminated are negatives of each other. Addition of these two equations will then eliminate that variable.

We illustrate the use of the two methods of solution by considering several verbal problems.

Problem. Mr. Jones desires fourteen ounces of a mixture of pipe tobacco made from Vitali tobacco (at $.38 an ounce) and Lebesgue Blend (at $.47 an ounce).

He plans to spend exactly $6.00 for the mixture. How much of each type of tobacco should he purchase?

Solution. We let letters denote the unknown quantities. For example, let

x = number of ounces of Vitali tobacco purchased and
y = number of ounces of Lebesgue Blend purchased.

The conditions stated verbally in the problem will now translate into linear equations involving x and y. Thus the conditions that he wants fourteen ounces of a mixture and wishes to spend $6.00 can be translated as

$$(i)\ x + y = 14$$
$$(ii)\ .38x + .47y = 6.$$

From (i) we have $x = 14 - y$. Substituting in (ii), we obtain

$$.38(14 - y) + .47y = 6,$$
$$5.32 - .38y + .47y = 6,$$
$$.09y = .68,$$
$$y = \frac{68}{9} = 7\frac{5}{9}\ \text{oz.}$$

Knowing the value of y, we can substitute into (i) to obtain

$$x = 14 - \frac{68}{9} = \frac{58}{9} = 6\frac{4}{9}\ \text{oz.}$$

Problem. Mrs. Brown plans to serve cheese and crackers at a party. She figures that a combined total of 5 pounds of cheese and crackers should suffice for her guests. Cheese costs $.98 per pound and crackers cost $.29 per pound. If she has only $4.00 to spend, how many pounds of each should she buy?

Solution. As in the preceding example, we let letters denote the unknown quantities. Let

x = number of pounds of cheese and
y = number of pounds of crackers.

The conditions on total weight and money spent on cheese and crackers mean that x and y satisfy the following equations:

$$(i)\ x + y = 5$$
$$(ii)\ .98x + .29y = 4.$$

From (i), $y = 5 - x$. Substituting into (ii), we have

$$.98x + .29(5 - x) = 4,$$
$$.98x + 1.45 - .29x = 4,$$
$$.69x = 2.55.$$

Rounding off to two decimal places, we have

$$x = \frac{2.55}{.69} = 3.70.$$

Again, by (ii), $y = 5 - 3.7 = 1.3$.

Problem. Find two numbers such that twice the larger is three times the smaller and such that five times the larger exceeds twice the smaller by eight.

Solution. Let x be the larger number and y the smaller number. Thus

$$2x = 3y \text{ (or } 2x - 3y = 0)$$

and
$$5x = 2y + 8 \text{ (or } 5x - 2y = 8).$$

Now, multiplying the first equation by 5 and the second equation by -2 gives us

$$10x - 15y = 0$$

and
$$-10x + 4y = -16.$$

Adding, we have $-11y = -16$, yielding $y = \frac{16}{11}$. Therefore, using the first equa-

tion, we obtain

$$2x = 3\left(\frac{16}{11}\right)$$

or $x = \frac{24}{11}$. So, our solution is $\left(\frac{24}{11}, \frac{16}{11}\right)$.

It is a good idea to check the solution by substituting into the original system of equations to make sure they are satisfied. This check will point out whether or not arithmetic mistakes have been made while trying to obtain the solution.

In closing this section, it should be noted that there is no reason to restrict our attention to a system of just two linear equations in two unknowns. In fact, in any reasonable types of applications one would be forced to consider larger systems of linear equations in more than two unknowns. A more detailed study of this topic will be made in Chap. 9.

Exercise Set 2.8

1. Express each of the following equations in the slope-intercept form $y = mx + B$:

   a. $7y = 3x - 14$ b. $-x - y = 4$ c. $\frac{5}{2}x - \frac{9}{2}y + 1 = 0$

   d. $\frac{2}{3}x = 3y + 2$ e. $118x + 95y - 436 = 0$ f. $4x - \sqrt{2}\,y = 0$

2. Find the slope of each of the lines determined by the equations in problem 1.

3. Given the line $y = -\dfrac{3}{2}x + \dfrac{5}{3}$:

   a. Write an equation of a line which is parallel but not identical to the given line.

   b. Write an equation of a line which is not parallel to the given line.

   c. Write a different equation which represents the same line as the given line.

4. Which of the following systems of equations are consistent? inconsistent? dependent?

   a. $x - y = -5$
      $-\sqrt{2}\,x + \sqrt{2}\,y = 5\sqrt{2}$

   b. $17x = 8 - 13y$
      $34x + 26y = 24$

   c. $4x + 3y = 12$
      $3x + 4y = 22$

   d. $3t - 19s = 25$
      $3t + 19s = 25$

   e. $x - y = 8$
      $y - x = \dfrac{1}{8}$

   f. $\dfrac{5}{2}x - \dfrac{12}{7}y = \dfrac{5}{3}$
      $35x - 24y = \dfrac{70}{3}$

* * *

Find the solution set of each of the following systems of two equations in two unknowns:

5. $3x + 4y = 6$
   $3x - 2y = 18$

6. $6W - t + 4 = 0$
   $2W + 4t - 1 = 0$

7. $\dfrac{1}{2}p - \dfrac{3}{2}q = \dfrac{2}{5}$
   $\dfrac{3}{2}p + \dfrac{7}{2}q = 1$

8. $\dfrac{3}{4}p + \dfrac{9}{4}q = \dfrac{7}{3}$
   $\dfrac{1}{4}p + \dfrac{13}{4}q = -\dfrac{8}{3}$

9. $5y + 6z = 0$
   $37y - 25z = 0$

10. $-3y - 12z = 0$
    $\dfrac{3}{2}y + 6z = 0$

11. $5x + 4y = 3x - y + 12$
    $-7y + 2x = 8y + 6$

12. $-4x - 18y + 2 = -3x - 1$
    $4x - 6y + 8 = 2x - 9$

13. $4s + \dfrac{2}{3}t = 5$

    $3s - 6 = 0$

14. $2t + s = s - 10$
    $s + 5t = -1$

15. $-\dfrac{2}{5}x + \dfrac{4}{3}y = \dfrac{8}{9}$

    $-\dfrac{3}{10}x + y = \dfrac{2}{3}$

16. $\dfrac{5}{6}x + \dfrac{4}{5}y = -\dfrac{12}{13}$

    $\dfrac{5}{9}x + \dfrac{8}{15}y = -\dfrac{8}{13}$

17. $at + bw - c = 0$
    $bt + aw - c = 0$

18. $ax + by = c$
    $dx + ey = f$

*19. $\sqrt{2}\, x + \sqrt{3}\, y = 5$
     $-\sqrt{3}\, x + \sqrt{3}\, y = 1$

*20. $x - \sqrt{5}\, y = 2 + \sqrt{5}$

     $-\sqrt{3}\, x + \dfrac{y}{6} = -\sqrt{6}$

21. The sum of two numbers is 10 and one of the numbers is 2 more than the other. Find the numbers.

22. Find two consecutive even integers whose sum is 762.

23. A man has $5.30 in change consisting of dimes and quarters. If he has 3 more quarters than dimes, how many dimes and quarters does he have?

24. A man has $4.00 in change consisting of nickels and quarters. If he has 36 coins in all, how many of each does he have?

25. A candy store has two types of candy, one selling for 60 cents a pound and the second for 75 cents a pound. If a customer wishes to buy 2 pounds of a mixture of these candies which costs $1.38, how many pounds of each type should be purchased?

26. A chemist wishes to mix an 80% solution of alcohol with a 90% solution of alcohol in order to obtain 25 gallons of an 84% solution of alcohol. How many gallons of the 80% solution and how many gallons of the 90% solution should be used?

27. A sum of $10,000 is invested in two mutual funds. Part is invested in a high-risk fund which yields a 7% return each year. The rest is invested in a less speculative fund which yields a 5% return each year. If the yearly income from the investments is $580, find the amount invested in each type of mutual fund.

28. At a college basketball game the price of admission was $1.00 for students and $2.50 for nonstudents. If the total gate receipts for 9,000 people in attendance amounted to $15,000, how many students and how many nonstudents attended the game?

29. A car and an airplane are traveling at constant speeds. The car travels 110 miles in the same time that the airplane travels 380 miles. If the airplane travels 135 miles per hour faster than the car, find the speed of each.

30. The sum of the digits in a 2 digit number is 10. If the digits are interchanged, the new number is 18 more than the original number. What is the original number?

*31. Interpret, geometrically, what is meant by a system of three linear equations in two unknowns being consistent.

*32. Solve (if possible) the system:

$$3x - 7y = -9$$
$$x + 8y = 3x + 11$$
$$5x + 11 = 9y$$

2.9 SYSTEMS OF LINEAR INEQUALITIES

We are now interested in solving systems of linear inequalities, i.e., finding all points (x_0, y_0) which satisfy two or more linear inequalities simultaneously. The set of all such points is, as usual, called the *solution set* of the system. To illustrate a graphic solution, which we can apply in all of our cases, let us look at the following problems:

Problem. Graph the solution set of the system

(i) $3x + 7y < 12,$

and   (ii) $4x - y < 3.$

Solution. We want to find

$$\{(x,y): \ 3x + 7y < 12\} \cap \{(x,y): \ 4x - y < 3\}.$$

We have, from inequality (i), that

$$7y < -3x + 12$$

or
$$y < -\frac{3}{7}x + \frac{12}{7}.$$

From inequality (ii) we have

$$-y < -4x + 3$$
or
$$y > 4x - 3.$$

Graphing the solution sets of these two inequalities yields Figure 2.23.

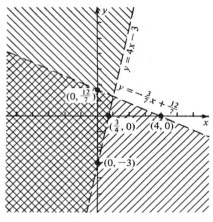

Figure 2.23

The intersection of the solution sets, and hence the graph of the solution set of the system, is the darkly shaded region of Figure 2.23.

Problem. Graph the solution set of the system

(i) $2x + 7y \geqslant 5$

and   (ii) $3x - 4y < 2.$

Solution. We want to find

$$\{(x,y): \; 2x + 7y \geqslant 5\} \cap \{(x,y): \; 3x - 4y < 2\}.$$

We have, from inequality (i), that

$$7y \geqslant -2x + 5$$
or
$$y \geqslant -\frac{2}{7}x + \frac{5}{7}.$$

From inequality (ii) we have

$$-4y < -3x + 2$$

or
$$y > \frac{3}{4}x - \frac{1}{2}.$$

Graphing the solution sets of these two inequalities, we have Figure 2.24.

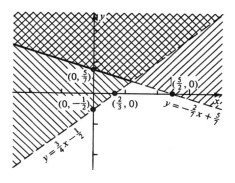

Figure 2.24

The desired region is the darkly shaded region in Figure 2.24. Recall that the solid line means we include the line and the dotted line indicates that we exclude the line.

Let us now consider how linear inequalities can be used to solve verbal problems. The procedure for setting up such problems is similar to that used in solving verbal problems by means of systems of linear equations.

Example. Mr. Jones, an appliance store owner, wishes to order washers and dryers from a factory. His storeroom is large enough to store at most a total of 40 washers and dryers. From his past experience Mr. Jones knows that many customers order a washer but not a dryer. Hence he feels that he must order at least twice as many washers as dryers. If x denotes the number of washers he orders and y the number of dryers he orders, then his order can be thought of as the ordered pair (x, y).

The order is subject to the previously stated conditions. Namely, the fact that he can store at most a total of 40 washers and dryers means that

$$x + y \leqslant 40.$$

Since he wishes to order at least twice as many washers as dryers, he wants

$$2y \leqslant x.$$

In addition, the number of washers and dryers are both nonnegative so that $0 \leqslant x$ and $0 \leqslant y$. Therefore any order (x, y) satisfying the conditions imposed by Mr. Jones satisfies the following system of linear inequalities:

$$x + y \leqslant 40,$$
$$2y \leqslant x,$$
$$0 \leqslant x,$$
$$0 \leqslant y.$$

The solution set of the system is the set of all points (x, y) which satisfy each of the inequalities and is nothing more than the intersection of the solution sets of each of the individual inequalities. The solution set of the system is the shaded region in Figure 2.25.

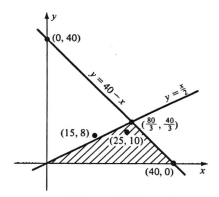

Figure 2.25

Any ordered pair (x, y) which satisfies the system of inequalities must be a point in the shaded region of Figure 2.25. Since the numbers of washers and dryers must both be integers, only points (x, y), where x and y are integers, which lie in the shaded region represent the possible orders Mr. Jones can make. Thus $(25, 10)$ (i.e., 25 washers and 10 dryers) is a possible order while $(15, 8)$ (i.e., 15 washers and 8 dryers) is not. Naturally, the order which is most advantageous for Mr. Jones to make would now be determined by other conditions such as consumer demands, profit per washer and dryer, and inventory costs.

Problem. Mr. Punt, owner of the Montana Grizzlies professional football team, leases a city municipal stadium which seats 80,000 people. He knows that at least 10,000 people purchase general admission tickets for each home game. Also, by league rule, at most 50,000 season tickets may be sold. Describe graphically the various possible combinations of types of ticket sales for a given game.

Solution. Let x denote the number of general admission tickets sold for a given game and let y denote the number of season tickets sold. Then x and y are subject to the following system of inequalities:

$$x + y \leqslant 80{,}000,$$
$$10{,}000 \leqslant x,$$
$$y \leqslant 50{,}000,$$
$$0 \leqslant y.$$

The solution set of this system is shown as the shaded region in Figure 2.26.

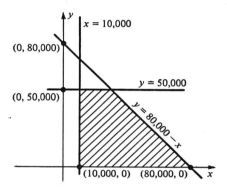

Figure 2.26

The various possible combinations of ticket sales are represented by the ordered pairs (x, y) corresponding to points which lie in the shaded region. It should be noted that the subset of the solution set which represents a sellout is that segment of the line $y = 80{,}000 - x$ which lies in the solution set.

The same remark applies here for linear inequalities that applied for systems of linear equations. Namely, one could consider large systems of linear inequalities in more than two unknowns. This would be necessary when considering more complex problems which arise in the real world.

Exercise Set 2.9

1. Graph the solution set for the system of inequalities:
$$3x + 4y < -7$$
$$-2x + y > 2$$

2. Graph the solution set for the system of inequalities:
$$x + y > 4$$
$$-x - y > 2$$

3. Graph the solution set for the system of inequalities:

$$2x + 3y \geqslant 6$$
$$x - 5y \leqslant 4$$
$$-2x + 9y \geqslant 0$$

* * *

Graph the solution set for each system of inequalities:

4. $-4x - 3y \geqslant 2$
   $5x - 2y < 1$

5. $4x + 3y \geqslant -2$
   $-5x + 2y < -1$

6. $\dfrac{1}{2}x - \dfrac{2}{3}y < \dfrac{1}{3}x - 7$

   $\dfrac{1}{2}x + y > -3$

7. $\dfrac{3}{2}x - \dfrac{1}{3}y < x + 2$

   $4x - 5 \geqslant -y$

8. $6x - y \leqslant 4$
   $2x + 3y \geqslant -1$

9. $5x - 2y \leqslant -4$
   $2x + 3y \geqslant -1$

10. $\dfrac{3}{4}x + 2 > \dfrac{7}{2}y - 1$

    $3x \leqslant 5$

11. $-\dfrac{5}{3}x - \dfrac{1}{3}y + 2 > -1$

    $2y \leqslant -1$

12. $3y \leqslant -4$
    $2x \geqslant 3y$

13. $3y \geqslant x$
    $3y \leqslant x$

14. $x + y + 1 > 15$
    $x + y - 1 > 15$

15. $4x - y > 7$
    $3x + 4y \leqslant 2$
    $5x + 2y > -1$

16. $2x - 4y \leqslant -3$
    $x + y > x - y + 3$
    $4x - 3y \geqslant 1$

17. $x + y > 0$
$-2x - y < 0$
$2x - 3y > 0$

18. $2x + y > 5$
$\frac{1}{4}x + \frac{1}{2}y < -2$
$-5x - \frac{5}{2}y > 3$

*19. $-\frac{51}{8}x - \frac{2}{15}y < \frac{1}{4}$
$\frac{3x}{7} + 5y > \frac{1}{4}$

*20. $\sqrt{2}\,x + \sqrt{2}\,y \geqslant 1$
$-\sqrt{3}\,x + \sqrt{3}\,y \leqslant \frac{1}{\sqrt{3}}$

21. A warehouse manager must order two types of automobile tires: black-walls and whitewalls. The warehouse can hold at most 5,000 tires and the manager must order at least three times as many whitewalls as blackwalls. Describe graphically the various possible orders the manager might make.

22. Do the same as in problem 21 if, in addition, the manager must order at least 500 blackwall tires.

23. Do the same as in problem 21 if, in addition, the manager must order at least 500 blackwalls but can order no more than 4,000 whitewalls.

24. A large corporation is taking a survey. It has a $2,000 maximum daily budget to hire consultants to help with the survey. A consultant earns $50 per day while a special consultant earns $100 per day. If at least one consultant of each type is hired each day, describe graphically the various possible combinations of types of consultants which can be hired on a given day.

25. Do the same as in problem 24 if, in addition, at least two consultants must be hired for each special consultant.

3 / *Functions and Their Graphs*

Playing a central role in all branches of mathematics and its applications is the idea of dependency of one quantity upon some other quantity. For example, the amount of interest accumulated on a fixed amount of capital depends upon the length of time of investment; the retail price of a product depends upon the cost of raw materials needed for the manufacture of the product; the total distance traveled by an automobile in three hours time depends upon the speed of travel.

This idea of dependency is discussed in terms of functions in the present chapter. We also introduce the notion of the graph of a function and discuss the sketching of graphs. This will give us a convenient way of obtaining a geometric "picture" of a function.

3.1 THE CONCEPT OF FUNCTION

From the examples mentioned in the introduction to this chapter we have

i) (The amount of interest accumulated) on a fixed amount of capital is dependent upon [the length of time of investment],

ii) (The retail price) of a product is dependent upon [the cost of raw materials] needed for the manufacture of the product, and

iii) (The total distance traveled) by an automobile in three hours is dependent upon [the speed of travel].

The quantities in parentheses and in brackets are called variables. It is seen that the variable in parentheses is determined by, and therefore dependent upon, the variable in brackets. For this reason, we refer to the determined variable, in parentheses, as the *dependent variable*. The determining variable, in brackets, is called the *independent variable*.

Usually variables are denoted by letters of the alphabet. The rule, according to which the value of the dependent variable is related to the value of the independent variable, is normally given by an equation. The following examples illustrate these conventions:

Example. Interest accrues at a rate of 4.5% per annum. An investment of $6240 is made at this rate. To write an equation describing the relationship between amount of interest accrued and the length of time of investment, we let t denote time (measured in years, since 4.5% is an annual rate), and i, the amount of interest accrued over a period of t years. It is known that

$$\text{interest} = (\text{principal}) \times (\text{rate}) \times (\text{time}).$$

Thus, the equation we seek is $i = (6240)(.045)t$ or $i = (280.8)t$.

Example. Lincoln Machine Co., manufacturer of hydraulic cylinders, finds that the average retail price of its product is 12.6 times the cost of raw materials used in the manufacture of this product. To write an equation describing this relationship, let p denote retail price and c the cost of materials. We are told that price = 12.6(cost). Hence $p = (12.6)c$ is the desired equation.

Example. Consider an equation that gives the distance traveled by a car in 3 hours time, in terms of the speed of travel. We can use this equation to determine how much distance is covered at a rate of 65 mph. It is known that distance = (speed) $\times$ (time). Letting d denote distance and s denote speed, we have $d = 3s$. In particular, if $s = 65$, then $d = 3(65) = 195$ miles.

In each of these examples, we see that to each value of the independent variable corresponds a unique value of the dependent variable. The following definition incorporates those properties which are common to the preceding examples.

3.1 Definition. *Let A and B be sets. A function f from A to B is a rule which assigns to each element x in A exactly one element y in B.*

In Definition 3.1, x plays the role of the independent variable and y the role of the dependent variable.

Example. Let A denote the set of universities and let B denote the set of states. Let f be the rule which assigns to each university the state in which the university is located. Then f is a function from A to B.

Example. Let A denote the set of cities. Let f be the rule which assigns to each city the number representing the population of the city. Then f is a function from A to $\mathcal{R}$.

Example. Let f be the rule which assigns to each real number x the number $-\frac{1}{2}x + 7$. In other words, f is the function from $\mathcal{R}$ to $\mathcal{R}$ defined by the equation $y = -\frac{1}{2}x + 7$. If $x = 8$, then $y = -\frac{1}{2}(8) + 7 = 3$. If $x = \frac{2}{3}$, then $y = -\frac{1}{2}\left(\frac{2}{3}\right) + 7 = \frac{20}{3}$.

Example. Let f be the rule which assigns to each real number x between -1 and 3 the number $-\frac{1}{2}x + 7$. Then f is the function from $[-1, 3]$ to $\mathcal{R}$ defined by the equation $y = -\frac{1}{2}x + 7$.

The set A appearing in Definition 3.1 is called the *domain* of f and is denoted by D_f. The domain may be thought of as the set of all values of the independent variable under consideration. Note that there is nothing special about the letters f, x, or y. Indeed, other letters will be used frequently to denote functions and the other variables.

Example. The first three examples of this section define functions given by the equations

$$i = (280.8)t,$$
$$p = (12.6)c,$$
$$d = 3s.$$

Usually we will be concerned with functions whose domains are sets of real numbers. If the function is defined by a formula and the domain is not explicitly stated, the domain is taken to be the set of all real numbers for which the formula makes sense. The following examples illustrate this convention:

Example. Let f be the function given by the equation

$$y = \frac{1}{x - 2}.$$

The domain is not explicitly stated. We see that the equation defining f makes sense so long as $x \neq 2$. If $x = 2$, then $x - 2 = 0$, and there is no corresponding value of y. Hence the domain of f is the set $D_f = \{x \in \mathcal{R}: \ x \neq 2\}$ and f is a function from D_f to $\mathcal{R}$.

Example. Let g be the function given by the equation $y = \sqrt{4 - 6x}$. Again, the domain of g is not explicitly specified. We observe, however, that the value of y will be well-defined provided the quantity $4 - 6x$ appearing under the radical sign is nonnegative. Using methods of Chap. 2, we obtain $4 - 6x \geqslant 0$ if and only if $x \leqslant \frac{2}{3}$. Thus $D_g = \left\{ x \in \mathfrak{R}: \ x \leqslant \frac{2}{3} \right\}$ and g is a function from D_g to $\mathfrak{R}$.

Example. Let a, b be real numbers and let h be the function given by the equation $y = ax + b$. Such a function is called a *linear* function. We see that $D_h = \mathfrak{R}$.

We conclude this section with the introduction of a notational device which simplifies statements involving functions. Given a function f, we shall denote by $f(x)$, read "f of x," the value of y which corresponds to x in the domain of f.

Problem. Let f be the function given by $y = 3x + 6$. Find $f(0), f(1), f\left(\frac{3}{2}\right)$, $f(-2), f(a)$, and $f(a + h)$.

Solution. $f(0)$ is the value of y corresponding to $x = 0$. Thus, $f(0) = 3 \cdot 0 + 6 = 6$. Similarly, $f(1)$ is the value of y assigned to $x = 1$ by f. We have $f(1) = 3 \cdot 1 + 6 = 9$. In the same fashion we obtain

$$f\left(\frac{3}{2}\right) = 3\left(\frac{3}{2}\right) + 6 = \frac{9}{2} + 6 = \frac{21}{2},$$
$$f(-2) = 3\,(-2) + 6 = -6 + 6 = 0,$$
$$f(a) = 3a + 6,$$
$$f(a + h) = 3(a + h) + 6 = 3a + 3h + 6.$$

In general, we have $f(x) = 3x + 6$.

For simplicity, we frequently refer to the equation which defines a function f as the function itself. For example, we might say, "Consider the function $f(x) = -3x + 10$," rather than "Consider the function f, given by the equation $y = -3x + 10$." It should be emphasized that, strictly speaking, this usage is technically incorrect. Indeed, f is a rule while $f(x)$ is a number (the value of y corresponding to x). However, no confusion will arise so long as we understand, at the outset, the basic definition of function.

Problem. Let $g(x) = -2 + x + x^2$. Find $g(0), g(1), g\left(-\frac{1}{2}\right)$, and $g(x + a)$.

Solution. We have

$$g(0) = -2 + 0 + 0^2 = -2,$$
$$g(1) = -2 + 1 + 1^2 = -2 + 2 = 0,$$

$$g\left(-\frac{1}{2}\right) = -2 + \left(-\frac{1}{2}\right) + \left(-\frac{1}{2}\right)^2 = -2 - \frac{1}{2} + \frac{1}{4} = -2 - \frac{1}{4} = -\frac{9}{4},$$
$$g(x + a) = -2 + (x + a) + (x + a)^2$$
$$= -2 + x + a + x^2 + 2ax + a^2.$$

Example. Let h be the function which assigns to each airport the city in which the airport is located. Then

$$h(\text{O'Hare}) = \text{Chicago}$$
and
$$h(\text{J. F. Kennedy}) = \text{New York}.$$

Exercise Set 3. 1

1. Let f be the function which assigns to each holiday the month in which it occurs. Find:

    $f(\text{Labor Day})$, $f(\text{Christmas})$, $f(\text{Valentine's Day})$.

2. Let g be the function which assigns to each integer n the integer $2n$. Find:
    $$g(-15), \quad g(1, 500, 369), \quad g(0).$$

3. Let h be the function which assigns to each real number x the number 100. Find:

    $$h(7), \quad h\left(\frac{1}{2}\right), \quad h(-\pi).$$

4. Use set notation to write the domain of f if:
    a. f is the function which assigns to each interstate highway a number.
    b. f is the rule which assigns to each professional baseball team the league to which it belongs.
    c. f is the function which assigns to each nonzero rational number $\frac{p}{q}$, the number $\frac{q}{p}$.
    d. f is the function given by the equation $y = \sqrt{-x}$.

* * *

5. Let $f(x) = x^2 - 1$. Find:
    $$f(-1), \quad f(0), \quad f(1), \quad f(200), \quad f(b), \quad f(a + b).$$

6. Let $f(x) = 1 - 2x^2$. Find:
    $$f(-2), \quad f\left(-\frac{1}{2}\right), \quad f(0), \quad f(30), \quad f(a), \quad f(a + b).$$

7. Let $f(x) = \frac{1}{x}$. Find:
    $$f(1), \quad f(2), \quad f\left(\frac{1}{2}\right), \quad f(10), \quad f\left(\frac{1}{10}\right), \quad f(500), \quad f\left(\frac{1}{500}\right).$$

8. Let $f(x) = (x - 3)^2$. Find:

$$f(-1), \quad f\left(-\frac{4}{3}\right), \quad f(-8), \quad f\left(\frac{4}{3}\right), \quad f\left(\frac{10}{3}\right), \quad f(75).$$

9. Let $g(x) = 2 - \frac{1}{2}x$. Find:

$$g(0), \quad g(2), \quad g\left(-\frac{1}{2}\right), \quad g(-3), \quad g(h), \quad g(x + h).$$

10. Let $g(x) = \sqrt{2}$. Find:

$$g(1), \quad g(\sqrt{2}), \quad g\left(\frac{35}{281}\right), \quad g(-2), \quad g(a), \quad g(a + h).$$

In problems 11–16, find the value of the dependent variable y if $x = 2, 1, -5, \frac{10}{3}, -\sqrt{2}$:

11. $y = 6x + 4$

12. $y = -5x - 7$

13. $y = \frac{-3}{8}$

14. $y = \frac{x - 2}{x + 2}$

15. $y = |x + 2|$

16. $y = |x| + 2$

In problems 17–22, find the domain of each of the given functions:

17. $f(x) = \frac{1}{x + 8}$

18. $y = 184.2$

19. $g(x) = 2x - 19$

20. $y = x^2$

21. $f(t) = |t|$

22. $g(t) = \frac{2t + 1}{t^2 + 1}$

23. Use the results of the first example of this section to find the amount of interest on $6240 at 4.5% per annum over a period of 5 years, 7 years, and 31 years.

24. Use the results of the second example of this section to find the retail price of the finished product if raw materials cost $18, $12.50, or $215.

25. Explain why each of the following rules fails to describe a function:
   a. r is the rule which assigns to each real number x all real numbers y such that $y < 2x$.
   b. r is the rule which assigns to each lock in the St. Lawrence Seaway the number of ships which pass through in an hour.
   c. r is the rule which assigns to each real number x the numbers which have the same absolute value as x.

26. Describe a function f such that:
   a. $D_f = \{2, 3, 5\}$
   b. $D_f = \{x: x \text{ is a triangle}\}$
   c. $D_f = \{t: t \text{ is an integer}\}$

3.2 THE GRAPH OF A FUNCTION

Consider a function f from one set of real numbers to another set of real numbers. With such a function f, we can associate a set G_f of ordered pairs of real numbers as follows: we let $(x, y) \in G_f$ if and only if $y = f(x)$. Then we see that

$$G_f = \{(x, f(x)): \ x \in D_f\}.$$

The set G_f is called the *graph* of f. From our work with the rectangular coordinate system we can assign to each pair $(x, f(x))$ in G_f a unique point in the plane. The set of all points in the plane corresponding to the ordered pairs in G_f will frequently form a curve. This curve is a pictorial representation of G_f and will reflect certain important properties of the function f. Because of the correspondence between points on the curve and ordered pairs in G_f, the curve itself is sometimes called the graph of f.

Example. Let $f(x) = 2x + \dfrac{1}{3}$. Then G_f consists of all ordered pairs of the form $\left(x, 2x + \dfrac{1}{3}\right)$ for $x \in D_f$. In order to determine some particular members of G_f, we may assign numerical values to x and compute the corresponding values of $2x + \dfrac{1}{3}$. Since $f(1) = 2(1) + \dfrac{1}{3} = \dfrac{7}{3}$, we have $\left(1, \dfrac{7}{3}\right) \in G_f$. Since $f(0) = 2(0) + \dfrac{1}{3} = \dfrac{1}{3}$, $\left(0, \dfrac{1}{3}\right) \in G_f$. Since $f\left(-\dfrac{5}{2}\right) = 2\left(-\dfrac{5}{2}\right) + \dfrac{1}{3} = -\dfrac{14}{3}$, we have $\left(-\dfrac{5}{2}, -\dfrac{14}{3}\right) \in G_f$. We see that $(2, 4) \notin G_f$ since $4 \neq f(2)$. Indeed $f(2) = 2(2) + \dfrac{1}{3} = \dfrac{13}{3}$, so that $\left(2, \dfrac{13}{3}\right) \in G_f$.

We now consider the problem of sketching the graph of a function. The steps to be followed are basically the same for all functions:

Step (1). Determine some particular ordered pairs belonging to the graph of f.

Step (2). Locate the points in the plane which correspond to the ordered pairs obtained in Step (1).

Step (3). Construct a "curve" which best seems to describe the array of points obtained in Step (2).

Before looking at examples, a few remarks are in order. First, the process of finding particular ordered pairs in the graph of f is best accomplished by means of a table of values. Construction of an adequate table of values is often quite difficult, however. We must select values of x which yield ordered pairs that give us a complete and accurate array of points with which to work. Second, the process of sketching a curve which best approximates the array of points given by Step (2) is complicated by the fact that there may be many different curves which seem to describe this set of points.

As we shall see, the curve may be disconnected, so that the word "curve" is inappropriate. For instance, in Chap. 4 we consider a function whose graph is a family of infinitely many disjoint line segments. Finally, we observe that while there is sometimes a tendency to think that all graphs are straight lines, this is not the case. For example, functions that contain terms where the independent variable is expressed with exponents greater than 1 (such as x^2; $x^3 + 1$; $x^4 + x^3 + x^2$, etc.) are nonlinear equations. The graphs of these functions are curved lines.

As we saw in Chap. 2, the graph of a linear equation $Ax + By + C = 0$, where not both A and B are zero, is a straight line. In particular, the linear function $y = ax + b$ may be interpreted as a linear equation $ax - y + b = 0$, so that the graph of any linear function is a straight line. We now illustrate the method of sketching the graphs of functions with a few examples:

Problem. Sketch the graph of the function $F(x) = -\frac{1}{2}x + 2$.

Solution. We observe that F is a linear function. Hence the graph of F will be a line with slope $-\frac{1}{2}$. Since two points determine a line, we need include only two values of x in our table:

x	0	2
$F(x)$	2	1

The reader should verify that the entries in the bottom row of this table are correct. Plotting the points $P(0, 2)$ and $Q(2, 1)$ we obtain the graph of F (see Figure 3.1).

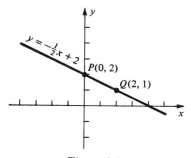

Figure 3.1

Problem. Sketch the graph of the function $g(x) = \dfrac{5}{2}$.

Solution. Again we are confronted by a linear function $y = 0x + \dfrac{5}{2}$. The graph of g will be a line with slope 0 (i.e., a horizontal line). The following is an adequate table of values for g:

x	1	2
$g(x)$	$\dfrac{5}{2}$	$\dfrac{5}{2}$

Locating the points $P\left(1, \dfrac{5}{2}\right)$ and $Q\left(2, \dfrac{5}{2}\right)$, and constructing the unique line through them, we obtain Figure 3.2.

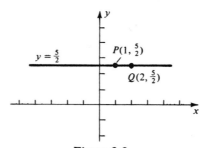

Figure 3.2

Problem. Sketch the graph of the function $h(x) = x^2 - 2$ for $x \geqslant 0$.

Solution. This is our first example of a nonlinear function. Note that the domain of h is explicitly given to be $D_h = \{x \in R : x \geqslant 0\}$. In setting up a table of values, we therefore select only nonnegative values for x. Since the graph of h is not a straight line, we must employ more than two entries in our table in order to obtain a reasonable sketch of the graph. The reader should check each of the entries in the table below:

x	0	1	2	$\dfrac{1}{4}$	$\dfrac{1}{2}$	$\dfrac{3}{2}$	$\dfrac{5}{2}$
$h(x)$	-2	-1	2	$-\dfrac{31}{16}$	$-\dfrac{7}{4}$	$\dfrac{1}{4}$	$\dfrac{17}{4}$

From the table we infer that

$$\left\{(0, -2), (1, -1), (2, 2), \left(\frac{1}{4}, -\frac{31}{16}\right), \left(\frac{1}{2}, -\frac{7}{4}\right), \left(\frac{3}{2}, \frac{1}{4}\right), \left(\frac{5}{2}, \frac{17}{4}\right)\right\} \subseteq G_h.$$

We now plot these points and draw a curve (see Figure 3.3) which seems to describe them.

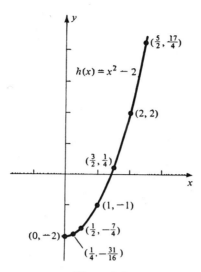

Figure 3.3

Suppose now that we had wished to have the graph of $h(x) = x^2 - 2$ for negative values of x. It should be apparent that the table of values obtained in the preceding example is incomplete in the sense that it provides us with absolutely no information as to the behavior of h for $x < 0$. Indeed, as far as we know, any one of the curves in Figure 3.4 could be the "correct" graph of h for $x \in \mathcal{R}$.

In order to determine which (if any) of these graphs is correct, we add to the table of values of the preceding example the following:

x	-1	-2	-3	$-\dfrac{1}{2}$	$-\dfrac{3}{2}$
$h(x)$	-1	2	7	$-\dfrac{7}{4}$	$\dfrac{1}{4}$

Plotting the additional points $(-1, -1), (-2, 2), (-3, 7), \left(-\dfrac{1}{2}, -\dfrac{7}{4}\right)$, and $\left(-\dfrac{3}{2}, \dfrac{1}{4}\right)$,
we obtain the curve in Figure 3.5 as the graph of h.

The solid portion of the curve in Figure 3.5 is an accurate sketch of the graph for values of x between -2 and 2. If larger values of x had been considered in the table, it would be seen that the remainder of the graph resembles the dotted portion of the curve in Figure 3.5. The problem of determining the shape of the graph for large values of $|x|$ will be discussed in more detail in Chap. 7.

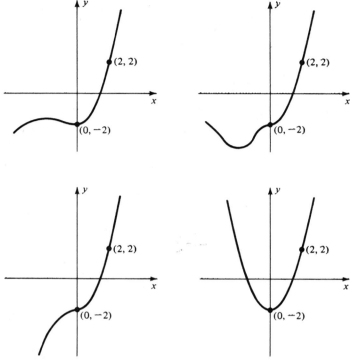

Figure 3.4

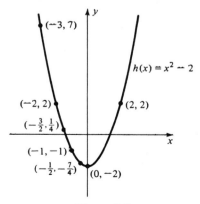

Figure 3.5

Problem. Sketch the graph of the function $T(x) = \begin{cases} 2x + 1, \text{ if } x \geqslant 1 \\ -x + 4, \text{ if } x < 1. \end{cases}$

Solution. Here we have two separate equations defining a single function T. If $x \geqslant 1$, we use $T(x) = 2x + 1$ to determine y. If $x < 1$, we employ the second equation, $T(x) = -x + 4$. We see that the function T may be thought of as a "combination" of two linear functions, each of which is defined over a specified domain. If we let $T_1(x) = 2x + 1$ for $x \geqslant 1$ and $T_2(x) = -x + 4$ for $x < 1$, we can obtain the graph T by graphing T_1 and T_2 on the same set of coordinate axes. That is, $G_T = G_{T_1} \cup G_{T_2}$.

x	1	3		x	0	-1
$T_1(x)$	3	7		$T_2(x)$	4	5

The graphs of T_1 and T_2 are shown in Figure 3.6.

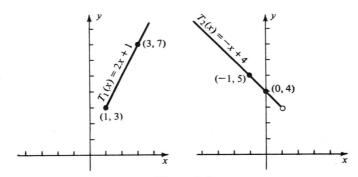

Figure 3.6

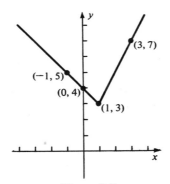

Figure 3.7

We observe that the point $(1, 3)$ does lie on the graph of T_1, but does not lie on the graph of T_2 (Why?). This is illustrated on the graph of T_2 by the little circle appearing at this point. We may now superimpose the graphs obtained above on the same pair of axes to obtain the graph of T (see Figure 3.7).

The following shows that not every curve is the graph of a function.

Example. Consider the circle drawn in Figure 3.8.

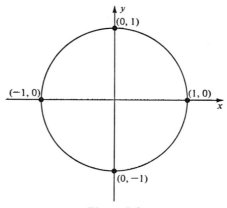

Figure 3.8

The circle is not the graph of a function f. If it were, then each of the points $(0, 1)$ and $(0, -1)$ would be of the form $(0, f(0))$. It would then follow that $f(0) = 1$ and $f(0) = -1$, contradicting the uniqueness of $f(0)$.

We close this section by considering the type of information we should like the graph of a function to reflect. Given a function $y = f(x)$, we should like to know where its graph crosses the x-axis if, in fact, it crosses at all. Such knowledge provides us with solutions of the equation $f(x) = 0$. We should also like to know where high points or low points appear on the graph of f. This information is easily obtained for the quadratic function to be examined next. However, in general, this information is difficult to obtain without deeper methods of analysis. In Chap. 7 we will examine techniques used in the sketching of graphs that are far superior to the methods we presently have at our disposal.

Exercise Set 3.2

1. Which of the following ordered pairs belong to G_f if f is the function which assigns to each real number x the number $3x + 2$?

$$(0, 0), \quad (1, 5), \quad (1, 3), \quad (-1, 1), \quad (-1, -1), \quad \left(\frac{8}{5}, \frac{34}{5}\right)$$

2. Which of the following points will belong to the graph of f, if $f(x) =$ $|x - 5|$?

$$(1, 4), \quad (1, -4), \quad (-1, 4), \quad (-1, 6), \quad (\sqrt{2}, \sqrt{2} - 5), \quad (\sqrt{2}, 5 - \sqrt{2})$$

3. If $F(x) = -x$, sketch the graph of $F(x)$.

4. If $g(x) = \dfrac{1}{x^2}$, complete the following table of values:

x	1	2	3	4	$\frac{1}{2}$	$\frac{1}{3}$	$\frac{1}{4}$	-1	-2	-3	-4	$-\frac{1}{2}$	$-\frac{1}{3}$	$-\frac{1}{4}$
$g(x)$														

5. On a set of coordinate axes, plot the points associated with the ordered pairs of G_g obtained in problem 4. Sketch G_g for the domain

$$D_g = \left[-4, -\frac{1}{4}\right] \cup \left[\frac{1}{4}, 4\right].$$

* * *

In problems 6–15, construct an adequate table of values and sketch the graph of the given function:

6. $f(x) = x + 3$

7. $f(x) = 2(x + 3)$

8. $G(x) = \dfrac{x + 1}{4}$

9. $G(x) = \dfrac{x + 1}{-4}$

10. $y = \dfrac{-2}{3}x + 2$

11. $y = \dfrac{3}{2}x + 2$

12. $k(x) = \begin{cases} x, \text{ if } x \geqslant 0 \\ x + 1, \text{ if } x < 0 \end{cases}$

13. $k(x) = \begin{cases} x + 2, \text{ if } x \leqslant 1 \\ 3x, \text{ if } x > 1 \end{cases}$

14. $h(x) = -x^2, -2 \leqslant x \leqslant 2$

15. $h(x) = x^2 + 2$

16. On the same set of axes sketch the graphs of f and g:

$$f(x) = \begin{cases} x, \text{ if } x \geqslant 0 \\ -x, \text{ if } x < 0 \end{cases} \qquad g(x) = |x|$$

17. On the same set of axes sketch the graphs of F and G:

$$F(x) = \frac{5}{2}x - 1 \qquad G(x) = -\frac{2}{5}x + 4$$

18. Let f be the function which assigns to each integer n the integer $2n$. Plot the graph of f for $D_f = \{-3, -2, -1, 0, 1, 2, 3, 4\}$.

19. Let f be the function which assigns to each nonzero rational number $\dfrac{p}{q}$, the number $\dfrac{q}{p}$. Sketch the graph of f for $D_f = \left(\dfrac{2}{5}, \dfrac{5}{2}\right)$.

20. Sketch the graph of F for $D_F = [-4, 4]$ if:

$$F(x) = \begin{cases} x, \text{ if } x \text{ is not an integer} \\ 0, \text{ if } x \text{ is an integer} \end{cases}$$

3.3 THE PRODUCT OF TWO LINEAR FACTORS

In the preceding sections we introduced the concept of a function and discussed some general aspects of the graph of a function. In this and subsequent sections, we turn our attention to the study of particular types of functions.

We begin by considering functions of the type

$$f(x) = A(x - a)(x - b), \quad A \neq 0,$$

where $A, a, b \in \mathcal{R}$. We are presently interested in finding the solution sets for $f(x) = 0, f(x) > 0$, and $f(x) < 0$, i.e.,

$$A(x - a)(x - b) = 0, \quad A(x - a)(x - b) > 0, \quad \text{and} \quad A(x - a)(x - b) < 0.$$

First, considering the equation

$$A(x - a)(x - b) = 0$$

we recall, from Chap. 1, that this equality holds if and only if

$$A = 0, \quad x - a = 0, \quad \text{or} \quad x - b = 0.$$

Since $A \neq 0$ we have that

$$x - a = 0 \quad \text{or} \quad x - b = 0,$$

i.e., $x = a$ or $x = b$.

Thus, we have proved

3.2 Theorem. *IF $A, a, b \in \mathcal{R}$ $(A \neq 0)$, THEN $A(x - a)(x - b) = 0$ if and only if $x = a$ or $x = b$.*

Problem. Find the solution set for each of the following:

a) $2(x - 2)\left(x - \dfrac{1}{2}\right) = 0$ c) $(2x + 1)\left(x - \dfrac{14}{3}\right) = 0$

b) $3\left(x - \dfrac{1}{4}\right)(x + 9) = 0$ d) $(3x - 5)\left(\dfrac{1}{2}x + 2\right) = 0.$

Solution.

a) By the preceding theorem, $2(x - 2)\left(x - \dfrac{1}{2}\right) = 0$ if and only if $x = 2$ or $x = \dfrac{1}{2}$.

Hence the solution set is $\left\{2, \dfrac{1}{2}\right\}$.

b) In place of $\left(x - \frac{1}{4}\right)(x + 9) = 0$ we may write $\left(x - \frac{1}{4}\right)(x - (-9)) = 0$. Thus

Theorem 3.2 implies that $x = \frac{1}{4}$ or $x = -9$. The solution set is $\left\{\frac{1}{4}, -9\right\}$.

c) Here we have $(2x + 1)\left(x - \frac{14}{3}\right) = 0$ if and only if $2\left(x + \frac{1}{2}\right)\left(x - \frac{14}{3}\right) = 0$.

This holds if and only if $x = -\frac{1}{2}$ or $x = \frac{14}{3}$. Thus the solution set is $\left\{-\frac{1}{2}, \frac{14}{3}\right\}$.

Note that $(2x + 1)\left(x - \frac{14}{3}\right) = 0$ if and only if $2x + 1 = 0$ or $x - \frac{14}{3} = 0$, i.e.,

$x = -\frac{1}{2}, \frac{14}{3}$.

d) Proceeding as in part c), we have that $(3x - 5)\left(\frac{1}{2}x + 2\right) = 0$ if and only if

$3x - 5 = 0$ or $\frac{1}{2}x + 2 = 0$. Now $3x - 5 = 0$ if and only if $3x = 5$, or, equiva-

lently, $x = \frac{5}{3}$; $\frac{1}{2}x + 2 = 0$ if and only if $\frac{1}{2}x = -2$, which implies that $x = -4$.

Hence the solution set is $\left\{\frac{5}{3}, -4\right\}$.

From parts (c) and (d) of the preceding example we can generalize Theorem 3.2.

3.3 Theorem. *IF A, a, b, c, $d \in \Re$ (A, a, c each nonzero), THEN*
$A(ax - b)(cx - d) = 0$ if and only if $x = \frac{b}{a}$ or $x = \frac{d}{c}$.

Let us now consider the inequalities

$$A(x - a)(x - b) > 0 \quad \text{and} \quad A(x - a)(x - b) < 0.$$

We indicate the method of finding the solution to these inequalities by examin-ing three problems.

Problem. Find the solution set of the inequality

$$-3(x - 2)(x + 1) > 0.$$

Solution. We see that the sign of the factor $x - 2$ depends upon whether $x > 2$ or $x < 2$. Similarly, the sign of the factor $x + 1$ depends upon whether $x > -1$ or $x < -1$. Thus, the sign of

$$-3(x - 2)(x + 1)$$

will be affected by the location of x and, in particular, will depend upon whether $x < -1$ (here $(x + 1) < 0$ and $(x - 2) < 0$), $-1 < x < 2$ (here $(x + 1) > 0$ and

$(x - 2) < 0$), or $x > 2$ (here $(x + 1) > 0$ and $(x - 2) < 0$). In tabular form we have

	-3	$x - 2$	$x + 1$	Sign of $-3(x - 2)(x + 1)$
$x < -1$	-	-	-	-
$-1 < x < 2$	-	-	+	+
$2 < x$	-	+	+	-

and, hence,

$$-3(x - 2)(x + 1) > 0 \text{ if and only if } x \in \{x: -1 < x < 2\}.$$

So the solution set of $-3(x - 2)(x + 1) > 0$ is $\{x: -1 < x < 2\}$.
Also, we can conclude that

$$-3(x - 2)(x + 1) < 0 \text{ if and only if } x \in \{x: x < -1\} \cup \{x: x > 2\}.$$

Problem. Find the solution set of the inequality

$$2\left(x + \frac{3}{2}\right)(x + 1) > 0.$$

Solution. The factor $x + \frac{3}{2}$ changes sign at $x = -\frac{3}{2}$ and the factor $x + 1$ changes sign at $x = -1$. Thus, we have

	$x + \frac{3}{2}$	$x + 1$	Sign of $2\left(x + \frac{3}{2}\right)(x + 1)$
$x < -\frac{3}{2}$	-	-	+
$-\frac{3}{2} < x < -1$	+	-	-
$-1 < x$	+	+	+

Therefore, from our table, we have that

$$2\left(x + \frac{3}{2}\right)(x + 1) > 0 \text{ if and only if } x \in \left\{x: x < -\frac{3}{2}\right\} \cup \{x: x > -1\}.$$

So the solution set is $\left\{x: x < -\frac{3}{2}\right\} \cup \{x: x > 1\}$.

Problem. Find the solution set of the inequality

$$(2 - 3x)(4x + 1) > 0.$$

Solution. The factor $(2 - 3x)$ changes sign at $x = \frac{2}{3}$ and the factor $(4x + 1)$ changes sign at $x = -\frac{1}{4}$. Thus, we have

	Sign of		
	$2 - 3x$	$4x + 1$	$(2 - 3x)(4x + 1)$
$x < -\frac{1}{4}$	+	−	−
$-\frac{1}{4} < x < \frac{2}{3}$	+	+	+
$\frac{2}{3} < x$	−	+	−

Therefore, the solution set is $\left\{ x: \ -\frac{1}{4} < x < \frac{2}{3} \right\}$.

Notice that solving the inequality

$$(2 - 3x)(4x + 1) > 0$$

is equivalent to solving the inequality

$$(-3)\left(x - \frac{2}{3}\right)(4)\left(x + \frac{1}{4}\right) > 0$$

or

$$-12\left(x - \frac{2}{3}\right)\left(x + \frac{1}{4}\right) > 0.$$

We now take up the problem of graphing functions of the type $f(x) = A(x - a)(x - b)$. As we have seen, $f(x) = 0$ if and only if $x = a$ or $x = b$. Thus, the graph of f can cross the x-axis at most two times. (When will the graph meet the x-axis only once?) Suppose that $a < b$ and $A > 0$. In tabular form we have

	Sign of		
	$x - a$	$x - b$	$A(x - a)(x - b)$
$x < a$	−	−	+
$a < x < b$	+	−	−
$b < x$	+	+	+

Thus, from the table we see that the graph of f will lie above the x-axis for $x \in \{x: \ x < a \text{ or } x > b\}$ and below the x-axis for $x \in \{x: \ a < x < b\}$. Using these facts together with a table of values, it is easy to obtain an accurate graph of f.

Problem. Sketch the graph of the function $f(x) = \left(x - \frac{1}{2}\right)(x - 3)$.

Solution. To relate this example to the prior discussion, we let $A = 1$, $a = \frac{1}{2}$ and $b = 3$. Observing that $a < b$, we conclude that $f(x) > 0$ if and only if $x < \frac{1}{2}$ or $x > 3$ and that $f(x) < 0$ for $\frac{1}{2} < x < 3$. Clearly, $f\left(\frac{1}{2}\right) = f(3) = 0$. We also have the following

x	$\frac{1}{2}$	3	1	$\frac{3}{2}$	2	$\frac{5}{2}$	$\frac{7}{2}$	4	5	0	-1	-2
$f(x)$	0	0	-1	$-\frac{3}{2}$	$-\frac{3}{2}$	-1	$\frac{3}{2}$	$\frac{7}{2}$	9	$\frac{3}{2}$	6	$\frac{25}{2}$

$\underbrace{\qquad\qquad}_{f(x) = 0} \quad \underbrace{\qquad\qquad\qquad}_{f(x) < 0} \quad \underbrace{\qquad\qquad\qquad\qquad}_{f(x) > 0}$

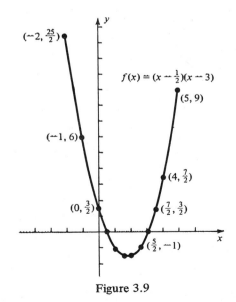

Figure 3.9

Notice that, in forming the table on page 108, we chose values of x for which $f(x)$ would be positive, negative, and zero. Such a choice contributes to the accuracy of the graph, which appears in Figure 3.9. This is an accurate sketch for values of x such that $-2 \leqslant x \leqslant 5$. It is a reasonable question to ask "what happens for $x < -2$ and $x > 5$, in particular, for very large values of $|x|$?". We will be able to answer this question in Chap. 7.

Now examine the case $f(x) = (x - a)(x - b)$, where $a = b$. Here we may write $f(x) = (x - a)(x - a) = (x - a)^2$. We immediately conclude that $f(x) \geqslant 0$ for all $x \in \mathcal{R}$, with $f(x) = 0$ if and only if $x = a$. Thus, the graph of f will lie entirely above the x-axis, actually touching the x-axis at $x = a$. Again, it is rather easy to obtain a graph of f as the following example shows.

Problem. Graph the function $f(x) = (x + 2)(x + 2)$.

Solution. We may write $f(x) = (x + 2)^2$. Hence $f(x) \geqslant 0$ for all $x \in \mathcal{R}$ with $f(x) = 0$ for $x = -2$.

In setting up a table for this function, we choose values of x "close to" -2.

x	-2	-3	-1	-4	0	-5	1	$-\dfrac{3}{2}$	$-\dfrac{5}{2}$
$y = h(x)$	0	1	1	4	4	9	9	$\dfrac{1}{4}$	$\dfrac{1}{4}$

The graph of f appears in Figure 3.10.

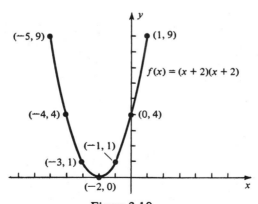

Figure 3.10

We conclude by observing that a knowledge of the solutions of $f(x) > 0$, $f(x) < 0$, and $f(x) = 0$ enabled us to choose relevant values of x in setting up our table and, thus, obtain a more accurate graph of the function f. Again, we can question the behavior of $f(x)$ for values of x for which $|x|$ is large.

Exercise Set 3.3

Consider the function $f(x) = 2(x + 2)(x - 1)$ in problems 1-4:

1. Find the solution set of the equation $f(x) = 0$.

2. Complete the following table:

	2	$x + 2$	$x - 1$	$2(x + 2)(x - 1)$
			Sign of	
$x < -2$				
$-2 < x < 1$				
$1 < x$				

3. Complete the following table of values:

x	-4	-3	-2.5	-2	-1	-.5	0	.5	1	1.5	2	3	4
$f(x)$													

4. Use the information obtained in problems 2 and 3 to sketch the graph of the function $f(x) = 2(x + 2)(x - 1)$.

5. Repeat problems 1-4 for $f(x) = -2(x + 2)(x - 1)$.

* * *

In problems 6-11, find the solution set of the equation $f(x) = 0$.

6. $f(x) = (x - 2)(x + 5)$

7. $f(x) = \left(x - \frac{2}{3}\right)(x + 4)$

8. $f(x) = -3(x + 6)\left(3x - \frac{1}{2}\right)$

9. $f(x) = (-5 - 4)(-x - 1)$

10. $f(x) = \frac{1}{16}(12 - 3x)(x + 2)$

11. $f(x) = \left(2x - \frac{3}{4}\right)^2$

In problems 12-17, find the solution set of the inequality $f(x) > 0$. Also, find the solution set of the inequality $f(x) < 0$.

12. $f(x) = (x - 2)(x + 5)$

13. $f(x) = \left(x - \frac{2}{3}\right)(x + 4)$

14. $f(x) = -3(x + 6)\left(3x - \frac{1}{2}\right)$

15. $f(x) = -5\left(x + \frac{4}{5}\right)(-x - 1)$

16. $f(x) = -\frac{1}{16}(12 - 3x)(x + 2)$

17. $f(x) = \left(2x - \frac{3}{4}\right)^2$

In problems 18–25, graph the function. Observe that the particular letter used to denote the independent variable is immaterial.

18. $f(x) = (x - 2)(x + 5)$

22. $f(x) = -5\left(x + \dfrac{4}{5}\right)(-x - 1)$

19. $g(x) = \left(x - \dfrac{2}{3}\right)(x + 4)$

23. $S(x) = 5\left(x + \dfrac{4}{5}\right)(x + 1)$

20. $F(t) = -3(t + 6)\left(3t - \dfrac{1}{2}\right)$

24. $f(W) = 2\left(W + \dfrac{1}{2}\right)\left(W - \dfrac{1}{2}\right)$

21. $G(z) = (z + 1)^2$

25. $g(t) = (t - 1)(1 - t)$

3.4 FACTORING QUADRATIC EXPRESSIONS

In the preceding section we considered the problem of graphing functions of the type $f(x) = A(x - r)(x - s)$. Information about solutions of the equation $f(x) = 0$ and of the inequalities $f(x) > 0$ and $f(x) < 0$ proved quite helpful in suggesting which values we should assign to the independent variable in order to obtain a graph of our function. Moreover, such information was rather easy to obtain for this type of function. If we multiply out the factors of $f(x)$ we see that

$$A(x - r)(x - s) = A[x^2 - (r + s)x + rs].$$

Thus, functions of this type are seen to fit the following definition:

3.4 Definition. *A function of the type $Q(x) = ax^2 + bx + c$, where a, b, $c \in \mathbb{R}$ with $a \neq 0$, is a quadratic function. The numbers a, b, and c are the coefficients of x^2, x, and 1, respectively. The coefficient of x^2 is the leading coefficient, while the coefficient of 1 is the constant term.*

We insist that the leading coefficient be different from zero since, otherwise, Q would take the form $Q(x) = bx + c$, which is the general form of the linear function. Although $a \neq 0$, it may well be the case that $b = 0$ or $c = 0$. The letter "x" possesses no special properties which make it particularly suitable for denoting the independent variable. Indeed, any letter of the alphabet may be used in this capacity.

Problem. Which of the following functions are quadratic functions?

a) $f(t) = 9t^2 - 2t + 4$

d) $h(x) = 4 - 29x^2$

b) $g(x) = -x^2 - x - \sqrt{2}$

e) $j(x) = (2x - 1)(x + 3)$

c) $K(w) = \dfrac{1}{w^2 + w + 1}$

f) $G(v) = v^2 + \sqrt{v} + \dfrac{1}{4}$

Solution.

a) $f(t) = 9t^2 - 2t + 4$ is quadratic with $a = 9$, $b = -2$, and $c = 4$.

b) $g(x) = -x^2 - x - \sqrt{2}$ is also a quadratic with $a = -1$, $b = -1$, and $c = -\sqrt{2}$.

c) $K(w) = \dfrac{1}{w^2 + w + 1}$ is not a quadratic since division by the variable occurs.

d) $h(x) = 4 - 29x^2$ is a quadratic. We may write $h(x) = -29x^2 + 0 \cdot x + 4$, thus it is apparent that $a = -29$, $b = 0$, and $c = 4$.

e) $j(x) = (2x - 1)(x + 3)$ is also a quadratic, but it appears in disguised form. If we multiply the factors of j, we obtain $j(x) = 2x^2 + 5x - 3$. Hence, $a = 2$, $b = 5$, and $c = -3$.

f) $G(v) = v^2 + \sqrt{v} + \dfrac{1}{4}$ is not a quadratic since the variable appears under a square root sign and, hence, cannot be expressed in the form $av^2 + bv + c$.

If $a, r, s \in \mathcal{R}$, then the function $f(x) = a(x - r)(x - s)$ is a quadratic function. Hence, all of the functions considered in Sec. 3.3 are quadratics. The question arises "is every quadratic function $Q(x) = ax^2 + bx + c$ expressible in factored form, $Q(x) = a(x - r)(x - s)$, for numbers r, $s \in \mathcal{R}$?" Unfortunately, such is not the case. We will show, in a later section, that many quadratics can be given in factored form while others can not be factored. We shall establish this fact using the method of *completing the square*, which we now introduce.

Let $g(x) = (x - r)^2$ where $r \in \mathcal{R}$. Multiplying, we obtain $g(x) = (x - r)(x - r) = x^2 - 2rx + r^2$. For the general form $ax^2 + bx + c$ we observe that $a = 1$, $b = -2r$, and $c = r^2$. Now $\dfrac{b}{2} = -r$, so that $\left(\dfrac{b}{2}\right)^2 = (-r)^2 = r^2$. Thus, $\left(\dfrac{b}{2}\right)^2 = c$, i.e., the square of one-half of the coefficient of x equals the constant term. This suggests the following.

3.5 Theorem. *Let* $g(x) = x^2 + bx + c$. *IF* $c = \left(\dfrac{b}{2}\right)^2$, *THEN* $g(x)$ *can be factored as* $g(x) = \left(x + \dfrac{b}{2}\right)\left(x + \dfrac{b}{2}\right) = \left(x + \dfrac{b}{2}\right)^2$.

In order to prove the theorem, we see that

$$\left(x + \frac{b}{2}\right)\left(x + \frac{b}{2}\right) = x^2 + 2\left(\frac{b}{2}\right)x + \left(\frac{b}{2}\right)^2 = x^2 + bx + \left(\frac{b}{2}\right)^2.$$

Since $\left(\dfrac{b}{2}\right)^2 = c$, we have $\left(x + \dfrac{b}{2}\right)\left(x + \dfrac{b}{2}\right) = x^2 + bx + c$. Hence, $g(x) = \left(x + \dfrac{b}{2}\right)^2$ and the theorem is proved.

Since $x^2 + bx + \left(\dfrac{b}{2}\right)^2 = \left(x + \dfrac{b}{2}\right)^2$, we call quadratics of the type $g(x) = x^2 + bx + \left(\dfrac{b}{2}\right)^2$ *perfect squares*.

Problem. Show that each of the following quadratics is a perfect square. Factor each one:

a) $f(t) = t^2 + 2t + 1$ c) $E(u) = u^2 + \dfrac{1}{2}u + \dfrac{1}{16}$

b) $H(x) = x^2 - 3x + \dfrac{9}{4}$ d) $k(x) = x^2 - \sqrt{2}\,x + \dfrac{1}{2}.$

Solution.

a) Here $b = 2$, and $c = 1$. We see that $\left(\dfrac{b}{2}\right)^2 = \left(\dfrac{2}{2}\right)^2 = 1^2 = 1$. Hence $\left(\dfrac{b}{2}\right)^2 = c$

and f is a perfect square. Factoring, we have $f(t) = \left(t + \dfrac{b}{2}\right)^2 = (t + 1)^2$.

b) In this case $b = -3$ and $c = \dfrac{9}{4}$. Hence, $\left(\dfrac{b}{2}\right)^2 = \left(\dfrac{-3}{2}\right)^2 = \dfrac{9}{4} = c$ so that H is a perfect square. We have $H(x) = \left(x + \dfrac{b}{2}\right)^2 = \left(x + \left(\dfrac{-3}{2}\right)\right)^2 = \left(x - \dfrac{3}{2}\right)^2.$

c) For $E(u) = u^2 + \dfrac{1}{2}u + \dfrac{1}{16}$, we see that $b = \dfrac{1}{2}$ and $c = \dfrac{1}{16}.$ Thus, $\left(\dfrac{b}{2}\right)^2 = \left(\dfrac{1/2}{2}\right)^2 = \left(\dfrac{1}{4}\right)^2 = \dfrac{1}{16} = c.$ We conclude that E is a perfect square and $E(u) = \left(u + \dfrac{b}{2}\right)^2 = \left(u + \dfrac{1}{4}\right)^2.$

d) Finally, for $k(x) = x^2 - \sqrt{2}\,x + \dfrac{1}{2}$, we have $b = -\sqrt{2}$ and $c = \dfrac{1}{2}.$ Now $\left(\dfrac{b}{2}\right)^2 = \left(-\dfrac{\sqrt{2}}{2}\right)^2 = \dfrac{2}{4} = \dfrac{1}{2} = c.$ Hence k is a perfect square, and $k(x) = \left(x + \dfrac{b}{2}\right)^2 = \left(x + \left(\dfrac{-\sqrt{2}}{2}\right)\right)^2 = \left(x - \dfrac{\sqrt{2}}{2}\right)^2.$

In each of the preceding examples, the leading coefficient, a, was 1. It is essential that this be the case in order that Theorem 3.5 may be applied. If $a \neq 1$, it may still be possible to write the given quadratic in factored form as the following example suggests.

Example. Consider the quadratic $q(x) = 4x^2 - 20x + 25$. We have $a = 4$ so that Theorem 3.5 is not directly applicable. However, we may write $q(x) = 4\left(x^2 - 5x + \dfrac{25}{4}\right) = 4g(x)$, where $g(x) = x^2 - 5x + \dfrac{25}{4}.$ For the quadratic $g(x)$ we have $a = 1$, $b = -5$, and $c = \dfrac{25}{4}.$ Hence $\left(\dfrac{b}{2}\right)^2 = \left(\dfrac{-5}{2}\right)^2 = \dfrac{25}{4} = c.$ We see that $g(x)$ is a perfect square with $g(x) = \left(x - \dfrac{5}{2}\right)^2.$ Therefore, we can write $q(x) = 4\left(x - \dfrac{5}{2}\right)^2.$

We now consider the method of completing the square for general quadratic functions. We will see, in the following sections, that this is an extremely important tool to have when we sketch graphs of general quadratic functions.

To illustrate the method of completing the square consider the quadratic $Q(x) = x^2 - 4x - 9$. Obviously Q is not a perfect square, since $\left(\dfrac{b}{2}\right)^2 = \left(\dfrac{-4}{2}\right)^2 =$ $(-2)^2 = 4$ and $c = -9$. However, we may add 4 $\left(\text{the value of } \left(\dfrac{b}{2}\right)^2\right)$ to $Q(x)$, and then subtract 4 from $Q(x)$ without altering the value of $Q(x)$. We obtain $Q(x) = (x^2 - 4x + 4) - 4 - 9 = (x^2 - 4x + 4) - 13$. The expression in parentheses is a perfect square, hence, $Q(x) = (x - 2)^2 - 13$, and the process is complete.

Problem. Complete the square for the following quadratic functions:

a) $q(x) = x^2 + 5x + 2$
b) $r(x) = x^2 + 3x + 1$
c) $s(t) = 3t^2 - 5t + 1$.

Solution.

a) For $q(x) = x^2 + 5x + 2$, we have $\left(\dfrac{b}{2}\right)^2 = \left(\dfrac{5}{2}\right)^2 = \dfrac{25}{4}$. Hence, we add and subtract $\dfrac{25}{4}$, obtaining

$$q(x) = \left(x^2 + 5x + \frac{25}{4}\right) - \frac{25}{4} + 2 = \left(x + \frac{5}{2}\right)^2 - \frac{17}{4}.$$

b) For $r(x) = x^2 + 3x + 1$, we see that $\left(\dfrac{b}{2}\right)^2 = \left(\dfrac{3}{2}\right)^2 = \dfrac{9}{4}$. Adding and subtracting $\dfrac{9}{4}$, we have

$$r(x) = \left(x^2 + 3x + \frac{9}{4}\right) - \frac{9}{4} + 1 = \left(x + \frac{3}{2}\right)^2 - \frac{5}{4}.$$

c) The function $s(t) = 3t^2 - 5t + 1$ differs from those previously discussed insofar as $a = 3$. We begin by first factoring 3 from each term of $s(t)$, obtaining $s(t) = 3\left(t^2 - \dfrac{5}{3}t + \dfrac{1}{3}\right)$. We now complete the square for $t^2 - \dfrac{5}{3}t + \dfrac{1}{3}$. Here $\left(\dfrac{b}{2}\right)^2 = \left(\dfrac{-5/3}{2}\right)^2 = \left(-\dfrac{5}{6}\right)^2 = \dfrac{25}{36}$. Thus, we have

$$t^2 - \frac{5}{3}t + \frac{1}{3} = \left(t^2 - \frac{5}{3}t + \frac{25}{36}\right) - \frac{25}{36} + \frac{1}{3} = \left(t - \frac{5}{6}\right)^2 - \frac{13}{36}.$$

Recalling that $s(t) = 3\left(t^2 - \dfrac{5}{3}t + \dfrac{1}{3}\right)$, we may write $s(t) = 3\left[\left(t - \dfrac{5}{6}\right)^2 - \dfrac{13}{36}\right] = 3\left(t - \dfrac{5}{6}\right)^2 - \dfrac{13}{12}$, and the example is complete.

In conclusion, we emphasize that completion of the square for a quadratic $Q(x) = ax^2 + bx + c$ does not directly provide us with a factorization of Q of the type $Q(x) = a(x - r)(x - s)$. We will show in the next section, however, that a factorization of this type can, in many cases, be obtained. Moreover, the means by which the factorization is accomplished depends upon completion of the square.

Exercise Set 3.4

1. A function of the type $Q(x) = ax^2 + bx + c$, $a \neq 0$, is a quadratic function. For each of the following quadratic functions, determine a, b, and c:

   a. $Q(x) = 3x^2 + 4x - 5$

   b. $Q(x) = -2x^2 + x + 8$

   c. $Q(x) = (2x - 5)(x + 3)$

   d. $Q(x) = x^2 - 16$

   e. $P(t) = \dfrac{21}{255} - \dfrac{3}{7}t - \dfrac{2}{3}t^2$

   f. $G(z) = (z - 1)^2$

2. a. Show that $x^2 - 3x + \dfrac{9}{4}$ is a perfect square.

   b. Factor $x^2 - 3x + \dfrac{9}{4}$.

3. a. Show that $4x^2 + 24x + 36$ is a perfect square.
   b. Factor $4x^2 + 24x + 36$.

4. Complete the square for the quadratic function

$$F(x) = x^2 + 4x - 3.$$

5. Complete the square for the quadratic function

$$G(x) = 2x^2 + 6x + 5.$$

* * *

In problems 6–15, determine whether or not the given quadratic is a perfect square. If it is, factor it.

6. $f(x) = x^2 - 6x + 9$

7. $g(x) = x^2 - 8x - 16$

8. $f(W) = W^2 - \dfrac{1}{3}W + \dfrac{1}{36}$

9. $g(t) = t^2 - \dfrac{4}{5}t + \dfrac{4}{25}$

10. $F(x) = 16 - 8x + x^2$

11. $h(x) = 4 + 4x - x^2$

12. $G(z) = z^2 + \dfrac{10}{3}z + \dfrac{50}{18}$

13. $f(x) = 9x^2 - 30x + 25$

14. $P(t) = 5t^2 - 3t + \dfrac{4}{25}$ 15. $H(y) = 72y^2 - 24y + 1$

In problems 16–25, complete the square for the given quadratic function.

16. $Q(t) = t^2 + 6t + 2$ 17. $j(x) = x^2 - 6x - 4$

18. $G(x) = x^2 - 8x - 16$ 19. $P(r) = r^2 + \dfrac{2}{3}r + \dfrac{1}{4}$

20. $f(y) = 3y + y^2 - \dfrac{9}{2}$ 21. $h(t) = t^2 - \dfrac{3}{7}t + \dfrac{7}{10}$

22. $F(W) = W^2 - \sqrt{3}\,W + 1$ 23. $f(x) = 2x^2 + 5x - 2$

24. $g(x) = 3 - 4x - 2x^2$ 25. $Q(s) = 9s^2 - 75s + 810$

*26. Let $Q(x) = ax^2 + bx + c$ be a quadratic function. Since $a \neq 0$, we may write $Q(x) = a\left[x^2 + \dfrac{b}{a}x + \dfrac{c}{a}\right]$. Let $g(x) = x^2 + \dfrac{b}{a}x + \dfrac{c}{a}$. Complete the square for this quadratic function.

3.5 THE QUADRATIC FORMULA

The method of completing the square, as discussed in Sec. 3.4, provides us with an effective technique for solving equations of the type $q(x) = 0$ where $q(x) = ax^2 + bx + c$.

Problem. Solve the equation $q(x) = 0$ where $q(x) = x^2 - 4x + 2$.

Solution. We first complete the square for q:

$$q(x) = (x^2 - 4x + 4) - 4 + 2 = (x - 2)^2 - 2.$$

Now $q(x) = 0$ if and only if $(x - 2)^2 - 2 = 0$. This occurs if and only if $(x - 2)^2 = 2$. Taking square roots of both sides of this equation, we have $x - 2 = \pm\sqrt{2}$. Note that $x - 2 = \pm\sqrt{2}$ means $x - 2 = \sqrt{2}$ or $x - 2 = -\sqrt{2}$. The two choices arise since $(\sqrt{2})^2 = 2$ and $(-\sqrt{2})^2 = 2$. Adding 2 to both sides yields $x = 2 \pm \sqrt{2}$. Hence, the solution set for the equation $q(x) = 0$ is

$$S = \{2 + \sqrt{2}, 2 - \sqrt{2}\}.$$

Problem. Solve the equation $r(t) = 0$ where $r(t) = 3t^2 + 9t + 4$.

Solution. Following the procedure used in the preceding example, we have $3t^2 + 9t + 4 = 3\left(t^2 + 3t + \dfrac{4}{3}\right)$. Now $3\left(t^2 + 3t + \dfrac{4}{3}\right) = 0$ if and only if $t^2 + 3t + \dfrac{4}{3} = 0$.

Completing the square, we have $t^2 + 3t + \dfrac{4}{3} = \left(t^2 + 3t + \dfrac{9}{4}\right) + \dfrac{4}{3} - \dfrac{9}{4} =$

$\left(t + \frac{3}{2}\right)^2 - \frac{11}{12}$. Now $\left(t + \frac{3}{2}\right)^2 - \frac{11}{12} = 0$ if and only if $\left(t + \frac{3}{2}\right)^2 = \frac{11}{12}$. Taking square roots, we obtain $t + \frac{3}{2} = \pm\sqrt{\frac{11}{12}}$; hence $t = -\frac{3}{2} \pm \sqrt{\frac{11}{12}}$. Thus the solution set is

$$S = \left\{-\frac{3}{2} + \sqrt{\frac{11}{12}}, -\frac{3}{2} - \sqrt{\frac{11}{12}}\right\}.$$

Problem. Let $s(x) = x^2 - 2x + 5$. Solve the equation $s(x) = 0$.

Solution. We have $s(x) = (x^2 - 2x + 1) - 1 + 5 = (x - 1)^2 + 4$. Now $(x - 1)^2 + 4 = 0$ if and only if $(x - 1)^2 = -4$. Since $(x - 1)^2 \geqslant 0$ for all $x \in R$, we see that there are no values of x for which $(x - 1)^2 = -4$, and therefore, the solution set of the equation $s(x) = 0$ is { }. That is, the equation $x^2 - 2x + 5 = 0$ has no solutions.

We recall that the general form of the quadratic function is $Q(x) = ax^2 + bx + c$, where $a \neq 0$. We may apply the method of completion of the square to $Q(x)$ just as we did in the previous examples. We have

$$ax^2 + bx + c = 0 \text{ if, and only if,}$$

$$a\left(x^2 + \frac{b}{a}x + \frac{c}{a}\right) = 0.$$

Now $\quad a\left(x^2 + \frac{b}{a}x + \frac{c}{a}\right) = a\left[\left(x^2 + \frac{b}{a}x + \left(\frac{b}{2a}\right)^2\right) - \left(\frac{b}{2a}\right)^2 + \frac{c}{a}\right]$

$$= a\left[\left(x + \frac{b}{2a}\right)^2 - \frac{b^2}{4a^2} + \frac{4ac}{4a^2}\right] = a\left[\left(x + \frac{b}{2a}\right)^2 - \frac{b^2 - 4ac}{4a^2}\right].$$

Hence $ax^2 + bx + c = 0$ if and only if

$$a\left[\left(x + \frac{b}{2a}\right)^2 - \frac{b^2 - 4ac}{4a^2}\right] = 0.$$

Since $a \neq 0$, this equation is equivalent to

$$\left(x + \frac{b}{2a}\right)^2 - \frac{b^2 - 4ac}{4a^2} = 0,$$

which occurs if and only if $\left(x + \frac{b}{2a}\right)^2 = \frac{b^2 - 4ac}{4a^2}$. Taking square roots, we have

$x + \frac{b}{2a} = \pm\sqrt{\frac{b^2 - 4ac}{4a^2}} = \pm\frac{\sqrt{b^2 - 4ac}}{2a}$. Subtracting $\frac{b}{2a}$ from both sides, we obtain

$$x = \frac{-b \pm \sqrt{b^2 - 4ac}}{2a}.$$

We have proven the following theorem which is known as the *quadratic formula*.

3.6 Theorem. *The solution set of the quadratic equation $ax^2 + bx + c = 0$ is*

$$\left\{ \frac{-b + \sqrt{b^2 - 4ac}}{2a}, \frac{-b - \sqrt{b^2 - 4ac}}{2a} \right\}.$$

The number $b^2 - 4ac$ appearing in the quadratic formula is called the *discriminant* of $Q(x)$. If $b^2 - 4ac < 0$, then the indicated square root cannot be taken. Thus the equation $Q(x) = 0$ has no solutions. If $b^2 - 4ac = 0$ then $\sqrt{b^2 - 4ac} = 0$ and, hence, the equation $Q(x) = 0$ has only one solution. The solution in this case is $x = -\dfrac{b}{2a}$. Finally, if $b^2 - 4ac > 0$ then $\dfrac{-b + \sqrt{b^2 - 4ac}}{2a}$ and $\dfrac{-b - \sqrt{b^2 - 4ac}}{2a}$ are distinct numbers. Thus the equation $Q(x) = 0$ has exactly two solutions. We state these results in the form of a theorem.

3.7 Theorem. *IF $Q(x) = ax^2 + bx + c$, where a, b, c, $\in \mathcal{R}$ and $a \neq 0$, THEN the equation $Q(x) = 0$*

    (i) *has no solutions if $b^2 - 4ac < 0$,*

    (ii) *has exactly one solution $\left(-\dfrac{b}{2a} \right)$ if $b^2 - 4ac = 0$, or*

    (iii) *has exactly two solutions $\left(\dfrac{-b + \sqrt{b^2 - 4ac}}{2a} \text{ and } \dfrac{-b - \sqrt{b^2 - 4ac}}{2a} \right)$*
    if $b^2 - 4ac > 0$.

It is customary to refer to the numbers $\dfrac{-b + \sqrt{b^2 - 4ac}}{2a}$ and $\dfrac{-b - \sqrt{b^2 - 4ac}}{2a}$ as the *zeros* of the quadratic function $Q(x)$ since $Q\left(\dfrac{-b + \sqrt{b^2 - 4ac}}{2a} \right) = Q\left(\dfrac{-b - \sqrt{b^2 - 4ac}}{2a} \right) = 0$. More generally we have the following definition:

3.8 Definition. *A number z belonging to the domain of the function f is a zero of f provided $f(z) = 0$.*

Let us now illustrate the use of the quadratic formula in determining the zeros of some particular quadratic functions.

Problem. Find the zeros of the quadratic $G(x) = 2x^2 + 4x + 7$.

Solution. By Theorem 3.6 the zeros of G are given by the formula $x = \dfrac{-b \pm \sqrt{b^2 - 4ac}}{2a}$. Here we have $a = -2$, $b = 4$, and $c = 7$. Hence, $f(x) = 0$

if, and only if,

$$x = \frac{-4 \pm \sqrt{4^2 - 4(-2)(7)}}{2(-2)} = \frac{-4 \pm \sqrt{16 + 56}}{-4}$$

$$= \frac{-4 \pm \sqrt{72}}{-4} = \frac{-4 \pm 6\sqrt{2}}{-4}.$$

Hence $z_1 = \frac{-4 + 6\sqrt{2}}{-4}$ and $z_2 = \frac{-4 - 6\sqrt{2}}{-4}$ are the zeros of G. Simplifying, we find $z_1 = 1 - \frac{3}{2}\sqrt{2}$ and $z_2 = 1 + \frac{3}{2}\sqrt{2}$.

Problem. What is the discriminant of the quadratic $q(x) = -5 + \frac{3}{2}x - \frac{1}{2}x^2$? What are the zeros of q?

Solution. The discriminant is given by $b^2 - 4ac$. Here we have $a = -\frac{1}{2}, b = \frac{3}{2}$, and $c = -5$ so that $b^2 - 4ac = \left(\frac{3}{2}\right)^2 - 4\left(-\frac{1}{2}\right)(-5) = \frac{9}{4} - 10 = -\frac{31}{4}$. Since the discriminant is negative, the quadratic q has no zeros.

Suppose now that $b^2 - 4ac \geq 0$ for the quadratic $Q(x) = ax^2 + bx + c$. Let $z_1 = \frac{-b + \sqrt{b^2 - 4ac}}{2a}$ and $z_2 = \frac{-b - \sqrt{b^2 - 4ac}}{2a}$ be the zeros of Q. Notice if $b^2 - 4ac = 0$, then $z_1 = z_2$.

Thus,

$$z_1 + z_2 = \frac{-b + \sqrt{b^2 - 4ac}}{2a} + \frac{-b - \sqrt{b^2 - 4ac}}{2a}$$

$$= \frac{-b + \sqrt{b^2 - 4ac} - b - \sqrt{b^2 - 4ac}}{2a}$$

$$= \frac{-2b}{2a} = -\frac{b}{a}.$$

In a similar fashion it can be shown that $z_1 z_2 = \frac{c}{a}$. Thus, we may write

$$Q(x) = ax^2 + bx + c = a\left(x^2 + \frac{b}{a}x + \frac{c}{a}\right) = a[x^2 - (z_1 + z_2)x + z_1 z_2].$$

Since $(x - z_1)(x - z_2) = x^2 - (z_1 + z^2)x + z_1 z_2$, it follows that $Q(x) = a(x - z_1)(x - z_2)$, and our next theorem is established.

3.9 Theorem. *IF $Q(x) = ax^2 + bx + c$ is a quadratic function for which $b^2 - 4ac \geq 0$ and if z_1 and z_2 are the zeros of Q, THEN $Q(x) = a(x - z_1)(x - z_2)$. Moreover, $z_1 + z_2 = \frac{b}{a}$ and $z_1 z_2 = \frac{c}{a}$.*

In the case where $Q(x)$ has no zeros (i.e., if $b^2 - 4ac < 0$) we say that the quadratic Q is *irreducible*. If $Q(x)$ has at least one zero then we say that Q is *reducible*.

Theorem 3.9 tells us that every quadratic with nonnegative discriminant can be factored. Moreover, the factorization may easily be obtained using the quadratic formula.

Problem. Determine if the given quadratic function is reducible or irreducible. If the function is reducible then find its factors.

(i) $f(x) = 4x^2 + 3x - 2$.

Solution. Since $b^2 - 4ac = (3)^2 - 4(4)(-2) = 9 + 32 = 41$, we have that f is reducible. Moreover, $f(x) = 0$ if and only if

$$x = \frac{-3 \pm \sqrt{3^2 - 4(4)(-2)}}{2(4)} = \frac{-3 \pm \sqrt{9 + 32}}{8} = \frac{-3 \pm \sqrt{41}}{8}.$$

Hence $z_1 = \frac{-3 + \sqrt{41}}{8}$ and $z_2 = \frac{-3 - \sqrt{41}}{8}$. By Theorem 3.9, we have

$$f(x) = 4\left(x - \frac{-3 + \sqrt{41}}{8}\right)\left(x - \frac{-3 - \sqrt{41}}{8}\right).$$

(ii) $g(t) = 3t^2 = 8t - 1$.

Solution. Since $b^2 - 4ac = (8)^2 - 4(3)(-1) = 64 + 12 = 76$, we have that g is reducible. Solving the equation $g(t) = 0$, we obtain

$$t = \frac{-8 \pm \sqrt{8^2 - 4(3)(-1)}}{2(3)} = \frac{-8 \pm \sqrt{64 + 12}}{6} = \frac{-8 \pm \sqrt{76}}{6} = \frac{-8 \pm 2\sqrt{19}}{6} = \frac{-4 \pm \sqrt{19}}{3}.$$

Hence, $z_1 = \frac{-4 + \sqrt{19}}{3}$ and $z_2 = \frac{-4 - \sqrt{19}}{3}$, so that

$$g(t) = 3\left(t - \frac{-4 + \sqrt{19}}{3}\right)\left(t - \frac{-4 - \sqrt{19}}{3}\right).$$

(iii) $h(w) = 2w^2 + w + 3$.

Solution. Since $b^2 - 4ac = (1)^2 - 4(2)(3) = -23$, we have that h is irreducible and, hence, cannot be factored.

Exercise Set 3.5

1. Let $f(x) = x^2 - 10x + 9$.
   a. Complete the square for $f(x)$.
   b. Find the solution set of the equation $f(x) = 0$.
   c. Express $f(x)$ as the product of two linear factors.

2. Use Theorem 3.6 to find the solution set of the quadratic equation $18x^2 - 9x - 20 = 0$.

3. Factor $18x^2 - 9x - 20$, using the results of problem 2.

4. For each of the following functions $Q(x)$, determine the number of solutions to the equation $Q(x) = 0$ by determining the value of the discriminant:
   a. $Q(x) = 4x^2 - 28x + 49$.
   b. $Q(x) = 3x^2 + x + 1$.
   c. $Q(x) = x^2 + x - 1$.

5. Which of the functions in problem 4 are reducible? Factor those which are reducible.

* * *

In problems 6–9, solve the given equation by completing the square.

6. $x^2 + \dfrac{1}{2}x - 7 = 0$

7. $u^2 - 5u + 3 = 0$

8. $4x^2 + x - 2 = 0$

9. $W^2 - \dfrac{11}{3}W = 2$

In problems 10–13, find the discriminant, $b^2 - 4ac$.

10. $Q(x) = x^2 + x + \dfrac{1}{2}$

11. $P(x) = 3x^2 + 9x + 5$

12. $F(y) = -2y^2 - 3y + 1$

13. $G(t) = 2t^2 + 2\sqrt{6}\,t + 3$

In problems 14–21, use the quadratic formula to find the zeros of each function. Then factor the function.

14. $f(x) = -3x^2 + x + 4$

15. $g(x) = \dfrac{1}{2}x^2 - 5x + 8$

16. $h(t) = 6 - 3t - t^2$

17. $R(x) = \dfrac{2}{3}x^2 - 3x - \dfrac{5}{3}$

18. $G(y) = \dfrac{3}{2}y^2 + y + \dfrac{1}{6}$

19. $U(V) = 10V^2 + 30V + 21$

20. $F(W) = W^2 - 25W + 2$

21. $f(t) = \sqrt{7}\,t^2 - t$

In problems 22–25, determine if the given function is reducible. If it is, find its factors.

22. $F(x) = x^2 - 2x - 1$

23. $h(x) = \dfrac{\sqrt{2}}{2}x^2 - \sqrt{5}\,x + \dfrac{\sqrt{2}}{3}$

24. $G(t) = 5t^2 + 5t + 5$

25. $P(z) = z^2 + 11z + 31$

*26. Find the zeros of the function.

$$P(x) = (x - a)^2 + 3(x - a) - 4.$$

3.6 THE GRAPH OF A QUADRATIC FUNCTION

In Chap. 2 we completely characterized the graph of the linear equation $Ax + By + C = 0$, where not both A and B are zero, as being a straight line. As a consequence, we see that the graph of every linear function $y = ax + b$ is a straight line. It shall be our purpose in this section to characterize the graph of the quadratic function $Q(x) = ax^2 + bx + c$. Again, the method of completing the square is our fundamental tool.

Consider the general quadratic Q. We may write

$$Q(x) = a\left[x^2 + \frac{b}{a}x + \frac{c}{a}\right] = a\left[x^2 + \frac{b}{a}x + \left(\frac{b}{2a}\right)^2 + \frac{c}{a} - \left(\frac{b}{2a}\right)^2\right]$$

$$= a\left[\left(x + \frac{b}{2a}\right)^2 + \frac{4ac}{4a^2} - \frac{b^2}{4a^2}\right]$$

$$= a\left(x + \frac{b}{2a}\right)^2 + \frac{4ac - b^2}{4a}.$$

We observe that $\left(x + \frac{b}{2a}\right)^2 \geqslant 0$ for all $x \in \Re$, with $\left(x + \frac{b}{2a}\right)^2 = 0$ if and only if $x = -\frac{b}{2a}$. Since $a \neq 0$, we have either $a > 0$ or $a < 0$. Suppose first that $a > 0$. Then $a\left(x + \frac{b}{2a}\right)^2 \geqslant 0$. Adding $\frac{4ac - b^2}{4a}$, we have

$$a\left(x + \frac{b}{2a}\right)^2 + \frac{4ac - b^2}{4a} \geqslant \frac{4ac - b^2}{4a},$$

or simply $Q(x) \geqslant \frac{4ac - b^2}{4a}$, with equality holding if, and only if, $x = -\frac{b}{2a}$, i.e., $Q\left(-\frac{b}{2a}\right) = \frac{4ac - b^2}{4a}$, and $Q(x) > \frac{4ac - b^2}{4a}$ for all $x \in \Re$ with $x \neq -\frac{b}{2a}$. Thus, we see that the number $\frac{4ac - b^2}{4a}$ is the smallest value of $Q(x)$. This means that the graph of Q will exhibit a "low point" at $\left(-\frac{b}{2a}, \frac{4ac - b^2}{4a}\right)$. Moreover, since $Q(x) \geqslant \frac{4ac - b^2}{4a}$, the graph of Q will lie *above* the horizontal line $y = \frac{4ac - b^2}{4a}$, actually touching this line for $x = -\frac{b}{2a}$. Since $\left(-\frac{b}{2a}, \frac{4ac - b^2}{4a}\right)$ is the lowest point on the graph of Q, the graph will rise as x increases from $-\frac{b}{2a}$ or as x decreases from $-\frac{b}{2a}$. We summarize these observations in Figure 3.11.

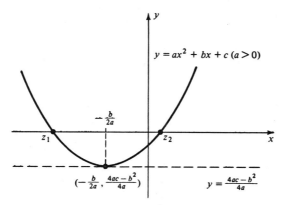

Figure 3.11

Turning now to the case where $a < 0$, we have $\left(x + \dfrac{b}{2a}\right)^2 \geqslant 0$ for all $x \in \mathcal{R}$, so that $a\left(x + \dfrac{b}{2a}\right)^2 \leqslant 0$ for all $x \in \mathcal{R}$. As before, equality holds if and only if $x = \dfrac{-b}{2a}$. Again we add $\dfrac{4ac - b^2}{4a}$ to both sides of the inequality to obtain $a\left(x + \dfrac{b}{2a}\right)^2 + \dfrac{4ac - b^2}{4a} \leqslant \dfrac{4ac - b^2}{4a}$, i.e.,

$$Q(x) \leqslant \frac{4ac - b^2}{4a} \text{ with } Q\left(-\frac{b}{2a}\right) = \frac{4ac - b^2}{4a}.$$

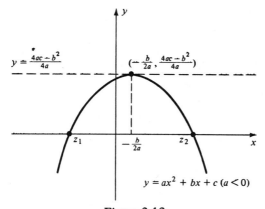

Figure 3.12

We see that $\dfrac{4ac - b^2}{4a}$ is the largest value of $Q(x)$. Thus, the graph of Q will be *below* the horizontal line $y = \dfrac{4ac - b^2}{4a}$, actually touching this line at the point $\left(-\dfrac{b}{2a}, \dfrac{4ac - b^2}{4a}\right)$. In this case, the graph of Q will fall as x decreases from $-\dfrac{b}{2a}$ or as x increases from $-\dfrac{b}{2a}$. These facts are reflected in Figure 3.12.

In applying the preceding discussion to the problem of graphing a particular quadratic function, we must answer the following four questions:

(i) What is the x-coordinate, $-\dfrac{b}{2a}$, of the high or low point?

(ii) What is the y-coordinate, $\dfrac{4ac - b^2}{4a}$, of the high or low point?

(iii) Is $a > 0$ (thus the graph opens upward) or is $a < 0$ (thus the graph opens downward)?

(iv) What are the zeros of the function (if any exist)?

The information provided by the answers to these four questions and a brief table of values of the function enables us to obtain an accurate graph of the function.

Problem. Graph the quadratic function $Q(x) = 2x^2 - 7x - 4$.

Solution. Here we have $a = 2$, $b = -7$, and $c = -4$. Thus

(i) $-\dfrac{b}{2a} = -\dfrac{(-7)}{2(2)} = \dfrac{7}{4}$

and

(ii) $\dfrac{4ac - b^2}{4a} = \dfrac{4(2)(-4) - (-7)^2}{4(2)} = -\dfrac{81}{8}$.

So $\left(\dfrac{7}{4}, -\dfrac{81}{8}\right)$ is the low point.

(iii) Since $a = 2 > 0$, we have that the graph opens upward.

(iv) $Q(x) = 2x^2 - 7x - 4 = (2x + 1)(x - 4) = 0$ if, and only if, $x = -\dfrac{1}{2}, 4$.

Using this information and the two additional points in the following table, we are able to sketch the graph of $Q(x)$.

x	$-\dfrac{3}{2}$	$-\dfrac{1}{2}$	$\dfrac{7}{4}$	4	5
$Q(x)$	11	0	$-\dfrac{81}{8}$	0	11

The graph of $Q(x)$ appears in Figure 3.13.

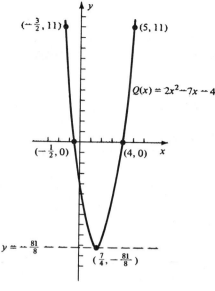

Figure 3.13

Problem. Graph the quadratic function $P(t) = -2t^2 + t - 1$.

Solution. Here we have $a = -2$, $b = 1$, and $c = -1$. Thus,

(i) $\quad -\dfrac{b}{2a} = -\dfrac{1}{2(-2)} = \dfrac{1}{4}$

and

(ii) $\quad \dfrac{4ac - b^2}{4a} = \dfrac{4(-2)(-1) - (1)^2}{4(-2)} = -\dfrac{7}{8}.$

So $\left(\dfrac{1}{4}, -\dfrac{7}{8}\right)$ is the high point.

(iii) Since $a = -2 < 0$, we have that the graph opens downward.

(iv) From parts (i), (ii), and (iii), we see that the largest value $P(t)$ can assume is $-\dfrac{7}{8}$. In particular, there are no values of t for which $P(t) = 0$, i.e., $P(t)$ has no zeros.

To obtain a reasonable table of values in order to sketch the graph of $P(t)$ we take values of t both greater than and less than the t-coordinate of the high point. In particular, we will take values of t on each side of the t-coordinate of the high point having equal distances from this coordinate. For example, let us choose $t = -\frac{1}{2}$, 1 $\left(-\frac{1}{2}\ \text{is}\ \frac{3}{4}\ \text{to the left of}\ \frac{1}{4}\ \text{and 1 is}\ \frac{3}{4}\ \text{to the right of}\ \frac{1}{4}\right)$ and $t = -2, \frac{5}{2}$ $\left(-2\ \text{is}\ \frac{9}{4}\ \text{to the left of}\ \frac{1}{4}\ \text{and}\ \frac{5}{2}\ \text{is}\ \frac{9}{4}\ \text{to the right of}\ \frac{1}{4}\right)$. Notice that $P(-2) = P\left(\frac{5}{2}\right)$ and $P\left(-\frac{1}{2}\right) = P(1)$.

t	-2	$-\frac{1}{2}$	$\frac{1}{4}$	1	$\frac{5}{2}$
$P(t)$	-11	-2	$-\frac{7}{8}$	-2	-11

Using the information we now have, we sketch the graph of $P(t)$ in Figure 3.14.

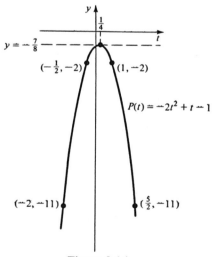

$$y = -\frac{7}{8}$$

$\left(-\frac{1}{2}, -2\right)$ $(1, -2)$

$$P(t) = -2t^2 + t - 1$$

$(-2, -11)$ $\left(\frac{5}{2}, -11\right)$

Figure 3.14

In the above problems and in the preceding discussion, we have seen that the graphs of quadratic functions all have the same general shape. The curve which arises as the graph of a quadratic function is called a *parabola*. The point with coordinates $\left(-\dfrac{b}{2a}, \dfrac{4ac - b^2}{4a}\right)$ is called the *vertex* of the parabola. The vertical line $x = -\dfrac{b}{2a}$ is called the *axis of symmetry* of the parabola since the curve is symmetric with respect to this line.

We now introduce the concept of the range of a function f. Let D_f be the domain of the function f. The *range of the function f*, denoted by R_f, is the set of values

$$R_f = \{y: \; y = f(x) \text{ for some } x \in D_f\}.$$

That is, given the domain D_f of a function, the range R_f of the function f is the collection of all values of the function $f(x)$ where the independent variable x is selected from the domain D_f.

In the case of the quadratic function, the domain of a quadratic function is the entire set of real numbers unless otherwise specified. The graph of the quadratic provides us with a means for determining the range. If $Q(x) = ax^2 + bx + c$ with $a > 0$, then the graph of Q rises from $\left(-\dfrac{b}{2a}, \dfrac{4ac - b^2}{4a}\right)$ without bound. Hence, the range of Q is

$$R_Q = \left\{x \in \mathbb{R}: \; x \geqslant \frac{4ac - b^2}{4a}\right\} = \left[\frac{4ac - b^2}{4a}, \infty\right).$$

If $a < 0$, the graph of Q falls from $\left(-\dfrac{b}{2a}, \dfrac{4ac - b^2}{4a}\right)$ without bound so that in this case

$$R_Q = \left\{x \in \mathbb{R}: \; x \leqslant \frac{4ac - b^2}{4a}\right\} = \left(-\infty, \frac{4ac - b^2}{4a}\right].$$

Our task is now accomplished. We have completely characterized the graphs of quadratic functions, though not without the expenditure of considerable effort. Hopefully, we have gained some respect for the difficulty encountered in the accurate graphing of functions.

Exercise Set 3.6

1. Let $Q(x) = 3x^2 - 5x + 4$. Determine $-\dfrac{b}{2a}$ and $\dfrac{4ac - b^2}{4a}$. What are the co-ordinates of the lowest point on the graph of $Q(x)$?

2. Let $P(x) = -4x^2 + 8x - 4$. Determine $-\dfrac{b}{2a}$ and $\dfrac{4ac - b^2}{4a}$. What are the coordinates of the highest point on the graph of $P(x)$?

3. Consider the function $F(x) = 3x^2 + x - 2$. Find the solution set for the equation $F(x) = 0$. Find the coordinates of the lowest point on the graph of $F(x)$.

4. Complete the following table of values for the function $F(x) = 3x^2 + x - 2$.

x			$-\dfrac{1}{6}$	0	$\dfrac{1}{2}$	$-\dfrac{1}{2}$	$+2$	-2
$F(x)$	0	0						

5. Sketch the graph of the function $F(x) = 3x^2 + x - 2$.

* * *

6. Show by direct substitution that $Q\left(\dfrac{-b}{2a}\right) = \dfrac{4ac - b^2}{4a}$ if $Q(x) = ax^2 + bx + c$.

In problems 7–16, construct a table of values for the given function. Then graph the function. Be sure to include the high (or low) point on the curve and the points, if any, at which the curve crosses or touches the x-axis.

7. $Q(x) = x^2 - 1$

8. $P(x) = 3x^2 + 2$

9. $G(t) = -t^2 + 2t + 3$

10. $h(x) = 4x^2 + 5x + 1$

11. $f(x) = \dfrac{1}{2}x^2 - 2x + \dfrac{1}{2}$

12. $Q(x) = 9x^2 + 3x - 56$

13. $P(W) = 4W^2 - 4W + 1$

14. $Q(t) = 25t^2 + 20t + 4$

15. $G(x) = -x^2 - 5x - 2$

16. $g(x) = \dfrac{2 - x^2}{3}$

In problems 17–20, determine, by examining the graph of the function $f(x)$, the solution set of the inequality $f(x) > 0$ and the solution set of the inequality $f(x) < 0$.

17. $f(x) = x^2 - 1$

18. $f(x) = 3x^2 + 2$

19. $f(x) = -x^2 + 2x + 3$

20. $f(x) = 4x^2 + 5x + 1$

For each of the functions in problems 21–24, find the domain D_P and the range R_P.

21. $P(x) = x^2 - x + 1$

22. $P(x) = -x^2 + 2x + 3$

23. $P(x) = 25x - 107$

24. $P(x) = 16 - x^2$

In problems 25–28, the graph of each of the given functions is a parabola. Determine the coordinates of the vertex of each parabola. Write an equation for the axis of symmetry for each parabola.

25. $P(x) = x^2 - x + 1$

26. $Q(x) = \dfrac{-4}{3}x^2 + \dfrac{1}{5}x + \dfrac{3}{4}$

27. $f(x) = 2x^2 - 1$

28. $g(t) = 5t^2 - 3t - 2$

3.7 EXPONENTS

Thus far we have considered in detail only two basic types of functions: the linear function $l(x) = ax + b$ and the quadratic function $Q(x) = ax^2 + bx + c$. Many other functions often arise in applications. In order to be adequately equipped to study these functions, it is necessary to know the basic definitions and rules of exponents.

We have already considered the numbers 1 and 2 as exponents. Indeed, we know that $a^1 = a$ and $a^2 = a \cdot a$ for any real number a. It is natural to extend this notation to include products of any number of factors of a.

3.10 Definition. *IF a is a real number and n is a positive integer, THEN*

$$a^n = \underbrace{a \cdot a \cdots a.}_{n \text{ factors}}$$

Example. According to Definition 3.10 we see that

$$2^3 = 2 \cdot 2 \cdot 2 = 8,$$

$$(-3)^4 = (-3)(-3)(-3)(-3) = 81,$$

$$\left(-\frac{1}{2}\right)^5 = \left(-\frac{1}{2}\right)\left(-\frac{1}{2}\right)\left(-\frac{1}{2}\right)\left(-\frac{1}{2}\right)\left(-\frac{1}{2}\right) = -\frac{1}{32}.$$

Suppose now that a is a real number and m, n are positive integers. Then

$$a^m \cdot a^n = \underbrace{(a \cdot a \cdots a)}_{m \text{ factors}} \underbrace{(a \cdot a \cdots a)}_{n \text{ factors}} = \underbrace{a \cdot a \cdots a}_{m + n \text{ factors}}.$$

Thus, we have $a^m \cdot a^n = a^{m+n}$. If $a \neq 0$, then

$$\frac{a^m}{a^n} = \frac{a \cdot a \cdots a(m \text{ factors})}{a \cdot a \cdots a(n \text{ factors})}.$$

If $m > n$, we may cancel all n of the factors of the denominator of this expression with the first n factors in the numerator, obtaining

$$\frac{a^m}{a^n} = \frac{a \cdot a \cdots a(m - n \text{ factors})}{1}.$$

Hence, for $m > n$, we have $\dfrac{a^m}{a^n} = a^{m-n}$. If $m < n$, then all m of the factors of the numerator may be cancelled with the first m factors of the denominator giving

$$\frac{a^m}{a^n} = \frac{1}{a \cdot a \cdots a(n - m \text{ factors})}.$$

Hence, we have $\dfrac{a^m}{a^n} = \dfrac{1}{a^{n-m}}$ if $m < n$. Finally, if $m = n$, then $a^m = a^n$ and

$\dfrac{a^m}{a^n} = 1$.

Consider now the expression $(a^m)^n$. Since $a^m = \underbrace{a \cdot a \cdots a}_{m \text{ factors}}$, it follows that

$$(a^m)^n = \underbrace{\underbrace{(a \cdot a \cdots a)}_{m \text{ factors}} \underbrace{(a \cdot a \cdots a)}_{m \text{ factors}} \cdots \underbrace{(a \cdot a \cdots a)}_{m \text{ factors}}}_{n \text{ factors of } (a \cdot a \cdots a)}$$

Counting up the total number of times a appears as a factor, we see that $(a^m)^n = a^{mn}$. We summarize these results in the following theorem:

3.11 Theorem. *IF a is a real number and m, n are positive integers, THEN*

(i) $a^m a^n = a^{m+n}$,

(ii) $(a^m)^n = a^{mn}$, *and*

(iii)
$$\dfrac{a^m}{a^n} = \begin{cases} a^{m-n}, & \text{if } m > n \\ 1, & \text{if } m = n \ (a \neq 0) \\ \dfrac{1}{a^{n-m}}, & \text{if } m < n. \end{cases}$$

Example. (a) $2^3 2^4 = 2^{3+4} = 2^7 = 128$

(b) $\dfrac{(-3)^5}{(-3)^2} = (-3)^{5-2} = (-3)^3 = -27$

(c) $\left[\left(\dfrac{1}{2}\right)^4\right]^2 = \left(\dfrac{1}{2}\right)^{4 \cdot 2} = \left(\dfrac{1}{2}\right)^8 = \dfrac{1}{256}$

(d) $\dfrac{5^2}{5^4} = \dfrac{1}{5^{4-2}} = \dfrac{1}{5^2} = \dfrac{1}{25}$

Suppose that a, b are real numbers and n is a positive integer. Then

$$(ab)^n = \underbrace{(ab)(ab) \cdots (ab)}_{n \text{ factors}}.$$

Grouping the a's and the b's together, we have

$$\underbrace{(ab)(ab) \cdots (ab)}_{n \text{ factors}} = \underbrace{(a \cdot a \cdots a)}_{n \text{ factors}} \underbrace{(b \cdot b \cdots b)}_{n \text{ factors}} = a^n b^n.$$

Thus, $(ab)^n = a^n b^n$. If $b \neq 0$, then

$$\left(\frac{a}{b}\right)^n = \left(\frac{a}{b}\right)\left(\frac{a}{b}\right) \cdots \left(\frac{a}{b}\right) = \frac{a \cdot a \cdots a(n \text{ factors})}{b \cdot b \cdots b(n \text{ factors})}$$

so that $\left(\frac{a}{b}\right)^n = \frac{a^n}{b^n}$. We summarize these observations in the following theorem:

3.12 Theorem. *IF a, b are real numbers and n is a positive integer, THEN*

(iv) $(ab)^n = a^n b^n$, *and*

(v) $\dfrac{a}{b}^n = \dfrac{a^n}{b^n}$ $(b \neq 0)$.

Example. (a) $(2x)^4 = 2^4 x^4 = 16x^4$

(b) $\left(\dfrac{2}{3}\right)^3 = \dfrac{2^3}{3^3} = \dfrac{8}{27}$

(c) $\left(\dfrac{y}{2}\right)^5 = \dfrac{y^5}{2^5} = \dfrac{y^5}{32}$

Properties (i)–(v) appearing in the preceding theorems are commonly called the basic laws of exponents. Until now, only positive integers have been used as exponents. We should like to extend our definitions so as to permit the use of any rational number as an exponent. We shall formulate our definitions of negative and fractional exponents in such a way that the five basic laws of exponents remain valid.

Let $a \neq 0$ and suppose m, n are positive integers with $m > n$. If rule (i) is to be satisfied, then $a^m \cdot a^{-n} = a^{m + (-n)} = a^{m-n}$. But $a^{m-n} = \dfrac{a^m}{a^n}$ according to rule (iii). Hence, if (i) and (iii) are to remain valid for negative exponents, we must have $a^m \cdot a^{-n} = \dfrac{a^m}{a^n}$. Dividing both sides of this equation by a^m, we obtain

$$a^{-n} = \frac{1}{a^n}.$$

Thus, we adopt the following definition:

3.13 Definition. *IF $a \neq 0$ and n is a positive integer, THEN*

$$a^{-n} = \frac{1}{a^n}.$$

Consider the expression a^0, where $a \neq 0$. If n is a positive integer, we want $a^n \cdot a^0 = a^{n+0} = a^n$. This means we must define $a^0 = 1$.

3.14 Definition. *IF a ≠ 0, THEN*

$$a^0 = 1.$$

Let us now reconsider rule (iii). If $a \neq 0$ and m, n are positive integers, then

$$\frac{a^m}{a^n} = \begin{cases} a^{m-n}, \text{ if } m > n \\ 1, \quad \text{ if } m = n \\ \frac{1}{a^{n-m}}, \text{ if } m < n. \end{cases}$$

Since we have now provided meaning for negative and zero exponents, this rule can be greatly simplified. Indeed, if $m = n$, then $\frac{a^m}{a^n} = 1$. But $a^{m-n} = a^0 = 1$ also. Thus we may write $\frac{a^m}{a^n} = a^{m-n}$. If $m < n$, then $\frac{a^m}{a^n} = \frac{1}{a^{n-m}}$. But $a^{m-n} = \frac{1}{a^{-(m-n)}} = \frac{1}{a^{n-m}}$ also. Hence, we again have $\frac{a^m}{a^n} = a^{m-n}$. In all three cases $(m > n, m = n, \text{ and } m < n)$ the same formula suffices:

$$\frac{a^m}{a^n} = a^{m-n} \text{ for all positive integers } m, n.$$

Problem. Rewrite each of the following, using exponents:

(a) $\frac{5^2}{5^4}$, (b) $\frac{3^{23}}{3^{19}}$, (c) $\frac{x^3}{x^{12}}$ $(x \neq 0)$.

Solution. (a) $\frac{5^2}{5^4} = 5^{2-4} = 5^{-2}$

(b) $\frac{3^{23}}{3^{19}} = 3^{23-19} = 3^4$

(c) $\frac{x^3}{x^{12}} = x^{3-12} = x^{-9}$

Problem. Rewrite each of the following, using exponents instead of fractions:

(a) $\frac{13x}{x^4}$, (b) $\frac{x^2 y}{xy^3}$, (c) $\frac{2a^3}{8a^4}$

Solution. (a) $\frac{13x}{x^4} = 13\left(\frac{x}{x^4}\right) = 13x^{1-4} = 13x^{-3}$

(b) $\frac{x^2 y}{xy^3} = \left(\frac{x^2}{x}\right)\left(\frac{y}{y^3}\right) = x^{2-1}y^{1-3} = x^{-1}y^{-2}$

(c) $\frac{2a^3}{8a^4} = \left(\frac{2}{2^3}\right)\left(\frac{a^3}{a^4}\right) = 2^{1-3}a^{3-4} = 2^{-2}a^{-1}$

Before introducing fractional exponents we consider the process of extraction of roots. If $a \in \mathcal{R}$, with $a \geqslant 0$, then $\sqrt{a}$ is that nonnegative real number having the property that $(\sqrt{a})^2 = a$. Extending the concept of square root, we define $\sqrt[n]{a}$ to be that real number having the property that $(\sqrt[n]{a})^n = a$. We call $\sqrt[n]{a}$ the n-th root of a. If n is even, then $\sqrt[n]{a}$ is defined only for $a \geqslant 0$. There are two n-th roots, one positive and the other negative. For the sake of exactness, the notation $\sqrt[n]{a}$ will be used only in reference to the positive n-th root of a. If n is odd, then $\sqrt[n]{a}$ is defined for every real number a. Moreover, $\sqrt[n]{a} > 0$ if, and only if, $a > 0$.

Example. (a) $\sqrt[3]{27} = 3$ since $3^3 = 27$

(b) $\sqrt[5]{32} = 2$ since $2^5 = 32$

(c) $\sqrt[4]{\dfrac{16}{81}} = \dfrac{2}{3}$ since $\left(\dfrac{2}{3}\right)^4 = \dfrac{16}{81}$

(d) $\sqrt[5]{-243} = -3$ since $(-3)^5 = -243$

(e) $\sqrt[4]{-16}$ is not defined

Returning now to the task of defining fractional exponents, we observe that if rule (ii) is to hold, then $\left(a^{\frac{1}{n}}\right)^n = \left(a^{\frac{1}{n} \cdot n}\right) = a^1 = a$ for each positive integer n. We know that $(\sqrt[n]{a})^n = a$. Hence, we have $\left(a^{\frac{1}{n}}\right)^n = (\sqrt[n]{a})^n$, and we are forced to define $a^{\frac{1}{n}}$ to be $\sqrt[n]{a}$, whenever this root exists. From here it is an easy matter to define $a^{\frac{m}{n}}$, where $\dfrac{m}{n}$ is any rational number.

3.15 Definition. *Let a be a nonzero real number and let m, n be nonzero integers with n positive. IF $\sqrt[n]{a}$ exists, THEN*

$$a^{\frac{m}{n}} = (\sqrt[n]{a})^m.$$

Example. (a) $16^{\frac{3}{4}} = (\sqrt[4]{16})^3 = 2^3 = 8$

(b) $27^{-\frac{2}{3}} = (\sqrt[3]{27})^{-2} = 3^{-2} = \dfrac{1}{3^2} = \dfrac{1}{9}$

(c) $(-32)^{\frac{2}{5}} = (\sqrt[5]{-32})^2 = (-2)^2 = 4$

(d) $(-125)^{-\frac{1}{3}} = (\sqrt[3]{-125})^{-1} = (-5)^{-1} = -\dfrac{1}{5}$

Problem. Write each of the following using fractional exponents:

(a) $\sqrt[7]{22}$, (b) $(\sqrt[3]{40})^2$, (c) $\dfrac{1}{\sqrt{13}}$, (d) $\left(\dfrac{\sqrt{13}}{\sqrt[5]{-12}}\right)^{-2}$

Solution. (a) $\sqrt[7]{22} = 22^{\frac{1}{7}}$

(b) $(\sqrt[3]{40})^2 = 40^{\frac{2}{3}}$

(c) $\dfrac{1}{\sqrt{13}} = (\sqrt{13})^{-1} = 13^{-\frac{1}{2}}$

(d) $\left(\dfrac{\sqrt{13}}{\sqrt[5]{-12}}\right)^{-2} = \dfrac{1}{\left(\dfrac{\sqrt{13}}{\sqrt[5]{-12}}\right)^2} = \dfrac{1}{\dfrac{(\sqrt{13})^2}{(\sqrt[5]{-12})^2}} = \dfrac{(\sqrt[5]{-12})^2}{(\sqrt{13})^2} = \dfrac{(-12)^{\frac{2}{5}}}{13}$

$= 13^{-1}(-12)^{\frac{2}{5}}$

Our definitions of negative and fractional exponents were governed by a desire to retain validity of the basic laws of exponents (i)–(v). The fact that (i)–(v) hold for fractional exponents could now be verified, but in a somewhat tedious fashion. Therefore we state this result without proof and use it in future work.

3.16 Theorem. *IF a, b are real numbers and r, s are rational numbers, THEN*

(i) $a^r a^s = a^{r+s}$,

(ii) $(a^r)^s = a^{rs}$,

(iii) $\dfrac{a^r}{a^s} = a^{r-s} \ (a \neq 0)$,

(iv) $(ab)^r = a^r b^r$, *and*

(v) $\left(\dfrac{a}{b}\right)^r = \dfrac{a^r}{b^r} \ (b \neq 0)$.

Example. (a) $5^{\frac{2}{3}} 5^{\frac{4}{3}} = 5^{\frac{2}{3} + \frac{4}{3}} = 5^2 = 25$

(b) $\left(2^{\frac{1}{2}}\right)^6 = 2^{\frac{6}{2}} = 2^3 = 8$

(c) $\dfrac{8^{\frac{2}{3}}}{8^{\frac{1}{3}}} = 8^{\frac{2}{3} - \frac{1}{3}} = 8^{\frac{1}{3}} = 2$

(d) $(250)^{\frac{1}{2}} = (25 \cdot 10)^{\frac{1}{2}} = (25)^{\frac{1}{2}}(10)^{\frac{1}{2}} = 5(10)^{\frac{1}{2}}$

(e) $\left(\dfrac{16}{625}\right)^{\frac{1}{4}} = \dfrac{16^{\frac{1}{4}}}{(625)^{\frac{1}{4}}} = \dfrac{2}{5}$.

Exercise Set 3.7

1. Find a^4 for each of the following values of a:

$$a = 1, \quad a = -1, \quad a = \frac{1}{2}, \quad a = 2, \quad a = \sqrt{2}, \quad a = 2\sqrt{2}$$

2. Find a^{-4} for each of the following values of a:

$$a = 1, \quad a = -2, \quad a = \frac{2}{3}, \quad a = \pi, \quad a = 100, \quad a = .01$$

3. Evaluate each of the following expressions:

$$(16)^{\frac{1}{4}}, \quad (16)^{\frac{3}{4}}, \quad (4)^{\frac{1}{4}}, \quad (4)^{\frac{3}{4}}, \quad (25)^{\frac{1}{2}}, \quad (25)^{-\frac{1}{2}}$$

4. Simplify each of the following expressions if $m = 2$ and $n = 3$:

$$3^m \cdot 3^n, \quad 3^m \cdot 3^{-n}, \quad (4^m)^n, \quad (4^n)^{-m}, \quad (4^m)^{-n}, \quad (-1)^m \cdot (-2)^n$$

5. Simplify each of the following expressions if $m = 2$ and $n = 3$:

$$5^m \cdot 6^m, \quad 27^n \cdot 27^{-n}, \quad \frac{15^m}{15^n}, \quad \frac{(107^m)^n}{(107^n)^m}, \quad \left(\frac{3}{2}\right)^m \cdot \left(\frac{8}{9}\right)^m, \quad \left(\frac{3}{2}\right)^m \cdot \left(\frac{8}{9}\right)^{-m}$$

* * *

6. Find $\left(\frac{2}{3}\right)^r$ for each of the following values of r:

$$r = 1, \quad r = -1, \quad r = 0, \quad r = 2, \quad r = \frac{1}{2}, \quad r = -6$$

7. Find $\left(-\frac{5}{4}\right)^r$ for each of the following values of r:

$$r = -1, \quad r = 2, \quad r = 0, \quad r = \frac{1}{3}, \quad r = \frac{3}{5}, \quad r = -3$$

8. Evaluate each of the following expressions:

$$2^2, \quad 2^5, \quad 3^3, \quad \left(\frac{1}{5}\right)^3, \quad \left(\frac{-2}{3}\right)^4, \quad (\sqrt{2})^4$$

9. Evaluate each of the following expressions:

$$3^3, \quad (-3)^5, \quad 4^{-2}, \quad 17^0, \quad \left(\frac{1}{5}\right)^{-1}, \quad (-1)^{\frac{2}{3}}$$

10. Evaluate each of the following expressions:

$$27^{\frac{1}{3}}, \quad 64^{\frac{1}{6}}, \quad 16^{-\frac{3}{4}}, \quad \left(\frac{9}{2}\right)^{-\frac{1}{2}}, \quad (-243)^{\frac{2}{5}}, \quad \left(-\frac{1}{8}\right)^{-\frac{1}{3}}$$

11. Evaluate each of the following expressions:

$$(-27)^{\frac{1}{3}}, \quad 64^{\frac{2}{3}}, \quad (1)^{-\frac{3}{4}}, \quad \left(\frac{9}{2}\right)^0, \quad (81)^{\frac{5}{8}}, \quad (-3)^{\frac{4}{3}}$$

12. Rewrite each of the following expressions using exponents:

$$\sqrt{19}, \quad \sqrt[3]{10}, \quad \frac{1}{\sqrt[4]{2}}, \quad \sqrt[7]{-13}, \quad (\sqrt[3]{4})^2, \quad \left(\frac{1}{\sqrt{2}}\right)^3$$

13. Rewrite each of the following expressions using exponents:

$$\frac{\sqrt{2}}{\sqrt{3}}, \quad \sqrt[5]{-5}, \quad \frac{\sqrt[3]{6}}{\sqrt[2]{6}}, \quad (\sqrt{10})^{10}, \quad \left(\frac{1}{\sqrt[3]{7}}\right), \quad \frac{2}{\sqrt{2}}$$

14. Rewrite each of the following expressions using exponents:

$$\sqrt{2} \cdot \sqrt{3}, \quad \sqrt{2} \cdot \sqrt[3]{2}, \quad 17^2 \cdot \sqrt[3]{17}, \quad \frac{1}{\sqrt{9}} \cdot \frac{1}{\sqrt{9}}, \quad \frac{\sqrt[5]{3}}{\sqrt{3}}, \quad \frac{\sqrt{5}}{2\sqrt{5}}$$

15. Rewrite each of the following expressions using exponents:

$$\sqrt{5}(\sqrt[3]{5}), \quad \sqrt[3]{-8}, \quad \sqrt[3]{-8} \cdot \sqrt[3]{-4}, \quad \frac{\sqrt{9}}{9\sqrt{9}}, \quad \sqrt{9} \cdot \sqrt{18}, \quad \sqrt{91} \cdot \sqrt{100}$$

16. Simplify the given expression:

$$(3x)^3, \quad \frac{1}{b^{-3}}, \quad (a^2 b)^4, \quad \frac{xy^4}{x^6 y^2}, \quad \left(\frac{t}{2}\right)^5, \quad \frac{(2x^2 y)^2}{x^2 y^2}$$

17. Simplify the given expression:

$$\left(\frac{1}{2}x\right)^{-2}, \quad \left(\frac{ab}{a+b}\right)^{-2}, \quad \left(x^{\frac{1}{2}}\right)^4, \quad \left(\frac{3x^3}{5xy}\right)^3, \quad \left(\frac{t}{2}\right)^{-5}, \quad \frac{(2x^2 y)^{\frac{1}{2}}}{x^2 y}$$

*18. Let a be a nonzero real number and let m, n be integers with n positive. If $a^{\frac{m}{n}}$ is defined, show that $\sqrt[n]{a^m}$ is defined and that $a^{\frac{m}{n}} = \sqrt[n]{a^m}$.

3.8 THE BINOMIAL THEOREM

In the previous section we defined exponents and discussed the rules of exponents. Now let us consider expressions of the form $(a + b)^n$, where a, b are real numbers and n is a positive integer. Frequently, it is convenient to display such expressions in expanded form. For example, consider the expression $(x + 1)^3$. By actually multiplying the terms $(x + 1)(x + 1)(x + 1)$, we obtain $(x + 1)^3 = x^3 + 3x^2 + 3x + 1$ (*verify*). In case $n \geq 3$, the process of actually multiplying out the terms of $(a + b)^n$ is quite tedious. Finding $(x + 1)^{49}$ in such a fashion, for example, would be a very time consuming task. Therefore, in this section, we shall present a method for expanding $(a + b)^n$ without having to resort to term by term multiplication. For this purpose, we will need some special notation.

3.17 Definition. *Let n be a positive integer. The number "n factorial," denoted by n!, is*

$$n! = 1 \cdot 2 \cdots n.$$

Example. (a) $1! = 1$
  (b) $2! = 1 \cdot 2 = 2$
  (c) $3! = 1 \cdot 2 \cdot 3 = 6$

(d) $4! = 1 \cdot 2 \cdot 3 \cdot 4 = 24$

(e) $5! = 1 \cdot 2 \cdot 3 \cdot 4 \cdot 5 = 120$

For convenience of notation we define $0! = 1$.

The preceding example suggests the fact that $(n + 1)! = (n!)(n + 1)$. The truth of this statement is shown as follows:

$$(n + 1)! = 1 \cdot 2 \cdots n(n + 1)$$
$$= (1 \cdot 2 \cdots n)(n + 1)$$
$$= (n!)(n + 1).$$

Another convenient notation is now defined.

3.18 Definition. *Let k and n be nonnegative integers with $k \leqslant n$. The binomial coefficient n by k, denoted by $\binom{n}{k}$, is*

$$\binom{n}{k} = \frac{n!}{k!(n - k)!}.$$

Example. (a) $\binom{4}{3} = \frac{4!}{3!(4 - 3)!} = \frac{4!}{3!1!} = \frac{1 \cdot 2 \cdot 3 \cdot 4}{(1 \cdot 2 \cdot 3)(1)} = 4$

(b) $\binom{5}{2} = \frac{5!}{2!(5 - 2)!} = \frac{5!}{2!3!} = \frac{1 \cdot 2 \cdot 3 \cdot 4 \cdot 5}{(1 \cdot 2)(1 \cdot 2 \cdot 3)} = 10$

(c) $\binom{3}{0} = \frac{3!}{0!(3 - 0)!} = \frac{3!}{0!3!} = \frac{1 \cdot 2 \cdot 3}{(1)(1 \cdot 2 \cdot 3)} = 1$

(d) $\binom{7}{7} = \frac{7!}{7!(7 - 7)!} = \frac{7!}{7!0!} = 1$

The preceding example shows that some cancellation will occur when computing $\binom{n}{k}$. More specifically,

$$\binom{n}{k} = \frac{n!}{k!(n - k)!} = \frac{1 \cdot 2 \cdots (n - k)(n - k + 1) \cdots (n - 1)n}{k! [1 \cdot 2 \cdots (n - k)]}$$
$$= \frac{(n - k + 1) \cdots (n - 1)n}{k!}.$$

This relation simplifies the computation of binomial coefficients.

Example. (a) $\binom{4}{3} = \frac{2 \cdot 3 \cdot 4}{1 \cdot 2 \cdot 3} = 4$

(b) $\binom{5}{2} = \frac{4 \cdot 5}{1 \cdot 2} = 10$

(c) $\binom{6}{1} = \frac{6}{1} = 6$

The principal concern of this section, called the *binomial theorem*, can now be stated.

3.19 Theorem *(Binomial theorem). IF a, b are real numbers and n is a positive integer, THEN*

$$(a + b)^n = a^n + \binom{n}{1}a^{n-1}b + \binom{n}{2}a^{n-2}b^2 + \cdots + \binom{n}{k}a^{n-k}b^k + \cdots + b^n.$$

The proof of the binomial theorem is omitted since it requires mathematical induction, a principle which we do not consider here. It should be noted that there are $n + 1$ terms in the expansion of $(a + b)^n$. The general term in the expansion is $\binom{n}{k}a^{n-k}b^k$. It should also be noted that the sum of the exponents of a and b in each term is n.

Problem. Expand $(a + b)^5$ using the binomial theorem.

Solution. We have $(a + b)^5 = a^5 + \binom{5}{1}a^4b + \binom{5}{2}a^3b^2 + \binom{5}{3}a^2b^3 + \binom{5}{4}ab^4 + b^5$
$$= a^5 + 5a^4b + 10a^3b^2 + 10a^2b^3 + 5ab^4 + b^5.$$

Note that there are 6 terms in the expansion and in each term the sum of the exponents of a and b is 5.

Problem. Expand $(x + 2)^4$.

Solution. Using the binomial theorem, we have

$$(x + 2)^4 = x^4 + \binom{4}{1}x^3 \cdot 2 + \binom{4}{2}x^2 \cdot 2^2 + \binom{4}{3}x \cdot 2^3 + 2^4$$
$$= x^4 + 8x^3 + 24x^2 + 32x + 16.$$

Problem. Expand $(2x - 3)^3$ using the binomial theorem.

Solution. $(2x - 3)^3$ is of the form $(a + b)^3$, where $a = 2x$ and $b = -3$. Thus

$$(2x - 3)^3 = (2x)^3 + \binom{3}{1}(2x)^2(-3) + \binom{3}{2}(2x)(-3)^2 + (-3)^3$$
$$= 8x^3 - 36x^2 + 54x - 27.$$

In some cases it may only be necessary to determine a particular term in the expansion of $(a + b)^n$.

Problem. Find the coefficient of y^4 in the expansion of $(y + 2)^7$.

Solution. The general term in the expansion of $(y + 2)^7$ is $\binom{7}{k}y^{7 - k}2^k$. The term y^4 occurs when $7 - k = 4$ or when $k = 3$. Hence, when $k = 3$

$$\binom{7}{k}y^{7 - k}2^k = 2^3\binom{7}{3}y^4$$
$$= 8\left(\frac{5 \cdot 6 \cdot 7}{1 \cdot 2 \cdot 3}\right)y^4$$
$$= 280y^4.$$

Therefore the coefficient of y^4 is 280.

Problem. Find the coefficient of x^5 in the expansion of $(3x - 1)^{10}$.

Solution. The general term of the expansion is

$$\binom{10}{k}(3x)^{10 - k}(-1)^k = (-1)^k3^{10 - k}\binom{10}{k}x^{10 - k}.$$

Therefore x^5 occurs when $k = 5$. In this case the coefficient of x^5 is

$$(-1)^5 3^5\binom{10}{5} = -3^5\left(\frac{6 \cdot 7 \cdot 8 \cdot 9 \cdot 10}{1 \cdot 2 \cdot 3 \cdot 4 \cdot 5}\right)$$
$$= -8748.$$

Exercise Set 3.8

1. Compute: $1!$, $8!$, $4! \cdot 2!$, $5 \cdot 4!$, $4! + 2!$, $6!$

2. Compute: $\binom{3}{2}$, $\binom{5}{2}$, $\binom{5}{3}$, $\binom{17}{0}$, $\binom{6}{5}$, $\binom{n}{1}$

3. Expand $(x + y)^3$
   a. By multiplying $(x + y)(x + y)(x + y)$;
   b. By using the binomial theorem.

4. Expand $(2x + 5)^3$
   a. By multiplying $(2x + 5)(2x + 5)(2x + 5)$;
   b. By using the binomial theorem.

* * *

In problems 5–20, expand the given expressions using the binomial theorem.

5. $(a + b)^5$

6. $(a - b)^5$

7. $(2a + b)^4$

13. $(x - 2)^6$

14. $(-2x + 1)^6$

15. $(3x - 4y)^4$

8. $(x - 2y)^4$

16. $\left(-\dfrac{1}{2}x + y\right)^4$

9. $(x + h)^3$

17. $(ax + by)^3$

10. $(x - a)^3$

18. $\left(t - \dfrac{2}{3}\right)^5$

11. $\left(\dfrac{x}{2} + 4\right)^4$

19. $\left(\dfrac{t}{2} - \dfrac{2}{t}\right)^4$

12. $\left(\dfrac{x}{3} + 3\right)^5$

20. $\left(\dfrac{1}{x} - \dfrac{1}{y}\right)^5$

21. Find the coefficient of t^5 in the expansion of $\left(2t - \dfrac{1}{2}\right)^8$.

22. Find the coefficient of y^4 in the expansion of $\left(\dfrac{2}{3}x - \dfrac{1}{5}y\right)^6$.

*23. Show that $\dbinom{n}{k} = \dbinom{n}{n - k}$.

*24. Use the expansion of $(1 + 1)^n$ to show that $\dbinom{n}{0} + \dbinom{n}{1} + \dbinom{n}{2} + \cdots + \dbinom{n}{n} = 2^n$.

3.9. THE ALGEBRA OF FUNCTIONS

Suppose now that we are given two functions, f and g, with respective domains D_f and D_g. Just as it is possible to combine two numbers by the operations of addition, subtraction, multiplication, and division, it is also possible to combine two functions. Let c be a real number.

a) The product of c and f, denoted by cf, is defined by the rule
   $(cf)(x) = cf(x)$.
b) The sum of f and g, denoted by $f + g$, is defined by the rule
   $(f + g)(x) = f(x) + g(x)$.
c) The difference of f and g, denoted by $f - g$, is defined by the rule
   $(f - g)(x) = f(x) - g(x)$.
d) The product of f and g, denoted by fg, is defined by the rule
   $(fg)(x) = f(x)g(x)$.
e) The quotient of f by g, denoted by $\dfrac{f}{g}$, is defined by the rule

$$\left(\dfrac{f}{g}\right)(x) = \dfrac{f(x)}{g(x)}.$$

The domain of cf is D_f. In order for $f(x) + g(x)$, $f(x) - g(x)$, and $f(x)g(x)$ to be defined, both $f(x)$ and $g(x)$ must be defined. Thus, the domain of $f + g$,

$f - g$, and fg is $D_f \cap D_g$. For $\dfrac{f(x)}{g(x)}$ to be defined we also need $g(x) \neq 0$. Thus the domain of $\dfrac{f}{g}$ is

$$\{x: \ x \in D_f \cap D_g \text{ and } g(x) \neq 0\}.$$

In the next few examples, the domain of each of the functions under consideration is clearly specified. In this way practice is gained in finding the domains of the various algebraic combinations of functions, as well as in determining the combinations themselves.

Example. Let $f(x) = 3x^2 + 2x + 1$ and $g(x) = -9x - 13$ with $D_f = [-2, 6]$ and $D_g = [0, 8]$. Then

$$
\begin{aligned}
(f + g)(x) = f(x) + g(x) &= (3x^2 + 2x + 1) + (-9x - 13) \\
&= 3x^2 + 2x - 9x + 1 - 13 \\
&= 3x^2 - 7x - 12, \\
(f - g)(x) = f(x) - g(x) &= (3x^2 + 2x + 1) - (-9x - 13) \\
&= 3x^2 + 2x + 9x + 1 + 13 \\
&= 3x^2 + 11x + 14, \text{ and} \\
(fg)(x) = f(x)g(x) &= (3x^2 + 2x + 1)(-9x - 13) \\
&= (3x^2 + 2x + 1)(-9x) + (3x^2 + 2x + 1)(-13) \\
&= -27x^3 - 18x^2 - 9x - 39x^2 - 26x - 13 \\
&= -27x^3 - 57x^2 - 35x - 13.
\end{aligned}
$$

The domain of each of these functions is

$$D_f \cap D_g = [-2, 6] \cap [0, 8] = [0, 6].$$

Finally, $\left(\dfrac{f}{g}\right)(x) = \dfrac{f(x)}{g(x)} = \dfrac{3x^2 + 2x + 1}{-9x - 13}$ for $x \in [0, 6]$ with $g(x) \neq 0$. Now $g(x) = -9x - 13 = 0$ if and only if $x = -\dfrac{13}{9}$. Since $-\dfrac{13}{9} \notin [0, 6]$, the domain of $\dfrac{f}{g}$ is $[0, 6]$.

Example. Let $F(w) = w^2 + 2w + 1$ and $g(t) = t - 2$ with $D_F = [-3, 4]$ and $D_g = [-2, 3]$. First observe that different letters have been used to denote the independent variable in these two functions. Since the particular letter used is irrelevant, we shall write $g(w) = w - 2$. Then

$$
\begin{aligned}
(F + g)(w) = F(w) + g(w) &= (w^2 + 2w + 1) + (w - 2) = w^2 + 3w - 1, \\
(F - g)(w) = F(w) - g(w) &= (w^2 + 2w + 1) - (w - 2) = w^2 + w + 3, \\
(Fg)(w) = (w^2 + 2w + 1)(w - 2) &= w^3 - 3w - 2, \text{ and} \\
\left(\dfrac{F}{g}\right)(w) &= \dfrac{w^2 + 2w + 1}{w - 2}.
\end{aligned}
$$

Now $D_F \cap D_g = [-3, 4] \cap [-2, 3] = [-2, 3]$. Thus the domain of $F + g$, $F - g$, and Fg is $[-2, 3]$. Since $g(2) = 0$, the domain of $\dfrac{F}{g}$ is

$$\{w: \ w \in D_F \cap D_g \text{ and } g(w) \neq 0\} = [-2, 2) \cup (2, 3].$$

The preceding ways in which functions were combined corresponded to operations with real numbers. There is a method of combining functions which does not correspond to operations with real numbers. A new function can sometimes be constructed by letting two functions operate on an element in succession. Suppose that f and g are functions and that $g(x) \in D_f$ for all $x \in D_g$. A new function, called the *composition of f and g* and denoted by $f \circ g$, is defined by the rule

$$(f \circ g)(x) = f(g(x)), \qquad x \in D_g.$$

A diagram can be used to give a schematic explanation of the composition of two functions. Consider four sets representing D_g, R_g, D_f, and R_f such that $R_g \subseteq D_f$ (see Figure 3.15). An arrow, drawn from an element x in the domain of g to the element $g(x)$ in the range of g, indicates the object assigned to x by the function g. In a similar manner, an arrow can be used to indicate the object assigned to each $y \in D_f$ by the function f. The object assigned to x by the composition $f \circ g$ is then indicated using an arrow obtained by combining the two given arrows as illustrated.

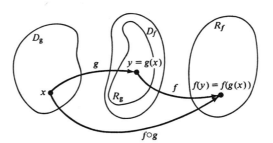

Figure 3.15

It is important to note that in order to form the composition of f and g it must be the case that $g(x) \in D_f$ for all $x \in D_g$. Equivalently, it is necessary to have $R_g \subseteq D_f$. The symbol $f \circ g$ indicates that the functions are applied successively in a particular order: first g, then f. Ordinarily, $g \circ f$ represents a different function when it is defined.

Example. Let g be the function, from the set of all states into the set of all cities, which assigns to each state the capital city of the state. Let f be the function, from the set of all cities into the set of real numbers, which assigns to each city the number representing the population of the city. Then $f \circ g$ is that function which assigns to each state the population of its capital city. For example,

$$g(\text{Ohio}) = \text{Columbus}$$
$$f(\text{Columbus}) = 540,000$$

so that

$$(f \circ g)(\text{Ohio}) = f(g(\text{Ohio}))$$
$$= f(\text{Columbus})$$
$$= 540,000.$$

Observe that in this example $f \circ g$ is defined. However, $g \circ f$ is not defined because $f(x)$ is not in D_g for any city x.

Example. Let m be the function, from the set of all people into the set of all people, which assigns to each person the mother of that person. Similarly, let f assign to each person the father of that person. Then $f \circ m$ assigns to each person that person's maternal grandfather. On the other hand, $m \circ f$ assigns to each person that person's paternal grandmother.

Our main concern is the composition of functions whose domains and ranges are sets of real numbers.

Problem. Let $f(x) = 2x + 4$ and $g(x) = x^2 + 3x - 4$. Find an equation which describes $f \circ g$.

Solution. Since no domains are specified, we assume that $D_f = D_g = \Re$. Clearly $R_g \subseteq \Re$, so that the range of g is contained in the domain of f. Now $(f \circ g)(x) = f(g(x)) = 2g(x) + 4 = 2(x^2 + 3x - 4) + 4 = 2x^2 + 6x - 8 + 4 = 2x^2 + 6x - 4$.

Problem. Let $g(x) = x^2 - 2$ for $x \geqslant 7$ and let $f(x) = \sqrt{3x - 9}$. Find the equation of $f \circ g$.

Solution. We must first determine D_f and R_g and be sure that $R_g \subseteq D_f$ in order that $f \circ g$ may be defined. Now D_f will consist of those values of x for which $3x - 9 \geqslant 0$, since we cannot take the square root of a negative number. Solving this inequality, we obtain $D_f = \{x: x \geqslant 3\}$. On the other hand, D_g is specified as being $[7, \infty)$. Now g is a quadratic with $a = 1$, $b = 0$, and $c = 2$. Hence, $-\dfrac{b}{2a} = 0$, and $\dfrac{4ac - b^2}{4a} = -2$. Since $a > 0$, the graph of g is a parabola, opening upward, with vertex $(0, -2)$. We have $g(7) = 7^2 - 2 = 49 - 2 = 47$. Since the graph of g is rising for $x > 0$, it is certainly also rising for $x \geqslant 7$. But this means that $g(x) \geqslant 47$ for $x \geqslant 7$. Hence, $R_g = [47, \infty)$, so that $R_g \subseteq D_f$. We may now compute $(f \circ g)(x)$.

$$(f \circ g)(x) = f(g(x)) = \sqrt{3g(x) - 9} = \sqrt{3(x^2 - 2) - 9} = \sqrt{3x^2 - 6 - 9}$$
$$= \sqrt{3x^2 - 15}.$$

The preceding example suggests that the difficult part of finding the composition of f and g is verifying that $R_g \subseteq D_f$. In this example, the time spent in developing the theory of graphs of quadratic functions paid off. Had g been a more complicated function, it might have been impossible to determine R_g with the tools that are currently available to us.

Exercise Set 3.9

1. For each pair of functions f and g, determine $f + g$, $f - g$, fg, and $\dfrac{f}{g}$. Specify the domain in each case.

   (a) $f(x) = x^2 + 2x - 1$ $\qquad$ $g(x) = \dfrac{x^2}{2} - 2x + 5$

   (b) $f(x) = 3x - 4$ $\qquad$ $g(x) = 3x + 4$

   (c) $f(x) = \sqrt{x} - x$ $\qquad$ $g(x) = \sqrt[3]{x} + \sqrt{x} + \sqrt{2}$

   (d) $f(x) = |x|$ $\qquad$ $g(x) = x + 1$

2. For each pair of functions f and g of problem 1, determine $g + f$, $g - f$, gf, and $\dfrac{g}{f}$. Specify the domain in each case.

3. For each pair of functions f and g, determine the composition $f \circ g$ in the following:

   a. Let g be the function which assigns to each test its score. Let f be the function which assigns to each test score a letter grade.

   b. Let g be the function which assigns to each telephone subscriber in the city of Cleveland a telephone number.
   Let f be the function which assigns to each telephone number in Cleveland the area code 216.

   c. Let g be the function which assigns to each day of the week the day which follows it.
   Let f be the function which assigns to each day of the week the day which precedes it.

   d. Let g be the function which assigns to each musical composition the key in which it is to be played.
   Let f be the function which assigns to each key the number of sharps or flats.

4. For each pair of functions f and g, determine whether or not $R_g \subseteq D_f$. If it is, form the composition $f \circ g$.

   a. $f(x) = 2x - 1$ $\qquad$ $g(x) = x^2$

   b. $f(x) = (x + 4)^5$ $\qquad$ $g(x) = 6 - x$

   c. $f(x) = 2 |x|$ $\qquad$ $g(x) = 33$

   d. $f(x) = \sqrt{x}$ $\qquad$ $g(x) = -x^2 + x - 1$

5. For each pair of functions f and g of problem 4, determine whether or not $R_f \subseteq D_g$. If it is, form the composition $g \circ f$.

* * *

6. Let $f(x) = 7x + 1$ and $g(x) = -2x - \dfrac{10}{3}$, with $D_f = [0, 9]$ and $D_g = [-2, 6]$.

   Find $f + g, f - g, fg$, and $\dfrac{f}{g}$. Specify the domain of each.

7. Repeat problem 6 for $f(x) = x^2 - 3$ and $g(x) = x^2 + 2x + 1$ with $D_f = \mathcal{R}$ and $D_g = (-10, 10)$.

8. Repeat problem 6 for $f(x) = \dfrac{1}{x}$ and $g(x) = \dfrac{3}{x^2 - 1}$, with $D_f = [2, 5]$ and

   $D_g = \left[\dfrac{3}{2}, 4 \right]$.

9. Repeat problem 6 for $f(x) = \dfrac{x}{x^2 + 1}$ and $g(x) = \dfrac{x - 1}{x + 1}$, with $D_f = \mathcal{R}$ and

   $D_g = [-4, -2]$.

In problems 10–30 find $(f \circ g)(x)$. *Assume that* $R_g \subseteq D_f$.

10. $f(x) = \dfrac{1}{2}x + 3$	$g(x) = -2x + \dfrac{7}{2}$
11. $f(x) = -x - 19$	$g(x) = x^2 + 1$
12. $f(x) = x^2 + 2x + 2$	$g(x) = \dfrac{1}{2}(x - 1)$
13. $f(x) = \dfrac{1}{x}$	$g(x) = x^2 + 1$
14. $f(x) = \dfrac{1}{x}$	$g(x) = \dfrac{1}{x}$
15. $f(x) = \dfrac{1}{x}$	$g(x) = \dfrac{3}{x^2 - 1}$
16. $f(x) = \dfrac{x}{x^2 + 1}$	$g(x) = \dfrac{x - 1}{x + 1}$
17. $f(x) = -3x + 1$	$g(x) = -\dfrac{1}{3}x + \dfrac{1}{3}$
18. $f(x) = x^2 - 9x + 2$	$g(x) = x^2$
19. $f(x) = 8 - x^3$	$g(x) = x^4$
20. $f(x) = x^5 + 3$	$g(x) = x + 2$ (*Hint:* Use the binomial theorem)
21. $f(x) = x^4$	$g(x) = -3x + 1$ (*Hint:* Use the binomial theorem)

22. $f(x) = x^{\frac{1}{3}} + x^{\frac{1}{2}}$ $g(x) = x^2$

23. $f(x) = x^{\frac{1}{3}} + x^{\frac{1}{2}}$ $g(x) = x^3$

24. $f(x) = |2x - 1|$ $g(x) = x + 2$

25. $f(x) = \dfrac{x}{x + 1}$ $g(x) = |x - 1|$

26. $f(x) = x^2 + 2x - 1$ $g(x) = x + h$

27. $f(x) = \sqrt{x^2 + 4}$ $g(x) = x + h$

28. $f(x) = 7$ $g(x) = x + h$

29. $f(x) = k$ $g(x) = x + h$

30. $f(x) = \dfrac{x}{x + 5}$ $g(x) = x + h$

*31. If $f(w) = \dfrac{2}{3}w + 7$ and $g(w) = \dfrac{3}{2}(w - 7)$ find $f \circ g$ and $g \circ f$. Are these functions the same? Does this prove that the operation "$\circ$" is commutative?

*32.   a. If $h(x) = (x^2 + 3)^5$, choose functions f and g such that $h = f \circ g$.
     b. If $h(x) = \sqrt{x - 12}$, choose functions f and g such that $h = f \circ g$.

3.10 POLYNOMIAL, RATIONAL, AND ALGEBRAIC FUNCTIONS

Let us return to our study of particular functions and their properties. We have noticed a similarity between linear and quadratic functions. Indeed, each is defined as a sum of terms of the type cx^n, where c is a constant and n is either 0, 1, or 2, with 2 appearing as an exponent only in the quadratic function. Both linear and quadratic functions fall into a more general class of functions, known as *polynomials*.

3.20 Definition. *A function of the type*

$$P(x) = a_n x^n + a_{n-1} x^{n-1} + \cdots + a_2 x^2 + a_1 x + a_0,$$

where $a_i \in \mathbb{R}$ for $i = 0, 1, 2, \ldots, n$, with $a_n \neq 0$, is a polynomial function of degree n.

The real numbers a_i $(i = 0, 1, 2, \ldots, n)$ are called *coefficients*. In particular, a_n is called the leading coefficient, and a_0 is called the constant term. The degree of a polynomial is seen to be the highest power to which the variable is raised. For example, the function

$$f(x) = 3x^6 - 2x^5 + x^2 + 5$$

is a sixth degree polynomial for which $a_6 = 3$, $a_5 = -2$, $a_4 = 0$, $a_3 = 0$, $a_2 = 1$, $a_1 = 0$, and $a_0 = 5$. We see that linear functions, $l(x) = ax + b\,(a \neq 0)$, are polynomials of degree 1, and quadratic functions, $Q(x) = ax^2 + bx + c\,(a \neq 0)$, are polynomials of degree 2. Unless otherwise specified, the domain of a polynomial function is the set $\Re$ of all real numbers.

Problem. Which of the following are polynomial functions?

(a) $f(x) = \dfrac{2}{3}x^4 - x^3 - x + 1052$

(b) $g(x) = (x - 3)\left(x^2 + 2x + \dfrac{1}{2}\right)$

(c) $h(x) = (x^4 - 3x^3 - 5x^2 + 8x + 1)^{1/2}$

Solution. (a) f is a polynomial of degree 4, for which $a_4 = \dfrac{2}{3}$, $a_3 = -1$, $a_2 = 0$, $a_1 = -1$, and $a_0 = 1052$.

(b) g is also a polynomial function. In order to see that g fits the form of the preceding definition, we need only carry out the indicated multiplication. We have

$$g(x) = (x - 3)\left(x^2 + 2x + \frac{1}{2}\right)$$

$$= x\left(x^2 + 2x + \frac{1}{2}\right) - 3\left(x^2 + 2x + \frac{1}{2}\right)$$

$$= x^3 + 2x^2 + \frac{1}{2}x - 3x^2 - 6x - \frac{3}{2}$$

$$= x^3 - x^2 - \frac{11}{2}x - \frac{3}{2}.$$

It is now clear that g is a third degree polynomial for which $a_3 = 1$, $a_2 = -1$, $a_1 = -\dfrac{11}{2}$, and $a_0 = -\dfrac{3}{2}$.

(c) h is not a polynomial since the variable appears under a square root sign.

In previous sections of this chapter we found that every linear function had a straight line as its graph, and that every quadratic function had a parabola as its graph. Unfortunately, it is not even the case that the graph of every third degree polynomial is a curve of one particular type. For example, the graphs of $f(x) = x^3 - x$ and $g(x) = x^3 - 4$ do not have the same general shape (see Figure 3.16).

As one might expect, a study of graphs of polynomial functions of degree $n > 3$ is even more difficult than is consideration of third degree polynomials.

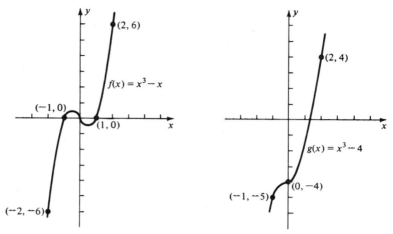

Figure 3.16

We consider the problem of graphing polynomial functions in more detail in Chap. 4 and again in Chap. 7, after we have developed the differential calculus.

In Sec. 3.5 we found that a quadratic function could have at most two zeros. It is possible to prove that a polynomial function

$$P(x) = a_n x^n + a_{n-1} x^{n-1} + \cdots + a_1 x + a_0$$

of degree n can have at most n zeros. It is also possible to derive formulas, similar to the quadratic formula, for determining the zeros of any third or fourth degree polynomial. We shall not consider these formulas here since they are complicated, and not of sufficient general interest to warrant the amount of time required for their study. It is interesting to note, however, that no formulas exist for finding zeros of polynomial functions of degree greater than or equal to 5, even though many such functions do have zeros! When it is necessary to determine the zeros of a polynomial, we usually resort to approximative methods. We shall reconsider the question of the existence of zeros of polynomials in Chap. 4.

We conclude Chap. 3 with an introduction to a class of functions closely related to polynomials. It is easy to show that the sum, difference, or product of two polynomial functions is again a polynomial function. However, if P and Q are polynomials, the function $R = \dfrac{P}{Q}$ need not be a polynomial. For example, if $P(x) = 3x^4 - 2x^3 + x + 9$, and $Q(x) = 2x^2 + 2$, then

$$R(x) = \left(\frac{P}{Q}\right)(x) = \frac{P(x)}{Q(x)} = \frac{3x^4 - 2x^3 + x + 9}{2x^2 + 2}.$$

Since division by the independent variable occurs, R is not a polynomial function, even though P and Q are.

3.21 Definition. *IF P and Q are polynomial functions, THEN the function*

$$R = \frac{P}{Q}$$

is a rational function.

Every polynomial may be thought of as a rational function with denominator $Q(x) = 1$. Also, the domain of the rational function $R = \dfrac{P}{Q}$ does not contain values of x for which $Q(x) = 0$. Note, too, that $R(z) = \dfrac{P(z)}{Q(z)} = 0$ if and only if $P(z) = 0$ and $Q(z) \neq 0$. Therefore the set of zeros of R is

$$\{z: \ P(z) = 0 \ \text{and} \ Q(z) \neq 0\}.$$

Problem. Find all the zeros of the rational function $R(x) = \dfrac{2x^2 + 3x - 2}{5x - 1}$.

Solution. We see that the domain of R is the set of all real numbers not equal to $\dfrac{1}{5}$, since $5\left(\dfrac{1}{5}\right) - 1 = 0$. Now $R(x) = 0$ if and only if $2x^2 + 3x - 2 = 0$. Applying the quadratic formula to $P(x) = 2x^2 + 3x - 2$, we obtain $P(x) = 0$ if and only if

$$x = \frac{-3 \pm \sqrt{9 - 4(2)(-2)}}{4} = \frac{-3 \pm \sqrt{25}}{4} = \frac{-3 \pm 5}{4}.$$

Hence, $z_1 = \dfrac{-3 + 5}{4} = \dfrac{1}{2}$ and $z_2 = \dfrac{-3 - 5}{4} = -2$, so the numbers $\dfrac{1}{2}$ and -2 are the zeros of R.

Problem. Find the zeros of the rational function $R(x) = \dfrac{2x^2 + 3x - 2}{x + 2}$.

Solution. From the preceding example, the zeros of the numerator are $\dfrac{1}{2}$ and -2. However, -2 is also a zero of the denominator and hence is not in the domain of R. It follows that $\dfrac{1}{2}$ is the only zero of R.

In constructing a rational function, we are permitted to add, subtract, multiply, and divide by the independent variable. If, in addition to these operations, the extraction of roots is also permitted, then the functions obtained are called *algebraic* functions. Every rational function is an algebraic function. However, there are algebraic functions such as

$$f(t) = t^2 - \sqrt{t + 4},$$

$$g(x) = \frac{6x - 7}{2 + \sqrt{x}}, \text{ and}$$

$$h(z) = z^{2/3} - 5z + 7$$

which are not rational functions.

We shall postpone our study of graphs of rational and algebraic functions until Chap. 4, in which certain facts about polynomial functions are derived that make the task of graphing such functions somewhat easier.

Exercise Set 3.10

1. Rewrite each of the following polynomial functions in the form $f(x) = a_n x^n + a_{n-1} x^{n-1} + \cdots + a_2 x^2 + a_1 x + a_0$:
   (a) $f(x) = 3x^2 - 8x^3 + 5x - 15$
   (b) $f(x) = \frac{2}{3}x^5 + \sqrt{2}\, x^3 - x^4 + 7002x^7$
   (c) $f(x) = (x^2 - 5x + 1)(4 - x)(x^3)$

2. Determine the degree of each of the polynomial functions listed in problem 1.

3. For each of the following polynomial functions, determine a_4, a_3, a_2, a_1, and a_0:
   (a) $g(x) = 2x^4 - x^3 + 9x^2 + x - \frac{7}{8}$
   (b) $h(x) = x^4 - x^2 + 1$
   (c) $F(x) = \sqrt{2}\, x + \sqrt{3}\, x^2 + \pi x^3 - \frac{13}{3}x^4$
   (d) $f(t) = \frac{t^4}{4} - \frac{2t^3}{3}$

4. Which of the following functions are polynomial functions? Which are rational functions? Which are algebraic functions?
   (a) $P(x) = 7x - 9$ (e) $F(y) = (y^3 + 1)^{20}$
   (b) $R(x) = -x^{-1}$ (f) $P(w) = -w^3 + w^2 + \sqrt{2w} - \sqrt{5}$
   (c) $T(x) = 25$ (g) $\dfrac{x + 1}{x^2 - 5x + 8}$
   (d) $G(x) = \sqrt{x^2 + 1}$ (h) $(x + 1)(3x - 4)^{1/2}$

* * *

5. Given an example of a polynomial function of degree 14; of degree 3; of degree 1; of degree 0.

6. Write a polynomial function such that
   (a) $a_n = 5$
   (b) $a_1 = 1, a_2 = 2, a_3 = -3, a_4 = -1$
   (c) $a_0 = 0$
   (d) $a_n = 3, a_0 = \sqrt{15}$
   (e) $a_{n-1} = a_{n-2} = 0$

7. For each polynomial function, state the degree and determine the coefficients $a_n, a_{n-1}, \ldots, a_0$. Specify the domain.
   (a) $F(t) = t^4 + \dfrac{15}{2}$
   (b) $g(x) = 2 - \sqrt{2}\,x + x^3 - x^{10}$
   (c) $h(u) = (u + 1)^5$ (*Hint*: Use the binomial theorem)
   (d) $f(x) = ax^2 + bx + c\ (a, b, c \in \mathcal{R})$
   (e) $h(x) = mx + b \qquad (m, b \in \mathcal{R})$
   (f) $G(p) = -\dfrac{12}{\sqrt{19}}$

8. Which of the following statements are always true?
   (a) The leading coefficient of a fifth degree polynomial is the coefficient of x^5.
   (b) The sum of a polynomial of degree 2 and a polynomial of degree 3 is a polynomial of degree 5.
   (c) The product of a polynomial of degree 2 and a polynomial of degree 3 is a polynomial of degree 5.
   (d) The sum of a polynomial of degree 3 and another polynomial of degree 3 is a polynomial of degree 3.
   (e) The polynomial $(3x + 2)^5$ is a polynomial of degree 15.

9. Explain why each of the following functions is a rational function, but not a polynomial. Specify the domain of each function.
   (a) $g(x) = \dfrac{x + 1}{x - 1}$
   (b) $g(t) = (3t^2 - 3)^{-1}$
   (c) $h(t) = t^2 + t + t^{-2}$
   (d) $R(y) = y^2\left(\dfrac{1}{y} + \dfrac{1}{y^3}\right)$

10. Determine the zeros of each of the following rational functions:
    (a) $G(x) = \dfrac{3x + 1}{x - 5}$
    (b) $H(x) = \dfrac{(2x - 3)(x^2 - 1)}{x^3 - 1}$
    (c) $R(z) = \dfrac{z^2 + 4z + 4}{3(z + 2)^2}$
    (d) $F(w) = \dfrac{4}{w - 1}$
    (e) $P(r) = \dfrac{r^2 + r + 3}{(r - 1)^5}$
    (f) $H(x) = \dfrac{(x^2 - 2)(x^2 + 5)}{x^2 + 1}$

11. Explain why each of the following functions is algebraic, but not rational:

   (a) $f(x) = \dfrac{(4x^2 - 9)}{\sqrt{4x^2 - 9}}$

   (c) $g(u) = (u + 1)^{1/2} (u^2 + 1)^{-1/2}$

   (b) $g(x) = \dfrac{x^2 + 2x + 5}{x^2 - 3x^{1/2} + 6}$

   (d) $f(t) = \dfrac{t}{4} + \dfrac{4}{t} + t^{1/4}$

12. Which of the following statements are always true?

   (a) $\{f: \ f \text{ is a quadratic function}\} \subseteq \{p: \ p \text{ is a polynomial function}\}$
   (b) $\{l: \ l \text{ is a linear function}\} \subseteq \{f: \ f \text{ is a quadratic function}\}$
   (c) $\{c: \ c \text{ is a constant function}\} \subseteq \{p: \ p \text{ is a polynomial function}\}$
   (d) $\{p: \ p \text{ is a polynomial function}\} \subseteq \{R: \ R \text{ is a rational function}\}$
   (e) $\{p: \ p \text{ is a polynomial function}\} \subseteq \{g: \ g \text{ is an algebraic function}\}$
   (f) $\{F: \ F \text{ is a rational function}\} \subseteq \{H: \ H \text{ is an algebraic function}\}$

4 / Algebraic Functions and Their Graphs

The main purpose of this chapter is to investigate in greater detail some of the functions we have seen before. Early discussion will be directed towards methods of finding zeros of polynomial functions. Using our knowledge of such zeros, we then give further consideration to the graphing of polynomial and related functions. The concluding sections of the chapter contain a discussion of the absolute value function and the greatest integer function.

Throughout this chapter the terms "polynomial" and "polynomial function" will be used synonymously. Moreover, we will frequently denote the polynomial function P by $P(x)$. This dual notation causes no confusion and allows us to state definitions and theorems in a more concise manner.

4.1 DIVISION OF POLYNOMIALS

Let us consider the process of dividing one polynomial $P(x)$ by another polynomial $D(x)$. We will call $P(x)$ the *dividend* and $D(x)$ the *divisor*. Suppose, for example, that the dividend is $P(x) = 2x^3 + 33x^2 - 16x - 3$ and that the divisor is $D(x) = x^2 + x + 1$. The division calculations are performed in tabular form as follows:

Quotient $\longrightarrow$ $2x + 31$

$$x^2 + x + 1 \enclose{longdiv}{2x^3 + 33x^2 - 16x + 4}$$

$\underline{2x^3 + 2x^2 \ + \ 2x}$ *Step 1.* Divide $2x^3$ by x^2 to obtain $2x$.

$31x^2 - 18x + 4$ *Step 2.* Multiply $x^2 + x + 1$ by $2x$.

Step 3. Subtract $2x^3 + 2x^2 + 2x$ from $2x^3 + 33x^2 - 16x + 4$.

153

Step 4. Divide $31x^2$ by x^2 to obtain 31.

$\underline{31x^2 + 31x + 31}$ *Step 5.* Multiply $x^2 + x + 1$ by 31.

Remainder $\longrightarrow$ $-49x - 34$ *Step 6.* Subtract $31x^2 + 31x + 31$ from $31x^2 - 18x + 4$.

Thus we may write

$$\frac{2x^3 + 33x^2 - 16x + 4}{x^2 + x + 1} = 2x + 31 + \frac{-49x - 34}{x^2 + x + 1},$$

or

$$2x^3 + 33x^2 - 16x + 4 = (x^2 + x + 1)(2x + 31) + (-49x - 34).$$

If we denote the *quotient* polynomial, $2x + 31$, by $Q(x)$ and the *remainder* polynomial, $-49x - 34$, by $R(x)$, then the preceding equation becomes

$$P(x) = D(x)\, Q(x) + R(x).$$

As another illustration, let $P(x) = 6x^2 + 17x - 14$ and $Q(x) = 2x - 5$. The division calculations are:

Quotient $\longrightarrow$ $3x + 16$

$2x - 5 \;\big|\; 6x^2 + 17x - 14$ *Step 1.* Divide $6x^2$ by $2x$ to obtain $3x$.

$\underline{6x^2 - 15x}$ *Step 2.* Multiply $2x - 5$ by $3x$.

$32x - 14$ *Step 3.* Subtract $6x^2 - 15x$ from $6x^2 + 17x - 14$.

Step 4. Divide $32x$ by $2x$ to obtain 16.

$\underline{32x - 80}$ *Step 5.* Multiply $2x - 5$ by 16.

Remainder $\longrightarrow$ 66   *Step 6.* Subtract $32x - 80$ from $32x - 14$.

Therefore

$$\frac{6x^2 + 17x - 14}{2x - 5} = 3x + 16 + \frac{66}{2x - 5},$$

or

$$6x^2 + 17x - 14 = (2x - 5)(3x + 16) + 66.$$

In this case $Q(x) = 3x + 16$ and $R(x) = 66$ so that the preceding equation becomes

$$P(x) = D(x)\, Q(x) + R(x).$$

The two examples we have just examined illustrate the fact that, if one polynomial is divided by another, then a quotient and remainder can always be found. This important fact is called the *division algorithm*.

4.1 Theorem. *For any two polynomials P(x) and D(x), D(x) not a constant, there exist unique polynomials Q(x) and R(x) such that*

$$P(x) = D(x)Q(x) + R(x)$$

and the degree of R(x) is less than the degree of D(x).

Problem. Find polynomials $Q(x)$ and $R(x)$ such that

$$2x^5 + x^4 - x^3 + 3x + 7 = (2x^3 + x^2 + 1)\, Q(x) + R(x),$$

where the degree of $R(x)$ is less than the degree of $2x^3 + x^2 + 1$.

Solution. If we let $P(x) = 2x^5 + x^4 - x^3 + 3x + 7$ and $D(x) = 2x^3 + x^2 + 1$, then the problem is to find the quotient and remainder when $P(x)$ is divided by $D(x)$. Note that the coefficient of x^2 in $P(x)$ is 0. Because of the tabular nature of the computations involved in dividing $P(x)$ by $D(x)$, space must be reserved for terms of degree 2. This is achieved by writing $P(x)$ as

$$P(x) = 2x^5 + x^4 - x^3 + 0 \cdot x^2 + 3x + 7$$

and proceeding with the usual computations as follows:

$$
\begin{array}{r}
x^2 - \dfrac{1}{2} \\
\hline
2x^3 + x^2 + 1 \,\big|\, 2x^5 + x^4 - x^3 + 0 \cdot x^2 + 3x + 7 \\
\underline{2x^5 + x^4 \qquad\quad + x^2} \\
- x^3 - x^2 + 3x + 7 \\
\underline{- x^3 - \dfrac{1}{2}x^2 \qquad - \dfrac{1}{2}} \\
- \dfrac{1}{2}x^2 + 3x + \dfrac{15}{2}
\end{array}
$$

Therefore $Q(x) = x^2 - \dfrac{1}{2}$ and $R(x) = -\dfrac{1}{2}x^2 + 3x + \dfrac{15}{2}$. Note that the degree of $R(x)$ is 2, the degree of $D(x)$ is 3 and that

$$2x^5 + x^4 - x^3 + 3x + 7 = (2x^3 + x^2 + 1)\left(x^2 - \frac{1}{2}\right) + \left(-\frac{1}{2}x^2 + 3x + \frac{15}{2}\right)$$

or

$$2x^5 + x^4 - x^3 + 3x + 7 = (2x^3 + x^2 + 1)\, Q(x) + R(x).$$

If $P(x)$ is divided by a divisor of the form $D(x) = ax + b$, where $a \neq 0$, then the division algorithm states that there are polynomials $Q(x)$ and $R(x)$ such that

$$P(x) = D(x)\, Q(x) + R(x)$$
$$= (ax + b)\, Q(x) + R(x),$$

where the degree of $R(x)$ is less than the degree of $D(x) = ax + b$. Since $D(x)$ is of degree 1 and the degree of $R(x)$ is less than 1, $R(x)$ must be a constant c. Thus, in this special case, we have

$$P(x) = (ax + b) Q(x) + c.$$

Problem. Find the quotient and constant remainder when $6x^3 - 19x^2 + 58x + 5$ is divided by $3x - 2$.

Solution. We want to divide $P(x) = 6x^3 - 19x^2 + 58x + 5$ by $D(x) = 3x - 2$. The computations are as follows:

$$
\begin{array}{r}
2x^2 - 5x \quad + 16 \\
3x - 2 \, \overline{\smash{)}\, 6x^3 - 19x^2 + 58x + 5} \\
\underline{6x^3 - 4x^2} \\
-15x^2 + 58x \\
\underline{-15x^2 + 10x} \\
48x + 5 \\
\underline{48x - 32} \\
37.
\end{array}
$$

Thus the quotient is $Q(x) = 2x^2 - 5x + 16$, the constant remainder is $R(x) = 37$, and we have

$$6x^3 - 19x^2 + 58x + 5 = (3x - 2)(2x^2 - 5x + 16) + 37.$$

Exercise Set 4.1

1. Divide $12x^2 - 17x - 40$ by $3x - 2$.
   Divide $12x^2 - 17x - 40$ by $4x + 5$.
   Divide $12x^2 - 17x - 40$ by $x + 1$.

2. Write an equation which illustrates the division algorithm (Theorem 4.1) for each of the pairs of polynomials in problem 1.

3. Divide $2x^5 + 4x^4 - x^3 - x - 1$ by $x^2 + x - 1$.
   Divide $2x^5 + 4x^4 - x^3 - x - 1$ by $2x^2 + 1$.
   Divide $2x^5 + 4x^4 - x^3 - x - 1$ by $4x^4 + x^3 - 2x^2 - x + 5$.

4. Write an equation which illustrates the division algorithm (Theorem 4.1) for each of the pairs of polynomials in problem 3.

5. If $P(x) = D(x)Q(x)$, then each of the polynomials $D(x)$ and $Q(x)$ is called a factor of $P(x)$. If the remainder $R(x)$ obtained when $P(x)$ is divided by $D(x)$ is zero, then $D(x)$ is a factor of $P(x)$. Determine whether $D(x)$ is a factor of $P(x)$.

(a) $P(x) = 12x^2 - 17x - 40$, $D(x) = 3x + 5$

(b) $P(x) = 2x^3 - 4x^2 - \dfrac{13}{2}x + 4$, $D(x) = x - \dfrac{1}{2}$

(c) $P(x) = x^3 + x^2 + x + 1$, $D(x) = x - 1$

(d) $P(x) = 3x^4 + 5x^3 + 15x^2 + 9x + 10$, $D(x) = x^2 + x + 2$

* * *

In problems 6–15, given polynomials P and D, find polynomials Q and R such that $P = DQ + R$ and the degree of R is less than the degree of D.

6. $P(x) = 2x^4 + 7x^3 - 18x^2 - 3x + 4$ $D(x) = 2x^2 - x + 5$

7. $P(r) = 6r^4 + 8r^3 + 15r^2 - 4r - 32$ $D(r) = 3r + 4$

8. $P(z) = 6z^4 + 8z^3 + 15z^2 - 4z - 32$ $D(z) = z + \dfrac{4}{3}$

9. $P(y) = 2y^3 - y^2 + 2y + 5$ $D(y) = 3y + 4$

10. $P(t) = -3t^5 + 8$ $D(t) = t - 6$

11. $P(x) = x^4 + 5x^2 - 8$ $D(x) = 3x$

12. $P(x) = x^4 + 5x^2 - 8$ $D(x) = 3x^4$

13. $P(t) = 8t^6 + 16t^4 + 8t^2 + 16$ $D(t) = 2t^3 + t^2 - t + 1$

14. $P(z) = z^7 - 1$ $D(z) = z^6 + z^5 + z^4 + z^3 + z^2 + 1$

15. $P(y) = 2y^3 - 3y^2 + 2y + 3$ $D(y) = \dfrac{1}{2}y - 1$

In problems 16–21, determine whether $D(x)$ is a factor of $P(x)$.

16. $P(x) = 4x^3 + 3x^2 - 6x - 5$ $D(x) = 4x - 5$

17. $P(x) = -24x^3 - 58x^2 + 85x - 21$ $D(x) = 3x - 1$

18. $P(x) = 30x^3 + 143x^2 + 117x - 56$ $D(x) = 2x + 7$

19. $P(x) = 30x^3 + 143x^2 + 117x - 56$ $D(x) = 10x^2 + 41x + 56$

20. $P(x) = -3x^4 - 5x^3 + 2x^2 + 5x + 1$ $D(x) = -x^2 + 1$

21. $P(x) = x^8 - 1$ $D(x) = x^7 + x^6 + x^5 + x^4 + x^3 + x^2 + x + 1$

*22. Show that $4t + \dfrac{1}{2}$ is a factor of the polynomial $P(t) = 8t^3 - 19t^2 - \dfrac{29}{2}t - \dfrac{3}{2}$. Use this fact and the quadratic formula to write $P(t)$ as the product of three *linear* factors.

*23. Do the same as in exercise 22 for $2x - \dfrac{1}{3}$ and $P(x) = 6x^3 + 5x^2 - 13x + 2$.

*24. Do the same as in exercise 22 for $3r + \dfrac{1}{4}$ and $P(r) = 24r^3 + 14r^2 - 11r - 1$.

4.2 SYNTHETIC DIVISION

It is not too difficult to see that, if we divide a polynomial of large degree by a polynomial of small degree using the usual long division process, then the tabular computations may get rather lengthy. This is certainly true if we divide a polynomial by a divisor of the form $x - a$. The division of a polynomial by a polynomial of the form $x - a$ can be simplified by the use of a technique called *synthetic division*. Essentially, synthetic division is a condensed version of the long division process.

Consider the format of the division of $3x^4 - 11x^3 + 14x^2 + 17x + 13$ by $x - 2$. The divisor is of the form $x - a$, where $a = 2$. We have

$$
\begin{array}{r}
\text{Quotient} \rightarrow \quad 3x^3 - 5x^2 + 4x + 25 \\
x - 2 \enclose{longdiv}{3x^4 - 11x^3 + 14x^2 + 17x + 13} \\
\underline{3x^4 - 6x^3} \\
-5x^3 + 14x^2 + 17x + 13 \\
\underline{-5x^3 + 10x^2} \\
4x^2 + 17x + 13 \\
\underline{4x^2 - 8x} \\
25x + 13 \\
\underline{25x - 50} \\
\text{Remainder} \rightarrow \quad 63
\end{array}
$$

For computational purposes it is the numerical coefficients of the x's which are important, not the x's themselves. In fact, the x's are only placeholders in our tabular method of computation. Thus the repetitious writing of the x's might be eliminated. Also, each number in the quotient appears several times in the course of the work. This repetition, too, might be superfluous. Let us consider steps which could be taken to shorten the process of dividing $3x^4 - 11x^3 + 14x^2 + 17x + 13$ by $x - 2$.

Simplification 1. Elimination of x's.

$$
\begin{array}{r}
\text{Quotient} \\
\text{coefficients} \rightarrow \quad 3 - 5 + 4 + 25 \\
1 - 2 \enclose{longdiv}{3 - 11 + 14 + 17 + 13} \\
\underline{3 - 6} \\
-5 + 14 + 17 + 13 \\
\underline{-5 + 10} \\
4 + 17 + 13 \\
\underline{4 - 8} \\
25 + 13 \\
\underline{25 - 50} \\
\text{Remainder} \rightarrow \quad 63
\end{array}
$$

Simplification 2. Omission of nonessential repetitions of numbers, recalling how each retained number is obtained.

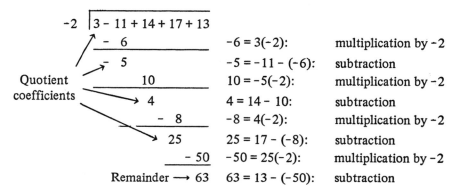

$-6 = 3(-2)$:	multiplication by -2
$-5 = -11 - (-6)$:	subtraction
$10 = -5(-2)$:	multiplication by -2
$4 = 14 - 10$:	subtraction
$-8 = 4(-2)$:	multiplication by -2
$25 = 17 - (-8)$:	subtraction
$-50 = 25(-2)$:	multiplication by -2
$63 = 13 - (-50)$:	subtraction

We can now easily observe that a simple pattern of computation is being followed: namely, multiplication by -2, followed by subtraction.

Simplification 3. Replace: Multiplication by -2, followed by subtraction.
By: Multiplication by 2, followed by addition.

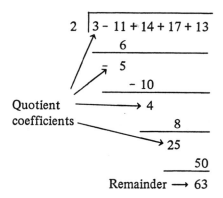

Simplification 4. Arrangement of work in a horizontal rather than a vertical line.

$$
\begin{array}{c|ccccc}
2 & 3 - 11 + 14 + 17 + 13 \\
& 6 - 10 \quad 8 \quad 50 \\
\hline
& 3 - 5 \quad 4 \quad 25 \quad 63
\end{array}
$$

Quotient   Remainder
coefficients

Simplification 5. Arrange in conventional form for synthetic division.

$$3 - 11 \quad 14 \quad 17 \quad 13 \qquad \underline{|2} = a$$
$$\underline{\quad 6 - 10 \quad 8 \quad 50 \quad}$$
$$3 - 5 \quad 4 \quad 25, \quad 63$$

Quotient   Remainder
coefficients

The last simplified form shows that the coefficients of the quotient and re-mainder can be found by bringing down the 3, to start the process, and then alternately multiplying by 2 and adding as indicated by the arrows in the following diagram.

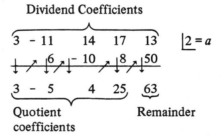

Dividend Coefficients

$$3 \ - 11 \qquad 14 \quad 17 \quad 13 \qquad \underline{|2} = a$$
$$6 \quad - 10 \quad 8 \quad 50$$
$$3 \ - \ 5 \qquad 4 \quad 25, \quad 63$$

Quotient   Remainder
coefficients

The interpretation of the numbers on the bottom line of the synthetic division diagram is quite important. We see that the last number represents the remainder while the preceding numbers represent the coefficients of the quotient, start-ing with the leading coefficient and ending with the constant term.

In general, the same technique can be used to divide any polynomial $P(x)$ by a polynomial of the form $x - a$. While the justification for this condensation of the long division process is somewhat lengthy, the process of synthetic division is not difficult and can be mastered rather quickly with practice.

Problem. Use synthetic division to divide $2x^3 + 9x^2 + 5x - 12$ by $x - 4$.

Solution. The divisor is of the form $x - a$, where $a = 4$. Using the experience gained in the earlier example, we see that the diagram to be completed is the following:

Dividend coefficients

$$2 \qquad 9 \qquad 5 \qquad - 12 \qquad \underline{|4} = a$$

If we now bring down the 2 and perform the synthetic division, we have

Dividend coefficients

$$
\begin{array}{ccccc}
2 & 9 & 5 & -12 & \underline{\;4} = a \\
 & 8 & 68 & 292 & \\
\hline
2 & 17 & 73 & 280 &
\end{array}
$$

Quotient
coefficients   Remainder

The quotient is $2x^2 + 17x + 73$ and the remainder is 280 so that

$$2x^3 + 9x^2 + 5x - 12 = (x - 4)(2x^2 + 17x + 73) + 280.$$

Problem. Use synthetic division to divide $x^4 + 6x^3 - 12x^2 + x - 5$ by $x + 3$.

Solution. In this case the divisor is $x + 3 = x - (-3)$ and, hence, is of the form $x - a$, where $a = -3$. The completed synthetic division diagram becomes

$$
\begin{array}{cccccc}
1 & 6 & -12 & 1 & -5 & \underline{\;-3} \\
 & -3 & -9 & 63 & -192 & \\
\hline
1 & 3 & -21 & 64 & -197 &
\end{array}
$$

The quotient is $x^3 + 3x^2 - 21x + 64$ and the remainder is -197 so that

$$x^4 + 6x^3 - 12x^2 + x - 5 = (x + 3)(x^3 + 3x^2 - 21x + 64) - 197.$$

In the long division process, terms missing in the dividend were written as having 0 as a coefficient. Recall that we wrote $2x^5 + x^4 - x^3 + 3x + 7$ as $2x^5 + x^4 - x^3 + 0 \cdot x^2 + 3x + 7$. Since synthetic division is merely a condensation of the tabular method for long division, 0 coefficients must be supplied as before.

Problem. Use synthetic division to divide $3x^5 - 6x^3 - 2x^2 + 1$ by $x + 2$.

Solution. The dividend has x^4 and x missing. Thus we supply the 0 coefficients and write

$$3x^5 - 6x^3 - 2x^2 + 1 = 3x^5 + 0 \cdot x^4 - 6x^3 - 2x^2 + 0 \cdot x + 1.$$

The divisor is $x + 2 = x - (-2)$ so that the completed synthetic division diagram is

$$
\begin{array}{ccccccc}
3 & 0 & -6 & -2 & 0 & 1 & \underline{\;-2} \\
 & -6 & 12 & -12 & 28 & -56 & \\
\hline
3 & -6 & 6 & -14 & 28 & -55 &
\end{array}
$$

The quotient is $3x^4 - 6x^3 + 6x^2 - 14x + 28$ and the remainder is -55.

Exercise Set 4.2

1. Divide $30x^3 + 143x^2 + 117x - 56$ by $x - 2$. Use synthetic division.

   Divide $30x^3 + 143x^2 + 117x - 56$ by $x + \dfrac{7}{2}$. Use synthetic division.

2. Write an equation which illustrates the division algorithm (Theorem 4.1) for each pair of polynomials in problem 1.

3. Is $x - 2$ a factor of $30x^3 + 143x^2 + 117x - 56$?

   Is $x + \dfrac{7}{2}$ a factor of $30x^3 + 143x^2 + 117x - 56$?

4. Find $P(2)$ if $P(x) = 30x^3 + 143x^2 + 117x - 56$.

   Find $P\left(-\dfrac{7}{2}\right)$ if $P(x) = 30x^3 + 143x^2 + 117x - 56$.

5. Compare the results of problem 4 with the results of problem 1.

* * *

In problems 6–15, divide the polynomial P by each of the polynomials D. Use synthetic division. In each case express P in the form $P = DQ + R$.

6. $P(x) = 4x^3 + 9x^2 - 2x + 21$;
$$D(x) = x + 3, \qquad D(x) = x - 4, \qquad D(x) = x + 1$$

7. $P(x) = 24x^3 + 28x^2 - 150x - 175$;
$$D(x) = x + \frac{5}{2}, \qquad D(x) = x - \frac{5}{2}, \qquad D(x) = x + \frac{7}{6}$$

8. $P(z) = z^5 - z^4 + z^3 - z^2 + z - 1$;
$$D(z) = z - 1, \qquad D(z) = z + 1, \qquad D(z) = z - 2$$

9. $P(z) = 6z^6 - 3z^2 + 8$;
$$D(z) = z - 2, \qquad D(z) = z + 4, \qquad D(z) = z + 5$$

10. $P(t) = t^4 - 2t^3 + 18t - 15$;
$$D(t) = t - 3, \qquad D(t) = t + 3, \qquad D(t) = t - \frac{1}{2}$$

11. $P(t) = \dfrac{1}{2}t^4 - \dfrac{1}{2}t^3 - \dfrac{3}{2}t^2 + \dfrac{1}{2}t + 1$;
$$D(t) = t - 1, \qquad D(t) = t + 4, \qquad D(t) = t - 2$$

12. $P(y) = 2y^3 - 7y^2 + 7y - 5$;
$$D(y) = y + 1, \qquad D(y) = y + \frac{5}{2}, \qquad D(y) = y - \frac{5}{2}$$

13. $P(y) = 2y^4 + 5y^3 + 2y + 5$;
$$D(y) = y + 1, \qquad D(y) = y + \frac{5}{2}, \qquad D(y) = y - \frac{5}{2}$$

14. $P(x) = 3x^6 - 1$;

$$D(x) = x + 1, \qquad D(x) = x - 2, \qquad D(x) = x + \frac{1}{3}$$

15. $P(x) = 3x^3 + x^2 - 6x - 2$;

$$D(x) = x + \sqrt{2}, \quad D(x) = x - \sqrt{2}, \quad D(x) = x + \frac{1}{3}$$

In problems 16–19, compute $P(b)$ for each of the polynomials P. Compare with problems 6–9 respectively.

16. $P(x) = 4x^3 + 9x^2 - 2x + 21$; $b = -3, \quad b = 4, \quad b = -1$

17. $P(x) = 24x^3 + 28x^2 - 150x - 175$; $b = -\dfrac{5}{2}, \ b = \dfrac{5}{2}, \ b = -\dfrac{7}{6}$

18. $P(z) = z^5 - z^4 + z^3 - z^2 + z - 1$; $b = -1, \quad b = -1, \ b = 2$

19. $P(z) = 6z^6 - 3z^2 + 8$; $b = 2, \quad b = -4, \ b = -5$

*20. Use synthetic division to find the quotient Q and the remainder R if $P(x) = 3x^4 - 6x^3 + 4x^2 - 2x + 4$ is divided by $D(x) = 2x + 6$.

$$\left[\text{Hint:} \quad \frac{P(x)}{D(x)} = \frac{P(x)}{2x + 6} = \frac{P(x)}{2(x + 3)} = \frac{1}{2}\left(\frac{P(x)}{x + 3}\right). \right]$$

*21. Repeat problem 20 for $P(x) = 5x^3 - 7x + 2$ and $D(x) = 2x - 1$.

4.3 REMAINDER THEOREM, FACTOR THEOREM

In this section we discuss two theorems concerning polynomials. These theorems are critical in the sense that they lead to conditions under which a polynomial can be factored.

4.2 Theorem *(Remainder theorem). IF a polynomial $P(x)$ is divided by $x - a$ until a constant remainder is obtained, THEN this remainder equals $P(a)$.*

Proof. We have previously observed that, in this case, there is a polynomial $Q(x)$ and a constant remainder $R(x) = c$ such that

$$P(x) = (x - a)Q(x) + c.$$

In particular, when $x = a$ it follows that

$$P(a) = (a - a)Q(a) + c,$$

or

$$P(a) = c.$$

A useful consequence of the remainder theorem is that when it is used in conjunction with synthetic division it gives us a fast tabular method for computing $P(a)$.

Problem. Find $P(2)$ if $P(x) = 6x^3 - 7x^2 - 18x - 5$.

Solution. Suppose $P(x)$ is divided by $x - 2$ using synthetic division. We have

$$
\begin{array}{rrrr|r}
6 & -7 & -18 & -5 & \underline{2} \\
 & 12 & 10 & -16 & \\
\hline
6 & 5 & -8 & -21 &
\end{array}
$$

giving us a remainder of -21. The remainder theorem tells us that this is precisely $P(2)$. In other words, $P(2) = -21$.

Problem. Find $P\left(-\dfrac{3}{2}\right)$ if $P(x) = 2x^5 + 4x^4 - x^3 + 5x - 1$.

Solution. This can be accomplished by dividing $P(x)$ by $x - \left(-\dfrac{3}{2}\right)$. If we use synthetic division, noting that the term $0 \cdot x^2$ must be supplied in $P(x)$, we have

$$
\begin{array}{rrrrrr|r}
2 & 4 & -1 & 0 & 5 & -1 & \left\lfloor -\dfrac{3}{2} \right. \\
 & -3 & -\dfrac{3}{2} & \dfrac{15}{4} & -\dfrac{45}{8} & \dfrac{15}{16} & \\
\hline
2 & 1 & -\dfrac{5}{2} & \dfrac{15}{4} & -\dfrac{5}{8} & -\dfrac{1}{16} &
\end{array}
$$

Since the remainder is $-\dfrac{1}{16}$, the remainder theorem tells us that $P\left(-\dfrac{3}{2}\right) = -\dfrac{1}{16}$.

There is a second important theorem which follows readily from the remainder theorem.

4.3 Theorem *(Factor theorem). IF $x - a$ is a factor of the polynomial $P(x)$, THEN $P(a) = 0$. Conversely, IF $P(a) = 0$, THEN $x - a$ is a factor of $P(x)$.*

Proof. Suppose, first, that $x - a$ is a factor of $P(x)$. This means that we can write $P(x) = (x - a)Q(x)$ for some polynomial $Q(x)$. It follows that $P(a) = (a - a)Q(a) = 0$.

On the other hand, assume $P(a) = 0$. By the division algorithm, there exist polynomials $Q(x)$ and a constant remainder $R(x)$ such that

$$P(x) = (x - a)Q(x) + R(x).$$

By the remainder theorem, $R(x)$ is the constant $P(a)$ so that

$$P(x) = (x - a)Q(x) + P(a).$$

Since $P(a) = 0$,

$$P(x) = (x - a)Q(x)$$

which shows that $x - a$ is a factor of $P(x)$.

The factor theorem can be used in conjunction with synthetic division to determine whether or not $x - a$ is a factor of $P(x)$.

Problem. Is $x - 1$ a factor of $P(x) = 3x^4 - 4x^2 - 2x + 10$?

Solution. We first divide $P(x)$ by $x - 1$ to find $P(1)$.

$$
\begin{array}{rrrrr|l}
3 & 0 & -4 & -2 & 10 & \underline{1} \\
  & 3 & 3 & -1 & -3 & \\
\hline
3 & 3 & -1 & -3 & 7 &
\end{array}
$$

Therefore $P(1) = 7 \neq 0$, implying $x - 1$ is not a factor of $P(x)$.

Problem. Is $x - \dfrac{3}{2}$ a factor of $P(x) = 4x^3 + 4x^2 - 21x + 9$?

Solution. Dividing $P(x)$ by $x - \dfrac{3}{2}$, we have

$$
\begin{array}{rrrr|l}
4 & 4 & -21 & 9 & \dfrac{3}{2} \\
  & 6 & 15 & 9 & \\
\hline
4 & 10 & -6 & 0 &
\end{array}
$$

which shows that $P\left(\dfrac{3}{2}\right) = 0$. By the factor theorem, $x - \dfrac{3}{2}$ divides $P(x)$.

Problem. Factor $P(x) = 4x^3 + 4x^2 - 21x + 9$.

Solution. The synthetic division diagram of the preceding problem shows that

$$P(x) = \left(x - \frac{3}{2}\right)(4x^2 + 10x - 6)$$

$$= 2\left(x - \frac{3}{2}\right)(2x^2 + 5x - 3).$$

In order to factor $P(x)$ further, it is necessary to factor $Q(x) = 2x^2 + 5x - 3$. The discriminant of $Q(x)$ is $5^2 - 4(2)(-3) = 49 > 0$. Therefore $Q(x)$ is reducible. Using the quadratic formula, we find that the zeros of Q are $\dfrac{-5 \pm \sqrt{49}}{4}$ or $\dfrac{1}{2}, -3$.

It follows that $Q(x) = 2\left(x - \dfrac{1}{2}\right)(x + 3)$. Hence,

$$P(x) = 4\left(x - \frac{3}{2}\right)\left(x - \frac{1}{2}\right)(x + 3).$$

Problem. Find a polynomial $P(x)$ such that -1, 2, and 3 are the zeros of $P(x)$.

Solution. By the factor theorem, if -1, 2, and 3 are zeros of $P(x)$, then $x + 1$, $x - 2$, and $x - 3$ must be factors of $P(x)$. Therefore, one such polynomial is

$$P(x) = (x + 1)(x - 2)(x - 3)$$
$$= x^3 - 4x^2 + x + 6.$$

Exercise Set 4.3

1. Let $P(x) = 5x^3 - 14x^2 - 23x - 4$.
   Use synthetic division to find $P(2)$, $P(-1)$, $P\left(\frac{1}{2}\right)$, $P\left(\frac{1}{5}\right)$, $P(3)$, and $P\left(-\frac{1}{5}\right)$.

2. Check the results of problem 1 by direct computation.

3. List the zeros of $P(x)$ which have been determined for
   $P(x) = 5x^3 - 14x^2 - 23x - 4$.

4. If $P(a) = 0$, then $x - a$ is a factor of $P(x)$. Express the polynomial
   $P(x) = 5x^3 - 14x^2 - 23x - 4$ as a product of three linear factors using the
   results of problems 1 and 2.

5. Determine which of the following polynomials $D(x)$ are factors of
   $P(x) = 6x^3 - 7x^2 - 18x - 5$. Use synthetic division.

   $$D(x) = x + 1, \quad D(x) = x - \frac{5}{2}, \quad D(x) = x - \frac{3}{2}, \quad D(x) = x + \frac{5}{2}$$

6. For the polynomials listed in problem 5, if $D(x)$ is a factor of $P(x)$, express
   $P(x)$ in the form $P(x) = D(x)Q(x)$.

* * *

7. If $P(x) = -5x^4 - 4x^3 + 3x^2 - 8$, use synthetic division to find $P(1)$, $P(-1)$,
   and $P(3)$.

8. If $P(y) = 3y^3 + 4y^2 - 3y + 1$, use synthetic division to find $P(-2)$, $P\left(\frac{1}{2}\right)$,
   and $P(-4)$.

9. If $P(t) = 6t^3 - 7t^2 - t + 2$, use synthetic division to find $P(0)$, $P\left(\frac{1}{2}\right)$, and
   $P\left(\frac{2}{3}\right)$.

10. If $P(z) = -8z^4 + 12z^3 + 2z^2 - z + 3$, use synthetic division to find
    $P(1)$, $P\left(-\frac{3}{2}\right)$, and $P\left(\frac{3}{4}\right)$.

11. If $P(w) = 2w^3 - 45w^2 - 100w + 147$, use synthetic division to find $P(5)$, $P(25)$, and $P(-33)$.

12. If $P(r) = 3r^4 - r + 10$, use synthetic division to find $P(1), P\left(-\dfrac{1}{2}\right)$, and $P(-5)$.

In problems 13–18, *use synthetic division* and the factor theorem to show that D is a factor of P. Express P as the product of three linear factors where possible.

13. $P(y) = -3y^3 + 4y^2 + 57y - 10$ $D(y) = y - 5$

14. $P(t) = t^3 + t^2 - 102t + 88$ $D(t) = t + 11$

15. $P(x) = x^3 - \dfrac{8}{27}$ $D(x) = x - \dfrac{2}{3}$

16. $P(t) = 5t^3 - 19t^2 + 36$ $D(t) = t + \dfrac{6}{5}$

17. $P(y) = 4y^3 - 4y^2 + 3y - 1$ $D(y) = y - \dfrac{1}{2}$

18. $P(x) = 2x^4 + x^3 + 2x + 1$ $D(x) = x + \dfrac{1}{2}$

19. Construct a polynomial with integer coefficients having $x - \dfrac{1}{2}, x + \dfrac{2}{3}$, and $x + 1$ as factors. Check your answer using synthetic division.

20. Construct a polynomial with integer coefficients such that the zeros of the polynomial are $\dfrac{1}{2}, -\dfrac{2}{3}$, and -1. Construct a different polynomial with the same zeros.

*21. Prove: If n is a positive integer and if a is any real number, then $x - a$ is a factor of $x^n - a^n$.

4.4 FINDING ZEROS OF POLYNOMIAL FUNCTIONS

The importance of finding zeros of functions should not be overlooked. In Chap. 7 we shall see that the maximization and minimization of functions and the accurate sketching of their graphs is dependent upon our ability to find zeros. The problem of finding zeros of functions is, in general, a difficult task. Our attention in this section is restricted to a discussion of zeros of polynomial functions. We will consider some theorems which restrict the set $\{x: P(x) = 0\}$ for a polynomial function P. We will accept the validity of these theorems without formal proof. The set of real zeros of a polynomial function will be denoted by Z. The following theorems relate to the number and kinds of elements which are contained in Z.

4.4 Theorem. *IF P is a polynomial of degree n, THEN Z contains at most n elements.*

Problem. Is there any limitation on Z if $P(x) = 3x^5 - \sqrt{2}\, x^4 + 7x^3 - x^2 + 5x - 12$?

Solution. P is a polynomial function of degree five. Therefore the set

$$Z = \{x:\ 3x^5 - \sqrt{2}\, x^4 + 7x^3 - x^2 + 5x - 12 = 0\}$$

can contain at most five elements.

The following theorem is useful in trying to locate intervals which contain zeros of a polynomial function.

4.5 Theorem. *IF P(a) and P(b) differ in sign for a polynomial function P, THEN Z contains at least one element z such that $a < z < b$.*

Problem. Does $P(x) = x^4 - 7x^2 + 10$ have a zero between 1 and 2?

Solution. We find $P(1)$ and $P(2)$ using synthetic division:

$$
\begin{array}{rrrrr|l}
1 & 0 & -7 & 0 & 10 & \underline{1} \\
  & 1 & 1 & -6 & -6 & \\
\hline
1 & 1 & -6 & -6 & 4 &
\end{array}
\qquad
\begin{array}{rrrrr|l}
1 & 0 & -7 & 0 & 10 & \underline{2} \\
  & 2 & 4 & -6 & -12 & \\
\hline
1 & 2 & -3 & -6 & -2 &
\end{array}
$$

Since $P(1) = 4$ and $P(2) = -2$ differ in sign, P has a zero between 1 and 2.

Theorem 4.5 can also be used to improve estimates of zeros.

Example. The preceding problem shows that $P(x) = x^4 - 7x^2 + 10$ has at least one zero between 1 and 2. Let us now consider a method for improving the precision of this estimation. We divide $[1, 2]$ into two subintervals of equal length and apply Theorem 4.5 to each subinterval. For example, the midpoint of $[1, 2]$ is $\dfrac{3}{2}$. Therefore the two subintervals are $\left[1, \dfrac{3}{2}\right]$ and $\left[\dfrac{3}{2}, 2\right]$. If we apply Theorem 4.5 to the endpoints of each subinterval, we will be able to determine which of the subintervals contains the zero. Using synthetic division

$$
\begin{array}{rrrrr|l}
1 & 0 & -7 & 0 & 10 & \dfrac{3}{2} \\[6pt]
  & \dfrac{3}{2} & \dfrac{9}{4} & -\dfrac{57}{8} & -\dfrac{171}{16} & \\[8pt]
\hline
1 & \dfrac{3}{2} & -\dfrac{19}{4} & -\dfrac{57}{8} & -\dfrac{11}{16} &
\end{array}
$$

From the preceding problem $P(1) = 4$ so that $P\left(\frac{3}{2}\right) = -\frac{11}{16}$ and $P(1)$ differ in sign. Thus P has a zero between 1 and $\frac{3}{2}$. It is clear that this procedure can now be repeated. The midpoint of $\left[1, \frac{3}{2}\right]$ is $\frac{5}{4}$. We could now apply this procedure to the intervals $\left[1, \frac{5}{4}\right]$ and $\left[\frac{5}{4}, \frac{3}{2}\right]$ and obtain an even better estimate of the location of this zero of P.

It has been noted in Chap. 3 that there is no general formula for finding zeros of polynomial functions. In fact, up to this point, we do not even have a general way of determining "possible zeros" of polynomial functions. The next result is very important because it points to a general method for determining certain kinds of "possible zeros."

4.6 Theorem. *Let $P(x) = a_n x^n + a_{n-1} x^{n-1} + \cdots + a_0$ be a polynomial with integer coefficients. IF Z contains a rational number $r = \frac{p}{q}$, where p and q are integers with no common divisors other than ± 1, THEN p is a divisor of a_0 and q is a divisor of a_n.*

The preceding theorem tells us that if $P(x) = a_n x^n + a_{n-1} x^{n-1} + \cdots + a_0$ has integer coefficients, then any rational zero of P must be of the form $\frac{p}{q}$, where p is a divisor of a_0 and q is a divisor of a_n. In other words, the set of possible rational zeros of P is

$$\left\{ \frac{p}{q} : \ p \text{ is a divisor of } a_0 \text{ and } q \text{ is a divisor of } a_n \right\}.$$

Problem. What are the possible rational zeros of $P(x) = 6x^3 - 13x^2 - 3x - 5$?

Solution. For the function P we have $a_0 = -5$ and $a_3 = 6$.

$$\text{divisors of } -5: \ \pm 1, \pm 5$$
$$\text{divisors of } 6: \ \ \ \pm 1, \pm 2, \pm 3, \pm 6$$

Therefore the set of possible rational zeros of P is

$$\left\{ \frac{p}{q} : \ p \in \{\pm 1, \pm 5\}, \ q \in \{\pm 1, \pm 2, \pm 3, \pm 6\} \right\}$$
$$= \left\{ \pm 1, \pm \frac{1}{2}, \pm \frac{1}{3}, \pm \frac{1}{6}, \pm 5, \pm \frac{5}{2}, \pm \frac{5}{3}, \pm \frac{5}{6} \right\}.$$

Problem. Find the rational zeros of $P(x) = x^5 + 3x^4 - 3x^3 - 9x^2 + 2x + 6$.

Solution. We again apply Theorem 4.6. In this case $a_0 = 6$ and $a_5 = 1$.

$$\text{divisors of 6: } \pm1, \pm2, \pm3, \pm6$$
$$\text{divisors of 1: } \pm1.$$

Therefore the set of possible rational zeros is $\{\pm1, \pm2, \pm3, \pm6\}$. We must now evaluate P at each of these possible zeros to determine whether or not they are zeros. Using synthetic division, we have

$$
\begin{array}{rrrrrr|l}
1 & 3 & -3 & -9 & 2 & 6 & \underline{1} \\
 & 1 & 4 & 1 & -8 & -6 & \\
\hline
1 & 4 & 1 & -8 & -6 & 0 &
\end{array}
$$

which shows that $P(1) = 0$. Similarly, synthetic division shows that $P(-1) = P(-3) = 0$ and that none of $P(-2)$, $P(2)$, $P(3)$, $P(-6)$, and $P(6)$ are zero. There-fore -1, 1, 3 are the only rational zeros of P. Since Z contains at most five elements, any other real zeros of P must be irrational numbers. The following problem bears out this fact.

Problem. Find the zeros of $P(x) = x^5 + 3x^4 - 3x^3 - 9x^2 + 2x + 6$.

Solution. By the preceding problem, the rational zeros of P are $-3, -1, 1$. Moreover, the synthetic division diagram used to show that $P(1) = 0$ also shows us that

$$P(x) = (x - 1)(x^4 + 4x^3 + x^2 - 8x - 6).$$

If we let $Q(x) = x^4 + 4x^3 + x^2 - 8x - 6$, then $P(-1) = -2Q(-1)$ and, since $P(-1) = 0$, it must be the case that $Q(-1) = 0$. Synthetic division can now be used to factor $x + 1$ from $Q(x)$. The diagram

$$
\begin{array}{rrrrr|l}
1 & 4 & 1 & -8 & -6 & \underline{-1} \\
 & -1 & -3 & 2 & 6 & \\
\hline
1 & 3 & -2 & -6 & 0 &
\end{array}
$$

shows that $Q(x) = (x + 1)(x^3 + 3x^2 - 2x - 6)$ hence

$$P(x) = (x - 1)(x + 1)(x^3 + 3x^2 - 2x - 6).$$

By a similar argument, $x^3 + 3x^2 - 2x - 6 = 0$ when $x = -3$ and $x^3 + 3x^2 - 2x - 6 = (x + 3)(x^2 - 2)$ so that

$$P(x) = (x - 1)(x + 1)(x + 3)(x^2 - 2).$$

Since $x^2 - 2 = (x + \sqrt{2})(x - \sqrt{2})$, the zeros of P are $\pm1, \pm\sqrt{2}, -3$ so that

$$Z = \{-3, -\sqrt{2}, -1, 1, \sqrt{2}\}.$$

Moreover, P can be factored completely as

$$P(x) = (x - 1)(x + 1)(x + 3)(x - \sqrt{2})(x + \sqrt{2}).$$

Problem. Find the zeros of $P(x) = x^3 + \frac{1}{6}x^2 - \frac{2}{3}x + \frac{1}{6}.$

Solution. Since the coefficients of P are not all integers, we cannot immediately apply Theorem 4.6. However, if we multiply P by the least common denominator of its coefficients, which in this case is 6, we have

$$6P(x) = 6x^3 + x^2 - 4x + 1.$$

If $Q(x) = 6x^3 + x^2 - 4x + 1$, then $6P(x) = Q(x)$, so that the zeros of P and Q are the same. Moreover, since Q has integer coefficients, Theorem 4.6 can be applied to find the rational zeros of Q. If this is done, it is seen that the rational zeros of Q are $-1, \frac{1}{3}, \frac{1}{2}$. Since Q is of degree 3, these are all the zeros of Q and hence of P.

Exercise Set 4.4

1. At most how many elements are contained in the set Z, if $P(x) = 8x^4 - 7x^3 + x - 17$?

2. If $P(x) = 2x^3 + 3x^2 - 5x - 3$, compute $P(-3)$, $P(-2)$, $P(-1)$, $P(0)$, $P(1)$, $P(2)$, and $P(3)$. Determine intervals of unit length which contain the zeros of P.

3. What are the possible rational zeros of P if $P(x) = 12x^3 + 44x^2 + 29x - 15$?

4. Find all the zeros of P if $P(x) = 12x^3 + 44x^2 + 29x - 15$.

5. Express $P(x) = 12x^3 + 44x^2 + 29x - 15$ as the product of three linear factors.

* * *

6. At most how many elements are contained in the set Z, if:
   a. $P(x) = x^5 - x^2 + 22x + 5$
   b. $P(x) = \frac{2}{3}x^7 - \frac{4}{5}x^2 + \sqrt{2}\,x + \frac{12}{7}$
   c. $P(x) = 1 + x + x^2 + x^3$

7. At most how many elements are contained in the set Z, if:
   a. $P(x) = x^{25} - 5x^{20} + 10x^{15} - 15x^{10} + 20x^5$
   b. $P(x) = x - \sqrt{35}$
   c. $P(x) = (3x^2 - 2x + 1)(4 - 5x - x^2)$

In problems 8–15, list the possible rational zeros of the given polynomial P. Determine which of the possible rational zeros are actually zeros of P.

8. $P(x) = x^3 + 5x^2 - 29x - 105$

9. $P(t) = 2t^3 + t^2 - t + 3$

10. $P(w) = 4w^4 - 4w^3 + 5w^2 - 4w + 1$

11. $P(r) = 2r^4 - 5r^2 + 7r - 3$

12. $P(t) = t^4 - 13t^2 + 36$

13. $P(x) = 35x^4 + 102x^3 - 46x^2 - 3x + 2$

14. $P(y) = 3y^6 + y^2 - 1$

15. $P(s) = 3s^6 + s^2 + 2$

In problems 16–19, factor the given polynomial as completely as possible.

16. $P(x) = 2x^3 + 3x^2 + 4x - 3$

17. $P(y) = 2y^3 + 3y^2 - 8y + 3$

18. $P(z) = 2z^3 + 3z^2 - 6z + 2$

19. $P(x) = x^4 - 5x^2 - 10x - 6$

20. Show that the function $P(t) = 2t^3 + t^2 - 5t + 2$ has a zero between $-\dfrac{5}{2}$ and -1.

21. Show that the function $Q(y) = y^5 + y^4 + y^3 - y^2 - y - 1$ has a zero between 0 and $\dfrac{3}{2}$.

22. Show that $x^3 - 2x^2 + 3x - 3$ has a real zero between 1 and 2. Show that $x^3 - 2x^2 + 3x - 3$ has a real zero between 1 and 1.5.

23. Show that $x^3 - x^2 - 5x + 3$ has no rational zeros. Isolate the real zeros of $x^3 - x^2 - 5x + 3$ between consecutive integers.

*24. Compute $P(1.1)$, $P(1.2)$, $P(1.3)$, $P(1.4)$ for $P(x) = x^3 - 2x^2 + 3x - 3$. Find a real zero of $x^3 - 2x^2 + 3x - 3$ correct to the nearest tenth.

25. Find all the rational zeros of the polynomial $P(x) = \dfrac{1}{3}x^3 - \dfrac{1}{2}x^2 - \dfrac{1}{2}x + \dfrac{1}{3}$.
(*Hint:* $P(x) = 0$ if and only if $6P(x) = 0$).

4.5 GRAPHING OF POLYNOMIAL FUNCTIONS

We now return to the problem of graphing functions which are carefully chosen polynomial functions of third and fourth degree. The functions are carefully chosen in order that we may apply Theorem 4.6 to locate some of the zeros.

Although the domain of every polynomial function is the set of all real numbers, we shall restrict the domain to an arbitrary interval for the purposes of graphing. The interval selected will be sufficiently large to exhibit the interesting features of the graph.

Problem. Sketch the graph of the polynomial function

$$P(x) = x^3 + 2x^2 - 5x - 6.$$

Solution. An application of Theorem 4.6 shows that the zeros of P are $-3, -1, 2$. The table of values should therefore include:

x	-3	-1	2
$P(x)$	0	0	0

Three important points on the graph will be the points whose coordinates are $(-3, 0)$, $(-1, 0)$, and $(2, 0)$. The graph will cross or touch the x-axis at no more than these three points because, by Theorem 4.4, P can have at most three real zeros.

We now want to determine the behavior of the graph for values of x between the zeros. Since -3, -2, and 2 are the zeros of $P(x)$, we know that $x + 3, x + 1$, and $x - 2$ are factors of $P(x)$. That is,

$$P(x) = x^3 + 2x^2 - 5x - 6 = (x + 3)(x + 1)(x - 2).$$

The following table of signs is constructed in a manner similar to those tables constructed in Sec. 3.3.

				Sign of
	$x + 3$	$x + 1$	$x - 2$	$P(x) = (x + 3)(x + 1)(x - 2)$
$x < -3$	$-$	$-$	$-$	$-$
$-3 < x < -1$	$+$	$-$	$-$	$+$
$-1 < x < 2$	$+$	$+$	$-$	$-$
$2 < x$	$+$	$+$	$+$	$+$

From this table we see that the graph of P must lie in the regions indicated between the dotted lines in Figure 4.1. We now extend our table of values, includ-

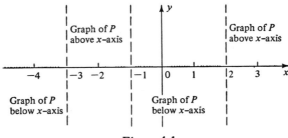

Figure 4.1

ing values of x for which $P(x)$ is positive, negative, and zero to obtain

x	-4	-3	-2	-1	$-\dfrac{3}{2}$	0	$\dfrac{1}{2}$	1	2	3
$P(x)$	-18	0	4	0	$\dfrac{21}{8}$	-6	$-\dfrac{63}{8}$	-8	0	24

Plotting these points and connecting them by a smooth curve we have the graph shown in Figure 4.2.

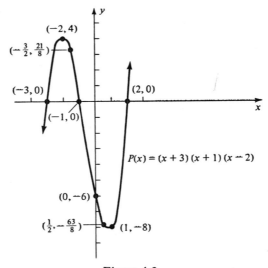

Figure 4.2

In sketching the graph, we have assumed that the graph has a peak near $(-2, 4)$ and a valley near $(1, -8)$. Obviously, it would be helpful to know exactly where such high points and low points occur, but methods for ascertaining these values will necessarily be postponed until techniques using differential calculus are available.

Problem. Sketch the graph of the polynomial function

$$P(x) = x^4 + x^3 + x^2 - x + 2.$$

Solution. An application of Theorem 4.6 shows that the rational zeros of P are -1 and 1. Thus, we can factor $P(x)$ as

$$P(x) = (x + 1)(x - 1)(x^2 + x + 2).$$

The factor $x^2 + x + 2$ has no real zeros and is positive for all real x. The table of signs of $P(x)$ is determined as before:

	$x + 1$	$x - 1$	$x^2 + x + 2$	Sign of $P(x) = (x + 1)(x - 1)(x^2 + x + 2)$
$x < -1$	$-$	$-$	$+$	$+$
$-1 < x < 1$	$+$	$-$	$+$	$-$
$1 < x$	$+$	$+$	$+$	$+$

Figure 4.3 describes the regions of the plane in which the graph of P must lie.

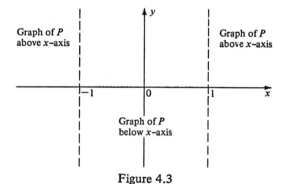

Figure 4.3

A table of values might include:

x	-2	$-\dfrac{3}{2}$	-1	$-\dfrac{1}{2}$	0	$\dfrac{1}{2}$	1	$\dfrac{3}{2}$	2
$P(x)$	12	$\dfrac{55}{16}$	0	$-\dfrac{21}{16}$	-2	$-\dfrac{33}{16}$	0	$\dfrac{115}{16}$	24

Plotting these points, a reasonable sketch of the graph of P is

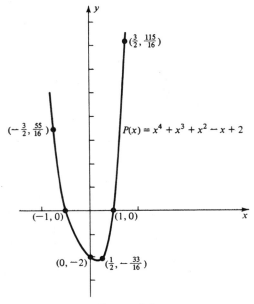

Figure 4.4

Again, we note that it would be desirable to have a technique which would help us to determine accurately the location of the high points and low points on graphs of this type.

Exercise Set 4.5

1. Construct a table of signs for the polynomial function
$P(x) = 2(x + 3)(x - 1)\left(x - \dfrac{3}{2}\right)$ as was done in the examples of this section.

2. Construct a table of values for the function of problem 1.

3. Sketch the graph of the function $P(x) = 2(x + 3)(x - 1)\left(x - \dfrac{3}{2}\right)$.

4. Sketch the graph of the polynomial function $P(x) = (x + 1)(x + 2)(x + 3)$.

5. Sketch the graph of the polynomial function $P(x) = x^3 - 2x^2 - 5x + 6$.
   (*Note:* It is necessary to determine the zeros of this function before sketching the graph.)

* * *

In problems 6–24, sketch the graph of the given polynomial.

6. $P(x) = -3x + 10$

7. $P(x) = -2$

8. $P(x) = x^3$

9. $P(x) = x^3 + 1$

10. $P(z) = z^3 - z$

11. $P(z) = 2z^2 - z + 1$

12. $P(t) = 2(t - 1)^2$

13. $P(t) = 2(t - 1)^2(t^2 + t + 1)$

14. $F(t) = \frac{1}{2}t(t - 2)(t + 1)$

15. $F(x) = 2x^3 + 5x^2 + 5x + 3$

16. $G(t) = 36t^4 - 25t^2 + 4$

17. $G(x) = 10x^4 - x^3 - 12x^2 + x + 2$

18. $P(t) = \dfrac{t^4}{10}$

19. $P(x) = (2x + 1)(x - 3)(x^2 - 2)$

20. $G(u) = 2u^3 + 3u^2 + 4u - 3$

21. $G(u) = 2u^3 + 3u^2 - 8u + 3$

22. $P(x) = 2x^3 + 3x^2 - 6x + 2$

*23. $P(z) = z^4 - 5z^2 - 10z - 6$

*24. $P(x) = x^4 - 5x^2 + 6$

4.6 GRAPHS OF RATIONAL FUNCTIONS

Now let us turn our attention to the graphs of certain rational functions. Any rational function R can be expressed as the quotient of two polynomials P/Q, so the zeros of Q determine those real numbers x which are not contained in the domain of R.

For instance, if $R(x) = \dfrac{1}{x}$, then the domain of R does not contain 0 and the graph of R will not include any point whose x-coordinate is 0, that is any point on the y-axis. If $R(x) = \dfrac{x - 1}{x^2 + 3x + 2}$ then the zeros of Q are -2 and -1 and R is not defined for $x = -2$ or for $x = -1$. Thus the graph of R will not include any point which lies on the line $x = -2$ nor any point on the line $x = -1$.

In sketching the graph of $R(x) = \frac{1}{x}$, we should then think of the line $x = 0$ (the y-axis) as a *barrier* which cannot be crossed by the graph of this function. Any such crossing would imply that there is a point $(0, R(0))$ belonging to the graph of $\frac{1}{x}$. This would be a contradiction.

Constructing a table of values for $R(x) = \frac{1}{x}$, we have:

x	-10	-4	-2	-1	$-\frac{1}{2}$	$-\frac{1}{4}$	$-\frac{1}{10}$	0	$\frac{1}{10}$	$\frac{1}{4}$	$\frac{1}{2}$	1	2	4	10
$R(x)$	$-\frac{1}{10}$	$-\frac{1}{4}$	$-\frac{1}{2}$	-1	-2	-4	-10		10	4	2	1	$\frac{1}{2}$	$\frac{1}{4}$	$\frac{1}{10}$

Plotting the points corresponding to each $(x, R(x))$ we have the results shown in Figure 4.5.

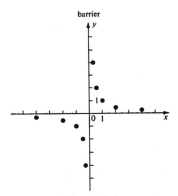

Figure 4.5

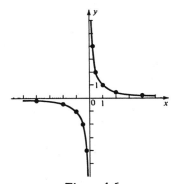

Figure 4.6

As the graph cannot cross the y-axis, no point which lies to the left of the barrier can be connected with any point which lies to the right of the barrier. Therefore, the graph consists of two totally separated curves as shown in Figure 4.6.

To sketch the graph of the rational function $R(x) = \dfrac{x-1}{x^2 + 3x + 2}$, we write it in the form $R(x) = \dfrac{x-1}{(x+2)(x+1)}$ and begin with a sign table as was done in Sec. 4.5 for polynomial functions. We note that the terms appearing in $R(x)$ have sign changes at $-2, -1$, and 1.

<table>
<tr><td></td><td colspan="4" style="text-align:center">Sign of</td></tr>
<tr><td></td><td>$x+2$</td><td>$x+1$</td><td>$x-1$</td><td>$R(x) = \dfrac{x-1}{(x+2)(x+1)}$</td></tr>
<tr><td>$x < -2$</td><td>$-$</td><td>$-$</td><td>$-$</td><td>$-$</td></tr>
<tr><td>$-2 < x < -1$</td><td>$+$</td><td>$-$</td><td>$-$</td><td>$+$</td></tr>
<tr><td>$-1 < x < 1$</td><td>$+$</td><td>$+$</td><td>$-$</td><td>$-$</td></tr>
<tr><td>$1 < x$</td><td>$+$</td><td>$+$</td><td>$+$</td><td>$+$</td></tr>
</table>

The barriers for $R(x)$ are the lines $x = -2$ and $x = -1$. Therefore the information already obtained can be concisely summarized in Figure 4.7.

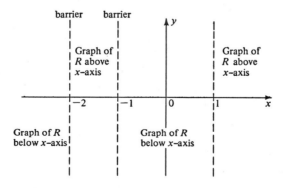

Figure 4.7

A table of values for $R(x) = \dfrac{x-1}{x^2 + 3x + 2}$ is

x	-4	-3	$-\dfrac{5}{2}$	$-\dfrac{9}{4}$	-2	$-\dfrac{5}{3}$	$-\dfrac{3}{2}$	$-\dfrac{4}{3}$	$-\dfrac{6}{5}$	-1	$-\dfrac{4}{5}$	$-\dfrac{1}{2}$	0	$\dfrac{1}{2}$	1	2	3	4
$R(x)$	$-\dfrac{5}{6}$	-2	$-\dfrac{14}{3}$	$-\dfrac{52}{5}$		12	10	$\dfrac{21}{2}$	$\dfrac{55}{5}$		$-\dfrac{15}{2}$	-2	$-\dfrac{1}{2}$	$-\dfrac{2}{15}$	0	$\dfrac{1}{12}$	$\dfrac{1}{10}$	$\dfrac{1}{15}$

Plotting these points and connecting them with curves which do not cross the barriers, we have the graph shown in Figure 4.8.

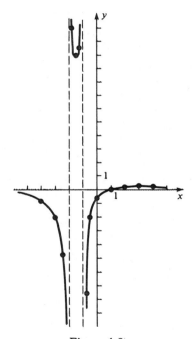

Figure 4.8

Notice that much more extensive tables of values are needed to make a reasonably accurate sketch of the graph of this function as compared to those needed for the graphs of the polynomials in Sec. 4.5. Too few points produce too little information.

Now, consider the rational function $R(x) = \dfrac{1}{x^2 + 1}$. The domain of this function is the set of all real numbers since the polynomial $x^2 + 1$ has no real zeros.

Furthermore, $R(x) > 0$ for all x and therefore the graph lies entirely above the x-axis. A table of values for $R(x)$ is

x	-3	-2	$-\dfrac{3}{2}$	-1	$-\dfrac{1}{2}$	0	$\dfrac{1}{2}$	1	$\dfrac{3}{2}$	2	3
$R(x)$	$\dfrac{1}{10}$	$\dfrac{1}{5}$	$\dfrac{4}{13}$	$\dfrac{1}{2}$	$\dfrac{4}{5}$	1	$\dfrac{4}{5}$	$\dfrac{1}{2}$	$\dfrac{4}{13}$	$\dfrac{1}{5}$	$\dfrac{1}{10}$

Connecting these points by a smooth curve, we obtain Figure 4.9.

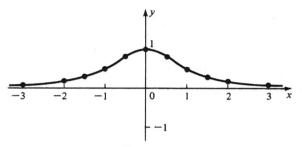

Figure 4.9

The function $R(x) = \dfrac{1}{x^2 + 1}$ has the property that, if the point whose co-ordinates are (a, b) lies on the graph, so does the point whose coordinates are $(-a, b)$. This property is called *symmetry* with respect to the y-axis. For any function f, if $f(x) = f(-x)$ for every x in the domain of f, then the graph of f is

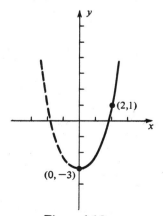

Figure 4.10

said to be symmetric with respect to the y-axis. Graphically this means that if the plane were folded on the y-axis, the portion to the right of the y-axis would be superimposed on the portion to the left of the y-axis. So, to sketch the graph of a function which possesses this type of symmetry, it is sufficient to plot points for positive values of x and sketch the remainder of the graph by symmetry. For example, if $f(x) = x^2 - 3$, then $f(-x) = (-x)^2 - 3 = x^2 - 3 = f(x)$ for all x. Therefore the graph of f is symmetric with respect to the y-axis. Thus to sketch the graph of f we need only sketch the graph for positive values of x (solid part of the curve in Figure 4.10). The rest of the graph is obtained from symmetry (dotted part of the curve in Figure 4.10).

Exercise Set 4.6

1. Given the rational function $R(x) = \dfrac{1}{x - 2}$, what is the domain of $R(x)$? What line is a barrier for $R(x)$?

2. Construct a sign table for the function $R(x) = \dfrac{1}{x - 2}$. Construct a table of values for $R(x)$.

3. Sketch the graph of the function $R(x) = \dfrac{1}{x - 2}$.

4. Given the rational function $R(z) = \dfrac{1}{(z + 1)(z - 2)}$, what is the domain of $R(z)$? What lines are barriers for $R(z)$?

5. Construct a sign table for the function $R(z) = \dfrac{1}{(z + 1)(z - 2)}$. Construct a table of values for $R(z)$.

6. Sketch the graph of the function $R(z) = \dfrac{1}{(z + 1)(z - 2)}$.

* * *

In problems 7–20, for each rational function R determine the domain of R, construct a sign table and a table of values for R, and sketch the graph of R.

7. $R(x) = \dfrac{1}{2x}$

8. $R(x) = \dfrac{2}{x}$

9. $R(z) = \dfrac{1}{z^2}$

10. $R(z) = \dfrac{4}{z^2 + 1}$

11. $R(t) = \dfrac{(t + 1)(t - 1)}{t + 1}$

12. $R(t) = \dfrac{(t + 1)^2 (t - 1)}{t - 1}$

13. $R(t) = \dfrac{t + 1}{t^2 - 1}$

14. $R(t) = \dfrac{t + 1}{t^2 - t - 2}$

15. $R(x) = \dfrac{1}{x^2 - 4x - 5}$

16. $R(x) = \dfrac{x + 2}{x^2 - 4x - 5}$

17. $R(s) = \dfrac{1}{s^3 - 8}$

18. $R(s) = \dfrac{2s^2 + 3s - 2}{s}$

19. $R(w) = \dfrac{w^2 - 9}{w}$

20. $R(w) = \dfrac{w}{w^2 - 9}$

21. Explain why the functions $f(x) = \dfrac{x^2 - 1}{x - 1}$ and $g(x) = x + 1$ are not the same.

22. Which of the functions in problems 7–20 possess the property of symmetry with respect to the y-axis?

4.7 GRAPHS OF ALGEBRAIC FUNCTIONS

Probably the simplest of the algebraic functions is the function $f(x) = \sqrt{x}$. This function is the rule which assigns to every number x, the nonnegative number whose square is x. Since the square of every real number is either positive or zero, x must necessarily be greater than or equal to zero. Another way of saying this, is that the domain of f is the set of nonnegative real numbers. A table of values for $f(x)$ is

x	0	$\frac{1}{9}$	$\frac{1}{4}$	$\frac{1}{2}$	1	$\frac{3}{2}$	2	4
$f(x)$	0	$\frac{1}{3}$	$\frac{1}{2}$	$\frac{1}{\sqrt{2}}$	1	$\frac{\sqrt{3}}{\sqrt{2}}$	$\sqrt{2}$	2

Figure 4.11 is a sketch of the graph of f.

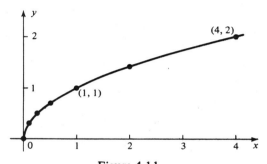

Figure 4.11

A variation of the function just described would be a function such as $g(x) = \sqrt{x^2 - 1}$. Again, the criterion for membership of a given x in the domain of g is an affirmative answer to the question: Is $x^2 - 1$ greater than or equal to zero? Rewriting $x^2 - 1$ as $(x + 1)(x - 1)$, we see that $x^2 - 1 = 0$ for $x = -1$ or $x = 1$. Constructing a sign table, we have

	Sign of			
	$x + 1$	$x - 1$	$x^2 - 1 = (x + 1)(x - 1)$	$g(x) = \sqrt{x^2 - 1}$
$x < -1$	−	−	+	+
$-1 < x < 1$	+	−	−	not defined
$1 < x$	+	+	+	+

Fom this diagram we can see that the domain of g is

$$D_g = \{x:\ x \leqslant -1 \text{ or } x \geqslant 1\}.$$

Therefore there will be no points on the graph of g which correspond to values of x in the interval $(-1, 1)$. This information is summarized in Figure 4.12.

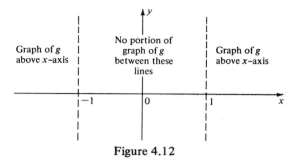

Figure 4.12

Since $x^2 - 1 = (-x)^2 - 1$, $g(x) = g(-x)$ for all x in the domain of g, g is symmetric with respect to the y-axis. Therefore it is sufficient to sketch that portion of the graph of g lying to the right of the y-axis and then duplicate it properly to the left of the y-axis using symmetry.

A table of values for g is

x	1	$\sqrt{2}$	$\dfrac{3}{2}$	2	3
$g(x)$	0	1	$\dfrac{\sqrt{5}}{2}$	$\sqrt{3}$	$\sqrt{8}$

Plotting these points and using symmetry yields Figure 4.13.

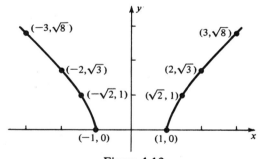

Figure 4.13

As one can see, the problem of sketching the graph of a more complicated algebraic function, such as

$$h(x) = 2x^3 - 5x + \sqrt{\frac{x-1}{x^4+10}} + (3x-5)^{\frac{4}{3}},$$

would present formidable difficulties if one uses only the techniques of sign tables, point plotting, and symmetry considerations. More general techniques for curve sketching are developed in Chap. 7.

Exercise Set 4.7

1. Given the algebraic function $f(x) = \sqrt{x + 1}$, what is the domain of f? In what interval will there be no points on the graph of f, either above, below, or on the x-axis?

2. Construct a table of values for the function $f(x) = \sqrt{x + 1}$.

3. Sketch the graph of the function $f(x) = \sqrt{x + 1}$.

4. Given the algebraic function $g(x) = \sqrt{x^2 - 4}$, determine the zeros of g. Construct a sign diagram for g. What is the domain of g?

5. Construct a table of values for the function $g(x) = \sqrt{x^2 - 4}$.

6. Sketch the graph of the function $g(x) = \sqrt{x^2 - 4}$.

* * *

In problems 7–20, for each algebraic function F, determine the zeros of F, construct a sign diagram for F, determine the domain of F, construct a table of values for F, and sketch the graph of F.

7. $F(x) = \sqrt{2x}$

8. $F(x) = 2\sqrt{x}$

9. $F(x) = \sqrt{-2x}$

10. $F(x) = -2\sqrt{-x}$

11. $F(z) = z^{\frac{1}{3}}$

12. $F(z) = z^{\frac{2}{3}}$

13. $F(x) = \sqrt{1 - x^2}$

14. $F(x) = \sqrt{4 - x^2}$

15. $F(t) = \sqrt{t^2}$

16. $F(t) = \sqrt{t^2 + 1}$

17. $F(x) = \sqrt{x^2 + x - 2}$

18. $F(x) = \sqrt{-x^2 - x + 2}$

19. $F(w) = \sqrt{2w^2 + 8w + 8}$

20. $F(w) = \sqrt{-2w^2 - 8w - 8}$

*21. Sketch the graph of the function $g(x) = (x - 1)^{\frac{2}{3}}$.

*22. Sketch the graph of the function $f(x) = x^2 + x^{\frac{1}{2}}$. (*Suggestion*: Sketch the graph of $f_1(x) = x^2$, then on the same set of axes, the graph of $f_2(x) = x^{\frac{1}{2}}$. Finally, sketch $f(x) = f_1(x) + f_2(x) = x^2 + x^{\frac{1}{2}}$ by summing graphically $f_1(x)$ and $f_2(x)$ for convenient choices of x.)

*23. Sketch the graphs of the functions $f(x) = \sqrt{9 - x^2}$ and $g(x) = -\sqrt{9 - x^2}$ on the same set of axes. What is the figure $G_f \cup G_g$?

4.8 ABSOLUTE VALUE FUNCTIONS

Let us consider now a table of values for the algebraic function $f(x) = \sqrt{x^2}$:

x	-2	$-\frac{3}{2}$	-1	$-\frac{1}{2}$	0	$\frac{1}{2}$	1	$\frac{3}{2}$	2
$f(x)$	2	$\frac{3}{2}$	1	$\frac{1}{2}$	0	$\frac{1}{2}$	1	$\frac{3}{2}$	2

Notice that $f(-x) = f(x)$ for every x. Notice also that $\sqrt{x^2}$ is the same as x if $x \geqslant 0$ and $\sqrt{x^2}$ is the negative of x if $x < 0$. This then, is another way of expressing the absolute value function. Recalling that

$$|x| = \begin{cases} x, & \text{if } x \geqslant 0 \\ -x, & \text{if } x < 0, \end{cases}$$

we see that the graph of $f(x) = |x|$ is composed of that part of the line $y = x$ where $x \geqslant 0$ together with that portion of the line $y = -x$ where $x < 0$ (see Figure 4.14).

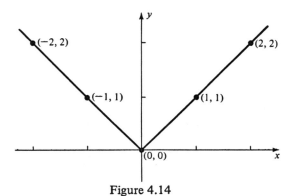

Figure 4.14

Let us examine in greater detail some of the properties of the absolute value function. Consider Figure 4.15 which shows the graph of $f(x) = |x|$ together with the line $y = 3$.

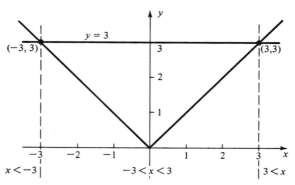

Figure 4.15

Inspection of Figure 4.15 reveals that
    (i) $|x| < 3$ *if and only if* $-3 < x < 3$,
    (ii) $|x| > 3$ *if and only if* $x > 3$ or $x < -3$,
    (iii) $|x| = 3$ *if and only if* $x = 3$ or $x = -3$.
A general statement to this effect is the following theorem:

4.7 Theorem. *IF c is a positive number, THEN*
    (i) $|x| < c$ *if and only if* $-c < x < c$,
    (ii) $|x| > c$ *if and only if* $x > c$ or $x < -c$,
and   (iii) $|x| = c$ *if and only if* $x = c$ or $x = -c$.

Example. In applying Theorem 4.7, we see that
    (i) $|x| < \dfrac{7}{3}$ if and only if $-\dfrac{7}{3} < x < \dfrac{7}{3}$,
    (ii) $|x| > \dfrac{7}{3}$ if and only if $x > \dfrac{7}{3}$ or $x < -\dfrac{7}{3}$,
    (iii) $|x| = \dfrac{7}{3}$ if and only if $x = \dfrac{7}{3}$ or $x = -\dfrac{7}{3}$.

Problem. For what values of x is $|x + 2| < \dfrac{7}{3}$?

Solution. From Theorem 4.7 we have $|x + 2| < \dfrac{7}{3}$ if and only if $-\dfrac{7}{3} < x + 2 < \dfrac{7}{3}$.

Adding -2 to both sides of the inequalities, we obtain

$$-\frac{7}{3} - 2 < x < \frac{7}{3} - 2.$$

Thus $|x + 2| < \frac{7}{3}$ if, and only if, $-\frac{13}{3} < x < \frac{1}{3}$. Therefore

$$\left\{ x: \ |x + 2| < \frac{7}{3} \right\} = \left(-\frac{13}{3}, \frac{1}{3} \right).$$

Problem. For what values of x is $|x + 2| > \frac{7}{3}$?

Solution. Again, an appeal to Theorem 4.7 yields $|x + 2| > \frac{7}{3}$ if, and only if,

$$x + 2 > \frac{7}{3} \quad \text{or} \quad x + 2 < -\frac{7}{3},$$

that is, $\qquad x > \frac{7}{3} - 2 \quad \text{or} \quad x < -\frac{7}{3} - 2.$

Hence, $|x + 2| > \frac{7}{3}$ if, and only if,

$$x > \frac{1}{3} \qquad \text{or} \quad x < -\frac{13}{3}.$$

Therefore $\left\{ x: \ |x + 2| > \frac{7}{3} \right\} = \left\{ x: \ x > \frac{1}{3} \text{ or } x < -\frac{13}{3} \right\}.$

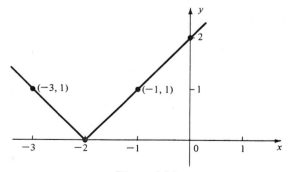

Figure 4.16

Problem. Sketch the graph of the function g defined by $g(x) = |x + 2|$.

Solution. Since

$$g(x) = |x + 2| = \begin{cases} x + 2, & \text{if } x \geqslant -2 \\ -(x + 2), & \text{if } x < -2, \end{cases}$$

the graph of g is composed of that part of the line $y = x + 2$ where $x \geqslant -2$ together with that portion of the line $y = -x - 2$ where $x < -2$ (see Figure 4.16).

Exercise Set 4.8

1. Find $|-5|$, $|2|$, $|2 + 5|$, $|2 - 5|$, $|2 + \sqrt{5}|$, and $|2 - \sqrt{5}|$.

2. a. For what values of x is $|x| < 8$?
   b. For what values of x is $|x| > 8$?
   c. For what values of x is $|x| = 8$?

3. a. For what values of x is $|2x + 3| < 8$?
   b. For what values of x is $|2x + 3| > 8$?
   c. For what values of x is $|2x + 3| = 8$?

4. Sketch the graph of the function $f(x) = |2x + 3|$.

5. Compare the answers for problem 3 with the graph of problem 4.

* * *

6. Determine each of the following:

   a. $|0|$ b. $\left|\dfrac{4}{3} - \dfrac{1}{2}\right|$ c. $\left|\dfrac{4}{3}\right| - \left|\dfrac{1}{2}\right|$ d. $|(2)(-3)|$ e. $|2| \cdot |-3|$

7. Determine each of the following:

   a. $|-7 + 7|$ b. $\left|-\dfrac{2}{5} - \dfrac{5}{4}\right|$ c. $\left|-\dfrac{2}{5}\right| - \left|-\dfrac{5}{4}\right|$

   d. $\left|-\dfrac{2}{5}\right| \cdot \left|-\dfrac{5}{4}\right|$ e. $-\left|\dfrac{2}{5}\right| \cdot \left|-\dfrac{5}{4}\right|$

8. Determine the values of y for which:

   a. $|y| < 5$ b. $|y| > 5$ c. $|y + 3| < 5$ d. $|y + 3| > 5$ e. $|y + 3| = 5$

9. Determine the values of y for which:

   a. $|y - 3| < 5$ b. $|y - 3| > 5$ c. $|y - 3| = 5$ d. $|3y| < 5$ e. $|-3y| < 5$

10. Express each of the following inequalities using absolute value notation:

   a. $-2 < x < 2$ b. $x < -2$ or $x > 2$ c. $-2 \leqslant x \leqslant 2$
   d. $-5 < x - 3 < -1$ e. $x - 3 < -5$ or $x - 3 > -1$

11. Express each of the following inequalities using absolute value notation.

a. $-\frac{1}{2} < x < \frac{1}{2}$　　　b. $x < -\frac{1}{2}$ or $x > \frac{1}{2}$　　　c. $-\frac{1}{2} \leqslant x \leqslant \frac{1}{2}$

d. $-1 < x + \frac{1}{2} < 2$　　　e. $x + \frac{1}{2} \leqslant -1$ or $x + \frac{1}{2} \geqslant 2$

12. Find all values of x for which:

a. $|x| = \frac{2}{3}$　　　b. $|x| < -\frac{2}{3}$　　　c. $|x| > -\frac{2}{3}$　　　d. $|x| > \frac{2}{3}$

13. Find all values of x for which:

a. $|x| = \sqrt{2}$　　　b. $|x| < \sqrt{2}$　　　c. $|x| < -\sqrt{2}$　　　d. $|x| > -\sqrt{2}$

In problems 14–21, sketch the graph of the given function.

14. $f(x) = |x - 2|$

15. $f(x) = |2x + 1|$

16. $f(x) = 3|x - 2|$

17. $f(x) = |3x - 2|$

18. $g(x) = x - |x|$

19. $g(x) = x + |x|$

20. $g(x) = |2 - x|$

21. $g(x) = |4 - x^2|$

22. Complete the following statements:
    a. If $t < 0$, then $|t| = $ _____.
    b. If $a < b$, then $|a - b| = $ _____.
    c. If $a < b$, then $|b - a| = $ _____.
    d. $|r| = 0$ if and only if _____.
    e. If $r \leqslant s$, then $-|s - r| = $ _____.

23. Solve the following inequalities given that ϵ is a positive number:

    a. $|g(t)| < \epsilon$　　　　　　　b. $|g(t)| < \frac{\epsilon}{2}$

    c. $|t - 1| < \epsilon$　　　　　　　d. $|t - 1| < \frac{\epsilon}{2}$

    e. $|2t + 5| < \epsilon$　　　　　　　f. $|-2t + 5| < \epsilon$

*24. a. Locate a point on the x-axis which satisfies the inequality $|x - 2| < \delta$ if $\delta = .25$.
     b. Locate a point $f(x)$ on the y-axis which satisfies $|f(x) - L| < \epsilon$ if $L = 1$ and $\epsilon = 0.3$.

4.9 GREATEST INTEGER FUNCTION

Another function of particular interest is one which belongs to the class of functions called *step functions*. If x is a real number, the symbol $[x]$ denotes the greatest integer n which is less than or equal to x. The number $[x]$ is usually called the "greatest integer in x." We see, for example, that $[2.1] = 2$ because $2 \leqslant 2.1 < 3$ and $[-\sqrt{3}] = -2$ because $-2 \leqslant -\sqrt{3} < -1$. Similarly, $[3] = 3$ and $\left[\dfrac{5}{8}\right] = 0$. Every real number x can be sandwiched in between two consecutive integers n and $n + 1$ such that $n \leqslant x < n + 1$. Therefore $[x] = n$ for all x in the interval $[n, n + 1)$.

If the function f is defined by $f(x) = [x]$, then f is called the *greatest integer function*. As we have just observed, $[x] = n$ for all x in $[n, n + 1)$ so that f is constant on each interval $[n, n + 1)$. Since $[n + 1] = n + 1$, we see that f jumps by one on the next successive interval $[n + 1, n + 2)$. From the short table of values

x	$-1 \leqslant x < 0$	$0 \leqslant x < 1$	$1 \leqslant x < 2$	$n \leqslant x < n + 1$
$f(x) = [x]$	-1	0	1	n

we see that the graph of the greatest integer function looks like a sequence of ascending steps (see Figure 4.17).

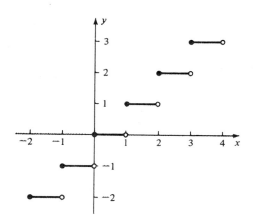

Figure 4.17

While the greatest integer function may, at first, seem a bit unusual, it is very useful in describing practical situations. For instance, consider how bonus stamps are given away in supermarkets. A single stamp is normally given for each 10-cent purchase, with any fractional part of a 10-cent purchase having no bonus stamp value. Suppose that a person spends an amount A in dollars. Then $10A$ represents the equivalent number of 10-cent purchases so that $[10A]$ is the number of bonus stamps the person receives. Thus, if a person spends \$2.67 at such a store, that person will receive

$$[10(2.67)] = [26.7] = 26$$

bonus stamps for such a purchase. We see that

$$s(A) = [10A]$$

is the function which determines the number of bonus stamps to be given to the customer making a purchase amounting to A dollars.

Let us close this section with an example of how the greatest integer function may be used to write a function which explicitly gives the cost of mailing a letter in terms of its weight. The cost of sending a letter first class is 8 cents per ounce or any fraction thereof. Let w denote the weight of a letter in ounces and let $P(w)$ denote the cost in cents of the postage required to send this letter first class. We assume $w > 0$. We see that $P(w)$ satisfies

$$P(w) = 8(n + 1) \quad \text{if } n < w \leqslant n + 1$$

for any nonnegative integer n. Let $O(w)$ denote the number of ounces of postage required for a letter of weight w. Then $O(w)$ satisfies

$$O(w) = n + 1 \quad \text{if } n < w \leqslant n + 1.$$

In this case, we have $P(w) = 8\,O(w)$. It therefore suffices to find an explicit expression for $O(w)$ in terms of w. Consider the function $\dfrac{[w]}{w}$. We see that $\dfrac{[w]}{w} = 1$ if w is an integer and $0 \leqslant \dfrac{[w]}{w} < 1$ if w is not an integer. It follows that the function $f(w) = \left[\dfrac{[w]}{w}\right]$ has the property that

$$f(w) = \begin{cases} 1, & \text{if } w \text{ is an integer} \\ 0, & \text{if } w \text{ is not an integer.} \end{cases}$$

We seek a function which will indicate when w is not an integer. If w is an integer, then $2^{-f(w)} = 2^{-1} = \dfrac{1}{2}$. If w is not an integer, then $2^{-f(w)} = 2^{0} = 1$. Thus the function $[2^{-f(w)}]$ has the property that

$$[2^{-f(w)}] = \begin{cases} 0, & \text{if } w \text{ is an integer} \\ 1, & \text{if } w \text{ is not an integer.} \end{cases}$$

A little thought will show that

$$O(w) = [w] + [2^{-f(w)}].$$

Exercise Set 4.9

1. Find each of the following:

   a. $\left[\dfrac{1}{2}\right]$ b. $\left[\dfrac{3}{2}\right]$ c. $\left[-\dfrac{1}{2}\right]$ d. $\left[\dfrac{7}{3}\right]$ e. $[-4]$

   f. $[7.51]$ g. $\left[-\dfrac{19}{2}\right]$ h. $\left[\dfrac{243}{4}\right]$ i. $[0]$ j. $\left[\dfrac{13}{2}-\dfrac{16}{3}\right]$

2. Find each of the following:

   a. $\left[4 \cdot \dfrac{5}{2}\right]$ b. $4[\sqrt{2}]$ c. $[6.4-6]$ d. $6.4-[6.4]$ e. $[6-6.4]$

   f. $[6]-[6.4]$ g. $[7.3]+[3.7]$ h. $[7.3+3.7]$

3. Find each of the following:

   a. $[x]$ if $\dfrac{1}{2} \leqslant x \leqslant \dfrac{3}{4}$ b. $[x]$ if $-\dfrac{3}{4} \leqslant x \leqslant -\dfrac{1}{2}$

   c. $[2x]$ if $\dfrac{1}{2} \leqslant x < 1$ d. $[2x]$ if $-1 \leqslant x < -\dfrac{1}{2}$

   e. $\left[\dfrac{x}{2}\right]$ if $4 \leqslant x < 6$ f. $\left[\dfrac{x}{2}\right]$ if $-6 \leqslant x < -4$

4. Construct a table of values for the function $f(x) = \left[\dfrac{x}{2}\right]$.

5. Sketch the graph of the function $f(x) = \left[\dfrac{x}{2}\right]$ in the interval $[-6, 6]$.

6. Sketch the graph of the function $s(A) = [2A]$.

* * *

In problems 7–18, construct a table of values and sketch the graph of the given function.

   7. $f(x) = [x + 1]$ 8. $f(x) = [x - 1]$

   9. $g(x) = [-x]$ 10. $g(x) = [-x - 1]$

   11. $f(t) = [3t]$ 12. $f(t) = \left[\dfrac{t}{3}\right]$

   13. $f(t) = 3[t]$ 14. $f(t) = -3[t]$

   15. $f(z) = [z] + 2$ 16. $f(z) = 2 - [z]$

17. $f(x) = x - [x]$

18. $f(x) = [x] + [2x]$

19. Use the postal rate function to determine the cost of mailing a letter weighing:

a. 3 oz.

b. 5 oz.

c. $2\frac{2}{3}$ oz.

d. $2\frac{5}{6}$ oz.

e. $3\frac{1}{4}$ oz.

f. 1.005 oz.

*20. Show that $[x + 1] = [x] + 1$ for every real number x.

*21. A function $f(x)$ is said to be periodic with period p if p is a positive real number having the property that $f(x + p) = f(x)$ for all x. Using the result of problem 20, show that the function $f(x) = x - [x]$ is periodic with period 1. What effect does the fact that f is periodic have on the graph of f?

5 | *The Exponential and Logarithm Functions*

In many applications of mathematics, two important functions frequently appear. These two functions are the *exponential* and *logarithm* functions. They are used, for example, in obtaining solutions to problems involving growth (such as population growth), decay (such as radioactive decay), and in computational problems. In this chapter we examine the basic properties of these functions and their relationship to each other.

5.1 EXPONENTIAL FUNCTIONS

In Chap. 3, we considered functions of the form

$$F(x) = x^a$$

where a was a rational number. For example, $F(x) = x^2$, $F(x) = x^{\frac{1}{2}} = \sqrt{x}$, or $F(x) = x^{-1} = \dfrac{1}{x}$. For such functions, the exponent is fixed and the number being raised to a power is variable.

Suppose now, however, that we interchange the roles of x and a. That is, we will consider functions of the form

$$E(x) = a^x$$

with the restriction that $a > 0$. For example, $E(x) = 2^x$, $E(x) = \left(\dfrac{1}{2}\right)^x$, or $E(x) = 3^x$. Note that now a is fixed and the exponent is variable. Functions of the form $E(x) = a^x$ are very different from functions of the form $F(x) = x^a$. A function such as E is called an *exponential* function with *base a*.

Let us consider, as an example, the function $E(x) = 2^x$. This equation defines the exponential function whose base is 2.

We recall from the rules of exponents the following:

(1) $2^0 = 1$,

(2) $2^{-n} = \dfrac{1}{2^n}$ for n a positive integer,

and

(3) $2^{\frac{m}{n}} = (\sqrt[n]{2})^m$ for $\dfrac{m}{n}$ a rational number.

Consequently, if $E(x) = 2^x$, then

$$E(3) \ = 2^3 = 8,$$
$$E(1) \ = 2^1 = 2,$$
$$E\left(\frac{1}{3}\right) = \sqrt[3]{2},$$
$$E(-3) = 2^{-3} = \frac{1}{2^3} = \frac{1}{8},$$
$$E\left(\frac{5}{2}\right) = 2^{\frac{5}{2}} = (\sqrt{2})^5 = 4\sqrt{2}.$$

There is no difficulty in defining 2^x when x is a rational number, but what if x is an irrational number? That is, how might we define a number such as $2^{\sqrt{3}}$?

It can be shown that if r and s are rational numbers and $r < s$, then $2^r < 2^s$. Moreover, it is also true, although it is difficult to prove, that there is precisely one real number, denoted by $2^{\sqrt{3}}$, such that $2^r < 2^{\sqrt{3}} < 2^s$ whenever r and s are rational numbers with $r < \sqrt{3} < s$. This enables us to approximate $2^{\sqrt{3}}$. For instance, we know that $1.5 < \sqrt{3} < 2$, $2^{1.5} = 2^{\frac{3}{2}} = 2.8$ (approximately) and $2^2 = 4$. Therefore

$$2.8 < 2^{\sqrt{3}} < 4.$$

In this manner we could approximate $2^{\sqrt{3}}$ to any desired degree of accuracy. In general, for any real number x the number 2^x is defined and it has the property that $2^r < 2^x < 2^s$ whenever r and s are rational numbers such that $r < x < s$.

A table of values of $E(x) = 2^x$ is

x	-3	$-\dfrac{1}{2}$	$-\dfrac{1}{3}$	0	$\dfrac{1}{3}$	$\dfrac{1}{2}$	1	2	$\dfrac{5}{2}$	3
$E(x) = 2^x$	$\dfrac{1}{8}$	$\dfrac{1}{\sqrt{2}}$	$\dfrac{1}{\sqrt[3]{2}}$	1	$\sqrt[3]{2}$	$\sqrt{2}$	2	4	$4\sqrt{2}$	8

If the points of the preceding table are used to sketch the graph of $E(x) = 2^x$, then we have the curve shown in Figure 5.1.

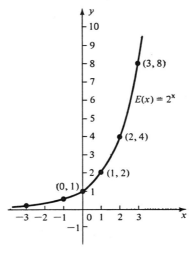

Figure 5.1

Problem. Sketch the graph of the function $E(x) = 5^x$.

Solution. This is the exponential function whose base is 5. We construct a short table of values:

x	-2	-1	$-\dfrac{1}{2}$	0	$\dfrac{1}{2}$	1	2
$E(x) = 5^x$	$\dfrac{1}{25}$	$\dfrac{1}{5}$	$\dfrac{1}{\sqrt{5}}$	1	$\sqrt{5}$	5	25

If the points obtained from the preceding table are plotted and connected by a smooth curve, the result is a reasonable sketch of the graph of $E(x) = 5^x$ (see Figure 5.2).

Notice that the graphs in Figures 5.1 and 5.2 both pass through the point (0, 1). In fact, the graph of every exponential function $E(x) = a^x$ will pass through the point (0, 1) because $E(0) = a^0 = 1$. Observe, too, that both of the preceding graphs are rising as we go from left to right. This is characteristic of the graphs of those exponential functions whose bases are greater than 1. That is, if $a > 1$, then the graph of $E(x) = a^x$ is rising as we go from left to right. The basic shape of the graph of $E(x) = a^x$, for $a > 1$, is shown in Figure 5.3.

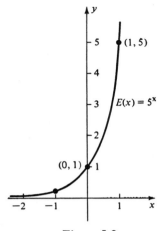

Figure 5.2

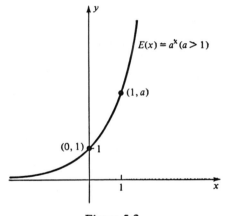

Figure 5.3

Problem. Sketch the graph of the function $E(x) = \left(\frac{2}{3}\right)^x$.

Solution. This exponential function has a base of $\frac{2}{3}$. Again, we construct a short table of values:

x	-3	-2	-1	0	1	2	3
$E(x) = \left(\frac{2}{3}\right)^x$	$\frac{27}{8}$	$\frac{9}{4}$	$\frac{3}{2}$	1	$\frac{2}{3}$	$\frac{4}{9}$	$\frac{8}{27}$

Using these points we obtain the graph of $E(x) = \left(\dfrac{2}{3}\right)^x$ (see Figure 5.4).

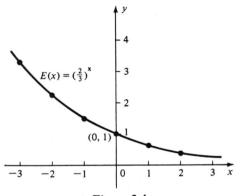

Figure 5.4

Note that the graph of $E(x) = \left(\dfrac{2}{3}\right)^x$ is falling as we go from left to right. This is a general property of those exponential functions whose bases are less than 1. That is, if $a < 1$, then the graph of $E(x) = a^x$ is falling as we go from left to right. Therefore if $a < 1$, the general shape of the graph of $E(x) = a^x$ is as shown in Figure 5.5.

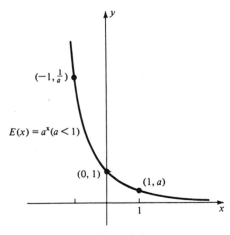

Figure 5.5

If $a > 0$, then a^x is defined for every real number x. Moreover, the usual rules for exponents still apply. Thus, if $x, y \in \mathcal{R}$, then

$$a^x a^y = a^{x+y},$$

$$a^{-x} = \frac{1}{a^x},$$

$$\frac{a^x}{a^y} = a^{x-y},$$

and

$$(a^x)^y = a^{xy}.$$

This means that we may continue to manipulate these expressions in the same manner as before.

Example. (a) $3^2 \, 3^{\sqrt{5}} = 3^{2 + \sqrt{5}}$

(b) $\dfrac{2^{\sqrt{6}}}{2^4} = 2^{\sqrt{6} - 4}$

(c) $\dfrac{7}{6}^{-\sqrt{3}} = \dfrac{1}{\left(\dfrac{7}{6}\right)^{\sqrt{3}}}$

(d) $(5^{\sqrt{2}})^3 = 5^{3\sqrt{2}}$

Exercise Set 5.1

1. a. If $x = 3$ find

$$2^x, x^3, x^{-3}, 2^{x+2}, (x + 2)^2, (2^x)^2, x^{\frac{3}{2}}, \left(\frac{3}{2}\right)^x.$$

   b. If $x = -3$ find

$$2^x, x^3, x^{-3}, 2^{x+2}, (x + 2)^2, (2^x)^2, x^{\frac{3}{2}}, \left(\frac{3}{2}\right)^x.$$

   c. If $x = \dfrac{1}{3}$ find

$$8^x, x^4, x^{-4}, 8^{-x}, 2^{2x}, \left(x + \frac{1}{3}\right)^x, (8^x)^3, \left(\frac{8}{27}\right)^x.$$

2. Let $E(x) = 3^x$.
   a. Construct a table of values for this function.
   b. Sketch the graph of this function.
   c. From the graph, estimate: $3^{\frac{1}{3}}, 3^{\frac{\sqrt{2}}{2}}, 3^{-\frac{1}{3}}$.

3. a. What is the domain of the function $E(x) = 3^x$?
   b. What is the range of the function $E(x) = 3^x$? (see Figure 5.3).

* * *

4. Let $E(x) = \left(\frac{1}{2}\right)^x$.

   a. Construct a table of values of this function.
   b. Sketch the graph of this function.
   c. Is there a real number x for which $E(x) = 0$?
   d. Find a real number x for which $E(x) < .005$.
   e. Find a real number x for which $E(x) < .00001$.

5. Let $J(t) = 10^t$.
   a. Construct a table of values for J.
   b. Sketch the graph of J.
   c. Is there a real number t for which $J(t) \leqslant 0$?
   d. Find a real number t for which $J(t) < .005$.
   e. Find a real number t for which $J(t) > 5000$.
   f. Use the graph to estimate $\sqrt[3]{10}$.. $\left(Hint: \sqrt[3]{10} = 10^{\frac{1}{3}} = J\left(\frac{1}{3}\right) \right)$.

6. Let $J(t) = 0.1^t$.
   a. Construct a table of values for J.
   b. Sketch the graph of J.
   c. Is there a real number t for which $J(t) \leqslant 0$?
   d. Find a real number t for which $J(t) < .002$.
   e. Find a real number t for which $J(t) > 2000$.
   f. Use the graph to estimate $\sqrt{0.1}, \sqrt{10}$.

7. If a is a positive real number and if $a > 1$, what is the domain of the function $E(x) = a^x$? What is the range?

8. If a is a positive real number and if $a < 1$, what is the domain of the function $E(r) = a^r$? What is the range?

9. Obtain bounds for each of the following:

$$3^{\sqrt{2}}, \ 3^{-\sqrt{2}}, \ 3^{\sqrt{10}}, \ 3^{-\sqrt{5}}, \ 3^{\pi}, \ 3^{-\pi}$$

10. Obtain bounds for each of the following:

$$\left(\frac{1}{2}\right)^{\sqrt{3}}, \ \left(\frac{1}{2}\right)^{-\sqrt{3}}, \ \left(\frac{1}{2}\right)^{\sqrt{10}}, \ \left(\frac{1}{2}\right)^{-\sqrt[3]{10}}, \ \left(\frac{1}{2}\right)^{\pi}, \ \left(\frac{1}{2}\right)^{1-\sqrt{2}}$$

11. Let $E(x) = (\sqrt{2})^x$.
   a. Construct a table of values for E.
   b. Sketch the graph of E.

12. Let $E(x) = \left(\dfrac{1}{\sqrt{2}}\right)^x$.
   a. Construct a table of values for E.
   b. Sketch the graph of E.

13. Let $E(x) = 4^x$.
   a. Find an x such that $E(x) = 2$.
   b. Is there an x such that $E(x) = 0$?
   c. Find an x such that $E(x) = 1$.
   d. Find an x such that $E(x) = \dfrac{1}{2}$.
   e. Is it possible to find more than one x such that $E(x) = 2$?
   f. Is it possible to find more than one x such that $E(x) = 1$?

14. Let $J(x) = \left(\dfrac{2}{3}\right)^x$.
   a. Find an x such that $J(x) = \dfrac{4}{9}$.
   b. Find an x such that $J(x) = \dfrac{9}{4}$.
   c. Find an x such that $J(x) = 1$.
   d. Is there an x such that $J(x) < 0$?
   e. What is the domain of $J(x)$?
   f. What is the range of $J(x)$?

15. Suggest a reason for the requirement $a > 0$, if $E(x)$ is a function of the form $E(x) = a^x$.

5.2 THE NUMBER e

The base a of the exponential function $E(x) = a^x$ can be any positive real number. However, in many applications it happens that one particular number appears as a base again and again. Because of the frequency with which it appears, this number merits special attention. In order to understand a little more about the nature of this special base, let us consider a problem pertaining to the computation of compound interest.

Suppose that money were in such great demand that banks were paying 100% interest annually on their savings accounts. At this rate a dollar deposited in a savings account for one year would earn one dollar in interest. Hence the amount in the account at the end of one year would be $1 + 1$ or 2 dollars. Suppose the bank compounds interest twice a year. The interest for one half year would be 50%. Thus a dollar left in a savings account would be worth $1 + \frac{1}{2}$ dollars at the end of six months. Therefore, at the end of the year, the account would be worth

$$\left(1 + \frac{1}{2}\right) + \left(1 + \frac{1}{2}\right)\frac{1}{2} = \left(1 + \frac{1}{2}\right)^2$$

dollars. If the bank compounds the interest three times a year, the account would be worth $\left(1 + \frac{1}{3}\right)^3$ dollars at the end of the year. In fact, if the bank compounds interest n times a year, the account would be worth $\left(1 + \frac{1}{n}\right)^n$ dollars at the end of the year. For example:

Interest compounded	Value of account at end of year	
annually	$(1 + 1)^1 = 2$	
semiannually	$\left(1 + \frac{1}{2}\right)^2 = 2.25$	
quarterly	$\left(1 + \frac{1}{4}\right)^4 = 2.44$	
monthly	$\left(1 + \frac{1}{12}\right)^{12} = 2.61$	approximate
daily	$\left(1 + \frac{1}{365}\right)^{365} = 2.67$	values
hourly	$\left(1 + \frac{1}{8760}\right)^{8760} = 2.713$	

The preceding list indicates that it is probably to the investor's advantage to have interest compounded n times per year for large values of n. What happens to the value of the account at the end of the year as n increases without bound? That is, what is the behavior of the number $\left(1 + \frac{1}{n}\right)^n$ as n increases without bound? There are two influences affecting $\left(1 + \frac{1}{n}\right)^n$. If n is large, $1 + \frac{1}{n}$ is close to 1 and $\left(1 + \frac{1}{n}\right)^n$ could therefore also be close to 1. However, if n is large, since $1 + \frac{1}{n}$ is greater than 1, the quantity $\left(1 + \frac{1}{n}\right)^n$ could conceivably grow without bound as n increases. Actually, the two influences offset each other to the extent that for sufficiently large n, $\left(1 + \frac{1}{n}\right)^n$ "approaches" a fixed number. This number is denoted by e and equals 2.718281828459 (to twelve decimal places).

It is the number e which appears as a base of the exponential function $E(x) = e^x$ in many applications. Because of its importance, the function $E(x) = e^x$ is called *THE* exponential function. For a given value of x, it would be difficult to compute e^x with the techniques presently at our disposal. Therefore, a table of values for this function can be found in Appendix I. It should be noted that, since e is larger than 1, the graph $E(x) = e^x$ has the basic shape indicated in Figure 5.3.

Exercise Set 5.2

1. Suppose a bank pays 100% interest, compounded quarterly, and suppose one dollar is deposited in a savings account.
   a. What is the interest rate per quarter?
   b. What is the interest to be paid at the end of the first quarter?
   c. What is the amount in the account at the end of the first quarter?
   d. What is the amount in the account at the end of the second quarter?
   e. What is the amount in the account at the end of the year?

2. a. Compute $\left(1 + \frac{1}{3}\right)^3$, $\left(1 + \frac{1}{5}\right)^5$, and $\left(1 + \frac{1}{6}\right)^6$.
   b. How do these numbers compare with e?

3. a. Use the tables in Appendix I to construct a table of values for the function $E(x) = e^x$.
   b. Sketch the graph of the function $E(x) = e^x$.

4. a. What is the domain of the function $E(x) = e^x$?
   b. What is the range of the function $E(x) = e^x$?

* * *

5. a. Use the tables in Appendix I to construct a table of values for the function $E(x) = e^{-x}$.
   b. Sketch the graph of $E(x) = e^{-x}$.

6. a. Use the tables in Appendix I to construct a table of values for the function $E(x) = e^{2x}$.
   b. Sketch the graph of this function.

7. a. Construct a table of values for the function $E(x) = e^{x+1}$.
   b. Sketch the graph of this function.

8. a. Construct a table of values for the function $E(x) = e^{x/3}$.
   b. Sketch the graph of this function.

9. a. Construct a table of values for the function $E(t) = e^{t^2}$.
   b. Sketch the graph of this function.
   c. What is the domain of this function?
   d. Use the graph to determine the range of this function.

10. a. Construct a table of values for the function $E(t) = e^{-t^2}$.
    b. Sketch the graph of this function.
    c. What is the domain of this function?
    d. What is the range of this function?

11. a. Compute $1 + \dfrac{1}{1!} + \dfrac{1}{2!} + \dfrac{1}{3!} + \dfrac{1}{4!} + \dfrac{1}{5!}$.
    b. Compare the result of the computation in part (a) with the number e.

12. Show that $\left(\dfrac{e^x + e^{-x}}{2}\right)^2 - \left(\dfrac{e^x - e^{-x}}{2}\right)^2 = 1$ for every real number x.

5.3 THE LOGARITHM FUNCTION

An examination of the graph of $E(x) = e^x$ points out two important properties possessed by the exponential function. First, if x_1 and x_2 are real numbers with $x_1 \neq x_2$, then $e^{x_1} \neq e^{x_2}$ (see Figure 5.6). It also appears that $e^x > 0$ for all x

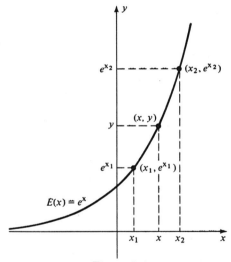

Figure 5.6

and that the range of E is the set of all positive real numbers. This means that for each positive real number y there is a unique real number x such that $y = e^x$. We now define a function L from the set of positive real numbers to the set of real numbers as follows:

$$L(y) = x \text{ if } y = e^x.$$

In other words, if y is a positive real number, then $L(y)$ is that exponent to which e must be raised to yield y.

Example. $L(e^3) = 3$ since 3 is the exponent to which e must be raised to yield e^3.

Example. $L\left(\dfrac{1}{\sqrt{e}}\right) = -\dfrac{1}{2}$ since $-\dfrac{1}{2}$ is the exponent to which e must be raised to yield $\dfrac{1}{\sqrt{e}}$.

Example. $L(1) = 0$ since 0 is the exponent to which e must be raised to yield 1.

Since $L(y)$ is that exponent to which e must be raised to yield y, we see that a fundamental property of the function L is the equality

$$e^{L(y)} = y$$

for all positive numbers y. In addition, since x is the exponent to which e must be raised to yield e^x, it follows that

$$L(e^x) = x.$$

The function L is commonly called the *natural logarithm* function and $L(y)$ is usually denoted by $\ln y$. The number $\ln y$ is called the *natural logarithm* of y. In terms of the ln notation, the equalities $e^{L(y)} = y$ and $L(e^x) = x$ become

$$e^{\ln y} = y \text{ and } \ln e^x = x.$$

Because of the latter equalities, the exponential and logarithm functions are said to be inverses of each other.

Problem. Find x if $\ln x^2 = \ln(2x - 1)$.

Solution. If x satisfies $\ln x^2 = \ln(2x - 1)$, then

$$e^{\ln x^2} = e^{\ln(2x-1)},$$

or

$$x^2 = 2x - 1.$$

Therefore $x^2 - 2x + 1 = 0$, or $(x - 1)^2 = 0$. It follows that $x = 1$.

Problem. Find x if $e^{7x} = 2$.

Solution. If x satisfies $e^{7x} = 2$, then, taking the natural logarithm of both sides, we have

$$\ln e^{7x} = \ln 2,$$

or

$$7x = \ln 2.$$

Therefore $x = \dfrac{\ln 2}{7}$.

If (x, y) is a point on the graph of the exponential function $E(x) = e^x$, then $y = e^x$ and it follows that $\ln y = x$. This means that the point (y, x) is on the graph of the logarithm function. Likewise, if (y, x) is on the graph of the logarithm function, then $x = \ln y$. This implies that $y = e^x$ which means that (x, y) is on the graph of the exponential function. In other words, because the exponential and logarithm functions are inverses of each other, the graph of the logarithm function consists of all points (y, x) such that (x, y) is on the graph of the exponential function. A moment's reflection on Figure 5.7 will show that the curve on the right is an accurate sketch of the graph of the logarithm function.

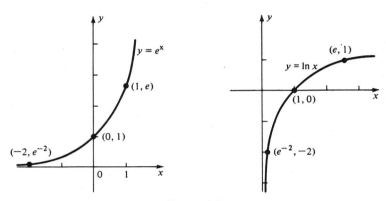

Figure 5.7

The logarithm function has three basic properties which can be used to simplify complicated arithmetic calculations (see Sec. 5.5) involving products, quotients, and roots of real numbers.

5.1 Theorem. *IF a and b are real numbers, THEN*
    (i) $\ln ab = \ln a + \ln b$ *(a and b both positive)*

    (ii) $\ln \dfrac{a}{b} = \ln a - \ln b$ *(a and b both positive)*

    (iii) $\ln a^b = b \ln a$ *(a positive)*

Proof. Let $u = \ln a$ and $v = \ln b$. This means that $e^u = a$ and $e^v = b$. Therefore

$$ab = e^u e^v = e^{u+v},$$

$$\frac{a}{b} = \frac{e^u}{e^v} = e^{u-v},$$

and
$$a^b = (e^u)^b = e^{bu}.$$

Since $u + v$ is the exponent to which e must be raised to obtain ab, it follows that

$$\ln ab = u + v = \ln a + \ln b$$

which proves (i).

Since $u - v$ is the exponent to which e must be raised to obtain $\dfrac{a}{b}$, it follows that

$$\ln \frac{a}{b} = u - v = \ln a - \ln b$$

which proves (ii).

Finally, bu is the exponent to which e must be raised to obtain a^b so that

$$\ln a^b = bu = b \ln a$$

which proves (iii).

We now examine how Theorem 5.1 is applied.

Problem. If $\ln 2 = .6931$ and $\ln 7 = 1.9459$, find $\ln 14$.

Solution. Since $14 = 2 \cdot 7$, we have, by part (i) of Theorem 5.1,

$$\begin{aligned}
\ln 14 = \ln (2 \cdot 7) &= \ln 2 + \ln 7 \\
&= .6931 + 1.9459 \\
&= 2.6390.
\end{aligned}$$

Problem. Find $\ln \dfrac{2}{7}$.

Solution. From the preceding problem, $\ln 2 = .6931$ and $\ln 7 = 1.9459$. Applying part (ii) of Theorem 5.1, we have

$$\begin{aligned}
\ln \frac{2}{7} &= \ln 2 - \ln 7 \\
&= .6931 - 1.9459 \\
&= -1.2528.
\end{aligned}$$

Problem. If $\ln 5 = u$ and $\ln 3 = v$, express $\ln (25 \sqrt[7]{9})$ in terms of u and v.

Solution. Note that $25 = 5^2$ and $\sqrt[7]{9} = 9^{\frac{1}{7}} = (3^2)^{\frac{1}{7}} = 3^{\frac{2}{7}}$. Thus

$$\begin{aligned}
\ln (25 \sqrt[7]{9}) &= \ln 25 + \ln \sqrt[7]{9} \\
&= \ln 5^2 + \ln 3^{\frac{2}{7}} \\
&= 2 \ln 5 + \frac{2}{7} \ln 3 \\
&= 2u + \frac{2}{7} v.
\end{aligned}$$

Problem. If $\ln 3 = u$ and $\ln 5 = v$, express $\ln \dfrac{27}{625}$ in terms of u and v.

Solution. We have $27 = 3^3$ and $625 = 5^4$. Therefore

$$
\begin{aligned}
\ln \frac{27}{625} &= \ln 27 - \ln 625 \\
&= \ln 3^3 - \ln 5^4 \\
&= 3 \ln 3 - 4 \ln 5 \\
&= 3u - 4v.
\end{aligned}
$$

Exercise Set 5.3

1. Use the definition of the natural logarithm function to determine each of the following:

   a. $\ln e^2$ b. $\ln e$ c. $\ln 1$

   d. $\ln \dfrac{1}{e}$ e. $\ln \sqrt{e}$ f. $\ln e^{15}$

   g. $\ln \dfrac{1}{\sqrt[3]{e}}$ h. $\ln e^x$ i. $\ln e^{m+n}$

2. a. Use the tables in Appendix II to construct a table of values for the function $L(x) = \ln x$.

   b. Sketch the graph of the function $L(x) = \ln x$.

   c. Compare your graph with the graph in Figure 5.7.

3. a. What is the domain of the function $L(x) = \ln x$?

   b. What is the range of the function $L(x) = \ln x$?

   c. How are the domain and range of the function $L(x) = \ln x$ related to the domain and range of the function $E(x) = e^x$?

4. Using Theorem 5.1, evaluate the following, given that $\ln 2 = .6931$ and $\ln 5 = 1.6094$:

   a. $\ln 10$ b. $\ln 4$ c. $\ln \dfrac{1}{2}$

   d. $\ln 2.5$ e. $\ln 25$ f. $\ln 0.4$

   g. $\ln \dfrac{1}{5}$ h. $\ln 8$ i. $\ln \dfrac{16}{25}$

   j. $\ln \sqrt{2}$ k. $\ln \sqrt{5}$ l. $\ln \sqrt{10}$

* * *

5. a. If $e^{3.15} = 23.34$, find $\ln 23.34$.

   b. If $e^{2.00} = 7.389$, find $\ln 7.389$.

   c. If $e^{-1.04} = 0.3535$, find $\ln 0.3535$.

   d. If $\ln .40 = -.9163$, find $e^{-.9163}$.

   e. If $\ln 1.5 = .4055$, find $e^{.4055}$.

   f. If $\ln 100 = 4.6052$, find $e^{4.6052}$.

g. If $e^u = v$, find $\ln v$.

h. If $\ln t = s$, find e^s.

6. Given $\ln 2 = .6931$, $\ln 3 = 1.0986$, and $\ln 7 = 1.9459$; find each of the following:

   a. $\ln 6$ b. $\ln \sqrt{7}$ c. $\ln 21$

   d. $\ln (14)^{-3}$ e. $\ln \dfrac{4}{3}$ f. $\ln 42$

   g. $\ln \dfrac{1}{6}$ h. $\ln \dfrac{49}{27}$ i. $\ln \left(\dfrac{3}{2}\right)^{\frac{2}{3}}$

7. Find x if:

   a. $\ln x^2 = \ln (6x - 9)$ b. $6e^{x+5} = 30$

   c. $3e^{-x} = 1$ d. $2 \ln x = \ln \left(3x - \dfrac{3}{4}\right)$.

8. Find x if:

   a. $e^{\frac{x}{2}} = 5$ b. $\ln \left(x^2 + \dfrac{1}{9}\right) = \ln \left(\dfrac{2}{3}x\right)$

   c. $\ln x^2 = \ln 8 + \ln (x - 2)$ d. $e^{3x-4} = 1.6$

9. If $\ln 4 = u$ and $\ln 3 = v$, express $\ln 2\sqrt[3]{12}$ in terms of u and v.

10. If $\ln 7 = u$ and $\ln 2 = v$, express $\ln \dfrac{49}{8}$ in terms of u and v.

11. Write each of the following as the natural logarithm of a single number:

    a. $\ln 3 + \ln 6 - \ln 12$

    b. $\ln 15 + \ln 25 + \ln 3 - 2 \ln 45 + \ln 27$

12. Write each of the following as the natural logarithm of a single number:

    a. $\ln 7 - 2 \ln 6 + \ln 9$

    b. $\dfrac{1}{2} \ln 64 + \dfrac{2}{3} \ln 27 - 5 \ln 2$

*13. Sketch the graphs of the functions $y = \ln x$, $y = e^x$, and $y = x$ on the same set of coordinate axes. How are these graphs related to each other?

*14. Sketch the graphs of the functions $y = x^2$ for $x \geqslant 0$, $y = \sqrt{x}$ for $x \geqslant 0$, and $y = x$ for $x \geqslant 0$ on the same set of coordinate axes. How are these graphs related to each other? (*Note*: The functions $f(x) = x^2$ and $g(x) = \sqrt{x}$ for $x \geqslant 0$ are inverses of each other since $(a, b) \in G_f$ if and only if $(b, a) \in G_g$.)

5.4 LINEAR INTERPOLATION

Before performing computations with logarithms, there is one computational procedure we must first examine. This procedure is known as linear interpolation. To interpolate means literally "to insert intermediate terms." In practice, interpolation pretty much has its literal meaning. Suppose we know that two points, $(a, f(a))$ and $(b, f(b))$, lie on the graph of a function f. If $a < c < b$, can we determine $f(c)$? It is clear that knowledge of $f(a)$ and $f(b)$ is not sufficient to determine $f(c)$. Can we reasonably approximate $f(c)$? One method is the use of linear interpolation to help us approximate the value of the function at the point c. Consider the line l passing through the points $(a, f(a))$ and $(b, f(b))$, as shown in Figure 5.8:

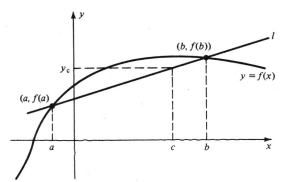

Figure 5.8

The line l has slope $\dfrac{f(b) - f(a)}{b - a}$ so that its equation is

$$\frac{y - f(a)}{x - a} = \frac{f(b) - f(a)}{b - a},$$

or

$$y = f(a) + \frac{f(b) - f(a)}{b - a}(x - a).$$

Let y_c denote the y-coordinate of the point on the line corresponding to $x = c$. Then

$$y_c = f(a) + \frac{f(b) - f(a)}{b - a}(c - a).$$

Note that the point (c, y_c) lies directly above or below the point $(c, f(c))$ on the graph of f. It is depicted as lying below $(c, f(c))$ in Figure 5.8. Therefore, if a and b are relatively close and the graph of f lies reasonably close to l over the interval $[a, b]$, then y_c is a reasonable approximation to $f(c)$. In other words, we can use y_c in place of $f(c)$. This procedure of approximation is known as *linear interpolation*.

Problem. If $f(1) = 2$ and $f(1.5) = 2.6$, use linear interpolation to approximate $f(1.3)$.

Solution. In this case, $a = 1$, $b = 1.5$, $f(a) = 2$, $f(b) = 2.6$, and $c = 1.3$. Therefore the points that are given to be on the graph of f are $(1, 2)$ and $(1.5, 2.6)$ (see Figure 5.9).

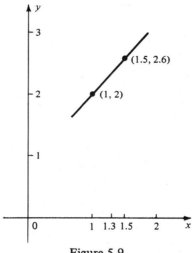

Figure 5.9

Thus, the equation of the line used in the interpolation procedure is

$$\frac{y - 2}{x - 1} = \frac{2.6 - 2}{1.5 - 1} = \frac{.4}{.5} = .8,$$

or

$$y = 2 + .8(x - 1).$$

When $x = c = 1.3$, we obtain

$$y_c = 2 + .8(1.3 - 1) = 2.24.$$

An approximation to $f(1.3)$ is 2.24.

Problem. Use linear interpolation to approximate $\sqrt{3}$.

Solution. We see that $\sqrt{3} = f(3)$, where $f(x) = \sqrt{x}$. The values of f at $x = 1$ and $x = 4$ are easily determined since $f(1) = \sqrt{1} = 1$ and $f(4) = \sqrt{4} = 2$. Let us take $a = 1$, $b = 4$, and $c = 3$. The interpolation formula gives

$$y_c = 1 + \frac{2-1}{4-1}(3-1)$$

$$= 1.67 \text{ (approximately).}$$

Therefore, an approximation to $\sqrt{3}$ is 1.67.

A table of natural logarithms is supplied in Appendix II. This table is a listing of ln x, rounded off to four decimal places, for selected values of x. To some extent the use of logarithms in computations will be restricted by the length of the table. However, we can use linear interpolation to approximate the logarithm of a number lying between two consecutive entries in the table. Suppose, for example, that a and b are two consecutive values of x listed in the table. Then ln a and ln b can be found in the table. If $a < c < b$, then ln c can be approximated by applying linear interpolation to $f(x) = \ln x$ on the interval $[a, b]$ (see Figure 5.10).

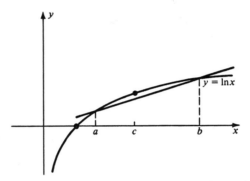

Figure 5.10

Our earlier work shows that

$$y_c = \ln a + \frac{\ln b - \ln a}{b-a}(c-a)$$

is an approximation to ln c.

Problem. Use linear interpolation to approximate ln 4.07.

Solution. From the table in Appendix II, we have ln 4.0 = 1.3863 and ln 4.1 = 1.4110. Therefore, the number whose logarithm we wish to find falls between two consecutive entries in the table. Thus $a = 4.0, b = 4.1$, and $c = 4.07$. Therefore, an approximation to ln 4.07 is

$$\ln 4.0 + \frac{\ln 4.1 - \ln 4.0}{4.1 - 4.0} (4.07 - 4.0) = 1.3863 + \frac{1.4110 - 1.3863}{.1} (.07)$$

$$= 1.4036$$

rounded off to four decimal places.

Problem. Use linear interpolation to approximate ln .76.

Solution. Again, from Appendix II, ln .7 = -.3567 and ln .8 = -.2231. There-fore, a = .7, b = .8, and c = .76. An approximation to ln .76 is

$$\ln .7 + \frac{\ln .8 - \ln .7}{.8 - .7} (.76 - .7) = -.3567 + \frac{-.2231 - (-.3567)}{.1} (.06)$$

$$= -.2765$$

rounded off to four decimal places.

Linear interpolation can also be used to find a number whose logarithm is known. Suppose that a and b are two consecutive values of x in the table of logarithms and that c is a number, to be determined, with the property that $\ln a < \ln c < \ln b$. Now $e^{\ln a} = a$ and $e^{\ln b} = b$. If we apply linear interpolation to the exponential function on the interval $[\ln a, \ln b]$, then $e^{\ln c} = c$ can be approximated. If this is done using the interpolation formula, it is seen that an approximation to c is

$$e^{\ln a} + \frac{e^{\ln b} - e^{\ln a}}{\ln b - \ln a} (\ln c - \ln a) = a + \frac{b - a}{\ln b - \ln a} (\ln c - \ln a).$$

The number c is sometimes called the *antilogarithm* of ln c. Therefore linear interpolation gives us a means for approximating antilogarithms.

Problem. Use linear interpolation to compute the antilogarithm of 1.9508.

Solution. From Appendix II we see that ln 7 = 1.9459 and ln 7.1 = 1.9601. Therefore,

$$\ln 7 < \ln c < \ln 7.1,$$

where ln c = 1.9508. By the interpolation formula, the antilogarithm of 1.9508 is approximately

$$7 + \frac{7.1 - 7}{1.9601 - 1.9459} (1.9508 - 1.9459) = 7 + \frac{.1}{.0142} (.0049)$$

$$= 7.03$$

rounded off to two decimal places. Therefore c approximately equals 7.03.

Problem. Use linear interpolation to compute the antilogarithm of .4441.

Solution. Using Appendix II we have ln 1.5 = .4055 and ln 1.6 = .4700. Therefore,

$$\ln 1.5 < \ln c < \ln 1.6,$$

where ln c = .4441. The antilogarithm of .4441 is approximately

$$1.5 + \frac{1.6 - 1.5}{.4700 - .4055} \; (.4441 - .4055) = 1.5 + \frac{.1}{.0645} \; (.0386)$$

$$= 1.56$$

rounded off to two decimal places. Thus, c is approximately 1.56.

Exercise Set 5.4

1. Suppose the following table of values is given:

x	0	$\frac{1}{2}$	1	$\frac{3}{2}$	2
$f(x)$	1	$\frac{5}{4}$	2	$\frac{13}{4}$	5

Use linear interpolation to approximate $f\left(\frac{1}{4}\right)$, $f\left(\frac{3}{4}\right)$, $f\left(\frac{5}{4}\right)$, and $f\left(\frac{7}{4}\right)$.

2. Given that the function $f(x)$ in problem 1 is determined by the equation $f(x) = x^2 + 1$, find the error incurred by using linear interpolation in each case.

3. For a function f, it is known that $f(2) = 3.43$ and $f(2.1) = 3.51$. Use linear interpolation to approximate
   a. $f(2.02)$ b. $f(2.06)$ c. $f(2.08)$

4. For a function g, it is known that $g(-1.8) = 2.46$ and $g(-1.7) = 2.08$. Use linear interpolation to approximate
   a. $g(-1.71)$ b. $g(-1.76)$ c. $g(-1.79)$

5. Use the tables in Appendix II and linear interpolation to approximate
   a. ln 4.03   b. ln 0.86

6. Use the tables in Appendix II and linear interpolation to approximate the antilogarithm of
   a. 0.7701   b. 1.3191

* * *

In problems 7–12, use linear interpolation and the tables in Appendix II to approximate the indicated logarithm.

7. ln 3.18   8. ln 4.44   9. ln 2.27

10. ln 1.56   11. ln 3.55   12. ln 0.73

In problems 13–16, use linear interpolation and the tables in Appendix II to approximate x.

13. ln $x = 1.6734$ 14. ln $x = 0.6435$

15. ln $x = -0.1983$ 16. ln $x = 2.1200$

In problems 17–20, use linear interpolation and the tables in Appendix I to approximate e^x for the specified x.

17. $e^{1.33}$ 18. $e^{3.77}$

19. $e^{-0.57}$ 20. $e^{-0.42}$

21. Use linear interpolation to approximate $\sqrt[3]{\dfrac{11}{15}}$ and $\sqrt[3]{\dfrac{1}{39}}$.

5.5 COMPUTATION WITH LOGARITHMS

In this section we examine how properties (i)-(iii) of logarithms, as stated in Theorem 5.1, can be used in computations. Recall that these properties were the following:

(i) $\ln ab = \ln a + \ln b$,

(ii) $\ln \dfrac{a}{b} = \ln a - \ln b$,

and   (iii) $\ln a^b = b \ln a$.

Notice that in (i) and (ii), the problem of finding the logarithm of a product and quotient is reduced to a problem of adding or subtracting logarithms. In a sense, the operations of multiplication and division can be replaced by the simpler operations of addition and subtraction. Likewise, (iii) shows that the problem of finding the logarithm of a number raised to a power can be reduced to a multiplication problem.

Problem. Use logarithms to find $(.9)(3.4)$.

Solution. Let $N = (.9)(3.4)$. From Appendix II we have $\ln .9 = -.1054$ and $\ln 3.4 = 1.2238$. Therefore

$$
\begin{aligned}
\ln N &= \ln (.9)(3.4) \\
&= \ln .9 + \ln 3.4 \\
&= -.1054 + 1.2238 \\
&= 1.1184.
\end{aligned}
$$

We are now in a situation where $\ln N$ is known. Our computations show that $\ln N = 1.1184$. In order to find N, we must find the antilogarithm of 1.1184.

Since $\ln 3.0 = 1.0986$ and $\ln 3.1 = 1.1314$, the antilogarithm of N is approximately

$$3 + \frac{3.1 - 3}{1.1314 - 1.0986}(1.1184 - 1.0986) = 3.06$$

rounded off to two decimal places. Therefore N is approximately 3.06. Actual multiplication shows that $N = 3.06$.

Problem. Use logarithms to find $\frac{3.5}{.93}$.

Solution. Let $N = \frac{3.5}{.93}$. Thus

$$\ln N = \ln \frac{3.5}{.93}$$
$$= \ln 3.5 - \ln .93$$
$$= 1.2528 - \ln .93.$$

The number $\ln .93$ is computed by linear interpolation. Since $\ln .9 = -.1054$ and $\ln 1 = 0$, an approximation to $\ln .93$ is

$$-.1054 + \frac{\ln 1 - \ln (.9)}{1 - .9}(.93 - .9) = -.0738$$

rounded off to four decimal places. Then, approximately,

$$\ln N = 1.2528 - (-.0738)$$
$$= 1.3266.$$

Finding the antilogarithm of 1.3266, we see that N is approximately

$$3.7 + \frac{3.8 - 3.7}{1.3350 - 1.3083}(1.3266 - 1.3083) = 3.77$$

rounded off to two decimal places. Actual division of 3.5 by .93 shows that $N = 3.763$ rounded off to three decimal places.

The preceding two problems could have been more easily solved simply by multiplying or dividing directly. This is not the case when it is necessary to extract roots of numbers.

Problem. Use logarithms to find $\sqrt[3]{2.5}$.

Solution. Let $N = \sqrt[3]{2.5} = (2.5)^{\frac{1}{3}}$. Thus $\ln 2.5 = .9163$ so that

$$\ln N = \ln (2.5)^{\frac{1}{3}}$$
$$= \frac{1}{3} \ln 2.5$$

$$= \frac{1}{3}(.9163)$$

$$= .3054$$

rounded off to four decimal places. Finding the antilogarithm of .3054 shows that N is approximately

$$1.3 + \frac{1.4 - 1.3}{.3365 - .2624}(.3054 - .2624) = 1.36$$

rounded off to two decimal places. It is easily checked that

$$(1.36)^3 = 2.515456.$$

In any given computation it may happen that most of the numbers involved appear between two consecutive entries in the table. In such cases it will be necessary to use linear interpolation to find these logarithms. Once this is done, the rest of the computation can proceed as usual.

Problem. Find $\dfrac{(4.07)\sqrt{.72}}{2.16}$.

Solution. If we let $N = \dfrac{(4.07)(\sqrt{.72})}{2.16}$, then

$$\ln N = \ln 4.07 + \frac{1}{2}\ln .72 - \ln 2.16.$$

Using linear interpolation and rounding off to four decimal places,

$$\ln 4.07 = 1.3863 + \frac{1.4110 - 1.3863}{4.1 - 4}(4.07 - 4) = 1.4036,$$

$$\ln .72 = -.3567 + \frac{-.2231 - (-.3567)}{.8 - .7}(.72 - .7) = -.3300,$$

and $\qquad \ln 2.16 = .7419 + \dfrac{.7885 - .7419}{2.2 - 2.1}(2.16 - 2.1) = .7699.$

Therefore,

$$\ln N = 1.4036 + \frac{1}{2}(-.3300) - .7699$$

$$= .4687$$

The antilogarithm of .4687 is approximately

$$1.5 + \frac{1.6 - 1.5}{.4700 - .4055}(.4687 - .4055) = 1.598 \text{ or } 1.60$$

rounded off to two decimal places. Therefore $\dfrac{(4.07)\sqrt{.72}}{2.16}$ is approximately equal to 1.60.

We now mention briefly a certain type of function which arises when considering mathematical models for simple kinds of population growth. The population could be the number of bacteria in a culture, the number of people in a country, etc. We let $P(t)$ denote the size of the population at time t. If the population increases at a rate which is proportional to the size of the population, then it can be shown (by use of the calculus) that $P(t)$ satisfies an equation of the form

$$P(t) = P_0 e^{kt},$$

where P_0 and k are positive constants. When $t = 0$,

$$P(0) = P_0 e^{k \cdot 0} = P_0 e^0 = P_0.$$

In other words, P_0 is the size of the population at time $t = 0$. Therefore P_0 is usually called the initial population. The constant k equals the constant of proportionality that determines the growth rate.

Example. Assume the world contains sufficient arable land to provide food for at most 40 billion people. Suppose the population of the world in 1965 was 3.15 billion and the population increases at a rate equal to 2.5% of the population. If we let $P(t)$ denote the population of the world at time t, then we will have

$$P(t) = P_0 e^{.025t}.$$

If we let $t = 0$ in the year 1965, then P_0 is the size of the population in 1965. Therefore, measured in billions,

$$P(t) = (3.15)e^{.025t}.$$

Assuming that the growth continues at the same rate, let us determine when the saturation point will be reached. That is, we wish to determine t so that $P(t) = 40$. For this value of t we have

$$40 = (3.15)e^{.025t}.$$

It follows that

$$\ln 40 = \ln 3.15 + \ln e^{.025t}$$
$$= \ln 3.15 + .025t.$$

We see that $\ln 40 = \ln 4 + \ln 10 = 1.3863 + 2.3026 = 3.6889$. Using linear interpolation, we have, approximately,

$$\ln 3.15 = 1.1314 + \frac{1.1632 - 1.1314}{3.2 - 3.1}(3.15 - 3.10)$$
$$= 1.1473.$$

Therefore,
$$.025t = 3.6889 - 1.1473$$
$$= 2.5416.$$

Solving for t, we have $t = \dfrac{2.5416}{.025}$, which is about 102. Therefore, under these assumptions, the saturation point would be reached in the year 2067.

Let us close this section by showing how logarithms can be used to compute the value of an investment on which compound interest is paid. Suppose, for instance, that a bank pays r percent interest compounded n times a year on its savings accounts. This means that there are n interest periods during the year. At the end of each interest period $\dfrac{r}{n}$ percent interest is paid on an investment left in a savings account for this interest period. Thus, if a person deposits an amount A_0 initially, then at the end of the first interest period the account is worth

$$A_0 + A_0 \frac{r}{100n} = A_0 \left(1 + \frac{r}{100n}\right).$$

If this amount is left in the account for the second interest period, the account is then worth

$$A_0 \left(1 + \frac{r}{100n}\right) + A_0 \left(1 + \frac{r}{100n}\right)\frac{r}{100n} = A_0 \left(1 + \frac{r}{100n}\right)^2.$$

In general, if an amount A_0 is deposited in a savings account and this account is left untouched for m interest periods, the account is worth an amount A given by the formula

$$A = A_0 \left(1 + \frac{r}{100n}\right)^m.$$

The number A can be computed by the use of logarithms as the following problem shows.

Problem. A bank pays 5% interest compounded quarterly (i.e., four times a year) on its savings accounts. If a person deposits three dollars in a savings account and leaves it there for two years, what is the account worth at that time?

Solution. Two years amounts to eight interest periods. Thus, the value A of the account at the end of two years is

$$A = 3\left(1 + \frac{5}{(100)4}\right)^8$$
$$= 3\left(1 + \frac{1}{80}\right)^8$$
$$= 3\left(\frac{81}{80}\right)^8.$$

Therefore,

$$\ln A = \ln\left[3\left(\frac{81}{80}\right)^8\right]$$

$$= \ln 3 + 8 \ln\left(\frac{81}{80}\right)$$

$$= \ln 3 + 8 \,[\ln 81 - \ln 80]$$

$$= \ln 3 + 8 \,[\ln (10)(8.1) - \ln (10)(8)]$$

$$= \ln 3 + 8 \,[\ln 10 + \ln 8.1 - \ln 10 - \ln 8]$$

$$= \ln 3 + 8 \,[\ln 8.1 - \ln 8]$$

$$= 1.0986 + 8 \,[2.0919 - 2.0794]$$

$$= 1.1986.$$

We see that $\ln 3.3 = 1.1939$ and $\ln 3.4 = 1.2238$. Thus, the antilogarithm A of 1.1986 is approximately

$$3.3 + \frac{3.4 - 3.3}{1.2238 - 1.1939}\,(1.1986 - 1.1939) = 3.32$$

rounded off to two decimal places. Hence, at the end of two years, the account is worth \$3.32.

Exercise Set 5.5

1. Use logarithms to find the indicated product:
   a. $(1.32)(2.6)$ b. $(2.89)(0.76)$

2. Use logarithms to find the indicated quotient:
   a. $\dfrac{1.32}{2.6}$ b. $\dfrac{2.89}{0.76}$

3. Check your answers to problems 1 and 2 by using ordinary multiplication and division.

4. Use logarithms to find the indicated cube root:
   a. $\sqrt[3]{1.32}$ b. $\sqrt[3]{3.75}$

5. Check your answers to problem 4 by cubing the result.

* * *

In problems 6–17, use logarithms to evaluate the given expression.

6. $(1.44)(1.68)$ 7. $(4.32)(0.84)$

8. $\dfrac{1.44}{1.68}$ 9. $\dfrac{0.84}{4.32}$

10. $\sqrt{3.14}$ 11. $\sqrt{5.21}$

12. $\dfrac{1}{3.38}$

13. $\dfrac{1}{4.69}$

14. $\dfrac{(0.88)\,(2.08)}{1.9}$

15. $\sqrt{\dfrac{1.9}{(0.88)\,(2.08)}}$

16. $\dfrac{2.2}{\sqrt{1.5}}$

17. $(3.44)^{\frac{2}{3}}$

18. If the number of bacteria in a culture at time t is given by the growth equation $P(t) = P_0 e^{4t}$, where t is time measured in days, when will the culture double its initial size?

19. A bank pays 5% interest on its savings accounts. If a person deposits three dollars in a savings account and leaves it there for two years, what is the account worth at that time if the interest is compounded:
    a) twice a year,
    b) three times a year,
    c) five times a year,
    d) six time a year?

20. A bank pays 7% interest on its savings accounts and the interest is compounded twice a year. If a person deposits five dollars in a savings account and leaves it there for six years, what is the account worth at that time?

6 / *Differential Calculus*

One of our main interests in this text is the study of functions and their graphs. The graphs of certain functions, such as linear or quadratic functions, are easily determined. We have previously seen that the graph of a linear function is a straight line while the graph of a quadratic function is a parabola. Although we have had some success in completely determining the graphs of a few special functions, we have also encountered some obstacles along the way. It has been noted, for example, that the accurate sketching of graphs of polynomials and, more generally, algebraic functions is a difficult task. This is the case because we simply do not have the mathematical tools needed to tackle this problem at the present time.

In this chapter the differential calculus is developed. For the time being we will be concerned primarily with the mechanical aspects of differentiation. While our discussion may not seem at all connected with the problem of curve sketching, we will actually be developing the standard machinery needed to sketch curves accurately. In Chap. 7 it will become apparent that the differential calculus is an indispensible tool which can be used to help us in graphing functions and in solving many other problems.

6.1 LIMITS

So far, the mathematical material we have examined has been primarily of an algebraic and geometric nature. Is calculus essentially more of the same? The answer is no. This is because the concept which plays the leading role in the calculus is that of a limit. It is the notion of a limit which distinguishes the calculus from algebra and geometry and lends an entirely different flavor to the remaining material of this text.

Given a function f and a number a, what is the behavior of $f(x)$ as x gets close to a? Consideration of this question will lead us to the limit concept. Let us take an informal look at this idea, using several specific examples.

Example. Let $f(x) = \frac{1}{2}x + 1$ and consider the following table of values of x and $f(x)$:

x	2.5	2.9	2.95	2.99	2.999	3.001	3.01	3.05	3.1	3.5
$f(x)$	2.25	2.45	2.475	2.495	2.4995	2.5005	2.505	2.525	2.55	2.75

The values of x in the table are close to 3. What is the behavior of the number $f(x)$ if x is close to 3? From the table we see that the closer x is to 3, the closer $f(x)$ is to 2.5. We might phrase this as follows: $f(x)$ approaches 2.5 as x approaches 3. In this case note that $f(3) = 2.5$. The graph of this function is shown in Figure 6.1 with some of the preceding points labeled.

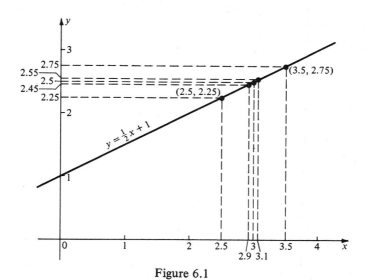

Figure 6.1

Example. Let

$$f(x) = \begin{cases} x^2 + 2, & \text{if } x \text{ is not an integer} \\ 1, & \text{if } x \text{ is an integer.} \end{cases}$$

Consider the following table of values for x and $f(x)$:

x	-.5	-.2	-.01	.01	.2	.5
$f(x)$	2.25	2.04	2.0001	2.0001	2.04	2.25

In examining the behavior of $f(x)$ as x gets close to 0 we see that $f(x)$ approaches 2. Simply stated, $f(x)$ approaches 2 as x approaches 0. Note that $f(0) = 1$, thus, the value of $f(x)$ at $x = 0$ presumably has no bearing on the fact that $f(x)$ approaches a value as x approaches 0. The graph of this function is shown in Figure 6.2.

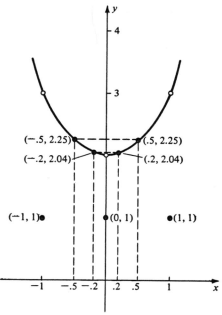

Figure 6.2

Example. Let $f(x) = \dfrac{-x^2 + 3x - 2}{x - 2}$. This function is not defined for $x = 2$. Hence, the domain of this function is $\{x: x \neq 2\}$. Note that $-x^2 + 3x - 2 = (x - 2)(1 - x)$. Consequently, when $x \neq 2$, we have $f(x) = \dfrac{(x - 2)(1 - x)}{x - 2} = 1 - x$.

Again, consider a table of values for x and $f(x)$:

x	1.5	1.8	1.9	1.99	1.999	2.001	2.01	2.1	2.2	2.5
$f(x)$	-.5	-.8	-.9	-.99	-.999	-1.001	-1.01	-1.1	-1.2	-1.5

As in the preceding examples, a reasonable interpretation of the data in the table is that $f(x)$ approaches -1 as x approaches 2. In this case, however, the function is not defined at $x = 2$. The graph of this function is shown in Figure 6.3.

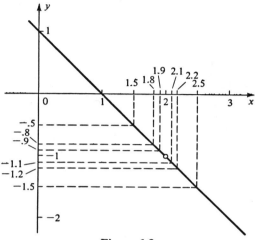

Figure 6.3

What are these examples really describing? In each case there are real numbers a and L such that $f(x)$ approaches L as x approaches a. Thus, in the first example $\left(\text{where } f(x) = \dfrac{1}{2}x + 1\right)$, $f(x)$ approaches $L = 2.5$ as x approaches $a = 3$. In the second example, $f(x)$ approaches $L = 2$ as x approaches $a = 0$. In the third example, we see that $f(x)$ approaches $L = -1$ as x approaches $a = 2$. A natural way to think of the number L is as a limiting value of the number $f(x)$ as x approaches a. In addition to giving us an intuitive look at this limiting process, the examples also point out that we must be careful in trying to determine L. Indeed, one might think that we always have $L = f(a)$. However, the second example shows that this is not the case. Moreover, the third example shows that the function may not even be defined at $x = a$. Thus, in trying to determine a limiting value of the number $f(x)$ as x approaches a, we can not necessarily let $x = a$ in the expression for $f(x)$.

Let us now try to make the idea of a limiting value of $f(x)$, as x approaches a, more precise. What does the phrase

"$f(x)$ approaches L as x approaches a"

really mean? To "approach" means "to get arbitrarily close to" in our particular setting. Hence, we mean "$f(x)$ get arbitrarily close to L as x gets sufficiently close to a." Note the dependency here. Namely, the closeness of $f(x)$ to L depends on the closeness of x to a. Let us examine this dependency more carefully.

Consider the first example again. We said that $\frac{1}{2}x + 1$ approaches 2.5 as x approaches 3. This means that $\frac{1}{2}x + 1$ gets arbitrarily close to 2.5 as x gets arbitrarily close to 3. Suppose we want $\frac{1}{2}x + 1$ to be within a distance $\frac{1}{10}$ of 2.5. That is, suppose we want

$$\left| \left(\frac{1}{2}x + 1 \right) - \frac{5}{2} \right| < \frac{1}{10}.$$

This is equivalent to

$$\left| \frac{1}{2}x - \frac{3}{2} \right| < \frac{1}{10},$$

which, in turn, is equivalent to

$$|x - 3| < \frac{1}{5}.$$

In other words, in order to have

$$\left| \left(\frac{1}{2}x + 1 \right) - \frac{5}{2} \right| < \frac{1}{10} \text{ we must have } |x - 3| < \frac{1}{5}.$$

Moreover, if $|x - 3| < \frac{1}{5}$ then $\left| \left(\frac{1}{2}x + 1 \right) - \frac{5}{2} \right| < \frac{1}{10}$. That is, if x is within a distance $\frac{1}{5}$ of 3, then $\frac{1}{2}x + 1$ is within a distance $\frac{1}{10}$ of 2.5. This is depicted in Figure 6.4, where the values of x within $\frac{1}{5}$ of 3 lie in the darkly shaded horizontal interval and the corresponding values of $\frac{1}{2}x + 1$ belong to the darkly shaded vertical interval, within $\frac{1}{10}$ of $\frac{5}{2}$.

In the second example, we said that $f(x)$ approaches 2 as x approaches 0. This means that $f(x)$ gets arbitrarily close to 2 as x gets close to 0. Suppose we require that $f(x)$ be within $\frac{1}{10000}$ of 2. That is, we want $|f(x) - 2| < \frac{1}{10000}$. Since $f(0) = 1$, we will not have $|f(0) - 2| < \frac{1}{10000}$. This goes back to the remark that we are not really interested in the behavior of $f(x)$ at $x = 0$ but only the behavior of $f(x)$ as x gets close to 0. Now if $x \neq 0$ and x is not an integer,

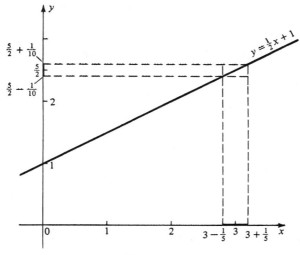

Figure 6.4

then we see from the way $f(x)$ is defined in this example that

$$|f(x) - 2| = |(x^2 + 2) - 2| = |x^2| = |x|^2.$$

In order to have $|x|^2 < \dfrac{1}{10000}$, it suffices to require $|x| < \dfrac{1}{100}$. Thus $|f(x) - 2| <$ $\dfrac{1}{10000}$ whenever $0 < |x| < \dfrac{1}{100}$. This situation is shown in Figure 6.5. Note that $0 < |x|$ merely guarantees that $x \neq 0$.

Before stating the formal definition of limit, let us summarize our observations. We have seen that the phrase

$$\text{``}f(x) \text{ approaches } L \text{ as } x \text{ approaches } a\text{''}$$

entails two main ideas. First, is the idea of dependency. In order to make $f(x)$ lie within a given distance from L, we have to describe how close x must be to a. Second, we do not want $x = a$. Thus, in the first example, given the number $\dfrac{1}{10}$, we were to find another positive number, call it δ (δ is the lower case Greek letter delta), such that $\left|\left(\dfrac{1}{2}x + 1\right) - \dfrac{5}{2}\right| < \dfrac{1}{10}$ whenever $0 < |x - 3| < \delta$. We found that we could let $\delta = \dfrac{1}{5}$. In the second example, given the number $\dfrac{1}{10000}$, we were to find another positive number δ such that $|f(x) - 2| < \dfrac{1}{10000}$ whenever $0 < |x - 0| = |x| < \delta$, where f is the function in the example. We found that we

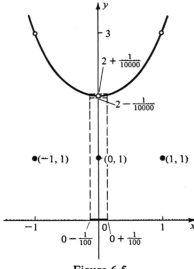

Figure 6.5

could let $\delta = \dfrac{1}{100}$. Of course, if the given numbers $\dfrac{1}{10}$ or $\dfrac{1}{10000}$ had been different, the number δ might also have been different. In general, when we assert
that $f(x)$ approaches L as x approaches a, we are stating that $f(x)$ can be made to
lie within a preassigned distance from L by taking x close enough to a (i.e., by
finding a suitable δ).

We now state the formal definition of limit. In what follows, the symbol ϵ denotes the lower case Greek letter epsilon.

6.1 Definition. *Let a and L be real numbers. IF for each $\epsilon > 0$, there exists a
$\delta > 0$ such that $|f(x) - L| < \epsilon$ whenever $0 < |x - a| < \delta$, THEN $\lim\limits_{x \to a} f(x) = L$,
read "the limit of $f(x)$, as x approaches a, is L."*

The formal definition of limit can be difficult to apply to particular functions.
For our purposes, it will not be necessary to resort to the strict ϵ, δ formulation
in order to find limits. That is, we will normally assert that $\lim\limits_{x \to a} f(x) = L$, when

this is the case, because of intuitive considerations.

Example. (i) We see now that $\lim\limits_{x \to 3} \left(\dfrac{1}{2} x + 1 \right) = \dfrac{5}{2}$. We read this as "the limit of

$\dfrac{1}{2} x + 1$, as x approaches 3, is $\dfrac{5}{2}$." This was our first example in

this section.

(ii) In our second example we let $f(x) = \begin{cases} x^2 + 2, \text{ if } x \text{ is not an integer} \\ 1, \text{ if } x \text{ is an integer.} \end{cases}$

Thus $\lim_{x \to 0} f(x) = 2$. This statement reads "the limit of $f(x)$, as x approaches 0, is 2."

(iii) For our third example, $\lim_{x \to 2} \dfrac{-x^2 + 3x - 2}{x - 2} = -1$, i.e., "the limit of $\dfrac{-x^2 + 3x - 2}{x - 2}$, as x approaches 2, is -1."

(iv) If $f(x) = x$ and a is any real number then

$$\lim_{x \to a} f(x) = \lim_{x \to a} x = a.$$

(v) If a and c are real numbers and $f(x) = c$ for all x then

$$\lim_{x \to a} f(x) = \lim_{x \to a} c = c.$$

Exercise Set 6.1

1. Let   a. $f(x) = x$,   b. $f(x) = -3x$,   and   c. $f(x) = \dfrac{(x + 3)(x + 1)}{x + 1}$.

    Complete the following statements for *each* of the given functions:

    If x is close to 2 but not equal to 2, then $f(x)$ is close to _____.

    If x is close to $\dfrac{4}{3}$ but not equal to $\dfrac{4}{3}$, then $f(x)$ is close to _____.

    If x is close to -1, but not equal to -1, then $f(x)$ is close to _____.

2. Let   a. $f(x) = x$,   b. $f(x) = -3x$,   and   c. $f(x) = \dfrac{(x + 3)(x + 1)}{x + 1}$.

    Complete the following statements for *each* of the given functions:

    $L = \lim_{x \to 2} f(x) = $ _____. Is this equal to $f(x)$ evaluated at $x = 2$?

    $L = \lim_{x \to \frac{4}{3}} f(x) = $ _____. Is this equal to $f(x)$ evaluated at $x = \dfrac{4}{3}$?

    $L = \lim_{x \to -1} f(x) = $ _____. Is this equal to $f(x)$ evaluated at $x = -1$?

3. Let   a. $f(x) = x$,   b. $f(x) = -3x$,   and   c. $f(x) = \dfrac{(x + 3)(x + 1)}{x + 1}$.

    For each of the given functions, by trial and error, determine an x_0, such that:
The difference between $f(x_0)$ and $\lim_{x \to 2} f(x)$ is less than .05.

The difference between $f(x_0)$ and $\lim\limits_{x \to \frac{4}{3}} f(x)$ is less than .05.

The difference between $f(x_0)$ and $\lim\limits_{x \to -1} f(x)$ is less than .05.

4. Let $f(x) = x^2$.

   a. Complete the following table of values:

x	1.5	1.6	1.7	1.71	1.73	1.732	1.734	1.74	1.75	1.8	1.9
$f(x)$											

   b. If x is close to $\sqrt{3}$, then $f(x)$ is close to _____.

   c. If $L = \lim\limits_{x \to \sqrt{3}} x^2$, then $L =$ _____.

5. Let $f(x) = \begin{cases} 2 & \text{if } x \text{ is rational} \\ 0 & \text{if } x \text{ is irrational.} \end{cases}$

   a. Complete the following table of values:

x	.5	$\dfrac{\sqrt{2}}{2}$	$\dfrac{4}{5}$	$\dfrac{\sqrt{3}}{2}$	$\dfrac{7}{8}$	$\dfrac{\sqrt[3]{6}}{2}$	$\dfrac{\pi}{3}$	$\dfrac{4}{3}$	$\dfrac{7}{5}$	$\sqrt{2}$	1.5
$f(x)$											

   b. Does this function have a limit as x approaches 1? If so, what is it?

* * *

In problems 6–17, find the limits, using intuitive considerations.

6. $\lim\limits_{x \to 3} (6x + 1)$

7. $\lim\limits_{x \to \frac{2}{3}} (-3x + 4)$

8. $\lim\limits_{x \to \frac{1}{2}} 7$

9. $\lim\limits_{x \to \frac{1}{2}} \sqrt{3}$

10. $\lim\limits_{x \to a} c$

11. $\lim\limits_{x \to b} k$

12. $\lim\limits_{x \to 1} \dfrac{x^2 - 1}{x - 1}$

13. $\lim\limits_{x \to -1} \dfrac{x^2 - 1}{x + 1}$

14. $\lim\limits_{x \to 1} \dfrac{x^2 + x - 6}{x - 2}$

15. $\lim\limits_{x \to 2} \dfrac{x^2 + x - 6}{x - 2}$

16. $\lim\limits_{x \to 1} f(x)$ if $f(x) = \begin{cases} x, & \text{if } x \neq 1 \\ -17, & \text{if } x = 1 \end{cases}$

17. $\lim_{x \to -3} g(x)$ if $g(x) = \begin{cases} x^2, \text{ if } x \text{ is an integer} \\ 0, \text{ if } x \text{ is not an integer} \end{cases}$

In problems 18–23, given f, L, ϵ, and a, find δ such that $|f(x) - L| < \epsilon$ whenever $0 < |x - a| < \delta$. Draw a picture which depicts each of these situations.

18. $f(x) = x + 1$, $L = 2$, $\epsilon = \dfrac{1}{3}$, $a = 1$

19. $f(x) = x + 1$, $L = 2$, $\epsilon = \dfrac{1}{1000}$, $a = 1$

20. $f(x) = 4 - 3x$, $L = 10$, $\epsilon = \dfrac{1}{100}$, $a = -2$

21. $f(x) = 4 - 3x$, $L = -2$, $\epsilon = \dfrac{1}{100}$, $a = 2$

22. $f(x) = x^2$, $L = 0$, $\epsilon = \dfrac{1}{4}$, $a = 0$

23. $f(x) = x^2$, $L = 0$, $\epsilon = \dfrac{1}{100}$, $a = 0$

24. In the definition of limit why is $|x - a| > 0$ rather than $|x - a| \geq 0$?

6.2 LEFT-HAND AND RIGHT-HAND LIMITS

Consider the function $f(x) = \begin{cases} x + 2, \text{ if } x \leq 1 \\ 3x + 1, \text{ if } x > 1 \end{cases}$ for values of x close to 1.

By examining the table of values

x	0	.5	.9	.99	.999	1.001	1.01	1.1	1.5
$f(x)$	2	2.5	2.9	2.99	2.999	4.003	4.03	4.3	5.5

we see that $\lim_{x \to 1} f(x)$ does not exist (see Figure 6.6). However, if we consider values of x such that $x > 1$ we have the table

x	1.001	1.01	1.1	1.5
$f(x)$	4.003	4.03	4.3	5.5

Thus, by restricting our consideration to values of x where $x > 1$, we see that the function $f(x)$ approaches 4 as x approaches 1. Here we say that we are considering values of x "to the right of 1" and we write

$$\lim_{x \to 1^+} f(x) = 4.$$

This statement reads "the limit of $f(x)$ as x approaches 1 from the right is 4."

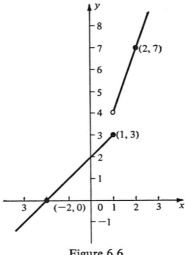

Figure 6.6

Now, considering values of x where $x < 1$, we have the table

x	0	.5	.9	.99	.999
$f(x)$	2	2.5	2.9	2.99	2.999

Hence, by restricting ourselves to values of x where $x < 1$, we have that the function $f(x)$ approaches 3 as x approaches 1. Here we say that we are considering values of x "to the left of 1" and we write

$$\lim_{x \to 1^-} f(x) = 3.$$

We read this as "the limit of $f(x)$ as x approaches 1 from the left is 3." Notice that the *right-hand limit*, $\lim_{x \to 1^+} f(x)$, and the *left-hand limit*, $\lim_{x \to 1^-} f(x)$, differ in this example.

In Sec. 6.1 we considered the behavior of the function

$$f(x) = \frac{1}{2}x + 1$$

for values of x close to 3 by examining the tables

x	2.5	2.9	2.95	2.99	2.999
$f(x)$	2.25	2.45	2.475	2.495	2.4995

and

x	3.001	3.01	3.05	3.1	3.5
$f(x)$	2.5005	2.505	2.525	2.55	2.75

In the first table values of x such that $x < 3$ are considered and in the second table we list values of $x > 3$. In order that

$$\lim_{x \to 3} f(x) = 2.5$$

we see that $f(x)$ must approach the number 2.5 when x gets close to 3 for both cases $x < 3$ and $x > 3$. That is,

$$\lim_{x \to 3} f(x) = 2.5$$

provided

$$\lim_{x \to 3^-} f(x) = 2.5 \text{ and } \lim_{x \to 3^+} f(x) = 2.5.$$

From our two tables, we conclude that this is the case, i.e., the left-hand limit, $\lim_{x \to 3^-} f(x)$, is 2.5 and the right-hand limit, $\lim_{x \to 3^+} f(x)$, is 2.5. Therefore, $\lim_{x \to 3} f(x) = 2.5$.

By combining the concepts of left-hand and right-hand limits with Definition 6.1 the following theorem can be proved.

6.2 Theorem. *IF a and L are real numbers, THEN*

$$\lim_{x \to a} f(x) = L$$

if, and only if,

$$\lim_{x \to a^+} f(x) = L \text{ and } \lim_{x \to a^-} f(x) = L.$$

Problem. Let $f(x) = \begin{cases} 2x + 3, \text{ if } x > 2 \\ 2 - x, \text{ if } x \leqslant 2. \end{cases}$

Find $\lim_{x \to 2^-} f(x)$, $\lim_{x \to 2^+} f(x)$, and determine if $\lim_{x \to 2} f(x)$ exists.

Solution. To find $\lim_{x \to 2^-} f(x)$ consider values of $x < 2$. For $x < 2$ we have that $f(x) = 2 - x$ and

x	1	1.5	1.9	1.99	1.999
$f(x)$	1	.5	.1	.01	.001

From this we conclude that $\lim_{x \to 2^-} f(x) = 0$.

To find $\lim_{x \to 2^+} f(x)$ consider values of $x > 2$. For $x > 2$ we have that $f(x) = 2x + 3$. By considering the table

x	3	2.5	2.1	2.01	2.001
$f(x)$	9	8	7.2	7.02	7.002

we conclude that $\lim_{x \to 2^+} f(x) = 7$.

Therefore, by Theorem 6.2, $\lim_{x \to 2} f(x)$ does not exist (since the left-hand and right-hand limits are unequal).

Problem. Find $\lim_{x \to -1^-} f(x)$, $\lim_{x \to -1^+} f(x)$, and determine $\lim_{x \to -1} f(x)$ (if this exists) where

$$f(x) = \begin{cases} x + 4, & \text{if } x \geqslant -1 \\ 4x + 7, & \text{if } x < -1. \end{cases}$$

Solution. For $x < -1$ we have $f(x) = 4x + 7$ and $\lim_{x \to -1^-} f(x) = \lim_{x \to -1^-} (4x + 7) = 3$. For $x > -1$ we have $f(x) = x + 4$ and $\lim_{x \to -1^+} f(x) = \lim_{x \to -1^+} (x + 4) = 3$. Since $\lim_{x \to -1^-} f(x) = 3$ and $\lim_{x \to -1^+} f(x) = 3$ we have, from Theorem 6.2, that $\lim_{x \to -1} f(x) = 3$ (see Figure 6.7).

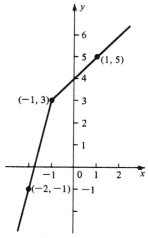

Figure 6.7

Exercise Set 6.2

1. Let $f(x) = \begin{cases} x + 4, & \text{if } x \leqslant 2 \\ 3x, & \text{if } x > 2. \end{cases}$

   a. Complete the following table of values for $f(x)$:

x	1.5	1.9	1.99	1.999	2.001	2.01	2.1	2.5
$f(x)$								

   b. Determine the following:
   $$\lim_{x \to 2^-} f(x), \quad \lim_{x \to 2^+} f(x), \quad \lim_{x \to 2} f(x), \quad f(2).$$

   c. Sketch the graph of $f(x)$.

2. Let $f(x) = \begin{cases} x - 4, & \text{if } x \leqslant 2 \\ -3x, & \text{if } x > 2. \end{cases}$

   a. Complete the following table of values for $f(x)$:

x	1.5	1.9	1.99	1.999	2.001	2.01	2.1	2.5
$f(x)$								

   b. Determine the following:
   $$\lim_{x \to 2^-} f(x), \quad \lim_{x \to 2^+} f(x), \quad \lim_{x \to 2} f(x), \quad f(2).$$

   c. Sketch the graph of $f(x)$.

3. Let $f(x) = \begin{cases} -1, & \text{if } x < 0 \\ 0, & \text{if } x = 0 \\ 1, & \text{if } x > 0. \end{cases}$

   a. Construct a table of values for $f(x)$.
   b. Determine $\lim_{x \to 0^-} f(x), \lim_{x \to 0^+} f(x), \lim_{x \to 0} f(x), \text{ and } f(0)$.
   c. Sketch the graph of $f(x)$.

4. Let $f(x) = x^2 - 3x$.
   a. Construct a table of values for $f(x)$.
   b. Determine $\lim_{x \to 0^-} f(x), \lim_{x \to 0^+} f(x), \lim_{x \to 0} f(x), \text{ and } f(0)$.

* * *

In problems 5–10, determine the indicated limits.

5. $\lim_{x \to -7^-} 2x, \quad \lim_{x \to -7^+} 2x, \quad \lim_{x \to 7} 2x.$

6. $\displaystyle\lim_{x\to 12^-}\frac{25}{23}$, $\displaystyle\lim_{x\to 12^+}\frac{25}{23}$, $\displaystyle\lim_{x\to 12}\frac{25}{23}$.

7. $\displaystyle\lim_{x\to 0^-}\frac{x^2}{x}$, $\displaystyle\lim_{x\to 0^+}\frac{x^2}{x}$, $\displaystyle\lim_{x\to 0}\frac{x^2}{x}$.

8. $\displaystyle\lim_{x\to 0^-}\frac{|x|}{x}$, $\displaystyle\lim_{x\to 0^+}\frac{|x|}{x}$, $\displaystyle\lim_{x\to 0}\frac{|x|}{x}$.

9. $\displaystyle\lim_{x\to -\frac{1}{2}^-}\frac{2}{x}$, $\displaystyle\lim_{x\to -\frac{1}{2}^+}\frac{2}{x}$, $\displaystyle\lim_{x\to -\frac{1}{2}}\frac{2}{x}$.

10. $\displaystyle\lim_{x\to 1^-}\left(\frac{1}{x}-x\right)$, $\displaystyle\lim_{x\to 1^+}\left(\frac{1}{x}-x\right)$, $\displaystyle\lim_{x\to 1}\left(\frac{1}{x}-x\right)$.

In problems 11–16, construct a table of values for the given function, determine the indicated limits, and sketch the graph of the function.

11. $g(x) = \begin{cases} x^2, & \text{if } x \leqslant 1 \\ 2-x, & \text{if } x > 1, \end{cases}$ $\qquad \displaystyle\lim_{x\to 1^-} g(x), \quad \lim_{x\to 1^+} g(x).$

12. $h(x) = \begin{cases} x^2, & \text{if } x < 0 \\ 10, & \text{if } x = 0 \\ x^3, & \text{if } x > 0, \end{cases}$ $\qquad \displaystyle\lim_{x\to 0^-} h(x), \quad \lim_{x\to 0^+} h(x).$

13. $f(x) = [x]$, $\qquad \displaystyle\lim_{x\to 5^-} f(x), \quad \lim_{x\to 5^+} f(x).$

14. $f(x) = x - [x]$, $\qquad \displaystyle\lim_{x\to 1^-} f(x), \quad \lim_{x\to 1^+} f(x).$

15. $f(z) = \dfrac{1}{z^2 + 1}$, $\qquad \displaystyle\lim_{z\to 0^-} f(z), \quad \lim_{z\to 0^+} f(z).$

16. $f(z) = \dfrac{4}{z^2 + 2}$, $\qquad \displaystyle\lim_{z\to -2^-} f(z), \quad \lim_{z\to -2^+} f(z).$

17. If $f(t) = \begin{cases} 2t - 3, & \text{if } t < 1 \\ -5t + 4, & \text{if } t > 1 \end{cases}$, find $\displaystyle\lim_{t\to 1^-} f(t)$ and $\displaystyle\lim_{t\to 1^+} f(t).$

18. If $H(w) = \begin{cases} 8w^2, & \text{if } w < -\dfrac{1}{2} \\ \\ 6w + 1, & \text{if } w \geqslant -\dfrac{1}{2} \end{cases}$, find $\displaystyle\lim_{w\to -\frac{1}{2}^-} H(w)$ and $\displaystyle\lim_{w\to -\frac{1}{2}^+} H(w).$

6.3 PROPERTIES OF LIMITS

There are many elementary facts concerning limits which will be extremely use-ful in reducing the work needed to calculate limits of more complicated func-tions. We state these facts without proof.

6.3 Theorem. *IF* $\lim_{x \to a} f(x) = L$, $\lim_{x \to a} g(x) = M$, *and c is a real number, THEN*

(a) $\lim_{x \to a} [cf(x)] = c \lim_{x \to a} f(x) = cL$,

(b) $\lim_{x \to a} [f(x) + g(x)] = \lim_{x \to a} f(x) + \lim_{x \to a} g(x) = L + M$,

(c) $\lim_{x \to a} [f(x) \cdot g(x)] = [\lim_{x \to a} f(x)] [\lim_{x \to a} g(x)] = LM$,

and (d) $\lim_{x \to a} \dfrac{f(x)}{g(x)} = \dfrac{\lim\limits_{x \to a} f(x)}{\lim\limits_{x \to a} g(x)} = \dfrac{L}{M}$ *(provided $M \neq 0$).*

We now illustrate the use of the different parts of the theorem just stated.

Example. (i) Since $\lim_{x \to 3} x = 3$, we see from (a) that

$$\lim_{x \to 3} 5x = 5 (\lim_{x \to 3} x) = 5 \cdot 3 = 15.$$

Similarly, we have

$$\lim_{x \to -4} (-\sqrt{2}\, x) = -\sqrt{2} (\lim_{x \to -4} x) = -\sqrt{2}(-4) = 4\sqrt{2}.$$

(ii) Since $\lim_{x \to -2} 3x = 3(\lim_{x \to -2} x) = 3(-2) = -6$ and $\lim_{x \to -2} 5 = 5$, we see from (b) that

$$\lim_{x \to -2} (3x + 5) = \lim_{x \to -2} 3x + \lim_{x \to -2} 5 = -6 + 5 = -1.$$

In a similar manner, we have

$$\lim_{x \to 1} \left(-\frac{1}{2} x + 4 \right) = \lim_{x \to 1} \left(-\frac{1}{2} x \right) + \lim_{x \to 1} 4 = -\frac{1}{2} + 4 = \frac{7}{2}.$$

(iii) From example (i) $\lim_{x \to -2} (3x + 5) = -1$. Also, $\lim_{x \to -2} x = -2$. Therefore, by (c),

$$\lim_{x \to -2} [3x^2 + 5x] = \lim_{x \to -2} [(3x + 5) \cdot x] = [\lim_{x \to -2} (3x + 5)] [\lim_{x \to -2} x]$$

$$= (-1) \cdot (-2) = 2.$$

Again, by (c), it follows that

$$\lim_{x \to -2} (9x^2 + 30x + 25) = \lim_{x \to -2} (3x + 5)^2$$

$$= \lim_{x \to -2} [(3x + 5)(3x + 5)]$$

$$= [\lim_{x \to -2} (3x + 5)] [\lim_{x \to -2} (3x + 5)]$$

$$= 1.$$

(iv) We know that $\lim\limits_{x \to -2} (3x + 5) = -1$ and $\lim\limits_{x \to -2} x = -2 \neq 0$. By (d), we have

$$\lim_{x \to -2} \frac{3x + 5}{x} = \frac{\lim\limits_{x \to -2} (3x + 5)}{\lim\limits_{x \to -2} x} = \frac{-1}{-2} = \frac{1}{2}.$$

Similarly,

$$\lim_{x \to -\frac{1}{2}} \frac{\sqrt{3}}{x^2 + 1} = \frac{\lim\limits_{x \to -\frac{1}{2}} \sqrt{3}}{\lim\limits_{x \to -\frac{1}{2}} (x^2 + 1)} = \frac{\sqrt{3}}{\lim\limits_{x \to -\frac{1}{2}} x^2 + \lim\limits_{x \to -\frac{1}{2}} 1}$$

$$= \frac{\sqrt{3}}{[\lim\limits_{x \to -\frac{1}{2}} x]^2 + 1}$$

$$= \frac{\sqrt{3}}{\dfrac{1}{4} + 1}$$

$$= \frac{4\sqrt{3}}{5}.$$

Notice that

$$\lim_{x \to a} x^2 = \lim_{x \to a} (x \cdot x) = [\lim_{x \to a} x] [\lim_{x \to a} x] = a \cdot a = a^2.$$

Thus

$$\lim_{x \to a} x^3 = \lim_{x \to a} (x^2 \cdot x) = [\lim_{x \to a} x^2] [\lim_{x \to a} x] = a^2 \cdot a = a^3.$$

Continuing in this manner, it is seen that

$$\lim_{x \to a} x^n = a^n$$

for every positive integer n. A very pleasant consequence of this fact is that limits of polynomial functions are quite simple to find. Suppose, for instance, that $p(x) = a_n x^n + a_{n-1} x^{n-1} + \ldots + a_1 x + a_0$. Then

$$\lim_{x \to b} p(x) = \lim_{x \to b} (a_n x^n + a_{n-1} x^{n-1} + \ldots + a_1 x + a_0)$$

$$= \lim_{x \to b} (a_n x^n) + \lim_{x \to b} (a_{n-1} x^{n-1}) + \ldots + \lim_{x \to b} (a_1 x) + \lim_{x \to b} a_0$$

$$= (a_n \lim_{x \to b} x^n) + (a_{n-1} \lim_{x \to b} x^{n-1}) + \ldots + (a_1 \lim_{x \to b} x) + (\lim_{x \to b} a_0)$$

$$= a_n b^n + a_{n-1} b^{n-1} + \ldots + a_1 b + a_0$$

$$= p(b).$$

The following example illustrates the usefulness of the preceding observation in finding limits involving polynomial functions.

Example. (i) $\lim_{x \to 3} \left(\frac{1}{2} x + 1 \right) = \left(\frac{1}{2} \cdot 3 \right) + 1 = \frac{5}{2}$ which agrees with the results of Sec. 6.1.

(ii) $\lim_{x \to -\frac{1}{2}} (x^3 - 2x + 1) = \left(-\frac{1}{2} \right)^3 - 2\left(-\frac{1}{2} \right) + 1 = \frac{15}{8}$.

(iii) $\lim_{x \to 1} [(7x^2 - x + 4) + (x^3 - 1)]$

$$= \lim_{x \to 1} (7x^2 - x + 4) + \lim_{x \to 1} (x^3 - 1)$$

$$= (7 - 1 + 4) + (1 - 1)$$

$$= 10.$$

(iv) $\lim_{x \to 5} [(x - 3)(x^2 + 6x - \sqrt{7})]$

$$= [\lim_{x \to 5} (x - 3)] [\lim_{x \to 5} (x^2 + 6x - \sqrt{7})]$$

$$= (5 - 3) [5^2 + 6 \cdot 5 - \sqrt{7}]$$

$$= 2(155 - \sqrt{7}).$$

Now let $p(x)$ and $q(x)$ be polynomial functions and assume $q(a) \neq 0$. We have

$$\lim_{x \to a} \frac{p(x)}{q(x)} = \frac{\lim_{x \to a} p(x)}{\lim_{x \to a} q(x)} = \frac{p(a)}{q(a)}.$$

Example. (i) $\lim_{x \to \sqrt{2}} \frac{x^2 - 1}{x^3 + 5x + 6} = \frac{\lim_{x \to \sqrt{2}} (x^2 - 1)}{\lim_{x \to \sqrt{2}} (x^3 + 5x + 6)}$

$$= \frac{\sqrt{2}^2 - 1}{\sqrt{2}^3 + 5\sqrt{2} + 6} = \frac{1}{7\sqrt{2} + 6}.$$

(ii) $\lim\limits_{x \to -1} \dfrac{x^4 + 16x^2 - 2x + 9}{7x^3 + x^2 - 3} = \dfrac{\lim\limits_{x \to -1}(x^4 + 16x^2 - 2x + 9)}{\lim\limits_{x \to -1}(7x^3 + x^2 - 3)}$

$$= -\frac{28}{9}.$$

Our last observation states that if p and q are polynomial functions and a is in the domain of $\dfrac{p}{q}$, then $\lim\limits_{x \to a} \dfrac{p(x)}{q(x)} = \dfrac{p(a)}{q(a)}$. In other words, the limit of $\dfrac{p(x)}{q(x)}$ as x approaches a is just the value of the function $\dfrac{p}{q}$ at $x = a$. This important property is given a special name.

6.4 Definition. *IF* $\lim\limits_{x \to x_0} f(x) = f(x_0)$, *THEN the function f is* **continuous** *at* $x = x_0$. *IF f is continuous at each point in an interval, THEN f is a continuous function on the interval.*

In particular, all polynomial functions are continuous. Continuous functions are those functions f for which $f(x)$ approaches $f(x_0)$ as x approaches x_0 for each number x_0 in the domain of f. Graphs of continuous functions have a very special geometric property. For instance, let f be a continuous function defined on an interval containing the number x_0. Then the distance from $(x, f(x))$ to $(x_0, f(x_0))$ is $d = \sqrt{(x - x_0)^2 + (f(x) - f(x_0))^2}$. As x approaches x_0 the terms $x - x_0$ and $f(x) - f(x_0)$ both approach 0. Therefore $(x - x_0)^2 + (f(x) - f(x_0))^2$ approaches 0 as x approaches x_0. Thus, it is a fact (see Sec. 6.4) that d approaches 0 as x approaches x_0. This means that the distance from $(x, f(x))$ to $(x_0, f(x_0))$ goes to 0 as x approaches x_0, so the point $(x, f(x))$ is "zeroing in" on $(x_0, f(x_0))$ as x approaches x_0. We can interpret this loosely to mean that $(x_0, f(x_0))$ cannot lie off the curve nor can there be a break in the curve as depicted in Figure 6.8.

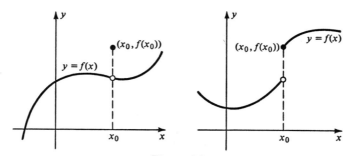

Figure 6.8

Thus, for our purposes, a function is continuous on an interval if its graph is an unbroken curve. A simple rule of thumb that is applicable here is the following: If the graph of a function can be sketched without lifting the pencil from the paper, then the function is continuous.

Exercise Set 6.3

1. Given $\lim\limits_{x \to 2} f(x) = \dfrac{7}{3}$ and $\lim\limits_{x \to 2} g(x) = -4$

    find: $\lim\limits_{x \to 2} 6 f(x)$, $\qquad$ $\lim\limits_{x \to 2} [f(x) + g(x)]$,

    $\lim\limits_{x \to 2} [f(x) - g(x)]$, $\qquad$ $\lim\limits_{x \to 2} [f(x) \cdot g(x)]$,

    $\lim\limits_{x \to 2} \dfrac{f(x)}{g(x)}$, $\qquad$ $\lim\limits_{x \to 2} [-2f(x) + 3g(x)]$.

2. Given $f(x) = 3x^2 + x - 5$ and $g(x) = 2x + 3$

    find: $\lim\limits_{x \to 2} f(x)$, $\qquad$ $\lim\limits_{x \to 2} g(x)$,

    $\lim\limits_{x \to 2} \dfrac{3x^2 + x - 5}{2x + 3}$, $\qquad$ $\lim\limits_{x \to 2} (3x^2 + x - 5)(2x + 3)$,

    $\lim\limits_{x \to 2} (3x^2 + x - 5)^5$, $\qquad$ $\lim\limits_{x \to 2} \dfrac{2x + 3}{3x^2 + x - 5}$.

3. Given $P(x) = x^5 - 4x^4 + 2x^3 + x - 12$

    find: $\lim\limits_{x \to 0} P(x)$, $\quad$ $\lim\limits_{x \to 1} P(x)$, $\quad$ $\lim\limits_{x \to -2} P(x)$, $\quad$ and $\lim\limits_{x \to 6} P(x)$

    (Use results of this section and synthetic division to find these limits.)

4. Which of the following functions are continuous at $x = 1$?
    a. $P(x) = x^{64} - 7x^{39} + x^{10} + x^9$

    b. $g(x) = \dfrac{1}{x + 1}$

    c. $h(x) = \dfrac{x^2 - 1}{x - 1}$

    d. $f(x) = [x]$

    e. $j(x) = \begin{cases} 2x^2, & \text{if } x \leqslant 1 \\ x + 1, & \text{if } x > 1 \end{cases}$

    f. $j(x) = \begin{cases} x^2, & \text{if } x \leqslant 1 \\ 2x, & \text{if } x > 1 \end{cases}$

** * **

In problems 5-18, find the indicated limits.

5. $\lim\limits_{x \to 7} (16 - 3x)$

6. $\lim\limits_{x \to -4} 16 - 3x^2$

7. $\lim\limits_{x \to -\frac{1}{2}} (2x^3 + 7x - 3)$

8. $\lim\limits_{x \to -1} x^4 + 2x^3 - x + 4$

9. $\lim\limits_{x \to 9} \dfrac{x + 5}{x + 4}$

10. $\lim\limits_{x \to 5} \dfrac{x^2 - 1}{x - 1}$

11. $\lim\limits_{x \to 1} \dfrac{x^2 - 1}{x - 1}$

12. $\lim\limits_{x \to -1} \dfrac{x^2 - 1}{x + 1}$

13. $\lim\limits_{x \to \sqrt{2}} (5 - \sqrt{2}x)$

14. $\lim\limits_{x \to \sqrt{2}} x^{12}$

15. $\lim\limits_{x \to 3} \dfrac{x^2 - x - 6}{x - 3}$

16. $\lim\limits_{x \to 3} \dfrac{x^2 - x - 6}{x + 3}$

17. $\lim\limits_{x \to a^2} \dfrac{1}{7x^2 + 4x + 3}$

18. $\lim\limits_{x \to b} \dfrac{4x^3 - 8x + 1}{x^4 - 16}$

In problems 19-24, determine whether the given function is continuous at $x = a$ for the given a. Sketch the graph in each case.

19. $f(x) = \begin{cases} 1, & \text{if } x \leqslant 0 \\ 2, & \text{if } x > 0 \end{cases} \qquad a = 0$

20. $f(x) = \begin{cases} x^2 + 2, & \text{if } x \text{ is not an integer} \\ 1, & \text{if } x \text{ is an integer} \end{cases} \qquad a = 1$

21. $g(x) = \begin{cases} x + 1, & \text{if } x \leqslant 1 \\ 2x, & \text{if } x > 1 \end{cases} \qquad a = 1$

22. $g(x) = \begin{cases} |x|, & \text{if } x \neq 0 \\ -2, & \text{if } x = 0 \end{cases} \qquad a = 0$

23. $h(x) = \begin{cases} x + 1, & \text{if } x \neq 1 \\ 2, & \text{if } x = 1 \end{cases} \qquad a = 1$

24. $h(x) = \begin{cases} x, & \text{if } x \geqslant 0 \\ -x, & \text{if } x < 0 \end{cases} \qquad a = 0$

*25. Let f and g be continuous functions each having the same domain. Use Theorem 6.3 to prove the following:
   (a) $f + g$ is continuous   (b) $f \cdot g$ is continuous.

*26. Let $G(x) = \begin{cases} x, & \text{if } x \text{ is a rational number} \\ 0, & \text{if } x \text{ is an irrational number.} \end{cases}$

Give a reasonable argument which shows that G is continuous at 0.

6.4 LIMITS OF COMPOSITE FUNCTIONS

In this section we consider limits of composite functions, exponential functions, and logarithm functions. This will enable us to find limits for all the functions which are important in our study of the calculus.

Let n be a positive integer and assume $\lim\limits_{x \to a} f(x)$ exists. By repeated application of Theorem 6.3 we have

$$\lim_{x \to a} [f(x)]^n = \lim_{x \to a} \underbrace{[f(x) \ldots f(x)]}_{n \text{ factors}}$$

$$= \underbrace{(\lim_{x \to a} f(x)) \ldots (\lim_{x \to a} f(x))}_{n \text{ factors}}$$

$$= [\lim_{x \to a} f(x)]^n.$$

Example. (i) $\lim\limits_{x \to -1} (x - 1)^7 = [\lim\limits_{x \to -1} (x - 1)]^7 = (-2)^7 = -128.$

(ii) $\lim\limits_{x \to 3} \dfrac{(2x^2 + 7)^3}{(x^3 - 12x + 14)^4} = \dfrac{\lim\limits_{x \to 3} (2x^2 + 7)^3}{\lim\limits_{x \to 3} (x^3 - 12x + 14)^4}$

$$= \dfrac{[\lim\limits_{x \to 3} (2x^2 + 7)]^3}{[\lim\limits_{x \to 3} (x^3 - 12x + 12)]^4}$$

$$= \dfrac{25^3}{54}.$$

Now suppose m is a negative integer and that $\lim\limits_{x \to a} f(x) \neq 0$. We can write $m = -n$, where n is a positive integer. Then, since $\lim\limits_{x \to a} f(x) \neq 0$,

$$\lim_{x \to a} [f(x)]^m = \lim_{x \to a} [f(x)]^{-n}$$

$$= \lim_{x \to a} \frac{1}{[f(x)]^n}$$

$$= \frac{\lim_{x \to a} 1}{\lim_{x \to a} [f(x)]^n}$$

$$= \frac{1}{[\lim_{x \to a} f(x)]^n}$$

$$= [\lim_{x \to a} f(x)]^{-n}$$

$$= [\lim_{x \to a} f(x)]^m.$$

Therefore the preceding result also holds for negative integers when $\lim_{x \to a} f(x) \neq 0$.

Example. (i) $\lim_{x \to -2} (x + 6)^{-3} = [\lim_{x \to -2} (x + 6)]^{-3} = 4^{-3} = \dfrac{1}{64}.$

(ii) $\lim_{x \to \frac{1}{3}} \left(3x^2 + 2x - \dfrac{4}{3}\right)^{-5} = \left[\lim_{x \to \frac{1}{3}} \left(3x^2 + 2x - \dfrac{4}{3}\right)\right]^{-5}$

$$= \left(-\frac{1}{3}\right)^{-5} = -243.$$

We have shown that if $\lim_{x \to a} f(x) \neq 0$ and n is an integer, then $\lim_{x \to a} [f(x)]^n = [\lim_{x \to a} f(x)]^n$. This is a limit theorem holding for integral exponents. This result also holds for rational exponents.

If r is a rational number, $\lim_{x \to a} f(x) = L$, $L \neq 0$, and L^r is defined, then

$$\lim_{x \to a} [f(x)]^r = [\lim_{x \to a} f(x)]^r = L^r.$$

This formula allows us to find limits of many more functions than before.

Example. (i) $\lim_{x \to 4} (x - 2)^{\frac{1}{3}} = [\lim_{x \to 4} (x - 2)]^{\frac{1}{3}} = 2^{\frac{1}{3}}.$

(ii) $\lim_{x \to -6} (x - 2)^{-\frac{1}{3}} = [\lim_{x \to -6} (x - 2)]^{-\frac{1}{3}} = (-8)^{-\frac{1}{3}} = -\dfrac{1}{2}.$

(iii) $\lim_{x \to 2} \sqrt{x^2 + 2x + 8} = \lim_{x \to 2} (x^2 + 2x + 8)^{\frac{1}{2}}$

$$= [\lim_{x \to 2} (x^2 + 2x + 8)]^{\frac{1}{2}}$$

$$= 16^{\frac{1}{2}}$$

$$= 4.$$

(iv) $\lim_{x \to 1} (2x^3 - x^2 + 3x - 1)(x^2 + 1)^{\frac{5}{4}}$

$$= [\lim_{x \to 1} (2x^3 - x^2 + 3x - 1)] [\lim_{x \to 1} (x^2 + 1)^{\frac{5}{4}}]$$

$$= 3 [\lim_{x \to 1} (x^2 + 1)]^{\frac{5}{4}}$$

$$= 3 \cdot 2^{\frac{5}{4}}$$

$$= 6 \cdot 2^{\frac{1}{4}}.$$

(v) $\lim_{x \to -1} \dfrac{(x^3 - 1)^{\frac{2}{7}}}{(5x^4 + 16x^2 - 1)^{\frac{1}{2}}}$

$$= \frac{\lim_{x \to -1} (x^3 - 1)^{\frac{2}{7}}}{\lim_{x \to -1} (5x^4 + 16x^2 - 1)^{\frac{1}{2}}}$$

$$= \frac{[\lim_{x \to -1} (x^3 - 1)]^{\frac{2}{7}}}{[\lim_{x \to -1} (5x^4 + 16x^2 - 1)]^{\frac{1}{2}}}$$

$$= \frac{(-2)^{\frac{2}{7}}}{20^{\frac{1}{2}}}.$$

Let us now examine the preceding observations from another point of view. Consider the function $h(x) = \sqrt{x^2 - 2x - 6}$. The number $\sqrt{x^2 - 2x - 6}$ is formed in two steps. First, we find $x^2 - 2x - 6$. Second, we take the square root of the number just found. If we write $u(x) = x^2 - 2x - 6$, then the two steps are

$$x \longrightarrow u(x) = x^2 - 2x - 6 \longrightarrow \sqrt{u(x)} = \sqrt{x^2 - 2x - 6} = h(x).$$

1st step   2nd step

We now see that the rule for finding $h(x)$ is really a composition of two separate rules in a specified order. This, of course, is the reason h is called a composite function. Thus $h(x) = \sqrt{u(x)}$, where $u(x) = x^2 - 2x - 6$. Suppose we want to find $\lim\limits_{x \to 4} h(x)$. If we write $g(u) = \sqrt{u}$, this amounts to asking the following question: What is the limiting value of $g(u(x))$ as x approaches 4, where $u(x) = x^2 - 2x - 6$? Note that as x approaches 4, $x^2 - 2x - 6$ approaches 2. Also, $\sqrt{u}$ approaches $\sqrt{4} = 2$ as u approaches 2. Therefore $g(u(x))$ approaches $\sqrt{2}$ as x approaches 4. This means that $\lim\limits_{x \to 4} h(x) = g(\lim\limits_{x \to 4} u(x))$. As one might suspect, this result is true in a more general setting. We state this important result without proof.

6.5 Theorem. *Let u be a function of x such that $\lim\limits_{x \to x_0} u(x) = u_0$. IF g is a continuous function on the interval (a, b) and $u_0 \in (a, b)$, THEN*

$$\lim_{x \to x_0} g(u(x)) = g(u_0) = g(\lim_{x \to x_0} u(x)).$$

In particular, IF u is continuous at x_0, THEN

$$\lim_{x \to x_0} g(u(x)) = g(u(x_0)).$$

In order to give us a large set of functions to which we can apply Theorem 6.5, we now make the following assumptions:

(a) *If r is a rational number then the function $g(u) = u^r$ is continuous.*
(b) *The exponential and logarithmic functions are continuous.*

Suppose $\lim\limits_{x \to x_0} u(x) = L \neq 0$ and L^r is defined. Then, by (a) and Theorem 6.5, we have $\lim\limits_{x \to x_0} [u(x)]^r = [\lim\limits_{x \to x_0} u(x)]^r$. This is our earlier observation. We now illustrate how Theorem 6.5 applies to the exponential and logarithmic functions.

Example. (i) Consider $\lim\limits_{x \to 2} e^{x^3}$. Let $g(u) = e^u$ and $u(x) = x^3$. Thus $e^{x^3} = g(u(x))$ and $\lim\limits_{x \to 2} u(x) = 8$. Hence,

$$\lim_{x \to 2} e^{x^3} = \lim_{x \to 2} g(u(x))$$

$$= g(\lim_{x \to 2} u(x))$$

$$= g(8)$$

$$= e^8.$$

(ii) Consider $\lim\limits_{x \to -1} \ln(x^2 + 4)$. Let $g(u) = \ln u$ and $u(x) = x^2 + 4$.

Then $\ln (x^2 + 4) = g(u(x))$ and $\lim\limits_{x\to-1} u(x) = 5$. Thus

$$\lim\limits_{x\to-1} \ln (x^2 + 4) = \lim\limits_{x\to-1} g(u(x))$$

$$= g(\lim\limits_{x\to-1} u(x))$$

$$= g(5)$$

$$= \ln 5.$$

(iii) Consider $\lim\limits_{x\to1} \dfrac{\sqrt{7 - 2x - 3x^2}}{e^{2x}}$. We have

$$\lim\limits_{x\to1} \sqrt{7 - 2x - 3x^2} = \sqrt{\lim\limits_{x\to1} (7 - 2x - 3x^2)} = \sqrt{2}$$

and

$$\lim\limits_{x\to1} e^{2x} = e^2.$$

Therefore,

$$\lim\limits_{x\to1} \frac{\sqrt{7 - 2x - 3x^2}}{e^{2x}} = \frac{\lim\limits_{x\to1} \sqrt{7 - 2x - 3x^2}}{\lim\limits_{x\to1} e^{2x}}$$

$$= \frac{\sqrt{2}}{e^2}.$$

(iv) Consider $\lim\limits_{x\to\frac{1}{2}} [-x^2 + \ln(2x + 4)]^{\frac{1}{3}}$. We know that $\lim\limits_{x\to\frac{1}{2}} \ln(2x + 4) = \ln 5$. Thus

$$\lim\limits_{x\to\frac{1}{2}} [-x^2 + \ln (2x + 4)]^{\frac{1}{3}} = [\lim\limits_{x\to\frac{1}{2}} [-x^2 + \ln (2x + 4)]]^{\frac{1}{3}}$$

$$= \left[-\frac{1}{4} + \ln 5\right]^{\frac{1}{3}}.$$

Exercise Set 6.4

1. Given $\lim\limits_{x\to a} f(x) = -1$ and $\lim\limits_{x\to a} g(x) = 4$

   find: $\lim\limits_{x\to a} [f(x)]^{25}$, $\quad \lim\limits_{x\to a} [g(x)]^{\frac{1}{2}}$,

   $\lim\limits_{x\to a} [f(x)]^{\frac{1}{3}}$, $\quad \lim\limits_{x\to a} [g(x)]^{\frac{3}{4}}$.

2. Given $f(x) = x^2 + 4$

 find: $\lim_{x \to 4} f(x)$, $\qquad \lim_{x \to 4} [f(x)]^{\frac{1}{2}}$,

 $\qquad \lim_{x \to -2} f(x)$, $\qquad \lim_{x \to -2} [f(x)]^{-\frac{1}{3}}$.

3. For the given function $h(x)$, choose a function g and a function u such that $h(x) = g(u(x))$.

 a. $h(x) = e^{2x}$
 b. $h(x) = \ln(x^2 + x - 1)$

 c. $h(x) = |4 - x^2|$
 d. $h(x) = \dfrac{1}{(x^3 + 1)^7}$

4. Use the results of problem 3 and Theorem 6.5 to find the following limits:

 a. $\lim_{x \to 0} e^{2x}$
 b. $\lim_{x \to 1} \ln(x^2 + x - 1)$

 c. $\lim_{x \to 3} |4 - x^2|$
 d. $\lim_{x \to -\frac{1}{2}} \dfrac{1}{\left(x^3 + \dfrac{1}{2}\right)^7}$

* * *

In problems 5–18, find the indicated limit.

5. $\lim_{x \to -1} (1 - 4x^2)^3$
6. $\lim_{x \to \frac{1}{4}} (3x - 1)(4x - 2)^5$

7. $\lim_{x \to 11} \sqrt{\dfrac{x + 4}{x + 7}}$
8. $\lim_{x \to 2} [x^2 + \sqrt{x^3 + 8}]^{\frac{2}{3}}$

9. $\lim_{x \to a} (x^2 + ax + a^2)$
10. $\lim_{x \to a} \sqrt{2x^2 - a^2}$

11. $\lim_{x \to 6} e^{\sqrt{x-2}}$
12. $\lim_{x \to 2} e^{x^2 - 2x - 1}$

13. $\lim_{x \to -1} \ln(x^2 + 1)$
14. $\lim_{x \to \frac{5}{2}} \ln(2x)$

15. $\lim_{x \to 0} \dfrac{\ln(x^3 - 3x^2 + x + 6)}{x^4 + 16}$
16. $\lim_{x \to 0} (x^2 + 1)e^{5x}$

17. $\lim_{x \to 4} \sqrt{x}\, e^{\sqrt{x-2}}$
18. $\lim_{x \to 2} (\ln x)\, e^{\sqrt{x}}$

In problems 19–24, find the limits. Consider the symbol Δx as representing a single quantity.

19. $\lim_{\Delta x \to 0} \Delta x$
20. $\lim_{\Delta x \to 0} 5\,\Delta x$

21. $\displaystyle\lim_{\Delta x \to 0} \; x\,(\Delta x)$

22. $\displaystyle\lim_{\Delta x \to 0} \; (x + \Delta x)$

23. $\displaystyle\lim_{\Delta x \to 0} \; [4x^3 + 7x^2 \Delta x - x(\Delta x)^2 + (\Delta x)^3]$

24. $\displaystyle\lim_{\Delta x \to 0} \; [3x^2 - 4x + x\Delta x + 8(\Delta x)^2]$

6.5 THE NEED FOR THE DERIVATIVE

We now examine several problems which arise quite naturally and formulate reasonable solutions to these problems. These solutions will require the notion of a limit. Indeed, this is the reason limits have already been discussed. The approach to the solution is the same in each case and will lead us to the concept of a derivative of a function.

Example. Let us consider a geometric problem. The tangent T to a circle at a point P on the circle is that line through P which touches the circle only at the point P (Figure 6.9). Is it possible to formulate this notion of a tangent line for curves other than circles?

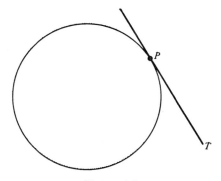

Figure 6.9

Given a point P on a curve ζ, our initial attempt might be to define the tangent line T to be a line which touches ζ only at P. However, if the curve has a "corner," there may be many lines touching the curve at that one point (Figure 6.10). Therefore this approach does not seem to be satisfactory. If we consider the circle once more, we see that T is a line which touches the circle at P and lies close to the curve at P. Hence our definition should reflect this idea of closeness.

We proceed as follows. Let P be a point on a curve ζ and let Q be any point on ζ which is distinct from P. Let PQ denote the line passing through P and Q.

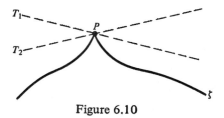

Figure 6.10

Now consider what happens as Q approaches P along ζ. The line PQ is shown for several positions of Q in Figure 6.11.

If the curve ζ is "smooth" at P, the line PQ would appear to approach a limiting position as Q approaches P. That is, we can think of PQ as rotating into coincidence with some "limiting line" T. Since the slope of a line is a measure of its inclinition to the x-axis, one would expect the slope of PQ to approach the slope of T as Q approaches P along ζ. In addition, since PQ is rotating into coincidence with T as Q approaches P, T would also appear to be close to the curve ζ at P. Intuitively, we see that the usual line T which is tangent to the circle at P can be obtained as a "limiting line" of PQ.

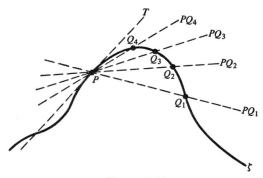

Figure 6.11

Our observations indicate that, under reasonable conditions, the tangent line to ζ at P should be that line passing through P whose slope equals the limiting value of the slope of PQ as Q approaches P along the curve, provided the latter limit exists. In order to be able to find this limit, we now consider curves obtained as graphs of functions.

Let f be a function defined on an interval. If x is a point in this interval, let P denote the point $(x, f(x))$ on the graph of f. If Q is any other point on the graph of f, Q is of the form $(z, f(z))$, where $z \neq x$. Letting $\Delta x = z - x$, we can write $z = x + \Delta x$. Then Q is the point $(x + \Delta x, f(x + \Delta x))$, where $\Delta x \neq 0$ (see Figure 6.12).

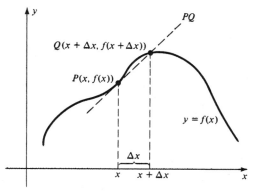

Figure 6.12

It follows that PQ has slope

$$\frac{f(x + \Delta x) - f(x)}{(x + \Delta x) - x} = \frac{f(x + \Delta x) - f(x)}{\Delta x}.$$

If f is continuous, then Q will approach P as Δx approaches 0. Therefore

$$\lim_{\Delta x \to 0} \frac{f(x + \Delta x) - f(x)}{\Delta x}$$

is defined to be the slope of the tangent line to the graph of f at the point $(x, f(x))$, provided this limit exists.

Example. Suppose a space ship is launched and we know that the vertical distance, in feet, of the space ship from the ground is given by the function

$$s(t) = t^4 + 100t,$$

where t represents the elapsed time, measured in seconds, since the launch. Computing $s(t)$ for $t = 1, 2, 3$ we have

$$s(1) = (1)^4 + 100(1) = 1 + 100 = 101,$$

$$s(2) = (2)^4 + 100(2) = 16 + 200 = 216,$$

and

$$s(3) = (3)^4 + 100(3) = 81 + 300 = 381.$$

Thus, the space ship is 101 feet off the ground after 1 second, 216 feet off the ground after 2 seconds, and 381 feet off the ground after 3 seconds.

We are interested in determining the velocity of the space ship at any time $t = t_1$. To do this let us consider the particular case when $t_1 = 2$ seconds. To determine the velocity of the space ship at the time $t_1 = 2$ seconds consider the following argument:

(a) The average velocity of the space ship for any time interval is given by

$$\frac{\text{distance traveled.}}{\text{time elapsed}}.$$

So, the average velocity from $t = 2$ seconds to $t = 3$ seconds is

$$\frac{s(3) - s(2)}{3 - 2} = \frac{381 - 216}{1} = 165 \text{ feet per second.}$$

This does not answer the question posed. It does, however, give us some idea as to what the answer is. Let us now diminish the time elapsed. For example, compute the average velocity from $t = 2$ seconds to $t = 2.1$ seconds. We have

$$\frac{s(2.1) - s(2)}{2.1 - 2} = \frac{229.4481 - 216}{.1} = \frac{13.4481}{.1}$$

$$= 134.481 \text{ feet per second.}$$

This gives us a better approximation to how fast the space ship is traveling at $t_1 = 2$ seconds.

To improve this approximation let us compute the average velocity from $t = 2$ seconds to $t = 2.01$ seconds. We have

$$\frac{s(2.01) - s(2)}{2.01 - 2} = \frac{217.32240801 - 216}{.01} = \frac{1.32240801}{.01}$$

$$= 132.240801 \text{ feet per second.}$$

(b) Now, let us consider the average velocity of the space ship from $t = t_2$ seconds to $t = 2$ seconds, where $t_2 < 2$. In this case we have

$$\frac{s(t_2) - s(2)}{t_2 - 2} = \frac{s(2) - s(t_2)}{2 - t_2}.$$

For the values $t_2 = 1, 1.9, 1.99$ we see that

$$\frac{s(2) - s(1)}{2 - 1} = \frac{216 - 101}{1} = 115 \text{ feet per second,}$$

$$\frac{s(2) - s(1.9)}{2 - 1.9} = \frac{216 - 203.0321}{.1} = \frac{12.9679}{.1}$$

$$= 129.679 \text{ feet per second,}$$

and

$$\frac{s(2) - s(1.99)}{2 - 1.99} = \frac{216 - 214.68239201}{.01} = \frac{1.31760799}{.01}$$

$$= 131.760799 \text{ feet per second.}$$

So, the space ship averages 115 feet per second from 1 second to 2 seconds, 129.679 feet per second from 1.9 seconds to 2 seconds, and 131.760799 feet per second from 1.99 seconds to 2 seconds.

(c) Combining the information from parts (a) and (b) in tabular form we have

t	1	1.9	1.99	2.01	2.1	3
$\dfrac{s(t) - s(2)}{t - 2}$	115	129.679	131.760799	132.240801	134.481	165

It appears that the space ship is traveling at 132 feet per second at time $t_1 = 2$ seconds. Therefore, from our earlier discussion of limits, we see that the velocity of the space ship at time $t_1 = 2$ seconds is given by

$$\lim_{t \to 2} \frac{s(t) - s(2)}{t - 2}.$$

(d) Generalizing, to find the velocity at time $t = t_1$, we must compute

$$\lim_{t \to t_1} \frac{s(t) - s(t_1)}{t - t_1}.$$

Problem. Assume that at each time t an object moving on a coordinatized line has coordinate $s(t)$. The function s is called the position function of the object. What is a reasonable way of defining the velocity of the object at any given time t?

Solution. As in the previous example, we define the average velocity of the object between the times t_1 and t_2 to be

$$\frac{s(t_2) - s(t_1)}{t_2 - t_1}.$$

For example, suppose $s(t) = t^3 - 2t - 1$ where the units of length and time are measured in feet and seconds, respectively. Then the average velocity of the object, in feet per second, between $t_1 = 1.2$ and $t_2 = 1.8$ is

$$\frac{s(1.8) - s(1.2)}{1.8 - 1.2} = \frac{1.232 - (-1.672)}{.6} = 4.84.$$

The average velocity of the particle, in feet per second, between $t_1 = 0$ and $t_2 = .5$ is

$$\frac{s(.5) - s(0)}{.5 - 0} = \frac{-1.875 - (-1)}{.5} = -1.75;$$

where the negative sign indicates a net change of position to the left in this time interval.

Our intuition tells us that if the time interval is small enough (i.e., t_1 and t_2 are sufficiently close), then the velocities of the object for times between t_1 and t_2 should not differ by much. It would be almost as if the particle were moving at a constant velocity in this interval. In this case the average velocity $\frac{s(t_2) - s(t_1)}{t_2 - t_1}$ would be a reasonable approximation to the velocity at time t_1.

Moreover, by making t_2 closer and closer to t_1, $\frac{s(t_2) - s(t_1)}{t_2 - t_1}$ should be getting closer and closer to the velocity of the object at time t_1. In other words, our intuitive ideas on how velocity should behave tell us how we should define velocity. That is, the velocity of the object at time t_1 is defined to be $\lim\limits_{t_2 \to t_1} \frac{s(t_2) - s(t_1)}{t_2 - t_1}$, provided this limit exists.

If we replace t_1 by t, then $\lim\limits_{t_2 \to t} \frac{s(t_2) - s(t)}{t_2 - t}$ is the velocity at time t. We can write $t_2 = t + \Delta t$; thus $\Delta t = t_2 - t$. Then, letting t_2 approach t is equivalent to letting Δt approach 0. Thus,

$$\lim_{t_2 \to t} \frac{s(t_2) - s(t)}{t_2 - t} = \lim_{\Delta t \to 0} \frac{s(t + \Delta t) - s(t)}{\Delta t}.$$

The latter limit is used to denote the velocity of the object at time t.

Example. Consider a tank into which a liquid is flowing. We will denote the volume of the liquid that has flowed into the tank at time t by $V(t)$. What is a reasonable way of defining the rate at which the liquid is flowing into the tank?

Our approach is the same as in the previous example. For $\Delta t \neq 0$, we define the average rate of flow of liquid between times t and $t + \Delta t$ to be

$$\frac{V(t + \Delta t) - V(t)}{\Delta t}.$$

For instance, let $V(t) = t^2$, where V is measured in gallons and t is measured in minutes. Then the average rate of flow of liquid in gallons per minute between times $t = 1$ and $t + \Delta t = 1.5$ (i.e., $\Delta t = .5$) is

$$\frac{V(1.5) - V(1)}{.5} = \frac{2.25 - 1}{.5} = 2.5.$$

Once again, we rely on our intuition as an aid in making a reasonable definition. If Δt is small, the rate of flow of liquid into the tank for times between t and $t + \Delta t$ should not differ by much. In other words, the average rate of flow, $\frac{V(t + \Delta t) - V(t)}{\Delta t}$, should be a reasonable approximation to the rate of flow at time t. By making Δt smaller, this approximation should improve. We would, therefore, be led to define the rate of flow at time t to be

$$\lim_{\Delta t \to 0} \frac{V(t + \Delta t) - V(t)}{\Delta t},$$

provided this limit exists.

What general conclusions can be drawn from our experience in the three pre-ceding examples? In each case the question posed was resolved by considering a particular limit. More precisely, each example involved a function, say f, and a limit of the form

$$\lim_{\Delta x \to 0} \frac{f(x + \Delta x) - f(x)}{\Delta x}.$$

In each case the solution to the problem was dependent on our ability to find such a limit. Thus, due to its apparent importance, we shall study this limit for the remainder of this chapter.

Exercise Set 6.5

1. a. Sketch the graph of the function $f(x) = x^2$ in the interval $[2, 4]$.
   b. Complete the following table of values where $f(x) = x^2$ and $x_0 = 3$:

Δx	.5	.3	.1	.01	−.01	−.1	−.3	−.5
$f(x_0 + \Delta x)$								
$\dfrac{f(x_0 + \Delta x) - f(x_0)}{\Delta x}$								

   c. Use the table to determine $\lim\limits_{\Delta x \to 0} \dfrac{f(x_0 + \Delta x) - f(x_0)}{\Delta x}$.
   d. What is the slope of the tangent line to the curve $f(x) = x^2$ at the point $(3, 9)$? What is the equation of this line?

2. Let $s(t) = 2t^2$ be the position function of an object moving on a coordi-natized line.
   a. Plot the position of the object for several values of $t \in [2, 4]$.
   b. Complete the following table of values for $s(t) = 2t^2$ and $t_0 = 3$:

Δt	.5	.3	.1	.01	−.01	−.1	−.3	−.5
$s(t_0 + \Delta t)$								
$\dfrac{s(t_0 + \Delta t) - s(t_0)}{\Delta t}$								

   c. Use the table to determine $\lim\limits_{\Delta t \to 0} \dfrac{s(t_0 + \Delta t) - s(t_0)}{\Delta t}$.
   d. What is the velocity of the object at time $t = 3$?

3. Water is leaking from a 1000 gallon tank. If the amount of water remaining at time t is given by the equation $V(t) = 1000 - t^2$, at what rate is the water flowing out when $t = 3$? To solve this problem:

a. Complete the table of values for $V(t) = 1000 - t^2$ and $t_0 = 3$:

Δt	.5	.3	.1	.01	-.01	-.1	-.3	-.5
$V(t_0 + \Delta t)$								
$\dfrac{V(t_0 + \Delta t) - V(t_0)}{\Delta t}$								

b. Use the table to determine $\lim\limits_{\Delta t \to 0} \dfrac{V(t_0 + \Delta t) - V(t_0)}{\Delta t}$.

c. Answer the question: At what rate is the water flowing out when $t = 3$?

* * *

4. Repeat problem 1 for the function $f(x) = x^3$ and $x_0 = 3$.

5. Repeat problem 2 for the function $s(t) = t^2 - t$ and $t_0 = 3$.

6. Repeat problem 3 for $t_0 = 4$.

In problems 7–24, find $\dfrac{f(x + \Delta x) - f(x)}{\Delta x}$ and simplify where possible.

7. $f(x) = 3x + 1$

8. $f(x) = x$

9. $f(x) = -6x + 7$

10. $f(x) = -4 - 3x$

11. $f(x) = 15$

12. $f(x) = \dfrac{1}{2}$

13. $f(x) = x^2$

14. $f(x) = 4x^2$

15. $f(x) = 4x^2 - x$

16. $f(x) = x - 3x^2$

17. $f(x) = \sqrt{x}$

18. $f(x) = -\sqrt{x}$

19. $f(x) = e^x$

20. $f(x) = e^{2x}$

21. $f(x) = \ln x$

22. $f(x) = \ln x^2$

23. $f(x) = \dfrac{1}{2x}$

24. $f(x) = \dfrac{1}{x^2}$

*25. Find the slope of the line tangent to the graph of $f(x) = x^2 + 1$ at the point (1, 2).

*26. Explain why the graph of the absolute value function, $f(x) = |x|$, does not have a tangent line at the point (0, 0).

6.6 THE DERIVATIVE AND Δ-PROCESS

We now start a detailed study of the important limit discussed in Sec. 6.5.

6.6 Definition. *Let f be a function defined in an interval and let x be a point in the interval. IF*

$$\lim_{\Delta x \to 0} \frac{f(x + \Delta x) - f(x)}{\Delta x}$$

*exists, THEN f is differentiable at x and this limit is called the **derivative** of f at x. IF f is differentiable at each x in the interval, THEN f is a differentiable function on the interval.*

Before proceeding, several remarks pertaining to notation are in order. The derivative of f is denoted by f', that is,

$$f'(x) = \lim_{\Delta x \to 0} \frac{f(x + \Delta x) - f(x)}{\Delta x},$$

whenever the latter limit exists. There are several other notations used to denote this limit. If we write $y = f(x)$, then we sometimes write $\Delta y = f(x + \Delta x) - f(x)$. In this case, $\lim_{\Delta x \to 0} \dfrac{f(x + \Delta x) - f(x)}{\Delta x}$ is written as $\lim_{\Delta x \to 0} \dfrac{\Delta y}{\Delta x}$ and denoted by $\dfrac{dy}{dx}$, y', $D_x y$, $D_x f$, or $\dfrac{d}{dx}[f(x)]$. In other words, all the symbols

$$f'(x), \frac{dy}{dx}, y', D_x y, D_x f, \frac{d}{dx}[f(x)]$$

represent the number $\lim_{\Delta x \to 0} \dfrac{f(x + \Delta x) - f(x)}{\Delta x}$. We will use the symbols $f'(x)$, $\dfrac{dy}{dx}$, y', and $\dfrac{d}{dx}[f(x)]$.

The process of finding the derivative of a function is called differentiation. At this time, we will use only the definition in finding derivatives. That is, we will find derivatives by finding the limit as Δx approaches 0 of the difference quotient

$$\frac{f(x + \Delta x) - f(x)}{\Delta x}.$$

This method for finding derivatives is commonly referred to as the Δ-*process*.

Problem. Find $f'(3)$ if $f(x) = 2x - 1$.

Solution. By Definition 6.6

$$f'(3) = \lim_{\Delta x \to 0} \frac{f(3 + \Delta x) - f(3)}{\Delta x}$$

$$= \lim_{\Delta x \to 0} \frac{[2(3 + \Delta x) - 1] - [2(3) - 1]}{\Delta x}$$

$$= \lim_{\Delta x \to 0} \frac{[6 + 2\Delta x - 1] - [6 - 1]}{\Delta x}$$

$$= \lim_{\Delta x \to 0} \frac{5 + 2\Delta x - 5}{\Delta x}$$

$$= \lim_{\Delta x \to 0} \frac{2\Delta x}{\Delta x} = \lim_{\Delta x \to 0} 2 = 2.$$

Problem. Find $g'(-2)$ where $g(x) = x^2 + 2x + 5$.

Solution. By Definition 6.6

$$g'(-2) = \lim_{\Delta x \to 0} \frac{g(-2 + \Delta x) - g(-2)}{\Delta x}$$

$$= \lim_{\Delta x \to 0} \frac{[(-2 + \Delta x)^2 + 2(-2 + \Delta x) + 5] - [(-2)^2 + 2(-2) + 5]}{\Delta x}$$

$$= \lim_{\Delta x \to 0} \frac{[4 - 4\Delta x + (\Delta x)^2 - 4 + 2\Delta x + 5] - [4 - 4 + 5]}{\Delta x}$$

$$= \lim_{\Delta x \to 0} \frac{-4\Delta x + (\Delta x)^2 + 2\Delta x + 5 - 5}{\Delta x}$$

$$= \lim_{\Delta x \to 0} \frac{-2\Delta x + (\Delta x)^2}{\Delta x} = \lim_{\Delta x \to 0} \frac{\Delta x(-2 + \Delta x)}{\Delta x}$$

$$= \lim_{\Delta x \to 0} (-2 + \Delta x) = -2.$$

Example. (i) To find $f'(x)$ where $f(x) = x$, we have

$$\frac{f(x + \Delta x) - f(x)}{\Delta x} = \frac{(x + \Delta x) - x}{\Delta x} = 1.$$

Thus,

$$f'(x) = \lim_{\Delta x \to 0} \frac{f(x + \Delta x) - f(x)}{\Delta x} = \lim_{\Delta x \to 0} 1 = 1.$$

If we write $y = x$, then we would also write $\dfrac{dy}{dx} = 1$ or $y' = 1$.

(ii) To find $g'(x)$ where $g(x) = 3x - 5$, we have

$$\frac{g(x + \Delta x) - g(x)}{\Delta x} = \frac{[3(x + \Delta x) - 5] - [3x - 5]}{\Delta x} = \frac{3\Delta x}{\Delta x} = 3.$$

Therefore

$$g'(x) = \lim_{\Delta x \to 0} \frac{g(x + \Delta x) - g(x)}{\Delta x} = \lim_{\Delta x \to 0} 3 = 3.$$

We also write $\dfrac{d}{dx} [3x + 5] = 3$.

(iii) In order to find $h'(x)$ where $h(x) = x^2 + 1$, we have that the difference quotient is

$$\frac{h(x + \Delta x) - h(x)}{\Delta x} = \frac{[(x + \Delta x)^2 + 1] - [x^2 + 1]}{\Delta x} = \frac{2x\Delta x + (\Delta x)^2}{\Delta x} = 2x + \Delta x.$$

By definition,

$$h'(x) = \lim_{\Delta x \to 0} \frac{h(x + \Delta x) - h(x)}{\Delta x} = \lim_{\Delta x \to 0} (2x + \Delta x) = 2x.$$

Problem. Find the equation of the tangent line to the curve $y = x^2 + 1$ at the point $(2, 5)$.

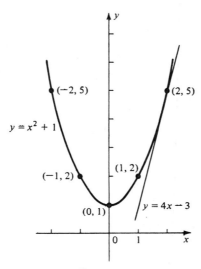

Figure 6.13

Solution. Using the notation of this section and the previous example the tangent line to the curve $y = x^2 + 1$ will be that line passing through $(2, 5)$ whose slope is the value of $\dfrac{dy}{dx}$ when $x = 2$. We know that $\dfrac{d}{dx}[x^2 + 1] = 2x$. Letting $x = 2$, we see that the desired tangent line has slope equal to 4. Since this line passes through $(2, 5)$, its equation is

$$\frac{y - 5}{x - 2} = 4$$

or

$$y = 4x - 3.$$

The tangent line and curve are shown in Figure 6.13.

Problem. Find $\dfrac{dy}{du}$ if $y = 2u^2 - 4u + 7$.

Solution. We have

$$\frac{\Delta y}{\Delta u} = \frac{[2(u + \Delta u)^2 - 4(u + \Delta u) + 7] - [2u^2 - 4u + 7]}{\Delta u}$$

$$= \frac{4u\,\Delta u + (\Delta u)^2 - 4\Delta u}{\Delta u}$$

$$= 4u - 4 + \Delta u.$$

Consequently,

$$\frac{dy}{du} = \lim_{\Delta u \to 0} \frac{\Delta y}{\Delta u} = \lim_{\Delta u \to 0}[4u - 4 + \Delta u] = 4u - 4.$$

Another notation is $\dfrac{d}{du}[2u^2 - 4u + 7] = 4u - 4$.

Problem. Find $\dfrac{dy}{dz}$ if $y = z^3$.

Solution. The difference quotient is

$$\frac{\Delta y}{\Delta z} = \frac{(z + \Delta z)^3 - z^3}{\Delta z}$$

$$= \frac{[z^3 + 3z^2\,\Delta z + 3z(\Delta z)^2 + (\Delta z)^3] - z^3}{\Delta z}$$

$$= 3z^2 + 3z\,\Delta z + (\Delta z)^2.$$

Hence,

$$\frac{dy}{dz} = \lim_{\Delta z \to 0} \frac{\Delta y}{\Delta z} = \lim_{\Delta z \to 0} [3z^2 + 3z\Delta z + (\Delta z)^2] = 3z^2.$$

Problem. Find $f'(w)$ if $f(w) = -4w^3 + 3w^2 - w + 11$.

Solution. In this case

$$\frac{f(w + \Delta w) - f(w)}{\Delta w}$$

$$= [-4(w + \Delta w)^3 + 3(w + \Delta w)^2 - (w + \Delta w) + 11] - [-4w^3 + 3w^2 - w + 11]/\Delta w$$

$$= [-4(w^3 + 3w^2\Delta w + 3w(\Delta w)^2 + (\Delta w)^3 + 3(w^2 + 2w\Delta w + (\Delta w)^2)$$

$$- (w + \Delta w) + 11] - [-4w^3 + 3w^2 - w + 11]/\Delta w$$

$$= -12w^2\Delta w - 12w(\Delta w)^2 - 4(\Delta w)^3 + 6w\Delta w + 3(\Delta w)^2 - \Delta w/\Delta w$$

$$= -12w^2 - 12w\Delta w - 4(\Delta w)^2 + 6w + 3\Delta w - 1.$$

Accordingly,

$$f'(w) = \lim_{\Delta w \to 0} \frac{f(w + \Delta w) - f(w)}{\Delta w}$$

$$= \lim_{\Delta w \to 0} [-12w^2 - 12w\Delta w - 4(\Delta w)^2 + 6w + 3\Delta w - 1]$$

$$= 12w^2 + 6w - 1.$$

The functions considered in the problems so far have been differentiable. One should not get the impression that all functions possess derivatives at all points in their domain.

Example. The function $f(x) = |x|$ does not possess a derivative at $x = 0$. For $x = 0$, the difference quotient is

$$\frac{f(0 + \Delta x) - f(0)}{\Delta x} = \frac{|\Delta x| - |0|}{\Delta x} = \frac{|\Delta x|}{\Delta x}.$$

If $\Delta x > 0$, then $\dfrac{|\Delta x|}{\Delta x} = \dfrac{\Delta x}{\Delta x} = 1$, i.e., $\lim_{\Delta x \to 0^+} \dfrac{|\Delta x|}{\Delta x} = \lim_{\Delta x \to 0^+} \dfrac{\Delta x}{\Delta x} = 1$. If $\Delta x < 0$,

then $\dfrac{|\Delta x|}{\Delta x} = \dfrac{-\Delta x}{\Delta x} = -1$, i.e., $\lim_{\Delta x \to 0^-} \dfrac{|\Delta x|}{\Delta x} = \lim_{\Delta x \to 0^-} \dfrac{-\Delta x}{\Delta x} = -1$. Consequently,

$\dfrac{|\Delta x|}{\Delta x}$ does not have a limit as Δx approaches 0. By definition of the derivative, f is not differentiable at $x = 0$.

Exercise Set 6.6

1. a. Use the Δ-process to find $f'(4)$ if $f(x) = x - 5$.
   b. Use the Δ-process to find $f'(x)$ if $f(x) = x - 5$.

2. a. Find $\dfrac{dy}{dx}$ if $y = x - 5$.
   b. Find y' if $y = x - 5$.
   c. Find $\dfrac{d}{du} [u - 5]$.

3. a. Use the Δ-process to find $f'(4)$ if $f(x) = 3x^2$.
   b. Use the Δ-process to find $f'(x)$ if $f(x) = 3x^2$.

4. a. Find $\dfrac{dy}{dz}$ if $y = 3z^2$.
   b. Find $g'(x)$ if $g(x) = 3x^2$.
   c. Find $\dfrac{d}{du} [3u^2]$.

5. Use the result of problem 3 to write the equation of the tangent line to the curve $y = 3x^2$ at the point $(4, 48)$.

* * *

In problems 6–15, use the Δ-process to find the derivative of the indicated function.

6. $f(x) = 4x + 7$

7. $g(u) = -6u + 14$

8. $h(z) = -4$

9. $y = -\dfrac{1}{2} x^2$

10. $y = -4w^2 + 7$

11. $w = 7z^2 - 2z + 10$

12. $y = x^4$

13. $f(x) = x^2(3x + 7)$

14. $z = -x^2 + 3x + 7$

15. $u = x^3 - x$

In problems 16–21, find the equation of the. line tangent to the graph of the given function at the specified point.

16. $f(x) = -3x + 5$, $(-1, 8)$

17. $f(x) = 4x + 7$, $(2, 15)$

18. $f(x) = -\dfrac{1}{2} x^2$, $\qquad$ $(2, -2)$

19. $f(x) = x^2 - 2x + 5$, $\quad$ $(1, 4)$

20. $f(x) = x^3$, $\qquad$ $(2, 8)$

21. $f(x) = -x^3 + 2$, $\qquad$ $(1, 1)$

22. Find the x-coordinate of the point on the curve, $y = x^2 - 6$, at which the tangent line is horizontal (that is, the slope is zero).

23. Find the x-coordinate of the point on the curve, $y = 3x^2 + x - 5$, at which the tangent line is horizontal.

24. Let $f(x) = c$ be a constant function. Show that $f'(x) = 0$ for all $x \in \mathcal{R}$.

6.7 DERIVATIVES OF SUMS, PRODUCTS, AND QUOTIENTS

We see that using the definition of the derivative can, indeed, be laborious. Therefore, let us attempt to derive some useful formulas so that we can avoid using the definition directly in future work. In this section we will prove several basic differentiation formulas and illustrate their use. These formulas are important because they show us how to differentiate sums, products, and quotients of differentiable functions. We summarize these formulas in the following theorem.

6.7 Theorem. *Let f and g be functions defined on the same interval and let c be a real number. IF f and g are differentiable at x, THEN*

(a) *cf is differentiable at x and* $\dfrac{d}{dx} [cf(x)] = c \dfrac{d}{dx} [f(x)]$,

(b) *f + g is differentiable at x and* $\dfrac{d}{dx} [f(x) + g(x)] = \dfrac{d}{dx} [f(x)] + \dfrac{d}{dx} [g(x)]$,

(c) *fg is differentiable at x and*

$$\frac{d}{dx} [f(x)g(x)] = \frac{d}{dx} [f(x)] \cdot g(x) + f(x) \cdot \frac{d}{dx} [g(x)],$$

(d) $\dfrac{f}{g}$ *is differentiable at x and*

$$\frac{d}{dx} \left[\frac{f(x)}{g(x)} \right] = \frac{g(x) \dfrac{d}{dx} [f(x)] - f(x) \dfrac{d}{dx} [g(x)]}{[g(x)]^2} \quad (provided\ g(x) \neq 0).$$

Proof. We will use the definition of the derivative and theorems concerning limits to prove (a), (b), and (c). The proof of (d) is left to the interested reader.

(a) The difference quotient for cf is

$$\frac{cf(x + \Delta x) - cf(x)}{\Delta x} = c\,\frac{f(x + \Delta x) - f(x)}{\Delta x}.$$

Since $\dfrac{d}{dx}[f(x)] = \lim\limits_{\Delta x \to 0} \dfrac{f(x + \Delta x) - f(x)}{\Delta x}$ exists, we have

$$\frac{d}{dx}[cf(x)] = \lim_{\Delta x \to 0} \frac{cf(x + \Delta x) - cf(x)}{\Delta x}$$

$$= \lim_{\Delta x \to 0} c\left[\frac{f(x + \Delta x) - f(x)}{\Delta x}\right]$$

$$= c \lim_{\Delta x \to 0} \frac{f(x + \Delta x) - f(x)}{\Delta x}$$

$$= c\,\frac{d}{dx}[f(x)].$$

(b) The difference quotient for $f + g$ is

$$\frac{(f + g)(x + \Delta x) - (f + g)(x)}{\Delta x} = \frac{f(x + \Delta x) + g(x + \Delta x) - f(x) - g(x)}{\Delta x}$$

$$= \frac{f(x + \Delta x) - f(x)}{\Delta x} + \frac{g(x + \Delta x) - g(x)}{\Delta x}.$$

Consequently,

$$\frac{d}{dx}[f(x) + g(x)] = \lim_{\Delta x \to 0} \frac{(f + g)(x + \Delta x) - (f + g)(x)}{\Delta x}$$

$$= \lim_{\Delta x \to 0} \frac{f(x + \Delta x) - f(x)}{\Delta x} + \lim_{\Delta x \to 0} \frac{g(x + \Delta x) - g(x)}{\Delta x}$$

$$= \frac{d}{dx}[f(x)] + \frac{d}{dx}[g(x)].$$

(c) The proof of this assertion is trickier. First, we examine the difference quotient for fg. We have

$$\frac{(fg)(x + \Delta x) - (fg)(x)}{\Delta x} = \frac{f(x + \Delta x)\,g(x + \Delta x) - f(x)\,g(x)}{\Delta x}.$$

Now, how are we going to use the fact that f and g are differentiable at x? The differentiability of f and g at x means that each of the difference quotients

$\dfrac{f(x + \Delta x) - f(x)}{\Delta x}$ and $\dfrac{g(x + \Delta x) - g(x)}{\Delta x}$ has a limit as Δx approaches 0. Hence it seems reasonable that the proof of (c) should use these facts just as (b) did. Therefore we should try to somehow introduce the difference quotients for f and g into the expression representing the difference quotient for fg. We achieve this by adding 0 in the appropriate form as follows:

$$\dfrac{f(x + \Delta x)g(x + \Delta x) - f(x)g(x)}{\Delta x}$$

$$= \dfrac{f(x + \Delta x)g(x + \Delta x) - f(x)g(x + \Delta x) + f(x)g(x + \Delta x) - f(x)g(x)}{\Delta x}$$

$$= \dfrac{[f(x + \Delta x) - f(x)]\,g(x + \Delta x) + f(x)\,[g(x + \Delta x) - g(x)]}{\Delta x}$$

$$= \left[\dfrac{f(x + \Delta x) - f(x)}{\Delta x}\right] g(x + \Delta x) + f(x)\left[\dfrac{g(x + \Delta x) - g(x)}{\Delta x}\right].$$

It is a fact, which we accept without proof, that if a function is differentiable at a point, it is continuous at that point. Therefore g is continuous at x so that $\lim\limits_{\Delta x \to 0} g(x + \Delta x) = g(x)$. Applying the results on limits of sums and products yields

$$\dfrac{d}{dx}\,[f(x)g(x)]$$

$$= \lim_{\Delta x \to 0} \dfrac{f(x + \Delta x)g(x + \Delta x) - f(x)g(x)}{\Delta x}$$

$$= \lim_{\Delta x \to 0}\left[\dfrac{f(x + \Delta x) - f(x)}{\Delta x}\right] \cdot \lim_{\Delta x \to 0} g(x + \Delta x) + f(x) \lim_{\Delta x \to 0}\left[\dfrac{g(x + \Delta x) - g(x)}{\Delta x}\right]$$

$$= \dfrac{d}{dx}\,[f(x)]\,g(x) + f(x)\,\dfrac{d}{dx}\,[g(x)].$$

Let us use the preceding differentiation formulas to show how easily poly-nomial functions can be differentiated. Recall that $\dfrac{d}{dx}\,[x] = 1$. Therefore, by (b),

$$\dfrac{d}{dx}\,[x^2] = \dfrac{d}{dx}\,[x \cdot x] = \dfrac{d}{dx}\,[x] \cdot x + x\,\dfrac{d}{dx}\,[x] = 1 \cdot x + x \cdot 1 = 2x.$$

Similarly,

$$\dfrac{d}{dx}\,[x^3] = \dfrac{d}{dx}\,[x^2 \cdot x] = \dfrac{d}{dx}\,[x^2] \cdot x + x^2 \cdot \dfrac{d}{dx}\,[x] = 2x \cdot x + x^2 \cdot 1 = 3x^2.$$

A similar argument shows $\dfrac{d}{dx} [x^4] = 4x^3$. Continuing in this manner, we see that, if n is a positive integer, then

$$\frac{d}{dx} [x^n] = nx^{n-1}.$$

Now let $p(x) = a_n x^n + a_{n-1} x^{n-1} + \ldots + a_1 x + a_0$ be a polynomial function. It can be easily shown that part (b) of Theorem 6.7 extends to any finite number of functions. That is, the derivative of the sum of any finite number of functions is the sum of the derivatives. It then follows that

$$\frac{d}{dx} [p(x)] = \frac{d}{dx} [a_n x^n] + \frac{d}{dx} [a_{n-1} x^{n-1}] + \ldots + \frac{d}{dx} [a_1 x] + \frac{d}{dx} [a_0]$$

$$= a_n \frac{d}{dx} [x^n] + a_{n-1} \frac{d}{dx} [x^{n-1}] + \ldots + a_1 \frac{d}{dx} [x] + \frac{d}{dx} [a_0]$$

$$= a_n \cdot nx^{n-1} + a_{n-1} (n-1)x^{n-2} + \ldots + a_1 x^0 + 0$$

$$= na_n x^{n-1} + (n-1)a_{n-1} x^{n-2} + \ldots + a_1.$$

Applications of the differentiation formulas developed in this section are presented in the following example.

Example.

(i) $\dfrac{d}{dx} [6x - 5] = \dfrac{d}{dx} [6x] + \dfrac{d}{dx} [-5] = 6\dfrac{d}{dx} [x] + \dfrac{d}{dx} [-5] = 6(1) + 0 = 6$

(ii) $\dfrac{d}{dx} [9x^2 - 6x + 7] = \dfrac{d}{dx} [9x^2] + \dfrac{d}{dx} [-6x] + \dfrac{d}{dx} [7]$

$$= 9\frac{d}{dx} [x^2] - 6\frac{d}{dx} [x] + \frac{d}{dx} [7] = 9(2x) - 6(1) + 0$$

$$= 18x - 6$$

(iii) $\dfrac{d}{dx} [-5x^3 + 7x^2 + 2x - 3] = 3(-5x^2) + 2(7x) + 2x^0$

$$= -15x^2 + 14x + 2$$

(iv) $\dfrac{d}{dx} [2x^4 + \sqrt{2} x^3 - x^2 + 5x + 8] = 4(2x^3) + 3(\sqrt{2} x^2) + 2(-x) + 5$

$$= 8x^3 + 3\sqrt{2} x^2 - 2x + 5$$

(v) $\dfrac{d}{du} [9u^{10} + 16u^7 - 11u^4 + 13] = 10(9u^9) + 7(16u^6) + 4(-11u^3)$

$$= 90u^9 + 112u^6 - 44u^3$$

(vi) $\dfrac{d}{dx} [(x^3 - 2x^2 + 3)(5x + 7)]$

$$= \dfrac{d}{dx} [x^3 - 2x^2 + 3] \cdot (5x + 7) + (x^3 - 2x^2 + 3) \dfrac{d}{dx} [5x + 7]$$

$$= (3x^2 - 4x)(5x + 7) + (x^3 - 2x^2 + 3) \cdot 5$$

(vii) $\dfrac{d}{dv} [(6v^4 - \sqrt{5}\,v + 9)(v^{20} + 3v^{15} - 11)]$

$$= \dfrac{d}{dv} [6v^4 - \sqrt{5}\,v + 9] \cdot (v^{20} + 3v^{15} - 11)$$

$$+ (6v^4 - \sqrt{5}\,v + 9) \dfrac{d}{dv} [v^{20} + 3v^{15} - 11]$$

$$= (24v^3 - \sqrt{5})(v^{20} + 3v^{15} - 11) + (6v^4 - \sqrt{5}\,v + 9)(20v^{19} + 45v^{14})$$

(viii) $\dfrac{d}{dx} \left[\dfrac{4x^2 - 11x + 2}{-x^3 + 6x^2 - 5x + 1} \right]$

$$= (-x^3 + 6x^2 - 5x + 11)(d/dx)[4x^2 - 11x + 2] - (4x^2 - 11x + 2)$$

$$\cdot (d/dx)[-x^3 + 6x^2 - 5x + 11]/(-x^3 + 6x^2 - 5x + 1)^2$$

$$= (-x^3 + 6x^2 - 5x + 11)(8x - 11) - (4x^2 - 11x + 2)$$

$$\cdot (-3x^2 + 12x - 5)/(-x^3 + 6x^2 - 5x + 1)^2$$

(ix) $\dfrac{d}{du} \left[\dfrac{\sqrt{3}\,u^5 - 2u^4 + u^3 + 7u^2 + 6}{u^2 + 1} \right]$

$$= (u^2 + 1)(d/dx)[\sqrt{3}\,u^5 - 2u^4 + u^3 + 7u^2 + 6] - (\sqrt{3}\,u^5 - 2u^4 + u^3$$

$$+ 7u^2 + 6)(d/du)[u^2 + 1]/(u^2 + 1)^2$$

$$= (u^2 + 1)(5\sqrt{3}\,u^4 - 8u^3 + 3u^2 + 14u) - (\sqrt{3}\,u^5 - 2u^4 + u^3 + 7u^2 + 6)$$

$$\cdot (2u)/(u^2 + 1)^2$$

Let m be a negative integer. There is a positive integer n such that $m = -n$. Therefore $x^m = \dfrac{1}{x^n}$. It follows that

$$\frac{d}{dx}[x^m] = \frac{d}{dx}\left[\frac{1}{x^n}\right]$$

$$= \frac{x^n \cdot \frac{d}{dx}[1] - 1 \cdot \frac{d}{dx}[x^n]}{(x^n)^2}$$

$$= \frac{-nx^{n-1}}{x^{2n}}$$

$$= -nx^{-n-1}$$

$$= mx^{m-1}.$$

Combining this with the result for positive integers, we see that

$$\frac{d}{dx}[x^n] = nx^{n-1}$$

for any nonzero integer n.

Example.

(i) $\dfrac{d}{dx}[x^{-1}] = -x^{-2}$

(ii) $\dfrac{d}{dx}[x^{-2}] = -2x^{-3}$

(iii) $\dfrac{d}{dx}[x^{-5}] = -5x^{-6}$

(iv) $\dfrac{d}{du}[3u^{-5} + 4u^{-2} + u - 7] = 3(-5u^{-6}) + 4(-2u^{-3}) + 1$

$$= -15u^{-6} - 8u^{-3} + 1$$

(v) $\dfrac{d}{dw}[(2w^{-6} - 3w)(w^2 - w^{-3})]$

$$= \frac{d}{dw}[2w^{-6} - 3w] \cdot (w^2 - w^{-3}) + (2w^{-6} - 3w)\frac{d}{dw}[w^2 - w^{-3}]$$

$$= [2(-6w^{-7}) - 3](w^2 - w^{-3}) + (2w^{-6} - 3w)[2w - (-3w^{-4})]$$

$$= (-12w^{-7} - 3)(w^2 - w^{-3}) + (2w^{-6} - 3w)(2w + 3w^{-4})$$

Exercise Set 6.7

1. Using the fact that $\dfrac{d}{dx}[x^n] = nx^{n-1}$, find

 a. $\dfrac{d}{dx}[x^2]$　 b. $\dfrac{d}{dx}[x^{12}]$　 c. $\dfrac{d}{dx}[x]$　 d. $\dfrac{d}{dx}[x^{-2}]$　 e. $\dfrac{d}{dx}\left[\dfrac{1}{x^3}\right]$

2. Use Theorem 6.7(a) and the results of problem 1 to find

 a. $\dfrac{d}{dx}[5x^2]$　 b. $\dfrac{d}{dx}\left[\dfrac{x^{12}}{5}\right]$　 c. $\dfrac{d}{dx}[-x]$　 d. $\dfrac{d}{dx}[\sqrt{2}\,x^{-2}]$ e. $\dfrac{d}{dx}\left[\dfrac{1}{3x^3}\right]$

3. Use Theorem 6.7(b) and the results of problem 2 to find

 a. $\dfrac{d}{dx}[5x^2 - x]$　　　　　 b. $\dfrac{d}{dx}\left[\dfrac{\sqrt{2}}{x^2} + \dfrac{1}{3x^3}\right]$

4. Use Theorem 6.7(c) and the results of problem 1 to find

 a. $\dfrac{d}{dx}[x^2 x^{12}]$　　　 b. $\dfrac{d}{dx}[x^{12} x^{-2}]$　　　 c. $\dfrac{d}{dx}\left[x^{12}\dfrac{1}{x^3}\right]$

5. Use Theorem 6.7(d) to find

 a. $\dfrac{d}{dx}\left[\dfrac{1}{x^3}\right]$　　 b. $\dfrac{d}{dx}\left[\dfrac{3x-4}{2x+1}\right]$　　 c. $\dfrac{d}{dx}\left[\dfrac{5x^2-x}{x^3}\right]$

* * *

In problems 6–19, find the indicated derivative.

6. $\dfrac{d}{dx}[x^2 - x + 4]$ 　　　　　 7. $\dfrac{d}{dx}[x^3 + 5x^2 - 6x + 3]$

8. $\dfrac{d}{du}[5u^4 - \sqrt{2}\,u + 10]$ 　　　 9. $\dfrac{d}{du}[u^6 - u^5 + u^4 - u^3 + u^2 - u + 1]$

10. $\dfrac{d}{dx}[(3x - 7)(x^2 - 5)]$ 　　　 11. $\dfrac{d}{dz}[(z + z^{-1})(z^{-2} + z^2)]$

12. $\dfrac{d}{dt}[(2t^2 - t + 3)(t^3 - t - 1)]$ 　　 13. $\dfrac{d}{dt}[(-t^2 + 5t - 8)(t^2 - 5t + 8)]$

14. $\dfrac{d}{dx}\left[\dfrac{8}{x^2 + 4}\right]$ 　　　　　 15. $\dfrac{d}{dx}\left[\dfrac{x^2 + 4}{8}\right]$

16. $\dfrac{d}{dv}[2v^{-2} - v^{-3}]$ 　　　　 17. $\dfrac{d}{dv}\left[\dfrac{2 - v}{v^2}\right]$

18. $\dfrac{d}{dz}\left[\dfrac{3z^{14}-z^7}{2z^2+3z-9}\right]$ 19. $\dfrac{d}{dz}\left[\dfrac{z^2-5z+6}{z-3}\right]$

20. If $f(x)=\dfrac{x^2-x^{-1}+5}{6x^{-1}+1}$, find $f'(x)$. What is the slope of the tangent line to

the curve $y=f(x)$ at $\left(1,\dfrac{5}{7}\right)$?

21. If $f(x)=3x^2+8x-39$, find a point on the curve $y=f(x)$ at which the slope of the tangent line is zero.

*22. Prove part (d) of Theorem 6.7.

*23. Prove: If the functions f, g, and h are all differentiable at x and $F(x)=f(x)g(x)\,h(x)$, then

$$F'(x)=f'(x)g(x)\,h(x)+f(x)g'(x)\,h(x)+f(x)\,g(x)\,h'(x).$$

24. Use problem 23 to find

$$\frac{d}{dt}\left[(5t^2-4t+6)(t^3+6t^2+t+1)(2t-4)\right].$$

6.8 THE CHAIN RULE

In the examples of the previous section Theorem 6.7 was used to differentiate sums, products, and quotients of polynomial functions. Of course, we do not want to restrict ourselves to the consideration of just polynomial functions. Basically, however, our only method of finding derivatives is the Δ-process. What happens if we try to apply this method to more complicated functions?

First, let us consider $f(x)=\sqrt{x^2+1}$ and use the Δ-process. The difference quotient is

$$\frac{f(x+\Delta x)-f(x)}{\Delta x}=\frac{\sqrt{(x+\Delta x)^2+1}-\sqrt{x^2+1}}{\Delta x}.$$

Both numerator and denominator approach 0 as Δx approaches 0. Hence, if the difference quotient is left in this form, we will not be able to determine its behavior as Δx approaches 0. We can, however, perform a useful, though somewhat tricky, manipulation as follows:

$$\frac{\sqrt{(x+\Delta x)^2+1}-\sqrt{x^2+1}}{\Delta x}=\frac{\sqrt{(x+\Delta x)^2+1}-\sqrt{x^2+1}}{\Delta x}$$

$$\cdot\frac{\sqrt{(x+\Delta x)^2+1}+\sqrt{x^2+1}}{\sqrt{(x+\Delta x)^2+1}+\sqrt{x^2+1}}$$

$$= \frac{[(x + \Delta x)^2 + 1] - [x^2 + 1]}{\Delta x \, [\sqrt{(x + \Delta x)^2 + 1} + \sqrt{x^2 + 1}]}$$

$$= \frac{2x \Delta x + (\Delta x)^2}{\Delta x \, [\sqrt{(x + \Delta x)^2 + 1} + \sqrt{x^2 + 1}]}$$

$$= \frac{2x + \Delta x}{\sqrt{(x + \Delta x)^2 + 1} + \sqrt{x^2 + 1}} \cdot$$

Consequently, since $\lim\limits_{\Delta x \to 0} \, [\sqrt{(x + \Delta x)^2 + 1} + \sqrt{x^2 + 1}] = 2\sqrt{x^2 + 1}$, we have

$$\frac{d}{dx} \, [\sqrt{x^2 + 1}] = \lim_{\Delta x \to 0} \left[\frac{2x + \Delta x}{\sqrt{(x + \Delta x)^2 + 1} + \sqrt{x^2 + 1}} \right]$$

$$= \frac{2x}{2\sqrt{x^2 + 1}}$$

$$= \frac{x}{\sqrt{x^2 + 1}} \cdot$$

The trick used to simplify the difference quotient for $f(x) = \sqrt{x^2 + 1}$ has a drawback. Namely, the same technique will not work for functions such as $f(x) = (x^2 + 1)^{\frac{5}{16}}$. It is for this reason that the Δ-process really has very limited usefulness. If we are going to be able to differentiate more complicated functions, then we will need more sophisticated facts with which to work. The following theorem, called the *chain rule*, provides us with these facts. The proof of the chain rule is omitted.

6.8 Theorem *(Chain Rule).* *Let $u(x)$ be a function of x such that u is differentiable at x_0. IF g is a differentiable function of u on the interval (a, b) and $u(x_0) \in (a, b)$, THEN*

$$\frac{d}{dx} \, [g(u(x))] = \frac{d}{du} \, [g(u)] \, \frac{d}{dx} \, [u(x)]$$

at $x = x_0$.

While the statement of the theorem is admittedly complicated, its application is not. The following problems illustrate the use of the chain rule.

Problem. Find $\dfrac{d}{dx} \, [(x^3 - 2x^2 + 4x + 1)^9]$.

Solution. If we expand this expression, we have a polynomial function which can be easily differentiated. However, just imagine the work needed to obtain such an expansion! This work can be avoided by recognizing that $(x^3 - 2x^2 + 4x + 1)^9$ is a composite function. If we write $u(x) = x^3 - 2x^2 + 4x + 1$ and $g(u) = u^9$, then

$$g(u(x)) = (x^3 - 2x^2 + 4x + 1)^9.$$

Since both g and u are polynomial functions, they are differentiable so that the chain rule applies. By the chain rule,

$$\frac{d}{dx}[(x^3 - 2x^2 + 4x + 1)^9] = \frac{d}{dx}[g(u(x))]$$

$$= \frac{d}{du}[g(u)]\frac{d}{dx}[u(x)]$$

$$= \frac{d}{du}[u^9]\frac{d}{dx}[x^3 - 2x^2 + 4x + 1]$$

$$= 9u^8(3x^2 - 4x + 4)$$

$$= 9(x^3 - 2x^2 + 4x + 1)^8(3x^2 - 4x + 4).$$

Problem. Find $\frac{d}{dx}[(7x^4 + 2x^3 - x^2 + 3x + 4)^{-5}]$.

Solution. Again, we may apply the chain rule. In order to do this, let $u(x) = 7x^4 + 2x^3 - x^2 + 3x + 4$ and $g(u) = u^{-5}$. Thus

$$\frac{d}{dx}[(7x^4 + 2x^3 - x^2 + 3x + 4)^{-5}] = \frac{d}{dx}[g(u(x))]$$

$$= \frac{d}{du}[g(u)]\frac{d}{dx}[u(x)]$$

$$= \frac{d}{du}[u^{-5}]\frac{d}{dx}[7x^4 + 2x^3 - x^2 + 3x + 4]$$

$$= -5u^{-6}(28x^3 + 6x^2 - 2x + 3)$$

$$= -5(7x^4 + 2x^3 - x^2 + 3x + 4)^{-6}$$

$$\cdot (28x^3 + 6x^2 - 2x + 3).$$

If n is a nonzero integer, then $\frac{d}{du}[u^n] = nu^{n-1}$. This formula is also true for nonzero rational numbers. That is, if r is any nonzero rational number, then

$$\frac{d}{du}[u^r] = ru^{r-1}.$$

We accept the preceding fact, called the *power formula,* without proof and now show how it enables us to differentiate a wide variety of functions.

Example.

(i) $\dfrac{d}{du} [u^{\frac{7}{8}}] = \dfrac{7}{8} u^{\frac{7}{8}-1} = \dfrac{7}{8} u^{-\frac{1}{8}}$

(ii) $\dfrac{d}{dx} [x^{-\frac{5}{3}}] = -\dfrac{5}{3} x^{-\frac{5}{3}-1} = -\dfrac{5}{3} x^{-\frac{8}{3}}$

(iii) $\dfrac{d}{dw} [(w^{\frac{1}{4}} + 2w - 3w^{\frac{3}{2}})(5w^{-1} + w^{\frac{5}{6}})]$

$= \dfrac{d}{dw} [w^{\frac{1}{4}} + 2w - 3w^{\frac{3}{2}}] \cdot (5w^{-1} + w^{\frac{5}{6}})$

$+ (w^{\frac{1}{4}} + 2w - 3w^{\frac{3}{2}}) \dfrac{d}{dw} [5w^{-1} + w^{\frac{5}{6}}]$

$= \left(\dfrac{1}{4} w^{-\frac{3}{4}} + 2 - \dfrac{9}{2} w^{\frac{1}{2}}\right)(5w^{-1} + w^{\frac{5}{6}})$

$+ (w^{\frac{1}{4}} + 2w - 3w^{\frac{3}{2}})(-5w^{-2} + \dfrac{5}{6} w^{-\frac{1}{6}})$

Example. Consider the function $f(x) = \sqrt{x^2 + 1}$ whose derivative was found using the Δ-process at the beginning of this section. Using the power formula and the chain rule, the derivative can be found quite easily. Let $u(x) = x^2 + 1$ and $g(u) = u^{\frac{1}{2}}$. Then $\sqrt{x^2 + 1} = g(u(x))$ so that

$$\dfrac{d}{dx} [\sqrt{x^2 + 1}] = \dfrac{d}{dx} [g(u(x))]$$

$$= \dfrac{d}{du} [g(u)] \dfrac{d}{dx} [u(x)]$$

$$= \dfrac{d}{du} [u^{\frac{1}{2}}] \dfrac{d}{dx} [x^2 + 1]$$

$$= \dfrac{1}{2} u^{-\frac{1}{2}} (2x)$$

$$= \dfrac{x}{u^{\frac{1}{2}}}$$

$$= \dfrac{x}{\sqrt{x^2 + 1}}.$$

Problem. Find the equation of the tangent line to the curve $y = (7x + 1)^{\frac{1}{3}}$ at the point $(1, 2)$.

Solution. The tangent line is that line passing through the point $(1, 2)$ whose slope is the value of $\dfrac{dy}{dx}$ when $x = 1$. If $u(x) = 7x + 1$ and $g(u) = u^{\frac{1}{3}}$, then $y = g(u(x))$. By the chain rule,

$$\frac{dy}{dx} = \frac{d}{du}\,[g(u)]\,\frac{d}{dx}\,[u(x)] = \frac{1}{3}(7x + 1)^{-\frac{2}{3}}(7) = \frac{7}{3}(7x + 1)^{-\frac{2}{3}}.$$

Letting $x = 1$, we see that the slope of the tangent line is

$$\frac{7}{3}(7 + 1)^{-\frac{2}{3}} = \frac{7}{12}.$$

Therefore, the equation of the desired tangent line is

$$\frac{y - 2}{x - 1} = \frac{7}{12}$$

or

$$y = \frac{7}{12}x + \frac{17}{12}.$$

In practice, the functions g and u are not usually explicitly written down when using the chain rule. Normally, it is quite easy to keep a mental record of the functions g and u and to apply the chain rule without an unnecessary amount of written work. In the following two examples, explicit mention of g and u is purposely omitted.

Example. $\dfrac{d}{dx}\,[(3x^4 + \sqrt{11}\,x^3 - 2x^2 - 7x + 4)\,(x^3 + 4)^{\frac{4}{3}}]$

$$= \frac{d}{dx}\,[(3x^4 + \sqrt{11}\,x^3 - 2x^2 - 7x + 4)]\,(x^3 + 4)^{\frac{4}{3}}$$

$$+ (3x^4 + \sqrt{11}\,x^3 - 2x^2 - 7x + 4)\frac{d}{dx}\,[(x^3 + 4)^{\frac{4}{3}}]$$

$$= (12x^3 + 3\sqrt{11}\,x^2 - 4x - 7)]\,(x^3 + 4)^{\frac{4}{3}}$$

$$+ (3x^4 + \sqrt{11}\,x^3 - 2x^2 - 7x + 4)\left[\frac{4}{3}(x^3 + 1)^{\frac{1}{3}}\,3x^2\right].$$

Notice that we first differentiated a product and, then, applied the chain rule.

Example.

$$\frac{d}{dx}\left[\frac{(3x^2 + x - 4)^{\frac{1}{3}}}{(x + 1)}\right] = \frac{(x + 1)\frac{d}{dx}[(3x^2 + x - 4)^{\frac{1}{3}}] - (3x^2 + x - 4)^{\frac{1}{3}}\frac{d}{dx}[(x + 1)]}{(x + 1)^2}$$

$$= \frac{(x + 1)\left[\frac{1}{3}(3x^2 + x - 4)^{-\frac{2}{3}}(6x + 1)\right] - (3x^2 + x - 4)^{\frac{1}{3}}}{(x + 1)^2}.$$

Here, we first differentiated a quotient and, then, applied the chain rule.

Exercise Set 6.8

1. Use the fact that $\frac{d}{du}[u^r] = ru^{r-1}$ to find:

   a. $\frac{d}{du}[u^{\frac{1}{2}}]$ b. $\frac{d}{du}[u^{-\frac{1}{2}}]$ c. $\frac{d}{du}[u^{\frac{3}{2}}]$ d. $\frac{d}{du}[u^{\frac{1}{3}}]$

2. Choose functions $u(x)$ and $g(u)$ such that $h(x) = g(u(x))$ if:

   a. $h(x) = (x^2 - 4)^{\frac{1}{2}}$ b. $h(x) = \frac{-8}{\sqrt{3x + 4}}$ c. $h(x) = \sqrt[3]{x^2 - x + 10}$

3. Use the chain rule together with the results of problems 1 and 2 to find:

   a. $\frac{d}{dx}[(x^2 - 4)^{\frac{1}{2}}]$ b. $\frac{d}{dx}\left[\frac{-8}{\sqrt{3x + 4}}\right]$ c. $\frac{d}{dx}\left[\sqrt[3]{x^2 - x + 10}\right]$

4. Find $f'(x)$, if $f(x) = \sqrt{x + 1}$, by:
   a. using the Δ-process;
   b. using the chain rule.

5. Find $g'(x)$, if $g(x) = (x^2 - 1)^3$, by:
   a. using the chain rule;
   b. by performing the multiplication and then finding the derivative.

* * *

In problems 6–11, use the chain rule to find the derivative of the given function.

6. $g(x) = \sqrt{5x^2 + 1}$

7. $g(x) = (5x^2 + 1)^{32}$

8. $y = \frac{1}{\sqrt{6 - x}}$

9. $y = (z^3 - 6)^{\frac{2}{5}}$

10. $s = (t + 1)^8$

11. $s = (3t - 12)^{-8}$

12. Find $\dfrac{d}{dx}(x^2 + 2x - 1)^4$.

13. Find $\dfrac{d}{dx}(x^3 + x^2 - x)^{-2}$.

14. Find $\dfrac{d}{dx}[(x - 3)^4(7x + 1)^3]$.

15. Find $\dfrac{d}{dx}[(x^4 - 3)(7x + 1)^{-3}]$.

16. Find $\dfrac{d}{dx}\left[\dfrac{3x^4}{(7x + 1)^3}\right]$.

17. If $y = (x^2 - 8x + 4)(x + 3)^{-\frac{4}{5}}$, find y'.

18. If $y = (3x^4 - x^3 + x^2 - 2x + 14)^{-\frac{1}{3}}$, find y'.

19. Find $\dfrac{d}{dw}[(w - 1)^{\frac{1}{3}}(2w^2 + 3)^{\frac{1}{2}}(4 - w)^{\frac{2}{9}}]$.

20. Find $\dfrac{d}{dx}\left[\dfrac{-7x}{\sqrt{x^3 + 2}}\right]$.

21. Find $\dfrac{d}{dx}[x^2\sqrt{x^2 - 11x - 3}]$.

22. Find the equation of the tangent line to the curve $y = (x^2 + 1)^{\frac{1}{3}}$ at the point $(-1, \sqrt[3]{2})$.

23. Find all values of t for which $h'(t) = 0$ if $h(t) = (t^3 - 4)^{\frac{5}{3}}$.

24. Show that the curve $y = \sqrt{x}$ has no tangent lines which are horizontal.

6.9 DERIVATIVE OF THE EXPONENTIAL FUNCTION

In many applications of the differential calculus, problems involving growth (such as population growth) and decay (such as radioactive decay) are present. Such applications are impossible to discuss without some knowledge of derivatives of exponential functions. The following theorem gives us the essential fact concerning the derivative of the exponential function.

6.9 Theorem. $\dfrac{d}{du}[e^u] = e^u$.

The preceding statement means that $\lim\limits_{\Delta u \to 0} \dfrac{e^{u + \Delta u} - e^u}{\Delta u} = e^u$. The following problems illustrate the use of Theorem 6.9.

Problem. Find $\dfrac{d}{dx} [e^{2x}]$.

Solution. We employ the chain rule once again. Let $u(x) = 2x$ and $g(u) = e^u$. Then $e^{2x} = g(u(x))$. Therefore

$$\frac{d}{dx} [e^{2x}] = \frac{d}{du} [g(u(x))]$$

$$= \frac{d}{du} [g(u)] \frac{d}{dx} [u(x)]$$

$$= \frac{d}{du} [e^u] \frac{d}{dx} [2x]$$

$$= e^u (2)$$

$$= 2e^{2x}.$$

Problem. Find $\dfrac{d}{dx} [e^{3x^2}]$.

Solution. Letting $u(x) = 3x^2$ and $g(u) = e^u$, we have $e^{3x^2} = g(u(x))$. Thus,

$$\frac{d}{dx} [e^{3x^2}] = \frac{d}{du} [g(u)] \frac{d}{dx} [u(x)]$$

$$= e^u (6x)$$

$$= 6xe^{3x^2}.$$

Problem. Find $\dfrac{d}{dx} [x^2 e^{\sqrt{x^3 - 7}}]$.

Solution. We have

$$\frac{d}{dx} [x^2 e^{\sqrt{x^3 - 7}}] = \frac{d}{dx} [x^2] e^{\sqrt{x^3 - 7}} + x^2 \frac{d}{dx} [e^{\sqrt{x^3 - 7}}]$$

$$= 2x e^{\sqrt{x^3 - 7}} + x^2 \frac{d}{dx} [e^{\sqrt{x^3 - 7}}].$$

In order to find $\dfrac{d}{dx} [e^{\sqrt{x^3 - 7}}]$, we let $u(x) = \sqrt{x^3 - 7}$ and $g(u) = e^u$. Then

$$\frac{d}{dx} [e^{\sqrt{x^3 - 7}}] = \frac{d}{du} [g(u)] \frac{d}{dx} [u(x)]$$

$$= e^{\sqrt{x^3-7}} \frac{d}{dx} [(x^3 - 7)^{\frac{1}{2}}]$$

$$= e^{\sqrt{x^3-7}} \cdot \frac{1}{2}(x^3 - 7)^{-\frac{1}{2}}(3x^2)$$

$$= \frac{3}{2} x^2 (x^3 - 7)^{-\frac{1}{2}} e^{\sqrt{x^3-7}}.$$

Therefore

$$\frac{d}{dx} [x^2 e^{\sqrt{x^3-7}}] = 2x e^{\sqrt{x^3-7}} + \frac{3x^4}{2}(x^3 - 7)^{-\frac{1}{2}} e^{\sqrt{x^3-7}}.$$

Problem. Find $\dfrac{d}{dx} \left[\dfrac{e^{x^3+7}}{\sqrt{2x^2 + 8}} \right]$.

Solution. In this solution, explicit mention of $u(x)$ and $g(u)$ is avoided:

$$\frac{d}{dx} \left[\frac{e^{x^3+7}}{\sqrt{2x^2 + 8}} \right]$$

$$= \frac{\sqrt{2x^2 + 8} \dfrac{d}{dx}(e^{x^3+7}) - e^{x^3+7} \dfrac{d}{dx}(\sqrt{2x^2 + 8})}{(\sqrt{2x^2 + 8})^2}$$

$$= \frac{\sqrt{2x^2 + 8}\,(3x^2)e^{x^3+7} - e^{x^3+7} \left[\dfrac{1}{2}(4x)(2x^2 + 8)^{-\frac{1}{2}} \right]}{2x^2 + 8}.$$

Problem. Find $\dfrac{d}{dx} [(1 + e^{x^3})^{\frac{5}{7}}]$.

Solution. We have

$$\frac{d}{dx} [(1 + e^{x^3})^{\frac{5}{7}}] = \frac{5}{7}(1 + e^{x^3})^{-\frac{2}{7}} \frac{d}{dx} [1 + e^{x^3}]$$

$$= \frac{5}{7}(1 + e^{x^3})^{-\frac{2}{7}} (e^{x^3}) \cdot 3x^2$$

$$= \frac{15x^2}{7}(1 + e^{x^3})^{-\frac{2}{7}} (e^{x^3}).$$

Problem. Find the equation of the tangent line to the curve $y = e^{x^3}$ at the point $(\sqrt[3]{2}, e^2)$.

Solution. We have

$$\frac{dy}{dx} = \frac{d}{dx}\,[e^{x^3}]$$

$$= 3x^2 e^{x^3}.$$

When $x = \sqrt[3]{2}$, $\dfrac{dy}{dx} = 3\sqrt[3]{4}\,e^2$. The equation of the tangent line is

$$\frac{y - e^2}{x - \sqrt[3]{2}} = 3\sqrt[3]{4}\,e^2$$

or

$$y - e^2 = 3\sqrt[3]{4}\,e^2\,(x - \sqrt[3]{2}).$$

Exercise Set 6.9

1. Find the indicated derivatives:

   a. $\dfrac{d}{du}\,[e^u]$ b. $\dfrac{d}{du}\,[2e^u]$ c. $\dfrac{d}{du}\,[e^{-u}]$ d. $\dfrac{d}{du}\,[e^{\frac{u}{2}}]$

2. Choose functions $u(x)$ and $g(u)$ such that $h(x) = g(u(x))$ where:

   a. $h(x) = e^{x^2 + x + 1}$ b. $h(x) = 2e^{\pi x}$ c. $h(x) = e^{-(x + \sqrt{2})}$ d. $h(x) = e^{\frac{x^2}{2}}$

3. Use the chain rule and the results of problems 1 and 2 to find:

   a. $\dfrac{d}{dx}\,[e^{x^2 + x + 1}]$ b. $\dfrac{d}{dx}\,[2e^{\pi x}]$ c. $\dfrac{d}{dz}\,[e^{-(z + \sqrt{2})}]$ d. $\dfrac{d}{dt}\,[e^{\frac{t^2}{2}}]$

4. Find $\dfrac{dy}{dx}$ where:

   a. $y = xe^x$ b. $y = \dfrac{x}{e^x}$ c. $y = \dfrac{e^x}{x}$ d, $y = (e^x + x)^{19}$

* * *

In problems 5–16, find the indicated derivative.

5. $\dfrac{d}{dx}\,[e^{3x}]$ 6. $\dfrac{d}{dx}\,[e^{-2x}]$

7. $\dfrac{d}{dx}\,[e^{x-1}]$ 8. $\dfrac{d}{dx}\,[e^{6x-7}]$

9. $\dfrac{d}{dt}[e^{t^3}]$

10. $\dfrac{d}{dt}[e^{4t^2-5t+11}]$

11. $\dfrac{d}{dt}[e^{\sqrt{t}}]$

12. $\dfrac{d}{dt}[e^{3\sqrt{t^2+1}}]$

13. $\dfrac{d}{dy}[\sqrt[3]{e^3y}]$

14. $\dfrac{d}{dv}[e^{\sqrt[3]{v^2+1}}]$

15. $\dfrac{d}{dx}[\sqrt{2}\,e^{x^2-3}]$

16. $\dfrac{d}{dy}\left[\dfrac{e^{y^2}}{\sqrt{1+y}}\right]$

17. Find y' if $y = (1 + e^x)^2$.

18. Find y' if $y = (1 + e^{2x})^{\frac{3}{2}}$.

19. Find $\dfrac{dy}{dz}$ if $y = \dfrac{e^{-2z} + e^{2z}}{2}$.

20. Find $\dfrac{dy}{dz}$ if $y = \dfrac{(e^{-2z})(e^{2z})}{2}$.

21. Find $f'(x)$ if $f(x) = e^{-x}(x^2 - 5x + 4)^{\frac{1}{3}}$.

22. Find $g'(z)$ if $g(z) = e^{-2z} + e^{z^2}$.

23. Write an equation of the tangent line to the curve $y = xe^x$ at the point $(1, e)$.

24. Find a point at which the tangent to the curve $y = e^x(x - 1)$ has a slope of zero.

6.10 DERIVATIVE OF THE LOGARITHM FUNCTION

As is the case with the exponential function, a knowledge of the derivative of the logarithm function is required when dealing with applications of the differential calculus. The following theorem is the basic fact we need.

6.10 Theorem. $\dfrac{d}{du}[\ln u] = \dfrac{1}{u}$.

The preceding statement means that $\displaystyle\lim_{\Delta u \to 0} \dfrac{\ln(u + \Delta u) - \ln u}{\Delta u} = \dfrac{1}{u}$. The use of Theorem 6.10 is illustrated in the following problems.

Problem. Find $\dfrac{d}{dx}[\ln(x^2 + 3x - 15)]$.

Solution. If we let $u(x) = x^2 + 3x - 5$ and $g(u) = \ln u$, then $\ln (x^2 + 3x - 15) = g(u(x))$.

Therefore

$$\frac{d}{dx} [\ln (x^2 + 3x - 15)] = \frac{d}{du} [g(u)] \frac{d}{dx} [u(x)]$$

$$= \frac{d}{du} [\ln u] \frac{d}{dx} [x^2 + 3x - 15]$$

$$= \frac{1}{u} \cdot (2x + 3)$$

$$= \frac{2x + 3}{x^2 + 3x - 15}.$$

Problem. Find $\dfrac{d}{dx} [\ln (1 + e^x)]$.

Solution. Letting $u(x) = 1 + e^x$ and $g(u) = \ln u$, we have

$$\frac{d}{dx} [\ln (1 + e^x)] = \frac{d}{du} [g(u)] \frac{d}{dx} [u(x)]$$

$$= \frac{1}{u} \cdot e^x$$

$$= \frac{e^x}{1 + e^x}.$$

Notice, by using the chain rule, we have

$$\frac{d}{dx} [\ln (f(x))] = \frac{1}{f(x)} \frac{d}{dx} [f(x)] = \frac{f'(x)}{f(x)}.$$

Using this in the preceding two problems would have given us the desired results immediately.

Problem. Find $\dfrac{d}{dx} [\sqrt{3x^2 - 5} \ln (x^3 + 7)]$.

Solution. We avoid explicit mention of g and u. Thus

$$\frac{d}{dx} [\sqrt{3x^2 - 5} \ln (x^3 + 7)]$$

$$= \frac{d}{dx} [\sqrt{3x^2 - 5}] \cdot \ln (x^3 + 7) + \sqrt{3x^2 - 5} \frac{d}{dx} [\ln (x^3 + 7)]$$

$$= \left[\frac{1}{2} (3x^2 - 5)^{-\frac{1}{2}} (6x) \right] \ln (x^3 + 7) + \sqrt{3x^2 - 5} \frac{3x^2}{x^3 + 7}.$$

Problem. Find $\dfrac{d}{dx}\left[\dfrac{1+4x^{\frac{1}{3}}}{\ln(5+2x)}\right]$.

Solution. Again, explicit mention of g and u is avoided. The computations proceed as follows:

$$\frac{d}{dx}\left[\frac{1+4x^{\frac{1}{3}}}{\ln(5+2x)}\right]$$

$$=\frac{\ln(5x+2x)\cdot\dfrac{d}{dx}[1+4x^{\frac{1}{3}}]-(1+4x^{\frac{1}{3}})\dfrac{d}{dx}[\ln(5+2x)]}{[\ln(5+2x)]^2}$$

$$=\frac{\ln(5+2x)\cdot\left(\dfrac{4}{3}x^{-\frac{2}{3}}\right)-(1+4x^{\frac{1}{3}})\dfrac{2}{5+2x}}{[\ln(5+2x)]^2}.$$

Problem. Find the equation of the tangent line to the curve $y=\ln(\sqrt{2}\,x-1)$ at the point $(2,\ln(2\sqrt{2}-1))$.

Solution. We have

$$\frac{dy}{dx}=\frac{d}{dx}[\ln(\sqrt{2}\,x-1)]$$

$$=\frac{\sqrt{2}}{\sqrt{2}\,x-1}.$$

When $x=2$, $\dfrac{dy}{dx}=\dfrac{\sqrt{2}}{2\sqrt{2}-1}$. The equation of the desired tangent line is

$$\frac{y-\ln(2\sqrt{2}-1)}{x-2}=\frac{\sqrt{2}}{2\sqrt{2}-1}$$

or

$$y-\ln(2\sqrt{2}-1)=\frac{\sqrt{2}}{2\sqrt{2}-1}(x-2).$$

Exercise Set 6.10

1. Find the indicated derivatives:

  a. $\dfrac{d}{du}[\ln u]$ b. $\dfrac{d}{du}[\ln 2u]$ c. $\dfrac{d}{du}2[\ln u]$ d. $\dfrac{d}{du}[\ln u]^2$

2. Choose functions $u(x)$ and $g(u)$ such that $f(x) = g(u(x))$ where:

a. $f(x) = \ln x^2$ b. $f(x) = \ln \dfrac{x+1}{x-1}$ c. $f(x) = 2 \ln x^3$ d. $f(x) = \ln \dfrac{1}{x}$

3. Use the chain rule and the results of problems 1 and 2 to find:

a. $\dfrac{d}{dx} [\ln x^2]$ b. $\dfrac{d}{dx} \left[\ln \dfrac{x+1}{x-1} \right]$ c. $\dfrac{d}{dx} [2 \ln x^3]$ d. $\dfrac{d}{dx} \left[\ln \dfrac{1}{x} \right]$

4. Find $\dfrac{dy}{dx}$ where:

a. $y = x \ln x$ b. $y = \dfrac{\ln x}{x}$ c. $y = \ln x + \ln x^2$ d. $y = (\ln x)(\ln x^2)$

* * *

In problems 5–12, find the indicated derivative.

5. $\dfrac{d}{dx} [\ln (1+x)]$

6. $\dfrac{d}{dx} [\ln (1+2x)]$

7. $\dfrac{d}{dx} [\ln (1+x^2)]$

8. $\dfrac{d}{dx} [\ln (1+x)^2]$

9. $\dfrac{d}{dt} [\ln (t^2 - 6t + 5)]$

10. $\dfrac{d}{dt} [\ln (1+\sqrt{t})]$

11. $\dfrac{d}{dx} [\ln (3 - 4e^x)]$

12. $\dfrac{d}{dx} [\ln (5 + e^{x^2})]$

Recall that if x and y are positive real numbers, then:

(i) $\ln xy = \ln x + \ln y$,

(ii) $\ln \dfrac{x}{y} = \ln x - \ln y$,

(iii) $\ln x^y = y \ln x$,

and

(iv) $\ln e^x = x$.

In problems 13–18, use these rules to simplify the given function. Then find the derivative of the resulting function.

13. $f(x) = \ln (xe^x)$

14. $g(x) = \ln x^{43}$

15. $h(u) = \ln \dfrac{u}{u+3}$

16. $F(t) = \ln t^{\frac{5}{2}}$

17. $F(x) = \ln \left(\dfrac{1}{\sqrt[3]{x}} \right)$

18. $H(w) = \ln \left(\dfrac{e^w}{w^2} \right)$

In problems 19–24 find y' where:

19. $y = x^2 \ln x$

20. $y = \dfrac{\ln x}{x^2}$

21. $y = \dfrac{\ln (2x + 1)}{1 + e^{-x}}$

22. $y = (\sqrt{1 + e^x}) (\ln \sqrt{x})$

23. $y = \dfrac{\ln (x^2 + 1)}{\ln (x^3 + 1)}$

24. $y = \sqrt{\ln (3x^2 + x + 5)}$

25. Find the equation of the tangent line to the curve $y = x \ln x$ at the point (e, e).

26. Find a point on the curve $y = x \ln x$ at which the tangent line is horizontal. Sketch the graph of $y = x \ln x$ on (e^{-1}, e).

6.11 HIGHER DERIVATIVES

As we have seen in preceding sections, the derivative of a function is, itself, a function. Consequently, the derivative of a function may also be differentiable. For example, if $y = 2x^3 - 8x^2 + 3x - 5$, then $\dfrac{dy}{dx} = 6x^2 - 16x + 3$. It follows that $\dfrac{dy}{dx}$ also has a derivative. In fact, $\dfrac{d}{dx}\left[\dfrac{dy}{dx}\right] = 12x - 16$ and the latter derivative is called the second derivative of y with respect to x. We write $\dfrac{d^2y}{dx^2}$ to denote the second derivative of y with respect to x. Thus $\dfrac{d^2y}{dx^2} = 12x - 16$, where $y = 2x^3 - 8x^2 + 3x - 5$. If we write $f(x) = 2x^3 - 8x^2 + 3x - 5$ and $f'(x) = 6x^2 - 16x + 3$, then the second derivative is denoted by $f''(x)$. That is, $f''(x) = 12x - 16$.

Problem. Find $\dfrac{d^2y}{dx^2}$ if $y = e^{3x}\sqrt{x + 1}$.

Solution. We have

$$\frac{dy}{dx} = 3e^{3x}(x + 1)^{\frac{1}{2}} + \frac{1}{2}e^{3x}(x + 1)^{-\frac{1}{2}}$$

so that

$$\frac{d^2y}{dx^2} = \frac{d}{dx}\left[\frac{dy}{dx}\right]$$

$$= 9e^{3x}(x + 1)^{\frac{1}{2}} + \frac{3}{2}e^{3x}(x + 1)^{-\frac{1}{2}} + \frac{3}{2}e^{3x}(x + 1)^{-\frac{1}{2}} - \frac{1}{4}e^{3x}(x + 1)^{-\frac{3}{2}}.$$

Problem. Find $f''(x)$ if $f(x) = \dfrac{x^2 + 1}{x - 1}$.

Solution. We have

$$f'(x) = \frac{(x - 1)\, 2x - (x^2 + 1)}{(x - 1)^2}$$

$$= \frac{x^2 - 2x - 1}{(x - 1)^2}.$$

Therefore

$$f''(x) = \frac{(x - 1)^2 (2x - 2) - (x^2 - 2x - 1)\, 2(x - 1)}{(x - 1)^4}$$

$$= \frac{4}{(x - 1)^3}.$$

Naturally, we can take derivatives of second derivatives to obtain third derivatives denoted by $\dfrac{d^3 y}{dx^3}$. Fourth derivatives, fifth derivatives, and so on, are defined in the obvious manner. Thus,

$$\frac{d^4 y}{dx^4}, \frac{d^5 y}{dx^5}, \ldots, \frac{d^n y}{dx^n}$$

denote the fourth, fifth, . . . , nth derivatives. Similarly, if we write

$$f'(x) = f^{(1)}(x), \quad f''(x) = f^{(2)}(x), \quad f'''(x) = f^{(3)}(x),$$

then

$$f^{(4)}(x), f^{(5)}(x), \ldots, f^{(n)}(x)$$

indicate the fourth, fifth, . . . , nth derivative.

Example. If $y = x^4 - 2x^3 - x^2 + 10$, then

$$\frac{dy}{dx} = 4x^3 - 6x^2 - 2x,$$

$$\frac{d^2 y}{dx^2} = 12x^2 - 12x - 2,$$

$$\frac{d^3 y}{dx^3} = 24x - 12,$$

$$\frac{d^4 y}{dx^4} = 24,$$

and
$$\frac{d^5 y}{dx^5} = 0.$$

Example. If $f(x) = e^{2x}$, then

$$f^{(1)}(x) = 2e^{2x},$$

$$f^{(2)}(x) = 4e^{2x},$$

and
$$f^{(3)}(x) = 8e^{2x}.$$

Exercise Set 6.11

1. Let $f(x) = 3x^2 - 7x + 8$. Find $f^{(1)}(x)$, $f^{(2)}(x)$, and $f^{(3)}(x)$.

2. If $g(x) = xe^x$, find $g'(x)$ and $g''(x)$.

3. If $y \sqrt{5x + 12}$, find $\dfrac{dy}{dx}$, $\dfrac{d^2 y}{dx^2}$, and $\dfrac{d^3 y}{dx^3}$.

4. If $f(x) = x^3 + \dfrac{1}{2}x$, find $f(2)$, $f'(2)$, and $f''(2)$.

5. If $g(x) = e^{x^2}$, find $g(0)$, $g'(0)$, and $g''(0)$.

* * *

In problems 6–15, find $\dfrac{d^2 y}{dx^2}$.

6. $y = 6x^2 + 5x + \sqrt{2}$ 7. $y = 6x^3 - \dfrac{11}{2}x^2 + 4x - \dfrac{1}{2}$

8. $y = x - 5$ 9. $y = 4e^x$

10. $y = x\sqrt{x^2 + 1}$ 11. $y = \dfrac{x + 1}{x - 1}$

12. $y = x^{-1}$ 13. $y = e^{-x}$

14. $y = \ln (x^2 + 2)$ 15. $y = x^2 e^{-x}$

In problems 16–19, find $f^{(1)}(x)$, $f^{(2)}(x)$, and $f^{(3)}(x)$.

16. $f(x) = 3 - x - 5x^2$ 17. $f(x) = 7x^4 - 5x^3 + 8x^2 + x - 5$

18. $f(x) = e^x \ln x$ 19. $f(x) = (3x - 4)^{\frac{4}{3}}$

In problems 20–23, find $f^{(1)}(a)$, $f^{(2)}(a)$, and $f^{(3)}(a)$ for the given value of a.

20. $f(x) = 4 + 8x - x^2$, $a = 1$

21. $f(x) = x^4 - 4x^3 + 6x^2 - 8$, $a = 0$

22. $f(x) = e^x \ln x,$ $\qquad\qquad a = 1$

23. $f(x) = (3x + 4)^{\frac{4}{3}},$ $\qquad\quad a = -1$

In problems 24–27, find a formula which gives $f^{(n)}(x)$ for each positive integer n.

*24. $f(x) = e^{-x}$ $\qquad\qquad$ *25. $f(x) = \dfrac{1}{x}$

*26. $f(x) = \dfrac{1}{1-x}$ $\qquad\qquad$ *27. $f(x) = \dfrac{1}{(1+x)^2}$

7 / Applications of the Derivative

In Chap. 6 we introduced the derivative of a function and explored various methods for evaluating derivatives. In this chapter we will investigate some important applications of the derivative. In particular, we will stress the use of the derivative in geometric problems, such as curve sketching, and in the solution of "max-min" problems.

In order to illustrate some of the difficulties which can arise in attempting to sketch the graph of a function, let us suppose that we want to sketch the graph of

$$f(x) = x^3 - 2x^2 - 5x + 6.$$

The first thing we do is to find the zeros of f. It is easily checked that the zeros of f are $-2, 1, 3$. Next, we would probably construct a simple table of values, including the zeros, such as

x	-2	-1	0	1	2	3
$f(x)$	0	8	6	0	-4	0

Our attempt to sketch the graph using the points in this table might result in a curve as shown in Figure 7.1.

Remember, the curve in Figure 7.1 is only an approximate sketch of the graph of f. We now ask the following question: How accurate a sketch of the graph of f is the curve drawn in Figure 7.1? This can be answered by considering the limitations of the point plotting method on which we have relied entirely in order to sketch the graph of f.

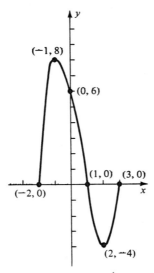

Figure 7.1

(1) The curve in Figure 7.1 is only a reflection of our table of values. En-largement of the table of values would result in more points and, hence, would probably produce a somewhat different sketch.

(2) The curve in Figure 7.1 shows a high point at $(-1, 8)$ and a low point at $(2, -4)$. There is no assurance that these points are actually high and low points on the graph of f.

(3) The pattern of the rising and falling of the graph of f cannot be precisely determined from a random table of values.

(4) The behavior of points $(x, f(x))$ for large values of $|x|$ is not indicated because such values are not included in the table.

Thus, the point plotting method may result in a sketch of the graph which is very rough indeed. In the preceding example it can be shown that the graph of f actually has a high point at $\left(\dfrac{2 - \sqrt{19}}{3}, \dfrac{52 + 40\sqrt{19}}{27}\right)$ and a low point at $\left(\dfrac{2 + \sqrt{19}}{3}, \dfrac{52 - 40\sqrt{19}}{27}\right)$. The complexity of these numbers clearly indicates that one is unlikely to include them in a table of values which is selected in a rather arbitrary manner. Hence, even if we enlarge our table of values in order to try and improve the accuracy of our sketch, the exact location of these high and low points is still likely to be missed. Likewise, a table of values will not give us precise information regarding the rising and falling of the graph.

Figure 7.2 is a comparison of the actual graph of f for the interval $[-2, 3]$ to the approximation just obtained. The solid curve in Figure 7.2 is the graph of f while the dotted curve is the approximation to the graph of f which we sketched in Figure 7.1

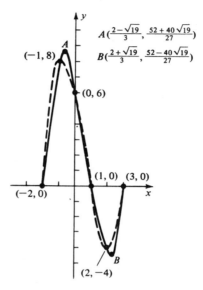

$A\left(\frac{2 - \sqrt{19}}{3}, \frac{52 + 40\sqrt{19}}{27}\right)$

$B\left(\frac{2 + \sqrt{19}}{3}, \frac{52 - 40\sqrt{19}}{27}\right)$

Figure 7.2

In order to draw a more accurate sketch of the graph of a function, one must be able to locate the high and low points precisely. Moreover, an exact description of the pattern of the rising and falling of the graph, together with other general information about the graph, is desirable. The derivative can be used to find this important information. Then a table of values can be constructed which includes the relevant information. It is this use of the derivative which occupies our attention for much of this chapter.

7.1. TANGENTS TO A CURVE

Let f be a function which is differentiable over the interval (a, b) so that $f'(x)$ exists for all $x \in (a, b)$. Let $x_0 \in (a, b)$ and let P be the point $(x_0, f(x_0))$ on the graph of f (see Figure 7.3).

In Chap. 6 we defined the tangent line to the curve at the point P. We found that the tangent line to the curve $y = f(x)$ at the point $x = x_0$ is the line, passing through the point $(x_0, f(x_0))$, which has a slope equal to $f'(x_0)$. Thus, we can determine the equation of the tangent line to the curve at $P(x_0, f(x_0))$ by find-

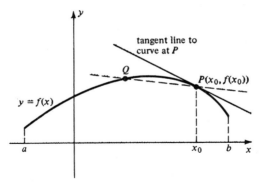

Figure 7.3

ing the slope of the line, $f'(x_0)$, and a point, $(x_0, f(x_0))$, through which it passes. We also say that $f'(x_0)$ is the slope of the curve $y = f(x)$ at $(x_0, f(x_0))$ or, simply at $x = x_0$.

Problem. Find the slope of the curve $y = \sqrt{x^2 + 1}$ at $x = 2, -1$.

Solution. Since $f(x) = \sqrt{x^2 + 1}$, we have

$$f'(x) = \frac{d}{dx}\left[(x^2 + 1)^{\frac{1}{2}}\right] = \frac{1}{2}(x^2 + 1)^{-\frac{1}{2}}\frac{d}{dx}[x^2 + 1]$$

$$= \frac{1}{2}(x^2 + 1)^{-\frac{1}{2}}(2x)$$

$$= \frac{x}{\sqrt{x^2 + 1}}.$$

So the slope of the curve at $x = 2$ is

$$f'(2) = \frac{2}{\sqrt{2^2 + 1}} = \frac{2}{\sqrt{5}} = \frac{2}{5}\sqrt{5}.$$

The slope of the curve at $x = -1$ is

$$f'(-1) = \frac{-1}{\sqrt{(-1)^2 + 1}} = -\frac{1}{\sqrt{2}} = -\frac{1}{2}\sqrt{2}.$$

Problem. Find the slope of the curve

$$y = x^3 - 2x^2 - 5x + 6$$

at $x = -\frac{3}{2}, 1$.

Solution. We have that

$$y' = 3x^2 - 4x - 5.$$

Therefore, the slope of the curve at $x = -\dfrac{3}{2}$ is

$$3\left(-\frac{3}{2}\right)^2 - 4\left(-\frac{3}{2}\right) - 5 = \frac{31}{4},$$

and the slope of the curve at $x = 1$ is

$$3(1)^2 - 4(1) - 5 = -6.$$

Notice that the signs of these slopes are different. We will soon see that the sign of the slope is very important in that it yields useful geometric information concerning the curve.

Problem. Find those values of x for which the slope of the curve $y = x^2 - x - 6$ is positive, zero, and negative.

Solution. First, $y' = 2x - 1$. We see that $y' > 0$ if and only if $2x - 1 > 0$, that is $x > \dfrac{1}{2}$. Also, $y' < 0$ if and only if $2x - 1 < 0$, that is, $x < \dfrac{1}{2}$. Finally, $y' = 0$ if and only if $2x - 1 = 0$, or $x = \dfrac{1}{2}$.

Problem. Find those intervals in which the function $y = 2x^3 - 9x^2 - 60x + 12$ has positive, negative, and zero slope.

Solution. The derivative is

$$y' = 6x^2 - 18x - 60 = 6(x^2 - 3x - 10)$$
$$= 6(x - 5)(x + 2).$$

So $y' > 0$ provided $x < -2$ or $x > 5$, $y' = 0$ if $x = 5, -2$, and $y' < 0$ provided $-2 < x < 5$.

Exercise Set 7.1

1. Let $f(x) = 2x^2 - x - 3$.
   a. Find $f'(x)$, $f'\left(-\dfrac{1}{2}\right)$, and $f'\left(\dfrac{1}{2}\right)$.
   b. What is the slope of the curve at the point $\left(-\dfrac{1}{2}, -2\right)$?

   What is the slope of the curve at the point $\left(\dfrac{1}{2}, -3\right)$?

c. Write an equation of the tangent line l_1 to the curve at the point $\left(-\dfrac{1}{2}, -2\right)$. Write an equation of the tangent line l_2 to the curve at the point $\left(\dfrac{1}{2}, -3\right)$.

2. a. On a set of coordinate axes, sketch the graph of the function $f(x) = 2x^2 - x - 3$.

b. On the same set of axes, sketch the tangent line to the curve at the point $\left(-\dfrac{1}{2}, -2\right)$ by using two points which lie on the line l_1 of problem 1c.

c. On the same set of axes, sketch the tangent line to the curve at the point $\left(\dfrac{1}{2}, -3\right)$ by using two points on the line l_2 of problem 1c.

3. Find those intervals on which the function $y = -x^2 + x - 2$ has positive slope. On what intervals does the function have negative slope? For what values of x does the function have zero slope?

4. Sketch the graph of the function $y = -x^2 + x - 2$. On what intervals does the graph appear to be rising or falling as x increases?

* * *

In problems 5–14, find the slope of the tangent line to the curve at each of the indicated values of x.

5. $f(x) = 4x^2 + 7x - 3$ $x = -1, \; x = 0, \quad x = 4$

6. $F(x) = -3x^2 + 2x + 4$ $x = \dfrac{1}{3}, \; x = -\dfrac{1}{3}, \; x = -3$

7. $g(x) = 3x - 7$ $x = -2, \; x = 0, \quad x = 2$

8. $G(x) = -3x + 21$ $x = -1, \; x = \dfrac{1}{2}, \quad x = \sqrt{2}$

9. $h(x) = \dfrac{4x + 3}{2 - 5x}$ $x = -4, \; x = \dfrac{4}{3}, \quad x = 115$

10. $H(x) = \dfrac{4x + 3}{2 + 5x^2}$ $x = -1, \; x = 0, \quad x = 1$

11. $f(x) = (x^2 - 1)^{\frac{2}{3}}$ $x = -3, \; x = 0, \quad x = 10$

12. $g(x) = \sqrt{3x^3 - x + 3}$ $x = \dfrac{1}{2}, \; x = -1, \; x = 2$

13. $h(x) = xe^x$ $x = -1, \; x = 0, \quad x = 1$

14. $t(x) = x \ln x$ $x = 1, \quad x = e, \quad x = e^2$

In problems 15–21, determine those intervals on which the function has positive slope, negative slope, and the points at which the slope is zero. Sketch the graph of the function. Note those intervals on which the graph appears to be rising or falling as x increases.

15. $f(x) = 7 - 2x$

16. $g(x) = -3x^2 + 7x - 1$

17. $h(x) = 2x^2 + 3x - 17$

18. $g(x) = x^3 - 8$

19. $p(x) = x^4$

20. $F(x) = e^{-x}$

21. $G(x) = \ln x^2$

In problems 22–25, find an equation of the line which is tangent to the graph of the given function at the given point.

22. $F(x) = (3x - 7) e^{5x}$, $\qquad x = 0$

23. $G(x) = \sqrt{3x - 1}$, $\qquad x = 5$

24. $g(x) = (x^2 + 3)^5 (2x - 7)^3$, $\qquad x = 3$

25. $f(x) = (2x^2 + 3x - 1) e^{-x}$, $\qquad x = -2$

In problems 26–30, find an equation of the line which is tangent to the graph of the given function at the given point. Sketch both the curve and the specified tangent line.

26. $F(x) = x^2 + 2x + 4$, $\qquad x = -1$

27. $H(x) = x^3$, $\qquad x = 1$

28. $P(x) = 3x^2 - 12x + 8$, $\qquad x = 2$

29. $f(x) = \sqrt{x^2 + 1}$, $\qquad x = 0$

30. $F(x) = \dfrac{1}{x}$, $\qquad x = \dfrac{1}{2}$

7.2 INCREASING AND DECREASING FUNCTIONS

We now start our discussion of those properties of the graph of a function which can be determined from the behavior of the derivatives of the function.

7.1 Definition. *Let f be a function defined on an interval I and let x_1, x_2 be arbitrary elements of I. IF $f(x_1) < f(x_2)$ whenever $x_1 < x_2$, THEN f is increasing on I. IF $f(x_1) > f(x_2)$ whenever $x_1 < x_2$, THEN f is decreasing on I.*

In general, if f is increasing on an interval I, then its graph is rising as we go from left to right over I (see Figure 7.4). Similarly, if f is decreasing on I, then its graph is falling as we go from left to right over I.

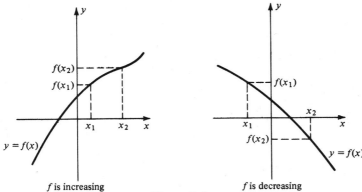

f is increasing   f is decreasing

Figure 7.4

For certain functions f it is relatively easy to determine whether f is increasing or decreasing on an interval I. For example, if $f(x) = x^2$ and $0 < x_1 < x_2$, then

$$(x_1)^2 < (x_2)^2$$

and so

$$f(x_1) < f(x_2).$$

Consequently, f is increasing on $(0, \infty)$.

On the other hand, if $g(x) = -3x + 1$, then $x_1 < x_2$ implies

$$g(x_1) > g(x_2)$$

since

$$-3x_1 + 1 > -3x_2 + 1.$$

Thus g is decreasing on $\mathcal{R}$.

However, for more complicated functions it may not be readily apparent that the function is increasing or decreasing on a particular interval. For instance, it is not readily apparent that

$$f(x) = 5 + 9x + 3x^2 - x^3$$

is increasing on $(-1, 3)$. Thus we want to develop an effective test which will allow us to determine whether a function is increasing or decreasing on an interval.

Some insight into this question can be obtained from the following argument. Assume that f is differentiable on an open interval I and suppose f is increasing on I. If $x \in I$, then for Δx sufficiently small, $x + \Delta x \in I$. If $\Delta x > 0$, then $x < x + \Delta x$ and $f(x) < f(x + \Delta x)$ because f is increasing on I. Since the quotient of two positive numbers is positive we have that

$$\frac{f(x + \Delta x) - f(x)}{\Delta x} > 0.$$

If $\Delta x < 0$, then $x + \Delta x < x$ so that $f(x + \Delta x) < f(x)$. Since the quotient of two negative numbers in positive we have that

$$\frac{f(x + \Delta x) - f(x)}{\Delta x} > 0.$$

It can then be shown that this implies

$$f'(x) = \lim_{\Delta x \to 0} \frac{f(x + \Delta x) - f(x)}{\Delta x} \geqslant 0.$$

In other words, if f is increasing on I, then $f'(x) \geqslant 0$ for all $x \in I$. A similar argument can be made to show that, if f is decreasing on I, then $f'(x) \leqslant 0$ for all $x \in I$. What is important to us is the fact that these conditions are almost sufficient to guarantee that the function is increasing or decreasing on I. The following theorem, stated without proof, gives us practical means with which to settle the question.

7.2 Theorem. *Let f be differentiable on an interval I.*
   (a) *IF $f'(x) > 0$ for all $x \in I$, THEN f is increasing on I.*
   (b) *IF $f'(x) < 0$ for all $x \in I$, THEN f is decreasing on I.*

The way in which Theorem 7.2 is applied is now shown by means of several solved problems.

Problem. Find those intervals on which $f(x) = x^2$ is increasing.

Solution. We have $f'(x) = 2x$. Notice that $f'(x) > 0$ provided $2x > 0$ and this is the case provided $x > 0$. Therefore, by Theorem 7.2, f is increasing on $(0, \infty)$. Note that this agrees with our earlier observation for the function f.

Problem. Find those intervals on which $g(x) = -3x + 1$ is decreasing.

Solution. Since $g'(x) = -3 < 0$ for all x, g is decreasing on $\mathcal{R}$ by Theorem 7.2. This, too, agrees with an earlier observation.

Problem. Find those intervals on which $f(x) = x^2 - 4x - 5$ is increasing or decreasing.

Solution. First,

$$f'(x) = 2x - 4 = 2(x - 2).$$

Therefore $f'(x) > 0$ provided $x - 2 > 0$, that is, $x > 2$. Similarly, $f'(x) < 0$ provided $x - 2 < 0$, that is, $x < 2$. Hence, by Theorem 7.2, f is increasing on

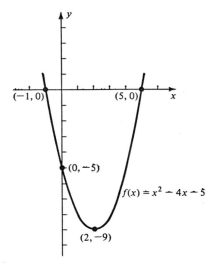

Figure 7.5

$(2, \infty)$ and decreasing on $(-\infty, 2)$. The graph of this function (Figure 7.5) bears out these facts.

Problem. Find those intervals on which $f(x) = 5 + 9x + 3x^2 - x^3$ is increasing or decreasing.

Solution. Note that we have previously mentioned that f is increasing on $(-1, 3)$. To see this we compute the first derivative of f:

$$f'(x) = 9 + 6x - 3x^2$$
$$= -3(x^2 - 2x - 3)$$
$$= -3(x + 1)(x - 3).$$

A sign table is useful for keeping track of the sign of f'.

	Sign of		
	$x + 1$	$x - 3$	$f'(x) = -3(x + 1)(x - 3)$
$x < -1$	$-$	$-$	$-$
$-1 < x < 3$	$+$	$-$	$+$
$3 < x$	$+$	$+$	$-$

By Theorem 7.2, f is decreasing on both $(-\infty, -1)$ and $(3, \infty)$ and is increasing on $(-1, 3)$.

In the previous examples Theorem 7.2 was applied to polynomial functions.

The theorem can be applied, however, to any differentiable function. This, of course, means Theorem 7.2 may be applied to algebraic functions as well as functions involving exponential and logarithm expressions, as we now illustrate.

Problem. Find those intervals on which $f(x) = (x^2 - 16)\sqrt{x^2 + 2}$ is increasing or decreasing.

Solution. The first derivative is

$$f'(x) = 2x(x^2 + 2)^{\frac{1}{2}} + (x^2 - 16)\frac{1}{2}(x^2 + 2)^{-\frac{1}{2}}(2x)$$

$$= 2x(x^2 + 2)^{\frac{1}{2}} + \frac{x(x^2 - 16)}{(x^2 + 2)^{\frac{1}{2}}}$$

$$= \frac{2x(x^2 + 2)}{(x^2 + 2)^{\frac{1}{2}}} + \frac{x(x^2 - 16)}{(x^2 + 2)^{\frac{1}{2}}}$$

$$= \frac{2x^3 + 4x + x^3 - 16x}{(x^2 + 2)^{\frac{1}{2}}} = \frac{3x^3 - 12x}{(x^2 + 2)^{\frac{1}{2}}}.$$

Now $(x^2 + 2)^{\frac{1}{2}} > 0$ for all $x \in \Re$; hence $f'(x) > 0$ if and only if $3x^3 - 12x > 0$, and $f'(x) < 0$ if and only if $3x^3 - 12x < 0$. Since

$$3x^2 - 12x = 3x(x^2 - 4)$$
$$= 3x(x + 2)(x - 2),$$

the following sign table determines the sign of f'.

Sign of

	x	$x + 2$	$x - 2$	$3x(x + 2)(x - 2)$
$x < -2$	–	–	–	–
$-2 < x < 0$	–	+	–	+
$0 < x < 2$	+	+	–	–
$2 < x$	+	+	+	+

Therefore, f is increasing on both $(-2, 0)$ and $(2, \infty)$ and is decreasing on both $(-\infty, -2)$ and $(0, 2)$.

Problem. Find those intervals on which $f(x) = xe^x$ is increasing or decreasing.

Solution. For this function the first derivative is

$$f'(x) = xe^x + e^x = (1 + x)e^x.$$

Since $e^x > 0$ for all x, the sign of f' depends on the sign of $1 + x$. Thus $f'(x) > 0$ if $1 + x > 0$, that is, $x > -1$. Also, $f'(x) < 0$ if $1 + x < 0$, that is, $x < -1$. Therefore f is increasing on $(-1, \infty)$ and decreasing on $(-\infty, -1)$.

Problem. Find those intervals on which $f(x) = \ln x$ and $g(x) = x \ln x$ are increasing.

Solution. Since $\ln x$ is defined only for $x > 0$, f and g are defined only for $x > 0$. Therefore, $f'(x) = \dfrac{1}{x}$ is positive for $x > 0$. Consequently, $f(x) = \ln x$ is increasing on $(0, \infty)$. For the function g we have $g'(x) = 1 + \ln x$ for $x > 0$. Therefore $g'(x) > 0$ provided $\ln x > -1$. But $\ln e^{-1} = -1$ and it has just been shown that the logarithm function is increasing. Hence $\ln x > -1$ if $x > e^{-1}$. It follows that g is increasing on (e^{-1}, ∞).

Let us now assess the usefulness of our ability to locate intervals on which a function is increasing or decreasing with regard to curve sketching. Suppose we wish to sketch the graph of a differentiable function f defined on $[a, b]$. Since f is differentiable, it is continuous, so that its graph is an unbroken curve. Assume that, in addition, we know f is increasing on (a, c) and decreasing on (c, b). We can think of the graph of f as the path of a moving point in the plane which starts at $(a, f(a))$ and stops at $(b, f(b))$. Because f is increasing on (a, c), this point must rise as it travels from left to right over (a, c). Since f is decreasing on (c, b), the point must fall as it travels from left to right over (c, b). Each of the different curves in Figure 7.6 exhibits these features.

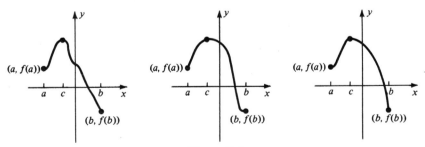

Figure 7.6

This illustrates the fact that, while knowing where the function is increasing or decreasing adds to our knowledge of its graph, we still need more information in order to ensure complete accuracy.

Exercise Set 7.2

1. Find those intervals on which the function $f(x) = 4x - 75$ is increasing (use Theorem 7.2).

2. Find those intervals on which the function $f(x) = 1 + 10x - 3x^2$ is decreasing (use Theorem 7.2).

3. Let $f(x) = x^4 - 4x^3 - 8x^2 + 7$.
   a. Find $f'(x)$.
   b. Complete the following sign table for f'.

Sign of f'

	x	$x - 4$	$x + 1$	$4x(x - 4)(x + 1)$
$x < -1$				
$-1 < x < 0$				
$0 < x < 4$				
$4 < x$				

   c. Use the sign table and Theorem 7.2 to determine those intervals on which f is increasing.
   d. Use the sign table and Theorem 7.2 to determine those intervals on which f is decreasing.

4. Use Theorem 7.2 to determine whether e^{-3x} is an increasing or a decreasing function.

* * *

In each of the problems 5–24, determine the intervals on which the function is increasing and the intervals on which it is decreasing.

5. $h(x) = -4x^2 + 7x - 1$

6. $h(x) = 4x^2 - 7x + 1$

7. $f(x) = x^3 - 3x$

8. $f(x) = (2x^2 + 5x - 3)^3$

9. $g(x) = 2x^3$

10. $g(x) = -37x^2$

11. $r(x) = \dfrac{x + 3}{x^2 + 2x + 2}$

12. $r(x) = \dfrac{2 - x}{x + 7}, x \neq -7$

13. $l(x) = \dfrac{5}{3} + \dfrac{1}{2}x$

14. $l(x) = 2 - \sqrt{2}\, x$

15. $f(x) = x^2 e^{2x}$

16. $f(x) = x^3 e^{-2x}$

17. $G(x) = \ln(x^2 + 1)$

18. $G(x) = \ln(1 - x^2),\, -1 < x < 1$

19. $H(x) = e^{\frac{3}{2}x}$

20. $H(x) = \ln 3x,\, x > 0$

21. $R(x) = (3x - 7)\sqrt{2x + 5},\, x > -\dfrac{5}{2}$

22. $R(x) = \sqrt{x^2 + 3x + 5}$

23. $F(x) = \dfrac{9}{x}$

24. $F(x) = \dfrac{9}{x^2}$

7.3 RELATIVE EXTREMA OF FUNCTIONS

In Sec. 7.2 we discussed a method used to determine the pattern of the rising and falling of the graph of a differentiable function. This section is devoted to a means for determining the high and low points on a graph. A high point is called a relative maximum and a low point is called a relative minimum.

7.3 Definition. *A function f has a relative maximum at $x = x_0$ and $f(x_0)$ is a relative maximum of f provided there exists an interval (c, d) containing x_0 such that $f(x) < f(x_0)$ for $c < x < x_0$ or $x_0 < x < d$.*

If f has a relative maximum at $x = x_0$, then the y-coordinate of the point $(x, f(x))$ is less than the y-coordinate of the point $(x_0, f(x_0))$ for all $x \neq x_0$ in an interval (c, d) which contains x_0. This means that $(x_0, f(x_0))$ is a high point on the graph of f. Figure 7.7 illustrates the graph of a function which has a relative maximum at $x = x_1$ and $x = x_2$.

We now give the analytic definition of a low point.

7.4 Definition. *A function f has a relative minimum at $x = x_0$ and $f(x_0)$ is a relative minimum of f provided there exists an interval (c, d) containing x_0 such that $f(x) > f(x_0)$ for $c < x < x_0$ or $x_0 < x < d$.*

If f has a relative minimum at $x = x_0$, the y-coordinate of the point $(x, f(x))$ is greater than the y-coordinate of the point $(x_0, f(x_0))$ for all $x \neq x_0$ in an

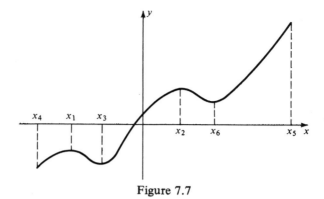

Figure 7.7

interval (c, d) which contains x_0. Therefore $(x_0, f(x_0))$ is a low point on the graph of f. The function whose graph is shown in Figure 7.7 has a relative minimum at $x = x_3$ and $x = x_6$.

Notice that from Figure 7.7 we have $f(x_2) < f(x_5)$ so that the phrase "$f(x_2)$ is a relative maximum" does not mean $f(x_2)$ is a "maximum" in the sense that $f(x_2)$ is larger than every other value of the function. Rather, it means that $f(x_2)$ is larger than values of the function computed for points close to x_2. That is why in our definition we specified some interval (c, d) and the word "relative" rather than "absolute." Similarly, the fact that $f(x_4) < f(x_3)$ indicates that $f(x_3)$ is not a "minimum." Thus "$f(x_3)$ is a relative minimum" means that $f(x_3)$ is less than values of the function computed for points close to x_3.

Figure 7.8 depicts the graph of a function f which has a relative maximum at $x = x_1$ and a relative minimum at $x = x_2$.

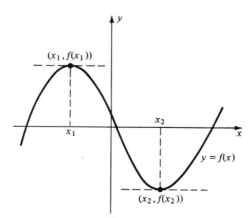

Figure 7.8

From the figure it appears that the tangent lines to the graph at the points $(x_1, f(x_1))$ and $(x_2, f(x_2))$ are horizontal. That is, the slope of each tangent line is zero. Since the slopes of the tangent lines to this curve at $(x_1, f(x_1))$ and $(x_2, f(x_2))$ are $f'(x_1)$ and $f'(x_2)$, respectively, we have $f'(x_1) = 0$ and $f'(x_2) = 0$. This is a fundamental connection between high and low points on the graph of a function and the derivative of the function. The following theorem is a formal statement of this observation.

7.5 Theorem. *Let f be a differentiable function. IF f has a relative maximum or a relative minimum at $x = x_0$, THEN $f'(x_0) = 0$.*

The importance of Theorem 7.5 lies in the fact that it gives us a practical means for determining those points at which a relative maximum or relative minimum can possibly occur. Namely, in order for a differentiable function f to have a relative maximum or relative minimum at $x = x_0$, we must have $f'(x_0) = 0$. Therefore, *the set of zeros of f' contains all those points at which a relative maximum or relative minimum can occur.*

Problem. What are the only possible values of x at which a relative maximum or relative minimum can occur for $f(x) = 3x^2 + 2x - 4$?

Solution. Since $f'(x) = 6x + 2$, $f'(x) = 0$ if, and only if, $x = -\frac{1}{3}$. Therefore, if f has a relative maximum or relative minimum, it must occur at $x = -\frac{1}{3}$.

Once the zeros of f' have been determined, it is then necessary to discover which of the zeros actually yields a relative maximum or relative minimum.

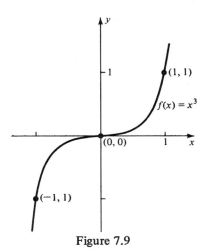

Figure 7.9

Caution is required at this point because it is not necessary that a relative maximum or relative minimum occur at every zero of f'. For instance, if $f(x) = x^3$, then $f'(x) = 3x^2$ and the only zero of f' is $x = 0$. However, we see from the graph of f (Figure 7.9) that $f(0) = 0$ is neither a relative maximum nor a relative minimum of f.

If $f'(x_0) = 0$, there is a simple way of showing that f has a relative maximum at $x = x_0$. Suppose we can find an interval (c, d) containing x_0 such that $f'(x) > 0$ for $c < x < x_0$ and $f'(x) < 0$ for $x_0 < x < d$. Then f is increasing on (c, x_0) and decreasing on (x_0, d) (see Figure 7.10). By continuity, since

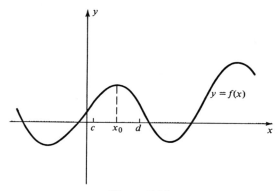

Figure 7.10

the graph of f must be an unbroken curve, this means $(x_0, f(x_0))$ is a high point on the graph of f and so $f(x_0)$ is a relative maximum of f. A similar argument holds for a relative minimum. These remarks are summarized in the following statement usually called the *first derivative test*.

7.6 Theorem. *Let f be a differentiable function defined at x_0.*

(a) *IF there exists an interval (c, d) containing x_0 such that $f'(x) > 0$ for $c < x < x_0$ and $f'(x) < 0$ for $x_0 < x < d$, THEN f has a relative maximum at $x = x_0$.*

(b) *IF there exists an interval (c, d) containing x_0 such that $f'(x) < 0$ for $c < x < x_0$ and $f'(x) > 0$ for $x_0 < x < d$, THEN f has a relative minimum at $x = x_0$.*

The relative maxima (plural of maximum) and relative minima (plural of minimum) are called the *relative extrema* of a function.

In practice Theorems 7.5 and 7.6 are used to locate the relative extrema of a function f as follows:

(1) Determine the set Z of zeros of f';

(2) For each $x_0 \in Z$ determine if condition (a) or condition (b) of the first derivative test (Theorem 7.6) holds or neither condition (a) nor condition (b) holds.

Problem. Find the relative extrema of $f(x) = 3x^2 + 2x - 4$.

Solution. We have $f'(x) = 6x + 2 = 6\left(x + \frac{1}{3}\right)$. The only zero of f is $x = -\frac{1}{3}$. Since $f'(x) < 0$ if $x < -\frac{1}{3}$ and $f'(x) > 0$ if $x > -\frac{1}{3}$, part (b) of Theorem 7.6 shows that f has a relative minimum at $x = -\frac{1}{3}$. The relative minimum is $f\left(-\frac{1}{3}\right) = -\frac{13}{3}$ and this is the only relative extremum of f.

Problem. Find the relative extrema of $f(x) = x^3 - 2x^2 - 5x + 6$.

Solution. Note that this is the function discussed in the introduction to this chapter. At the time, we stated that the graph of f had a high point at $\left(\frac{2 - \sqrt{19}}{3}, \frac{52 + 40\sqrt{19}}{27}\right)$ and a low point at $\left(\frac{2 + \sqrt{19}}{3}, \frac{52 - 40\sqrt{19}}{27}\right)$. We are now in a position to verify this. We have $f'(x) = 3x^2 - 4x - 5$. By the quadratic formula, the zeros of f' are

$$x_1 = \frac{2 - \sqrt{19}}{3}, \quad x_2 = \frac{2 + \sqrt{19}}{3}.$$

These are the only possible values of x at which a relative maximum or relative minimum can occur. Note that $f'(x) = 3(x - x_1)(x - x_2)$ and $x_1 < x_2$. Thus, a sign table is useful here.

	Sign of		
	$x - x_1$	$x - x_2$	$f'(x) = 3(x - x_1)(x - x_2)$
$x < x_1$	−	−	+
$x_1 < x < x_2$	+	−	−
$x_2 < x$	+	+	+

Using Theorem 7.6, the first two entries in the last column of the sign table show that f has a relative maximum at $x = x_1$. The last two entries in the last column show that f has a relative minimum at $x = x_2$. Therefore $f(x_1) =$

$\dfrac{52 + 40\sqrt{19}}{27}$ is a relative maximum of f and $f(x_2) = \dfrac{52 - 40\sqrt{19}}{27}$ is a relative minimum of f.

Problem. Find the relative extrema of $g(x) = (3x + 2)e^{-2x}$.

Solution. The first derivative is

$$g'(x) = (3x + 2)(-2e^{-2x}) + 3e^{-2x}$$
$$= (-6x - 1)e^{-2x}.$$

From Chap. 5, $e^{-2x} > 0$ for all x so that $g'(x) = 0$ if, and only if, $-6x - 1 = 0$. Therefore, $x = -\dfrac{1}{6}$ is the only possible point at which a relative maximum or relative minimum can occur. If $x < -\dfrac{1}{6}$, then $-6x - 1 > 0$ so that $g'(x) > 0$. Consequently, since g is increasing on $\left(-\infty, -\dfrac{1}{6}\right)$ and decreasing on $\left(-\dfrac{1}{6}, \infty\right)$, g has a relative maximum at $x = -\dfrac{1}{6}$. The relative maximum is $g\left(-\dfrac{1}{6}\right) = \dfrac{3}{2}e^{\frac{1}{3}}$.

Problem. Find the relative extrema of $h(x) = \dfrac{4x - 7}{2x + 3}$.

Solution. The first derivative of h is

$$h'(x) = \frac{(2x + 3)(4) - (4x - 7)(2)}{(2x + 3)^2}$$

$$= \frac{26}{(2x + 3)^2}.$$

Therefore $h'(x) \neq 0$ for any value of x. By Theorem 7.5, h has no relative extrema.

We are now able to locate exactly the high and low points on the graph of a differentiable function. While this enables us to be even more accurate than before when sketching the graph of a function, more information is still needed. For example, each of the curves in Figure 7.6 is determined by a function which is increasing on (a, c), decreasing on (c, b), and has a relative maximum at $x = c$. Since the graphs in Figure 7.6 appear so different, information concerning more than the behavior of the first derivative, as stated in Theorems 7.5 and 7.6, is needed.

Exercise Set 7.3

1. If the following sketch represents the graph of a function f, determine those values of x at which f has a relative minimum and those values of x at which f has a relative maximum.

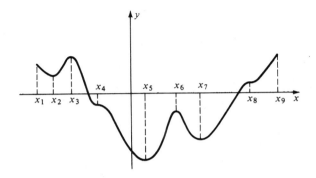

2. a. If $g'(x) = x^2(x + 1)(2x - 9)$, for what values of x might the function g attain a relative maximum or minimum value?

   b. If $G'(x) = \dfrac{3}{x^2 + 1}$, for what values of x might the function G attain a relative maximum or minimum value?

3. a. If $f(x) = x^2 + 4x - 12$, use Theorem 7.6 to show that f has a relative minimum at $x = -2$ and find its value.

   b. If $f(x) = -x^2 + 4x - 12$, use Theorem 7.6 to show that f has a relative maximum at $x = 2$ and find its value.

4. Let $f(x) = 4x^3 - 3x^2 - 6x + 5$.
   a. Find $f'(x)$.
   b. Determine the set Z of zeros of f'.
   c. For each $x_0 \in Z$, determine if either condition (a) or (b) of Theorem 7.6 holds.
   d. Determine the relative extrema of f.

* * *

In problems 5–22, determine the intervals on which the given function is increasing or decreasing and the relative extrema.

5. $h(x) = 9x^2 - 36x + 5$

6. $g(x) = -4x^2 + 5x - 8$

7. $f(x) = ax^2 + bx + c, a \neq 0$ (compare with results of Chap. 3, Sec. 6)

8. $g(x) = ax + b, a \neq 0$

9. $H(x) = 7x - 1$

10. $F(x) = 49$

11. $G(x) = (2 - 7x)\sqrt{4 + 3x}, x > -\dfrac{4}{3}$

12. $H(x) = (x^2 + 3)\sqrt{x^2 + 5}$

13. $P(x) = -\dfrac{4}{3}x^3 + \dfrac{33}{2}x^2 - 35x + 13$

14. $P(x) = 3x^4 - 2x^3 - 3x^2 - 18x + 4$

15. $G(x) = \dfrac{2x + 3}{x^2 + 4}$

16. $g(x) = \dfrac{15}{x^2 + 1}$

17. $G(x) = (-2x^2 - 3x + 12)\, e^{-x}$

18. $h(x) = (7 - 4x)\, e^{4x}$

19. $g(x) = (x^2 + 2x + 1)^5 (3x - 7)^4$

20. $f(x) = (2x + 3)^{11} (4 - 9x)^{12}$

21. $h(x) = \sqrt{x^2 + 4x + 5}$

22. $R(x) = \sqrt{4 - x^2},\ -2 < x < 2$

*23. Determine conditions on A such that the function $f(x) = x^3 + Ax + B$ has (a) two relative extrema; (b) no relative extrema. Is it possible for this function to have exactly one relative extremum?

*24. Use the results of problem 23 to determine the number of relative extrema for each of the following functions:
(a) $F(x) = x^3 - 7x + 10$
(b) $G(x) = x^3 + 3x - 9$
(c) $H(x) = x^3$

7.4 THE SECOND DERIVATIVE TEST

If f'' exists, there is a simple and useful test which employs the second derivative in determining the relative extrema of the function f. In order to see how the conditions of the test are derived, let us assume that f is differentiable and $f'(x_0) = 0$. This means that f may have a relative maximum or a relative minimum at $x = x_0$. Assume there exists an interval (c, d) containing x_0 such that $f''(x) < 0$ for all $x \in (c, d)$. Then f' is decreasing on (c, d). Therefore $f'(x_1) > f'(x_2)$ whenever $c < x_1 < x_2 < d$. In particular,

$$f'(x) > f'(x_0) \text{ for } c < x < x_0 \text{ and } f'(x_0) > f'(x) \text{ for } x_0 < x < d.$$

Since $f'(x_0) = 0$, we have

$$f'(x) > 0 \text{ for } c < x < x_0 \text{ and } 0 > f'(x) \text{ for } x_0 < x < d.$$

By the first derivative test, f has a relative maximum at $x = x_0$.

If f'' is continuous and $f''(x_0) < 0$, then it can be shown that there exists an interval (c, d) containing x_0 such that $f''(x) < 0$ for all $x \in (c, d)$. A similar statement holds in the case of a relative minimum. Both statements are summarized in the following theorem, known as the *second derivative test*.

7.7 Theorem. *Let f' and f'' be continuous on an interval (c, d) containing the point x_0 and assume $f'(x_0) = 0$.*
  (a) *IF $f''(x_0) < 0$, THEN f has a relative maximum at $x = x_0$.*
  (b) *IF $f''(x_0) > 0$, THEN f has a relative minimum at $x = x_0$.*

If f'' can be easily calculated and $f''(x_0) \neq 0$, then the second derivative test has one main advantage when compared to the first derivative test. Namely, it only requires that we determine the sign of the single number $f''(x_0)$ rather than the sign of f' over an entire interval which contains x_0.

Problem. Find the relative extrema of $f(x) = 3x^2 + 2x - 4$.

Solution. (Note that this problem was solved by means of the first derivative test in Sec. 7.3.) We have $f'(x) = 6\left(x + \dfrac{1}{3}\right)$ and $f''(x) = 6$ for all x. In particular, $f'\left(-\dfrac{1}{3}\right) = 0$ and $f''\left(-\dfrac{1}{3}\right) = 6 > 0$. Therefore f has a relative minimum at $x = -\dfrac{1}{3}$.

Problem. Find the relative extrema of $f(x) = x^3 - 2x^2 - 5x + 6$.

Solution. (This problem, too, was solved by means of the first derivative test in Sec. 7.3.) The first two derivatives of f are

$$f'(x) = 3x^2 - 4x - 5,$$

and
$$f''(x) = 6x - 4 = 2(3x - 2).$$

The zeros of f' are $x_1 = \dfrac{2 - \sqrt{19}}{3}$ and $x_2 = \dfrac{2 + \sqrt{19}}{3}$. Since

$$f''(x_1) = 2\left[3\left(\frac{2 - \sqrt{19}}{2}\right) - 2\right] = 2(4 - \sqrt{19}) < 0$$

and
$$f''(x_2) = 2\left[3\left(\frac{2 + \sqrt{19}}{2}\right) - 2\right] = 2(4 + \sqrt{19}) > 0,$$

f has a relative maximum at $x = x_1$ and a relative minimum at $x = x_2$.

Problem. Find the relative extrema of $g(x) = e^x - x$.

Solution. We have $g'(x) = e^x - 1$ and the only zero of g' is $x = 0$. The second derivative is $g''(x) = e^x$ and $g''(0) = 1$ is positive. Therefore g has a relative minimum at $x = 0$ and this relative minimum is $g(0) = 1$.

Problem. Find the relative extrema of $h(x) = \dfrac{4x^2 + 1}{x}$.

Solution. The first two derivatives of h are

$$h'(x) = \frac{x(8x) - (4x^2 + 1)}{x^2} = \frac{4x^2 - 1}{x^2},$$

and

$$h''(x) = \frac{x^2(8x) - (4x^2 - 1)(2x)}{x^4} = \frac{2}{x^3}.$$

The zeros of h' are $x = \pm\dfrac{1}{2}$. We have

$$h''\left(\frac{1}{2}\right) = \frac{2}{\left(\frac{1}{2}\right)^3} = 16 > 0$$

and

$$h''\left(-\frac{1}{2}\right) = \frac{2}{\left(-\frac{1}{2}\right)^3} = -16 < 0.$$

By the second derivative test, h has a relative minimum at $x = \dfrac{1}{2}$ and a relative maximum at $x = -\dfrac{1}{2}$. The relative minimum is $h\left(\dfrac{1}{2}\right) = 4$ and the relative maximum is $h\left(-\dfrac{1}{2}\right) = -4$.

Exercise Set 7.4

1. a. If $f(x) = x^2 + 4x - 12$, use Theorem 7.7 to show that f has a relative minimum at $x = -2$.
   b. If $f(x) = -x^2 + 4x - 12$, use Theorem 7.7 to show that f has a relative maximum at $x = 2$.

2. Let $f(x) = 4x^3 - 3x^2 - 6x + 5$.
   a. Find $f'(x)$.
   b. Find $f''(x)$.
   c. Determine the set Z of zeros of $f'(x)$.

   d. For each $x_0 \in Z$, find $f''(x_0)$.

   e. Use Theorem 7.7 to determine those values of x_0 for which $f(x_0)$ is a relative maximum or relative minimum of $f(x)$.

   f. Determine the relative extrema of f.

3. Let $g(x) = x^3 - 3x^2 - 9x + 2$.

   a. Find $g'(x)$ and $g''(x)$.

   b. Determine the set Z of zeros of $g'(x)$.

   c. Use the second derivative test to locate the relative extrema of $g(x)$.

   d. Find the relative extrema of g.

4. Find the relative extrema of the function $E(x) = e^{2x} - 2x$ (use Theorem 7.7).

* * *

In problems 5-14, locate the relative extrema of the given function by using the second derivative test.

5. $h(x) = 9x^2 - 36x + 5$

6. $g(x) = -4x^2 + 5x - 8$

7. $P(x) = -\frac{4}{3}x^3 + \frac{33}{2}x^2 - 35x + 13$

8. $P(x) = 3x^4 - 2x^3 - 3x^2 - 18x + 4$

9. $f(z) = z - 3z^3$

10. $f(z) = z^3 - 2z^2 - 5z + 6$

11. $L(x) = x \ln x, x > 0$

12. $L(x) = x - \ln x, x > 0$

13. $G(x) = (7 - 4x)e^{4x}$

14. $H(x) = \sqrt{x^2 + 4x + 5}$

In problems 15-24, determine the relative extrema for each of the given functions (use either Theorem 7.6 or Theorem 7.7).

15. $f(x) = x^4 - 9$

16. $F(x) = x^3 + \frac{2}{3}$

17. $P(x) = x^4 - x^3 - x^2 + x - 1$

18. $P(x) = x^4 + x^3 + x^2 - 9x + 2$

19. $h(x) = x^2 e^x$

20. $H(x) = x^2 e^{-x}$

21. $f(x) = (x^2 - 16)\sqrt{x^2 + 2}$

22. $F(x) = (3x - 7)\sqrt{2x + 5}$

23. $g(x) = \ln(x^2 + 1)$

24. $G(x) = (\ln x)^2, x > 0$

7.5 CONCAVITY AND POINTS OF INFLECTION

By using either the first or second derivative tests, we are now able to locate high and low points on the graph of a function, and by examining the sign of the first derivative we can tell where the graph is rising and where it is falling. It seems like this information should be sufficient to allow us to accurately sketch the graph and yet we have stated that it is not. An example quickly points out the difficulty. It can be shown that $f(x) = -x^3 + 3x^2 - 1$ is increasing on $(0, 2)$. Note that $f(0) = -1$ and $f(2) = 3$. Therefore the graph of f over $[0, 2]$ can be thought of as the path of a point which starts at $(0, -1)$ and rises to $(2, 3)$. Two such possibilities are shown in Figure 7.11.

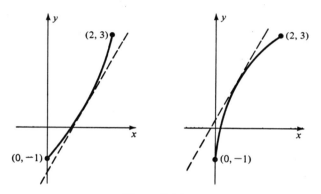

Figure 7.11

However, the curves in Figure 7.11 are basically different. If we wanted to describe the difference between them loosely, we might say that the first "curves upward" while the second "curves downward." We make this difference precise in the following definition:

7.8 Definition. *Let f be a differentiable function defined on (c, d).*

(a) *IF for each $x_0 \in (c, d)$ the graph of f lies above the tangent line to the curve at $(x_0, f(x_0))$, THEN the graph of f is **concave upward** on (c, d).*

(b) *IF for each $x_0 \in (c, d)$ the graph of f lies below the tangent line to the curve at $(x_0, f(x_0))$, THEN the graph of f is **concave downward** on (c, d).*

In Figure 7.12 the graph of f is concave upward on (a, b) and concave downward on (b, c).

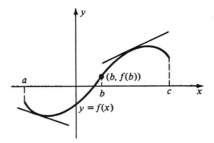

Figure 7.12

Notice, from Figure 7.12, that the concavity of the graph changes at the point $(b, f(b))$. In other words $(b, f(b))$ is a point at which the basic shape of the graph changes. Such points are very important and are called *points of inflection*.

7.9 Definition. *Let f be defined on the interval (c, d) and let $x_0 \in (c, d)$. IF the graph of f is concave upward on (c, x_0) and concave downward on (x_0, d) (or concave downward on (c, x_0) and concave upward on (x_0, d)), THEN f has a point of inflection at $x = x_0$ and $(x_0, f(x_0))$ is a point of inflection of the graph.*

Our opening remarks concerning Figure 7.12 show that a basic problem which still remains in our attempt to sketch graphs accurately is one of determining the concavity of the graph of a function. The following theorem is most useful in this determination.

7.10 Theorem. *Let f'' exist on (c, d).*
  (a) *IF $f''(x) > 0$ for all $x \in (c, d)$, THEN the graph of f is concave upward on (c, d).*
  (b) *IF $f''(x) < 0$ for all $x \in (c, d)$, THEN the graph of f is concave downward on (c, d).*

While a rigorous proof of Theorem 7.10 is somewhat complicated, we can argue geometrically to convince ourselves of its plausibility. For example, if the condition stated in part (a) holds, then $f''(x) > 0$ for all $x \in (c, d)$. Therefore f' is increasing on (c, d). For each point $x \in (c, d)$ visualize the tangent line to the graph at $(x, f(x))$. The slope of this line is $f'(x)$. Thus the fact that f' is increasing on (c, d) means that, as x goes from left to right in (c, d), the inclination of the tangent line to the graph must increase. This, in turn, means that the tangent line must be turning in a counterclockwise manner. It is geometrically apparent that this can only happen in a situation such as depicted in Figure 7.13, that is, when the graph is concave upward on (c, d). A similar geometric argument can be given for part (b).

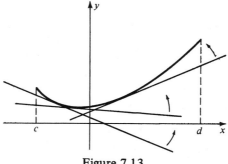

Figure 7.13

The following problems demonstrate how Theorem 7.10 is applied.

Problem. Determine those intervals on which the function $f(x) = 5 + 9x + 3x^2 - x^3$ is concave upward or concave downward.

Solution. The first two derivatives of f are

$$f'(x) = 9 + 6x - 3x^2,$$

and

$$f''(x) = 6 - 6x = 6(1 - x).$$

Therefore $f''(x) > 0$ if and only if $1 - x > 0$. Consequently, the graph of f is concave upward on $(-\infty, 1)$. Similarly, $f''(x) < 0$ if and only if $1 - x < 0$. Thus the graph of f is concave downward on $(1, \infty)$. It should also be noted that the concavity of the graph of f changes at $x = 1$. This means that f has a point of inflection at $x = 1$ and that $(1, f(1)) = (1, 16)$ is a point of inflection of the graph.

Problem. Determine those intervals on which the function $g(x) = 5(4x + 1)e^{2x}$ is concave upward or concave downward.

Solution. The first two derivatives of g are

$$g'(x) = \frac{d}{dx}[5(4x + 1)e^{2x}] = 5[4e^{2x} + (4x + 1)2e^{2x}] = 10e^{2x}(4x + 3),$$

and $g''(x) = \frac{d}{dx}[10e^{2x}(4x + 3)]$

$$= 10[4e^{2x} + (4x + 3)2e^{2x}]$$

$$= 20e^{2x}(4x + 5).$$

Since $e^{2x} > 0$ for all x, we see that $g''(x) > 0$ if, and only if, $4x + 5 > 0$, that is, $x > -\frac{5}{4}$. Hence, the graph of g is concave upward on $\left(-\frac{5}{4}, \infty\right)$. Likewise $g''(x) < 0$

if, and only if, $4x + 5 < 0$, that is, $x < -\frac{5}{4}$. This means that the graph of g is concave downward on $\left(-\infty, -\frac{5}{4}\right)$. Because of the change in concavity at $x = -\frac{5}{4}$, the point $\left(-\frac{5}{4}, -20e^{-\frac{5}{2}}\right)$ is a point of inflection on the graph of g.

From the preceding problems it should be observed that both second derivatives vanish at those values of x which yield a point of inflection. This property, which holds in general, is stated in the following theorem:

7.11 Theorem. *Let f'' exist at $x = x_0$. IF f has a point of inflection at $x = x_0$, THEN $f''(x_0) = 0$.*

Problem. What are the only possible values of x at which a point of inflection can occur for $f(x) = 3x^5 - 5x^4 + 1$?

Solution. The first two derivatives of f are

$$f'(x) = 15x^4 - 20x^3$$

and
$$f''(x) = 60x^3 - 60x^2 = 60x^2(x - 1).$$

We see that $f''(x) = 0$ if and only if $x = 0$ or $x = 1$. By Theorem 7.11 the only possible points of inflection must occur at $x = 0$ or $x = 1$.

Once the zeros of f'' have been determined, it is necessary to determine which of the zeros actually yields a point of inflection. One way of doing this is to check the concavity of the graph of f on both sides of a zero of f'' and apply Definition 7.9. There is also another method which can be used to quickly locate points of inflection. For this purpose assume that $f''(x_0) = 0$ and that f''' is continuous on an open interval containing x_0. If $f'''(x_0) > 0$, then there exists an interval (c, d) containing x_0 such that $f'''(x) > 0$ for all $x \in (c, d)$. Therefore f'' is increasing on (c, d). Since $f''(x_0) = 0$, this means that $f''(x) < 0$ for all $x \in (c, x_0)$ and $f''(x) > 0$ for all $x \in (x_0, d)$. Therefore the graph of f is concave downward on (c, x_0) and concave upward (x_0, d). It follows that f has a point of inflection at $x = x_0$. A similar argument in the case $f'''(x_0) < 0$ leads to a similar conclusion. Hence we have proved the following result sometimes called the *third derivative test.*

7.12 Theorem. *Let $f''(x_0) = 0$ and assume f''' is continuous on an open interval containing x_0. IF $f'''(x_0) \neq 0$, THEN f has a point of inflection at $x = x_0$.*

Problem. Locate the points of inflection on the graph of $f(x) = 3x^5 - 5x^4 + 1$.

Solution. From the preceding problem we found that $f''(x) = 60x^2(x - 1)$ so that the only possible points of inflection must occur for $x = 0$ or $x = 1$. Computing the third derivative we have

$$f'''(x) = 180x^2 - 120x = 60x(3x - 2).$$

Since $f'''(1) = -60 \neq 0$, the graph of f has a point of inflection at $x = 1$ by the third derivative test. On the other hand, $f''(0) = 0$ so that the third derivative test does not apply for the value $x = 0$. However, $f''(x) < 0$ for all $x \in (-1, 0)$ and for all $x \in (0, 1)$. Therefore the graph of f is concave downward on both $(-1, 0)$ and $(0, 1)$ so that the concavity does not change at $x = 0$. It follows that the graph of f does not have a point of inflection at $x = 0$.

Exercise Set 7.5

1. If the following sketch represents the graph of a function f, determine those intervals on which the graph of f is concave downward and those intervals on which the graph of f is concave upward. Determine the points of inflection.

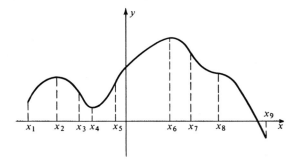

2. Let f be a function such that $f''(x) = 3x - 1$. Use Theorem 7.10 to determine those intervals on which the graph of f is concave downward and those intervals on which it is concave upward. Are there any points of inflection?

3. Let g be a function such that $g''(x) = (e^x - 1)(x^2 - 2x - 3)$. Use Theorem 7.11 to determine the possible points of inflection.

4. Let $f(x) = x^3 - 2x^2 + 5x - 25$.
   a. Find $f'(x), f''(x)$, and $f'''(x)$.
   b. Use Theorem 7.12 to determine the points of inflection.

* * *

In problems 5-20, determine those intervals on which the function is concave upward, those intervals on which the function is concave downward, and any points of inflection.

5. $f(x) = 5x^4 - 1$

6. $f(x) = x^3 + 2$

7. $g(x) = xe^x$

8. $G(x) = x - e^x$

9. $F(x) = x \ln x, x > 0$

10. $H(x) = x^2 + \ln x, x > 0$

11. $P(z) = z^2 + 3z + 2$

12. $g(t) = -\dfrac{2}{3}t^3 - \dfrac{5}{2}t^2 + 3t + 10$

13. $F(t) = \dfrac{3t + 7}{1 - 4t}, t \neq \dfrac{1}{4}$

14. $G(z) = \dfrac{z}{z + 1}, z \neq -1$

15. $f(x) = \sqrt{x}, x > 0$

16. $h(x) = \sqrt{x^2 + 1}$

17. $p(x) = x^4 - 4x^3 + 10$

18. $P(t) = 20t^5 - 2t^3 + t$

19. $G(x) = 2x^2 e^{-x}$

20. $f(z) = (3z + 2)\sqrt{1 - 2z}, z < \dfrac{1}{2}$

7.6 LIMITS INVOLVING INFINITY

Given a function f and real numbers a and L, we considered in Chap. 6 the notion of $f(x)$ approaching L as x approaches a. There are several other types of limits that occur quite frequently. We briefly examine some of these limits.

Let L be a real number. The function $f(x)$ approaches L as x approaches infinity, written $\lim\limits_{x \to \infty} f(x) = L$, if $f(x)$ can be made arbitrarily close to L by taking x sufficiently large. This situation is depicted in Figure 7.14. More precisely, $\lim\limits_{x \to \infty} f(x) = L$ if and only if for each $\epsilon > 0$ there is a number M such that $|f(x) - L| < \epsilon$ whenever $x > M$. We prefer, however, not to use the formal definition. Instead, we will rely on our intuition.

Example. For large values of x, the number $\dfrac{1}{x}$ becomes arbitrarily small. Thus we have $\lim\limits_{x \to \infty} \dfrac{1}{x} = 0$. Similarly, $\lim\limits_{x \to \infty} \dfrac{1}{x^2} = 0$. In fact, $\lim\limits_{x \to \infty} \dfrac{1}{x^n} = 0$ for every positive integer n. If c is any real number and n is a positive integer, we have

$$\lim_{x \to \infty} \dfrac{c}{x^n} = 0.$$

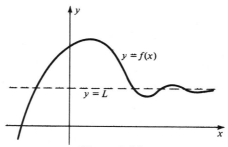

Figure 7.14

Now consider two polynomials $p(x)$ and $q(x)$. For example, let $p(x) = a_n x^n + a_{n-1} x^{n-1} + \cdots + a_0$ and $q(x) = b_m x^m + b_{m-1} x^{m-1} + \cdots + b_0$ be polynomials of degree n and m respectively, where $m \geqslant n$. First, let us assume that $n = m$. Thus, for $x \neq 0$,

$$\frac{p(x)}{q(x)} = \frac{a_n x^n + a_{n-1} x^{n-1} + \cdots + a_0}{b_n x^n + b_{n-1} x^{n-1} + \cdots + b_0} = \frac{a_n + \dfrac{a_{n-1}}{x} + \dfrac{a_{n-2}}{x^2} + \cdots + \dfrac{a_0}{x^n}}{b_n + \dfrac{b_{n-1}}{x} + \dfrac{b_{n-2}}{x^2} + \cdots + \dfrac{b_0}{x^n}},$$

where the final expression is obtained by dividing both numerator and denominator by x^n. As x increases without bound, all terms in the numerator and denominator approach zero, except a_n and b_n. Thus, for the case $n = m$,

$$\lim_{x \to \infty} \frac{p(x)}{q(x)} = \frac{a_n}{b_n}.$$

Example.

(i) $\displaystyle \lim_{x \to \infty} \frac{x}{2x + 1} = \frac{1}{2}$

and

(ii) $\displaystyle \lim_{x \to \infty} \frac{3x^3 - 6x^2 + 11x - 17}{-4x^3 - x^2 + 5x - 2} = -\frac{3}{4}.$

Now assume $n < m$. Then

$$\frac{p(x)}{q(x)} = \frac{\dfrac{a_n}{x^{m-n}} + \dfrac{a_{n-1}}{x^{m-n-1}} + \cdots + \dfrac{a_0}{x^m}}{b_m + \dfrac{b_{m-1}}{x} + \cdots + \dfrac{b_0}{x^m}}.$$

In this case, as x increases without bound, all terms in the numerator and denominator approach zero, except b_m. Therefore, when $n < m$,

$$\lim_{x \to \infty} \frac{p(x)}{q(x)} = 0.$$

Example.

(i) $\lim\limits_{x\to\infty} \dfrac{x}{2x^2 + 1} = 0$

and

(ii) $\lim\limits_{x\to\infty} \dfrac{3x^3 - 6x^2 + 11x - 17}{7x^4 - 4x^3 - x^2 + 5x - 2} = 0.$

In a similar fashion, we write $\lim\limits_{x\to-\infty} f(x) = L$ if $f(x)$ approaches L as x decreases without bound. For similar reasons the same conclusions of our previous examples concerning rational functions hold if we replace ∞ by $-\infty$.

Example. (i) If c is any real number and n is a positive integer, then

$\lim\limits_{x\to-\infty} \dfrac{c}{x^n} = 0.$

(ii) $\lim\limits_{x\to-\infty} \dfrac{x}{2x + 1} = \dfrac{1}{2},$

(iii) $\lim\limits_{x\to-\infty} \dfrac{3x^3 - 6x^2 + 11x - 17}{-4x^3 - x^2 + 5x - 2} = -\dfrac{3}{4},$

(iv) $\lim\limits_{x\to-\infty} \dfrac{x}{2x^2 + 1} = 0,$

and

(v) $\lim\limits_{x\to-\infty} \dfrac{3x^3 - 6x^2 + 11x - 17}{7x^4 - 4x^3 - x^2 + 5x - 2} = 0.$

One should note that $\lim\limits_{x\to-\infty} e^x = 0$ as pictured in Figure 7.15.

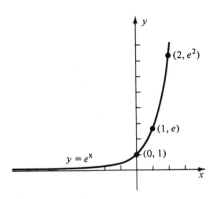

Figure 7.15

The symbol ∞ does not represent a number and thus should not be treated as such. Rather, the symbol $\lim_{x \to \infty} f(x) = L$ merely indicates that $f(x)$ approaches a limiting value L as x increases without bound. Since ∞ does not represent a number, one *should not* substitute ∞ for x when $f(x)$ is given by a formula. The same statements hold for the symbol $-\infty$.

The function $f(x)$ approaches infinity as x approaches infinity, written $\lim_{x \to \infty} f(x) = \infty$, if $f(x)$ can be made arbitrarily large by taking x sufficiently large. Formally, $\lim_{x \to \infty} f(x) = \infty$ if for each positive number M there is a number N such that $f(x) > M$ if $x > N$. For example, $\lim_{x \to \infty} x^2 = \infty$ because given a positive number M we will have $x^2 > M$ if $x > \sqrt{M}$ (see Figure 7.16).

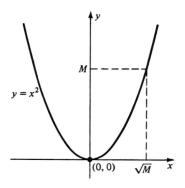

Figure 7.16

In general, if $c > 0$ and n is a positive integer, then $\lim_{n \to \infty} cx^n = \infty$.

The function $f(x)$ approaches minus infinity as x approaches infinity, written $\lim_{x \to \infty} f(x) = -\infty$, if $f(x)$ decreases without bound as x increases without bound.

Example. If $c < 0$ and n is a positive integer, then $\lim_{n \to \infty} cx^n = -\infty$.

Let $p(x) = a_n x^n + a_{n-1} x^{n-1} + \cdots + a_0$ be a polynomial function. Thus, for $x \neq 0$,

$$p(x) = x^n \left[a_n + \frac{a_{n-1}}{x} + \cdots + \frac{a_0}{x^n} \right].$$

As x increases without bound all terms in brackets approach 0, except a_n. Therefore the factor in brackets approximates a_n as x increases without bound. Thus $p(x)$ is approximated by $a_n x^n$, which approaches infinity as x increases without bound. Hence, if $a_n > 0$, then $\lim\limits_{x \to \infty} p(x) = \infty$ and if $a_n < 0$, then $\lim\limits_{x \to \infty} p(x) = -\infty$.

Example. (i) $\lim\limits_{x \to \infty} (6x - 10) = \infty$,

(ii) $\lim\limits_{x \to \infty} (x^4 - 167x^3 - 11x^2 + 4x - 70) = \infty$,

(iii) $\lim\limits_{x \to \infty} (-6x + 10) = -\infty$,

and

(iv) $\lim\limits_{x \to \infty} (-x^4 + 167x^3 - 11x^2 + 4x + 70) = -\infty$.

Consider the rational function $\dfrac{p(x)}{q(x)} = \dfrac{a_n x^n + \cdots + a_0}{b_m x^m + \cdots + b_0}$ as before, except now assume that $n > m$. If $x \neq 0$, then

$$\frac{p(x)}{q(x)} = x^{n-m} \left[\frac{a_n + \dfrac{a_{n-1}}{x} + \cdots + \dfrac{a_0}{x^n}}{b_m + \dfrac{b_{m-1}}{x} + \cdots + \dfrac{b_0}{x^m}} \right].$$

As x increases without bound all terms in the numerator and denominator approach 0, except a_n and b_m. Therefore the factor in brackets approximates $\dfrac{a_n}{b_m}$ as x increases without bound. If a_n and b_m have the same sign, then $\dfrac{a_n}{b_m}$ is positive. Thus the quotient $\dfrac{p(x)}{q(x)}$ approximates $x^{n-m} \dfrac{a_n}{b_m}$, which approaches infinity as x increases without bound. Hence, if a_n and b_m have the same sign and $n > m$, then

$$\lim_{x \to \infty} \frac{p(x)}{q(x)} = \infty.$$

Example. (i) $\lim\limits_{x \to \infty} \dfrac{x^2}{2x + 1} = \infty$,

(ii) $\lim\limits_{x \to \infty} \dfrac{-x^2}{-3x + 1} = \infty$,

and

(iii) $\lim\limits_{x \to \infty} \dfrac{-2x^4 + 3x^3 - 6x^2 + 11x - 17}{-4x^2 - x + 5} = \infty$.

Again, consider the rational function $\dfrac{p(x)}{q(x)} = \dfrac{a_n x^n + \cdots + a_0}{b_m x^m + \cdots + b_0}$, where $n > m$.

If a_n and b_m differ in sign, then $\dfrac{p(x)}{q(x)}$ approaches minus infinity as x increases without bound. That is, if a_n and b_m differ in sign and $n > m$, then

$$\lim_{x \to \infty} \frac{p(x)}{q(x)} = -\infty.$$

Example. (i) $\displaystyle\lim_{x \to \infty} \frac{-x^2}{2x + 1} = -\infty$,

(ii) $\displaystyle\lim_{x \to \infty} \frac{x^2}{-3x + 1} = -\infty$,

and

(iii) $\displaystyle\lim_{x \to \infty} \frac{-2x^4 + 3x^3 - 6x^2 - 11x - 17}{4x^2 - x + 5} = -\infty$.

Note that $\displaystyle\lim_{x \to \infty} e^x = \infty$ and $\displaystyle\lim_{x \to \infty} \ln x = \infty$ (see Figure 7.17).

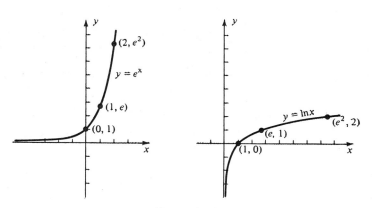

Figure 7.17

Another important type of limit is a *one-sided limit*. The function $f(x)$ approaches ∞ as x approaches a from the right, written $\displaystyle\lim_{x \to a^+} f(x) = \infty$, if $f(x)$ can be made arbitrarily large by taking x close to but greater than a. Formally, $\displaystyle\lim_{x \to a^+} f(x) = \infty$ if for each number M there is a $\delta > 0$ such that $f(x) > M$ if $a < x < a + \delta$ (see Figure 7.18).

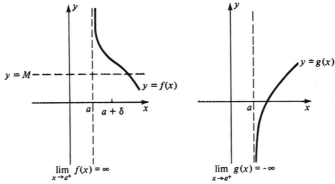

Figure 7.18

Example. For values of x close to but greater than 0, $\dfrac{1}{x}$ can be made arbitrarily large. Therefore $\displaystyle\lim_{x\to0^+} \dfrac{1}{x} = \infty$.

Example. For values of x close to but greater than 1, $\dfrac{x^2}{x-1}$ can be made arbitrarily large. Thus $\displaystyle\lim_{x\to1^+} \dfrac{x^2}{x-1} = \infty$.

The function $g(x)$ approaches $-\infty$ as x approaches a from the right, written $\displaystyle\lim_{x\to a^+} g(x) = -\infty$, if $g(x)$ decreases without bound by taking x close to but greater than a (see Figure 7.18).

Example. $\displaystyle\lim_{x\to2^+} \dfrac{x}{4-x^2} = -\infty$.

It should be noted that $\displaystyle\lim_{x\to0^+} \ln x = -\infty$ (see Figure 7.19).

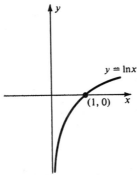

Figure 7.19

We also consider one-sided limits as x approaches a from the left. The function $f(x)$ approaches ∞ as x approaches a from the left, written $\lim\limits_{x \to a^-} f(x) = \infty$, if $f(x)$ can be made arbitrarily large by taking x close to but less than a. Formally, $\lim\limits_{x \to a^-} f(x) = \infty$ if for each number M there is a $\delta > 0$ such that $f(x) > M$ if $a - \delta < x < a$ (see Figure 7.20).

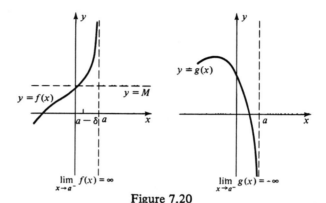

$$\lim_{x \to a^-} f(x) = \infty \qquad \lim_{x \to a^-} g(x) = -\infty$$

Figure 7.20

Example. (i) $\lim\limits_{x \to -3^-} \dfrac{x^2}{3 - x} = \infty.$

(ii) $\lim\limits_{x \to -2^-} \dfrac{-1}{x + 2} = \infty.$

In similar manner as before, the function $g(x)$ approaches $-\infty$ as x approaches a from the left, written $\lim\limits_{x \to a^-} g(x) = -\infty$, if $g(x)$ decreases without bound by taking x close to but less than a (see Figure 7.20).

Example. (i) $\lim\limits_{x \to 0^-} \dfrac{1}{x} = -\infty.$

(ii) $\lim\limits_{x \to 1^-} \dfrac{x^2}{1 - x} = -\infty.$

Exercise Set 7.6

1. a. If $f(x) = \dfrac{2}{x}$, find

$f(5), f(10), f(2000), f(8{,}000{,}000), \lim\limits_{x \to \infty} \dfrac{2}{x}.$

b. If $f(x) = \dfrac{2}{x}$, find

$f(-5), f(-10), f(-2000), f(-8,000,000), \displaystyle\lim_{x\to-\infty} \dfrac{2}{x}$.

2. a. If $f(x) = 2^x$, find
$f(1), f(5), f(10), f(100), \displaystyle\lim_{x\to\infty} 2^x$.

b. If $f(x) = 2^x$, find
$f(-1), f(-5), f(-10), f(-100), \displaystyle\lim_{x\to-\infty} 2^x$.

3. a. If $f(x) = \dfrac{x+1}{x-1}$, find

$f(2), f(5), f(100), f(4{,}999{,}999), \displaystyle\lim_{x\to\infty} \dfrac{x+1}{x-1}$.

b. If $f(x) = \dfrac{x+1}{x-1}$, find

$f(-2), f(-5), f(-100), f(-4{,}999{,}999), \displaystyle\lim_{x\to-\infty} \dfrac{x+1}{x-1}$.

4. a. If $f(x) = \dfrac{x+1}{x-1}$, find

$f(1.5), f(1.1), f(1.01), f(1.00004), \displaystyle\lim_{x\to1^+} \dfrac{x+1}{x-1}$.

b. If $f(x) = \dfrac{x+1}{x-1}$, find

$f(0.5), f(0.9), f(0.999), f(0.9999994), \displaystyle\lim_{x\to1^-} \dfrac{x+1}{x-1}$.

5. a. Find $\displaystyle\lim_{x\to\infty} \dfrac{2x^2 + x - 12}{3x^3 - x^2 + 15}$.

b. Find $\displaystyle\lim_{x\to-\infty} \dfrac{-3x^3 - x^2 + 15}{2x^2 + x - 12}$.

c. Find $\displaystyle\lim_{x\to-\infty} \dfrac{4x^3 - 3x^2 + 15}{3x^3 - x^2 - 12}$.

* * *

In addition to the limits already defined in the text, we say $f(x)$ approaches minus infinity as x approaches minus infinity and write $\displaystyle\lim_{x\to-\infty} f(x) = -\infty$, provided that $f(x)$ decreases without bound as x decreases without bound. In problems 6–26, find the indicated limits.

6. $\displaystyle\lim_{x\to\infty} \dfrac{-100}{x^2}$, $\qquad\qquad\qquad \displaystyle\lim_{x\to-\infty} \dfrac{-100}{x^2}$

7. $\displaystyle\lim_{x\to\infty} \dfrac{-3x+4}{2x-1}$, $\qquad\qquad\quad \displaystyle\lim_{x\to-\infty} \dfrac{-3x+4}{2x-1}$

8. $\lim\limits_{x \to \infty} -5x^3$, $\lim\limits_{x \to -\infty} -5x^3$

9. $\lim\limits_{x \to \infty} \dfrac{1}{x^2 - 1}$, $\lim\limits_{x \to -\infty} \dfrac{1}{x^2 - 1}$

10. $\lim\limits_{x \to \infty} \dfrac{2x^2 - 5x + 6}{x^3 + 1}$, $\lim\limits_{x \to \infty} \dfrac{-2x^2 + 5x - 6}{x^3 + 1}$

11. $\lim\limits_{x \to -\infty} \dfrac{x^3 + 1}{2x^2 - 5x + 6}$, $\lim\limits_{x \to -\infty} \dfrac{1 - x^3}{2x^2 - 5x + 6}$

12. $\lim\limits_{x \to \infty} \dfrac{6x^5 - 2x^3 + 11}{3 + 5x - 10x^5}$, $\lim\limits_{x \to -\infty} \dfrac{6x^5 - 2x^3 + 11}{10x^5 - 5x - 3}$

13. $\lim\limits_{x \to -\infty} \dfrac{\sqrt{2}\,x^2 - 6x + 11}{7x^2}$, $\lim\limits_{x \to \infty} \dfrac{\sqrt{2}\,x^2 - 6x + 11}{-7x^2}$

14. $\lim\limits_{x \to \infty} x^2$, $\lim\limits_{x \to -\infty} x^2$

15. $\lim\limits_{x \to \infty} 47x^3$, $\lim\limits_{x \to -\infty} 47x^3$

16. $\lim\limits_{x \to \infty} (5x - 350{,}000)$, $\lim\limits_{x \to -\infty} (350{,}000 - 5x)$

17. $\lim\limits_{x \to \infty} \dfrac{x^3}{x^2 + 1}$, $\lim\limits_{x \to -\infty} \dfrac{x^3}{x^2 + 1}$

18. $\lim\limits_{x \to 3^+} \dfrac{5}{x - 3}$, $\lim\limits_{x \to 3^-} \dfrac{5}{x - 3}$

19. $\lim\limits_{x \to 3^+} \dfrac{5}{(x - 3)^2}$, $\lim\limits_{x \to 3^-} \dfrac{5}{(x - 3)^2}$

20. $\lim\limits_{x \to 0^+} \dfrac{5}{x - 3}$, $\lim\limits_{x \to 1^-} \dfrac{5}{x - 3}$

21. $\lim\limits_{x \to 1^+} \dfrac{x + 1}{x^2 - 1}$, $\lim\limits_{x \to 1^-} \dfrac{x + 1}{x^2 - 1}$

22. $\lim\limits_{x \to 1^+} \dfrac{x^2 - 1}{x + 1}$, $\lim\limits_{x \to 1^-} \dfrac{x^2 - 1}{x + 1}$

23. $\lim\limits_{x \to -\infty} (2x^4 - x^3 + 5x^2 + 4x + 7)$, $\lim\limits_{x \to -\infty} (x^3 + 5x^2 + 4x + 7)$

24. $\lim\limits_{x \to 1^+} \dfrac{x^2 - 1}{x - 1}$, $\lim\limits_{x \to 1^+} \dfrac{x^2 + 1}{x - 1}$

25. $\lim\limits_{x \to -2^+} \dfrac{x^2 + 5x + 6}{x + 2}$, $\lim\limits_{x \to 2^-} \dfrac{x^2 + 5x + 6}{x - 2}$

26. $\lim\limits_{x \to 0^+} \dfrac{-2}{x^2}$, $\lim\limits_{x \to 0^-} \dfrac{-2}{x^2}$

7.7 CURVE SKETCHING

In this section all the machinery developed in Secs. 7.1-7.6 is brought to bear on the problem of curve sketching. Let us suppose we wish to sketch the graph of a given function f. In order to have a systematic way for doing this, it is recommended that the order of procedure be as follows:

Step 1. Find the zeros of f if they can be easily determined by techniques learned previously.

Step 2. Determine those intervals on which the function is increasing or decreasing by examining the sign of the first derivative of f (Sec. 7.2).

Step 3. Locate the relative extrema of f by either the first derivative test or the second derivative test (Sec. 7.3 and 7.4).

Step 4. Determine those intervals on which the graph of the function is concave upward or concave downward by examining the sign of the second derivative of f (Sec. 7.5).

Step 5. Locate the points of inflection by using the information in Step 4 or the third derivative test (Sec. 7.5).

Step 6. If possible, determine any limits involving infinity when applicable (Sec. 7.6).

Step 7. Sketch the graph of f by sketching a curve which accurately reflects the information obtained in Steps 1-6.

The procedure just outlined uses our knowledge of the first three derivatives of f. In practice it is usually wise to determine and list the first three derivatives of f before following the steps outlined above.

The techniques of curve sketching are shown in the following solved problems.

Problem. Sketch the graph of $f(x) = x^3 - 6x^2 + 9x$.

Solution. The first three derivatives of f are

$$f'(x) = 3x^2 - 12x + 9 = 3(x^2 - 4x + 3) = 3(x - 1)(x - 3),$$

$$f''(x) = 6x - 12 = 6(x - 2),$$

and

$$f'''(x) = 6.$$

Step 1. We can write

$$f(x) = x^3 - 6x^2 + 9x = x(x^2 - 6x + 9) = x(x - 3)^2.$$

Thus, the zeros of f are $x = 0, 3$.

Step 2. The following sign table indicates the sign of f':

		Sign of	
	$x-1$	$x-3$	$f'(x) = 3(x-1)(x-3)$
$x < 1$	−	−	+
$1 < x < 3$	+	−	−
$3 < x$	+	+	+

Hence, f is increasing on $(-\infty, 1)$, decreasing on $(1, 3)$, and increasing on $(3, \infty)$.

Step 3. The zeros of f' are $x = 1, 3$. Since $f''(1) = -6 < 0$, f has a relative maximum at $x = 1$. Thus, $(1, 4)$ is a high point on the graph of f. Since $f''(3) = 6 > 0$, f has a relative minimum at $x = 3$ so that $(3, 0)$ is a low point on the graph of f.

Step 4. From the fact that $f''(x) = 6(x - 2)$, we see that the graph of f is concave downward on $(-\infty, 2)$ and concave upward on $(2, \infty)$.

Step 5. The only zero of f'' is $x = 2$. Since $f'''(2) = 6 \neq 0$, f has a point of inflection at $x = 2$. Thus $(2, 2)$ is a point of inflection on the graph of f.

Step 6. We see that

$$\lim_{x \to \infty} f(x) = \lim_{x \to \infty} (x^3 - 6x^2 + 9x) = \infty$$

and

$$\lim_{x \to -\infty} f(x) = \lim_{x \to -\infty} (x^3 - 6x^2 + 9x) = -\infty.$$

Step 7. The information obtained in Steps 1-6 is now used to sketch the graph of f shown in Figure 7.21.

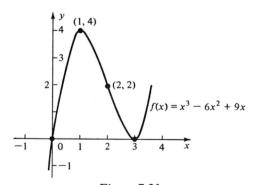

Figure 7.21

Problem. Sketch the graph of $g(x) = \dfrac{2x + 1}{5 - 3x}$.

Solution. Before computing the derivatives of g, it should be noted that g is not defined at $x = \frac{5}{3}$. Therefore the line $x = \frac{5}{3}$ will serve as a barrier for the graph of g. The derivatives of g are

$$g'(x) = \frac{13}{(5 - 3x)^2},$$

$$g''(x) = \frac{78}{(5 - 3x)^3},$$

and

$$g'''(x) = \frac{702}{(5 - 3x)^4}.$$

Step 1. $g(x) = 0$ if, and only if, $2x + 1 = 0$ so that the only zero of g is $x = -\frac{1}{2}$.

Step 2. For all $x \neq \frac{5}{3}$ we have $g'(x) > 0$. Thus g is increasing on $\left(-\infty, \frac{5}{3}\right)$ and $\left(\frac{5}{3}, \infty\right)$.

Step 3. As we have observed in Step 2, the first derivative of g is never zero. Hence, g has no relative extrema.

Step 4. We see that $g''(x) > 0$ if, and only if, $(5 - 3x)^3 > 0$. This is the case if, and only if, $5 - 3x > 0$, that is, $x < \frac{5}{3}$. Therefore, the graph of g is concave upward on $\left(-\infty, \frac{5}{3}\right)$. Similarly, the graph of g is concave downward on $\left(\frac{5}{3}, \infty\right)$.

Step 5. The function g'' has no zeros. Consequently, there are no points of inflection.

Step 6. The denominator of g vanishes at $x = \frac{5}{3}$, so it is at this point that we should examine right and left hand limits. For this function we have

$$\lim_{x \to \frac{5}{3}^-} \frac{2x + 1}{5 - 3x} = \infty,$$

and

$$\lim_{x \to \frac{5}{3}^+} \frac{2x + 1}{5 - 3x} = -\infty.$$

Taking limits as x goes to plus or minus infinity, we have

$$\lim_{x \to \infty} \frac{2x + 1}{5 - 3x} = -\frac{2}{3},$$

and

$$\lim_{x \to -\infty} \frac{2x + 1}{5 - 3x} = -\frac{2}{3}.$$

Step 7. The graph of g is now sketched using the information found in Steps 1-6 (see Figure 7.22).

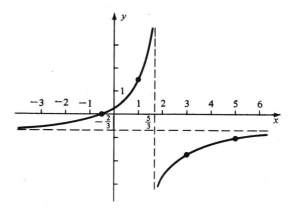

Figure 7.22

Problem. Sketch the graph of the function $h(x) = x - \sqrt{x - 1}$.

Solution. First, note that this function is defined only for $x \geqslant 1$. The derivatives of h are

$$h'(x) = 1 - \frac{1}{2\sqrt{x - 1}} = \frac{2\sqrt{x - 1} - 1}{2\sqrt{x - 1}},$$

$$h''(x) = \frac{1}{4}(x - 1)^{-\frac{3}{2}},$$

and

$$h'''(x) = -\frac{3}{8}(x - 1)^{-\frac{5}{2}}.$$

Step 1. The function h has no zeros because there are no real numbers x such that $x = \sqrt{x - 1}$. This is the case because there are no real solutions of the equation $x^2 = x - 1$.

Step 2. For $x > 1$ we have $\sqrt{x - 1} > 0$. Therefore, $h'(x) > 0$ if, and only if, $2\sqrt{x - 1} - 1 > 0$. But this is the case if, and only if, $\sqrt{x - 1} > \frac{1}{2}$, that is, $x - 1 > \frac{1}{4}$. It follows that h is increasing on $\left(\frac{5}{4}, \infty\right)$. A similar argument shows that h is decreasing on $\left(1, \frac{5}{4}\right)$.

Step 3. We see that $h'(x) = 0$ if, and only if, $2\sqrt{x - 1} - 1 = 0$ that is $x = \frac{5}{4}$. Since

$$h''\left(\frac{5}{4}\right) = \frac{1}{4}\left(\frac{5}{4} - 1\right)^{-\frac{3}{2}} = 2 > 0,$$

h has a relative minimum at $x = \frac{5}{4}$. This relative minimum is $h\left(\frac{5}{4}\right) = \frac{3}{4}$.

Step 4. $h''(x) > 0$ for all $x > 1$ so that the graph of h is concave upward on $(1, \infty)$.

Step 5. The function h'' has no zeros so there can be no points of inflection. Therefore, in this problem, information regarding the third derivative is not needed.

Step 6. We now wish to determine the behavior of $h(x)$ as x approaches infinity. Since $h(x) = x - \sqrt{x - 1}$ and both numbers in this difference become large as x becomes large, it will be difficult to determine the behavior of $h(x)$ as x becomes large. Therefore, we try to write the expression for $h(x)$ in a form which will more clearly reveal its behavior as x approaches infinity. We write

$$h(x) = x - \sqrt{x - 1}$$
$$= x\left(1 - \sqrt{\frac{1}{x} - \frac{1}{x^2}}\right).$$

The expression inside the parentheses approaches 1 as x approaches infinity. Consequently, $\lim\limits_{x \to \infty} h(x) = \infty$.

Step 7. Using the information determined in Steps 1-6, we now sketch the graph of h (see Figure 7.23).

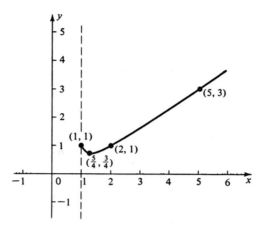

Figure 7.23

Exercise Set 7.7

1. Let $f(x) = 4x^3 + 3x^2 - \dfrac{5}{2}$.

So $f'(x) = 12x^2 + 6x$,
$f''(x) = 24x + 6$,
and $f'''(x) = 24$.

The only zero of f is $\dfrac{1}{2}$.

The function f is increasing on $\left(-\infty, -\dfrac{1}{2}\right)$ and $(0, \infty)$.

The function f is decreasing on $\left(-\dfrac{1}{2}, 0\right)$.

The function f has a relative minimum at $x = 0$ and $f(0) = -\dfrac{5}{2}$.

The function f has a relative maximum at $x = -\dfrac{1}{2}$ and $f\left(-\dfrac{1}{2}\right) = -1$.

The graph of f is concave upward on $\left(-\dfrac{1}{4}, \infty\right)$.

The graph of f is concave downward on $\left(-\infty, -\dfrac{1}{4}\right)$.

The graph of f has a point of inflection at $\left(-\dfrac{1}{4}, -\dfrac{19}{8}\right)$.

The $\lim\limits_{x \to -\infty} f(x) = -\infty$ and the $\lim\limits_{x \to \infty} f(x) = \infty$.

a. Construct a brief table of values for f using the given information.
b. Sketch the graph of the function f.
c. Label the relative extrema, zeros, and point of inflection.

2. Let $f(x) = x^3 - x^2 - 8x + 6$.
   a. Find f', f'', and f'''.
   b. Find the zeros of f.
   c. Determine those intervals on which the function f is increasing and those intervals on which it is decreasing.
   d. Find the relative extrema of f.
   e. Determine those intervals on which the graph of f is concave upward and those intervals on which it is concave downward.
   f. Find the points of inflection.
   g. Find $\lim\limits_{x \to \infty} f(x)$ and $\lim\limits_{x \to -\infty} f(x)$.
   h. Construct a brief table of values.
   i. Sketch the graph of the function.

* * *

In problems 3–21, determine the zeros of the function, the relative extrema, the intervals on which the function is increasing or decreasing, concavity, points of inflection, and infinite limits. Use this information to sketch the graph of the function.

3. $f(x) = 2x^2 - 3x - 2$

4. $g(x) = -2x^2 + 3x - 5$

5. $h(x) = -4x^3 - 3x^2 + 18x + 6$

6. $p(x) = 2x^3 + x^2 - 4x - 3$

7. $P(x) = x^3 + 2x - 3$

8. $H(x) = 2x^3 + 3x^2 + 2x + 1$

9. $F(x) = \dfrac{2x + 7}{3 - 4x}; \ x \neq \dfrac{3}{4}$

10. $G(x) = \dfrac{4x + 7}{(3x - 1)(x + 4)}; \ x \neq \dfrac{1}{3}, -4$

11. $f(x) = 3xe^{-2x}$

12. $H(x) = e^{3x}$

13. $F(x) = \dfrac{x^2}{e^x}$

14. $f(z) = z^3 - 33z^2 + 362z - 1320$

15. $h(x) = 4x^4 - 12x^2 + 9$

16. $T(z) = \sqrt{4 - z^2}$

17. $T(z) = \sqrt{z^2 - 4}$

18. $L(x) = x \ln x; \ x > 0$

19. $L(x) = x^2 + \ln x; \ x > 0$

20. $f(t) = \dfrac{t}{t+1}$

21. $f(x) = x^4 + x^2 - 6$

7.8 MAXIMA AND MINIMA OF FUNCTIONS

The derivative has many other applications in addition to its use in curve sketching. One particularly interesting class of problems to which the derivative may be applied is the class of *max-min problems.* These are practical problems which arise from a need to determine either the maximum or minimum value for a given function. We examine these in Sec. 7.9. In order to study these problems, we must first find out what is meant by the maximum value and the minimum value of a function. Following is a definition of these two terms.

7.13 Definition. *Let f be a function defined on an interval I.* (a) *f achieves its maximum value at $x = x_0$ on I and $f(x_0)$ is the maximum value of f on I provided $f(x_0) \geqslant f(x)$ for all $x \in I$.* (b) *f achieves its minimum value at $x = x_0$ on I and $f(x_0)$ is the minimum value of f on I provided $f(x_0) \leqslant f(x)$ for $x \in I$.*

Let us be careful at this point to distinguish between the terms "maximum" and "relative maximum," and "minimum" and "relative minimum." If $f(x_0)$ is a relative maximum of f, then $(x_0, f(x_0))$ is a *high* point on the graph of f. However, if $f(x_0)$ is the maximum value on I, then $(x_0, f(x_0))$ is the *highest* point on the graph of f for values of x in I. If $f(x_0)$ is a relative minimum of f, then $(x_0, f(x_0))$ is a *low* point on the graph of f. On the other hand, if $f(x_0)$ is the minimum value of f on I, then $(x_0, f(x_0))$ is the *lowest* point on the graph of f for values of x in I. For instance, Figure 7.24a represents the graph of a function f defined on $[c,d]$. We see that f attains its maximum value at $x = c$ and its minimum value at $x = x_3$. Note, too, that f has a relative minimum at

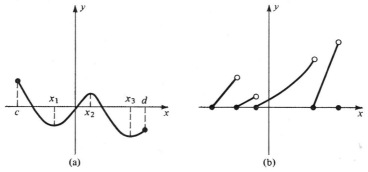

(a) (b)

Figure 7.24

$x = x_1$ and a relative maximum at $x = x_2$. Note further that the minimum $f(x_3)$ is also a relative minimum while the maximum $f(c)$ is not a relative maximum.

Figure 7.24b represents the graph of a function f defined on $[c, d]$ which does not attain a maximum value (it does attain a minimum value however).

It is natural to ask: "What kind of functions actually achieve their maximum or minimum value?" Fortunately, it can be shown that every function which is continuous on a closed interval $[c, d]$ achieves both its maximum and minimum values on this interval. If, in addition, f is differentiable on (c, d) and attains either its maximum or minimum value at a point x_0 such that $c < x_0 < d$, then $f'(x_0) = 0$. It follows that such a function must achieve its maximum value and minimum value at one of the end points or at a zero of the first derivative which lies strictly between the end points. These conditions are summarized in the following statement.

7.14 Theorem. *Let f be continuous on the closed interval $[c, d]$ and differentiable on (c, d). IF f attains its maximum or minimum value at $x = x_0$, THEN either x_0 is an end point of the interval $[c, d]$ or $f'(x_0) = 0$.*

Problem. Determine the maximum and minimum values of the function $f(x) = x^2 - x - 2$ on the closed interval $[-2, 4]$.

Solution. Since f is a polynomial function, it is continuous. Therefore, we are guaranteed that f achieves its maximum value and minimum value on $[-2, 4]$.

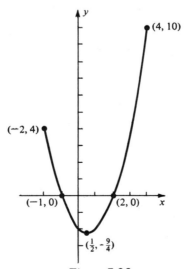

Figure 7.25

We note that $f'(x) = 2x - 1$ so that the only zero of f' which lies between -2 and 4 is $x = \frac{1}{2}$. According to Theorem 7.14, f must achieve its maximum value at one of the points $x = -2, \frac{1}{2}, 4$ and its minimum at one of the points $x = -2, \frac{1}{2}, 4$. Computing the value of the function at each of these points, we have $f(-2) = 4$, $f\left(\frac{1}{2}\right) = -\frac{9}{4}$, and $f(4) = 10$. Thus the maximum value of f on $[-2, 4]$ is 10 and the minimum value of f on $[-2, 4]$ is $-\frac{9}{4}$. The graph of f is shown in Figure 7.25.

Problem. Determine the maximum and minimum values of the function $g(x) = x + \frac{1}{x}$ on the closed interval $\left[\frac{1}{2}, \frac{3}{2}\right]$.

Solution. The derivative of g is

$$g'(x) = 1 - \frac{1}{x^2} = \frac{x^2 - 1}{x^2}.$$

Therefore the only zero of g' which lies between $\frac{1}{2}$ and $\frac{3}{2}$ is $x = 1$. By Theorem 7.14, g achieves its maximum value at one of the points $x = \frac{1}{2}, 1, \frac{3}{2}$ and its minimum value at one of these points. Since $g\left(\frac{1}{2}\right) = \frac{5}{2}, g(1) = 2$, and $g\left(\frac{3}{2}\right) = \frac{13}{6}$, the maximum value of g on $\left[\frac{1}{2}, \frac{3}{2}\right]$ is $\frac{5}{2}$ and the minimum value of g on $\left[\frac{1}{2}, \frac{3}{2}\right]$ is 2.

Problem. Determine the maximum and minimum values of the function $g(x) = x + \frac{1}{x}$ on the closed interval $[3, 5]$.

Solution. Note that this is essentially the same function as in the preceding problem, except that now it is defined on a different interval. Again, $g'(x) = \frac{x^2 - 1}{x^2}$, so that g' has no zeros in $[3, 5]$. By Theorem 7.14, g must attain its maximum and minimum values at the end points of the interval $[3, 5]$. We have $g(3) = \frac{10}{3}$ which is less than $g(5) = \frac{26}{5}$. Therefore the maximum value of g on $[3, 5]$ is $\frac{26}{5}$ and the minimum value of g on $[3, 5]$ is $\frac{10}{3}$.

The graph of the function g over each of the intervals is shown in Figure 7.26. This graph illustrates that the maximum and minimum value of a function on an interval I may change if I changes.

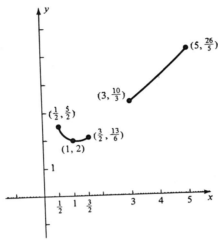

Figure 7.26

Problems involving the maximization and minimization of functions defined on intervals which are not closed can normally be solved by locating the relative extrema of the function involved and determining those intervals on which the function is increasing or decreasing. The following problems illustrate some typical situations.

Problem. Determine the minimum value of $f(x) = \dfrac{(x^2 + 9)^2}{x}$ on $(0, \infty)$.

Solution. We have

$$f'(x) = \frac{x\left[4x(x^2 + 9)\right] - (x^2 + 9)^2}{x^2}$$

$$= \frac{3(x^2 + 9)(x^2 - 3)}{x^2}.$$

Since $x^2 > 0$ and $x^2 + 9 > 0$ for $x \in (0, \infty)$, we see that $f'(x) = 0$ if and only if $x^2 - 3 = 0$. Thus, the only zero of f' in $(0, \infty)$ is $x = \sqrt{3}$. Moreover, if $0 < x < \sqrt{3}$, then $f'(x) < 0$. It follows that f is decreasing on $(0, \sqrt{3})$. Similarly,

f is increasing on $(\sqrt{3}, \infty)$. Therefore f has a minimum value at $x = \sqrt{3}$ and this minimum value is $f(\sqrt{3}) = \dfrac{144}{\sqrt{3}}$.

Problem. Determine the maximum value of the function $f(x) = xe^{-x}$. Does this function have a minimum value?

Solution. Note that the domain of this function is the set of all real numbers. The first derivative of f is $f'(x) = e^{-x}(1 - x)$. Since $e^{-x} > 0$ for all x, $f'(x) > 0$ if and only if $1 - x > 0$ and $f'(x) < 0$ if and only if $1 - x < 0$. Therefore, f is increasing on $(-\infty, 1)$ and decreasing on $(1, \infty)$. This means f achieves its maximum value at $x = 1$ and this maximum value is $f(1) = e^{-1}$. The function does not have a minimum value because $\lim\limits_{x \to -\infty} xe^{-x} = -\infty$.

In the next section we shall see how our ability to find the maximum and minimum values of a function on a given interval can be applied in the solution of verbal max-min problems.

Exercise Set 7.8

1. a. Sketch the graph of a function which has both a maximum and a minimum value on the interval $[-1, 1]$.
   b. Sketch the graph of a function which has neither a maximum nor a minimum value on the interval $[-1, 1]$.
   c. Sketch the graph of a function which has a maximum value but not a minimum value on the interval $[-1, 1]$.

2. Let g be a function which is continuous on $[-2, 5]$ and differentiable on $(-2, 5)$. Let Z be the set of zeros of g' and suppose $Z = \{-1, 2, 3\}$.
   a. Is g guaranteed to have a minimum value on $[-2, 5]$? A maximum value?
   b. According to Theorem 7.14, where must such values occur?

3. Let $f(x) = 4x^2 - 2x + 5$. Thus $f'(x) = 8x - 2$.
   a. Let Z be the set of zeros of $f'(x)$. List the elements of Z.
   b. For $x_0 \in Z$, compute $f(x_0)$.
   c. Compute $f(-1)$ and $f(1)$.
   d. What is the maximum value of $f(x)$ on $[-1, 1]$?
   e. What is the minimum value of $f(x)$ on $[-1, 1]$?
   f. What is the maximum value of $f(x)$ on $(-1, 1)$?
   g. What is the minimum value of $f(x)$ on $(-1, 1)$?

4. Suppose $f(x) = 3x^4 + 2x^3 - 15x^2 + 12x - 2$. So, $f'(x) = 12x^3 + 6x^2 - 30x + 12 = 6(x - 1)(x + 2)(2x - 1)$.
   a. Let Z be the set of zeros of $f'(x)$. List the elements of Z.
   b. For each $x_0 \in Z$, compute $f(x_0)$.

   c. Compute $f(-3)$ and $f(3)$.

   d. What is the maximum value of $f(x)$ on $[-3, 3]$?

   e. What is the minimum value of $f(x)$ on $[-3, 3]$?

   f. What is the maximum value of $f(x)$ on $(-\infty, \infty)$?

   g. What is the minimum value of $f(x)$ on $(-\infty, \infty)$?

* * *

In problems 5–22, find the maximum and the minimum value of the function in the indicated interval, or show that such values do not exist.

5. $f(x) = 4x^2 + 7x + 8$, $[-1, 1]$ and $(-1, 1)$

6. $f(x) = -x^2 + 2x + 1$, $[0, 4]$ and $[4, 12]$

7. $f(x) = x^3 - 3x^2 + 3x + 1$, $[0, 3]$ and $(-\infty, \infty)$

8. $g(x) = 3x^4 + 4x^3$, $[-2, 0]$ and $\left(-\dfrac{1}{2}, \dfrac{1}{2}\right)$

9. $g(x) = x^3 + \dfrac{48}{x}$, $(-\infty, 0)$ and $(0, \infty)$

10. $g(x) = x^3 + \dfrac{48}{x}$, $(-\infty, -1)$ and $[1, \infty)$

11. $l(x) = 4x - 12$, $[-10, 10]$ and $(-10, 10)$

12. $g(x) = -2$, $[-10, 10]$ and $(-\infty, \infty)$

13. $f(t) = \dfrac{t^2}{t + 2}$, $(-\infty, -2)$ and $(-2, \infty)$

14. $f(t) = \dfrac{4t}{t^2 + 4}$, $[-4, 4]$ and $(-\infty, \infty)$

15. $y = (x + 3)^5$, $[0, 1]$ and $[-4, 1]$

16. $y = \dfrac{x + 5}{x - 1}$, $[2, 5]$ and $[2, \infty)$

17. $y = \dfrac{4}{x}$, $(-1, 1)$ and $(-\infty, \infty)$

18. $y = \dfrac{1}{x^2 - 5x + 7}$, $[-3, 0]$ and $[0, 3]$

19. $f(x) = x - e^x$, $[-10, 10]$ and $(-\infty, \infty)$

20. $g(x) = \dfrac{1}{x} \ln x$, $\left[\dfrac{1}{e^2}, e^2\right]$ and $(0, e^2)$

21. $f(x) = x\sqrt{4 - x^2}$, $[-1, 1]$ and $[-2, 2]$

22. $R(x) = x^{\frac{1}{2}} + x^{-\frac{1}{2}}$, $(0, 1)$ and $\left[\dfrac{1}{2}, \dfrac{3}{2}\right]$

7.9 APPLICATIONS TO VERBAL PROBLEMS

We now consider several practical situations in which the differential calculus may be used to solve problems involving maximum or minimum values of a function and problems involving instantaneous rates of change. Though the max-min problems we present may seem to be unrelated to each other, the same basic steps are used in the solution of each one. They are as follows:

1) Represent the verbal problem mathematically, that is, obtain a function which indicates the relationship between the quantities given in the statement of the problem.
2) Using the methods developed in the preceding sections, maximize (or minimize, depending upon the nature of the problem) the function obtained in the first step.
3) Interpret verbally the mathematical result obtained in the second step.

We now illustrate this approach by solving a number of typical max-min problems.

Problem. A man has a rectangular sheet of metal, 20 inches wide and 32 inches long. From this piece of metal he wishes to construct a rectangular box without a top. In order to make this box, he intends to cut four square pieces, all of the same size, from the corners of the sheet, and bend up the sides (see Figure 7.27). What size piece should he cut from each corner in order that the box have a maximum volume?

Solution. Let us begin by expressing the volume of the box as a function of some other variable quantity. To do this, let x denote the length of a side of one of the square pieces to be removed.

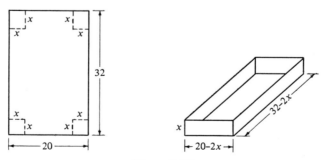

Figure 7.27

From Figure 7.27 we see that the box will then have sides of length $32-2x$, width $20-2x$, and height x, where $0 \leqslant x \leqslant 10$ (if x were larger, the box would have no width). Thus the volume of the box is given by

$$V(x) = (32-2x)(20-2x)x = 4(x^3 - 26x^2 + 160x).$$

This is the function we wish to maximize over the interval $[0, 10]$. Now the derivative of V is

$$V'(x) = 4(3x^2 - 52x + 160).$$

The zeros of V' may be found using the quadratic formula, thus, $V'(x) = 0$ if, and only if,

$$x = \frac{52 \pm \sqrt{2704 - 4(3)(160)}}{2(3)} = \frac{52 \pm \sqrt{784}}{6} = \frac{52 \pm 28}{6}.$$

Thus $x = 4$ or $x = \frac{40}{3}$. Since $\frac{40}{3} > 10$, the only zero of V' in the interval $[0, 10]$ is $x = 4$. By Theorem 7.14, V must achieve its maximum value for $x = 0, 4,$ or 10. Now $V(0) = 0$, $V(4) = 1152$, and $V(10) = 0$. Thus, the maximum volume of a rectangular box formed in this fashion is 1152 cubic inches, and such a box is obtained when squares measuring 4 inches on a side are cut from the corners of the metal sheet.

Problem. It is known that a given environment can support only a limited population of any particular species. It is also known that the normal development of a species involves a growth in number to a certain maximal point, followed either by a gradual leveling of population growth or by a marked decrease in population. In a certain environmental setting it is theorized that the number $N(t)$ of people present at time t (measured in years) is approximated by the function

$$N(t) = e^{-t^2 + 200t - 9984} + 4,000,000.$$

Letting the present time correspond to $t = 0$, what is the maximum human population this environment can support? In how many years will the population reach this maximum level? Assuming no measures are taken to curtail the growth of the population before its maximum level is attained, what will be the size of the population 4 years after the maximum level is attained?

Solution. We are provided with a mathematical formulation of the problem in the form of the function

$$N(t) = e^{-t^2 + 200t - 9984} + 4,000,000.$$

The time interval over which this function is to be maximized is $[0, \infty)$.
  Now

$$N'(t) = (-2t + 200) e^{-t^2 + 200t - 9984}.$$

Since $e^{-t^2+200t-9984} > 0$ for all values of t, we see that $N'(t) > 0$ for $-2t + 200 > 0$, $N'(t) < 0$ for $-2t + 200 < 0$, and $N'(t) = 0$ for $-2t + 200 = 0$. Therefore, N is increasing on $(0, 100)$, and decreasing on $(100, \infty)$. Consequently, N achieves its maximum value at $t = 100$, and this maximum value is

$$N(100) = e^{-10,000+20,000-9984} + 4,000,000$$

$$= e^{16} + 4,000,000$$

which is approximately 12,872,000.

Therefore, a maximum population of 12,872,000 will occur in 100 years. Assuming there has been no population control, the number of people present 4 years later will be

$$N(104) = e^{-10,816+20,800-9984} + 4,000,000$$

$$= e^1 + 4,000,000$$

which is approximately 4,000,003. We see that there will be a cataclysmic reduction in population if the maximum level is allowed to occur.

Problem. A man wishes to build a fence which will enclose a rectangular region of area 40,000 square feet. If fencing material costs $1.50 per running foot, what are the dimensions of the rectangular region of the desired area which can be enclosed at minimum cost? What is the cost of materials in this case?

Solution. Let x and y denote the lengths of two adjacent sides of the region under consideration (see Figure 7.28).

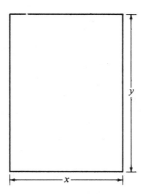

Figure 7.28

Since the area of the region is 40,000 square feet, we must have $xy = 40,000$, so that

$$y = \frac{40,000}{x}.$$

The total length of fence needed to enclose the region is given by

$$2x + 2y = 2x + \frac{80,000}{x}.$$

Since the cost per running foot of material is $1.50, we see that total cost is given by the function

$$c(x) = 1.5 \left[2x + \frac{80,000}{x} \right]$$

$$= 3 \left[x + \frac{40,000}{x} \right]$$

$$= 3[x + 40,000x^{-1}].$$

The interval over which this function is to be minimized is $(0, \infty)$. Now

$$c'(x) = 3[1 - 40,000x^{-2}],$$

so that $c'(x) = 0$ if and only if $\dfrac{40,000}{x^2} = 1$.

Thus, $c'(x) = 0$ if $x^2 = 40,000$, or $x = \pm 200$. Since $-200 \notin (0, \infty)$, the only possibility is $x = 200$. We may write

$$c'(x) = 3 \left[\frac{x^2 - 40,000}{x^2} \right] = \frac{3}{x^2} (x - 200)(x + 200).$$

From this form it is seen that $c'(x) < 0$ for $0 < x < 200$ and $c'(x) > 0$ for $x > 200$. Thus c is decreasing on $(0, 200)$ and increasing on $(200, \infty)$, so that c has a minimum at $x = 200$. For minimum cost then, we should choose $x = 200$ and $y = \dfrac{40,000}{200} = 200$. Thus the rectangular region is actually a square, 200 feet on a side, and the total cost of materials is

$$c(200) = 3 \left[200 + \frac{40,000}{200} \right]$$

$$= 3(400) = \$1200.$$

Problem. A recent study of the advertising policy of the Cole Corporation revealed that the profit P per item produced is related to the promotional expenditure s per item according to the formula

$$P = -s^3 + 7s^2 - 8s + 192,$$

where both quantities are measured in dollars. How much should the company invest in advertising during the coming year in order to maximize profit if 100,000 items are to be produced?

Solution. The mathematical problem is that of maximizing the profit function

$$P(s) = -s^3 + 7s^2 - 8s + 192$$

over the interval $[0, \infty]$. The derivative of this function is

$$P'(s) = -3s^2 + 14s - 8.$$

Using the quadratic formula, we find that $P'(s) = 0$ if and only if

$$s = \frac{-14 \pm \sqrt{196 - 96}}{2(3)} = \frac{-14 \pm 10}{-6}.$$

Hence, $s = 4$ or $s = \frac{2}{3}$. Therefore $P'(s) = -3\left(s - \frac{2}{3}\right)(s - 4)$. If $0 < s < \frac{2}{3}$, then $P'(s) < 0$, implying that P is decreasing on $\left(0, \frac{2}{3}\right)$. Similarly, P is increasing on $\left(\frac{2}{3}, 4\right)$, and decreasing on $(4, \infty)$. Thus, the maximum value of P on $[0, \infty)$ must occur either at $s = 0$ or $s = 4$. Since $P(0) = 192$ and $P(4) = 208$, P attains its maximum value at $s = 4$. Therefore, $\$4.00$ should be spent on advertising for each unit produced. An anticipated production of 100,000 units suggests that $\$400,000$ should be spent on advertising during the next year.

Problem. A psychologist is conducting the following experiment on reaction time in rats. A specimen is placed upon an electrified grid, a current is introduced into the grid, and the time required for the specimen to remove itself from the grid is measured and recorded. The psychologist is interested in the effect of altering the duration of the stimulus, while leaving its strength fixed. Specifically, he discovers that a current t seconds in duration results in a reaction time S according to the rule

$$S(t) = 2t^3 - \frac{3}{2}t^2 - \frac{9}{8}t + \frac{59}{32}.$$

What should the duration of the current be in order to minimize reaction time?

Solution. We wish to minimize the function

$$S(t) = 2t^3 - \frac{3}{2}t^2 - \frac{9}{8}t + \frac{59}{32}$$

on the interval $[0, \infty)$. The derivative of this function is

$$S'(t) = 6t^2 - 3t - \frac{9}{8}.$$

Using the quadratic formula, it is seen that the zeros of S' are $t = -\frac{1}{4}, \frac{3}{4}$. Therefore,

$$S'(t) = 6\left(t + \frac{1}{4}\right)\left(t - \frac{3}{4}\right).$$

As in the preceding problem, we see that S is decreasing on $\left(0, \frac{3}{4}\right)$ and increasing

on $\left(\frac{3}{4}, \infty\right)$. Thus S has a relative minimum at $t = \frac{3}{4}$. The reaction time corre-

sponding to a stimulus $\frac{3}{4}$ seconds in duration is

$$S\left(\frac{3}{4}\right) = 2\left(\frac{3}{4}\right)^3 - \frac{3}{2}\left(\frac{3}{4}\right)^2 - \frac{9}{8}\left(\frac{3}{4}\right) + \frac{59}{32}$$

$$= 2\left(\frac{27}{64}\right) - \frac{3}{2}\left(\frac{9}{16}\right) - \frac{27}{32} + \frac{59}{32} = 1 \text{ second.}$$

Let us now turn to problems involving instantaneous rates of change. In
Sec. 6.3 we found that if s denotes the position of an object on a straight line
and if t denotes time, then $\frac{ds}{dt}$ represents the instantaneous velocity of the object
at time t. Now velocity may be interpreted as the rate of change in position with
respect to time. In problems arising in the physical world, it is often necessary
to determine the instantaneous rate at which one quantity is changing with
respect to another quantity. Specifically, if $y = f(x)$ is a function relating some
quantity y to some other quantity x, then $\frac{dy}{dx}$ represents the *instantaneous
rate at which y is changing with respect to x.* In this regard we consider the
following problem.

Problem. A conical water tank is 20 feet high and 8 feet in diameter at the
top (see Figure 7.29).

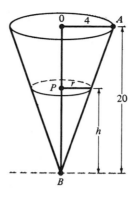

Figure 7.29

If the tank is initially full of water and if the water is allowed to flow out at the bottom at a constant rate of 2 cubic feet per minute, find the rate at which the water level is falling when the level is 10 feet from the bottom of the tank.

Solution. The mathematical problem is that of determining the instantaneous rate at which height h is changing with respect to time t, at that instant when $h = 10$. Thus, we must find $\dfrac{dh}{dt}$ when $h = 10$. It is known that the volume of a right circular cone of height h and base radius r is given by

$$V = \frac{1}{3}\pi r^2 h.$$

From similarity of triangles $\triangle AOB$ and $\triangle CPB$ (in Figure 7.29), we have the proportion

$$\frac{4}{20} = \frac{r}{h},$$

so that $r = \dfrac{1}{5}h.$

Substitution of $\dfrac{1}{5}h$ for r in the volume formula yields

$$V = \frac{1}{3}\pi\left(\frac{1}{5}h\right)^2 h$$

$$= \frac{\pi}{75}h^3.$$

Thus, $h = \left(\dfrac{75}{\pi}V\right)^{\frac{1}{3}}.$

By the chain rule,

$$\frac{dh}{dt} = \frac{dh}{dV}\frac{dV}{dt}$$

$$= \frac{1}{3}\left(\frac{75}{\pi}\right)\left(\frac{75}{\pi}V\right)^{-\frac{2}{3}}\frac{dV}{dt}.$$

We are given that $\dfrac{dV}{dt} = 2$, so that

$$\frac{dh}{dt} = \frac{1}{3}\left(\frac{75}{\pi}\right)\left(\frac{75}{\pi}V\right)^{-\frac{2}{3}} \quad (2)$$

$$= \frac{50}{\pi}\left(\frac{75}{\pi}V\right)^{-\frac{2}{3}}.$$

When $h = 10$, $r = \dfrac{1}{5}(10) = 2$, so that

$$V = \frac{1}{3}\pi(2)^2(10) = \frac{40\pi}{3}.$$

Thus, for $h = 10$, we have

$$\frac{dh}{dt} = \frac{50}{\pi}\left[\left(\frac{75}{\pi}\right)\left(\frac{40\pi}{3}\right)\right]^{-\frac{2}{3}}$$

$$= \frac{50}{\pi}(1000)^{-\frac{2}{3}}$$

$$= \frac{50}{\pi}\left(\frac{1}{100}\right) = \frac{1}{2\pi}.$$

Hence the rate at which the water level is falling at the instant when the water level is 10 feet from the bottom of the tank is $\dfrac{1}{2\pi}$ feet per second.

Exercise Set 7.9

1. The Big Corporation is producing baking pans from 8×13 rectangular sheets of aluminum by cutting equal squares from the four corners and turning up the sides to form the pan. What size squares should be removed to produce pans of maximum volume? What is the maximum volume?

2. A farmer wishes to construct a rectangular grazing area adjacent to a river, the total area of which must be 60,000 square feet. In purchasing fencing material for this project, he decides that the river will impose a natural barrier to his livestock, and that no fencing need be erected along its bank. How should he choose the dimensions of the pasture in order to minimize the amount of material needed (assuming that the bank of the river is straight)?

3. The rabbit population in a certain environmental setting at time t is given by the function $u(t) = 5000 + e^{-2t^2 + 48t - 296}$, where $t = 0$ corresponds to the present time. If t is measured in years, how many years from now will the rabbit population reach a maximum level? What will be the maximum population?

4. A telephone company has 10,000 phones in a certain city. The basic rate charge is $8.00 per month. If the rate is raised, the number of phones in use will be reduced by 300 for each $.35/month increase. What rate will produce the maximum gross income?

* * *

5. A man wishes to construct a rectangular box out of a sheet of metal 38 inches wide and 84 inches long by cutting a square piece out of each corner and bending up the sides. How should he choose the dimensions of the removed squares in order that the volume of the resulting box be as large as possible? What is the volume of the box so constructed?

6. A certain item sells for 50¢. A market analysis on this item reveals that for each increase of 2¢ in the price, the number of items sold decreases by 80 per day. If daily sales at 50¢ per item number 5460, what should the price of the item be in order to maximize profit?

7. Each page of a particular publication is to contain 24 square inches of printed material. The page must have margins of $1\frac{1}{2}$ inches at the top and bottom and margins of 1 inch on each side. What should be the dimensions of the page in order to make the total page area as small as possible?

8. A group of students plans to charter a bus to travel to a football game. If 40 persons buy tickets, the price per person is $14.00. For each additional passenger, the fare decreases by 8¢ for all. What number of passengers provides the maximum revenue for the bus line?

9. The sum of two positive numbers is 44. Find the numbers, if their product is to be as large as possible.

10. A man is enjoying a Sunday afternoon in his yacht, 4 miles off the Florida coast, when he receives word that his daughter has just eloped with a mathematician. The man's home is approximately 10 miles down

the straight shoreline from the nearest point on the shore. His yacht can travel a maximum speed of 18 mi/hr, and a taxi, which he can catch anywhere on the shoreline, can travel 30 mi/hr. At what point should the man land in order to reach his home in the shortest length of time?

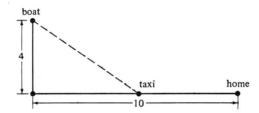

11. Referring to the elopement of the preceding problem, in anticipation of his daughter's indiscretion, the father had constructed a wall 8 feet in height around his house, leaving a distance of 1 foot between the wall and the house. What is the length of the shortest ladder that can reach over the wall to the house?

12. A man wishes to build a restaurant with a seating capacity of no fewer than 90 chairs. He knows that weekly profit per chair is $70 for the first 90 chairs and that each additional chair results in a decrease in profit of 30¢ for every chair in the restaurant. How many chairs should he install to maximize profit?

13. A physician is building a new office with the following floor plan:

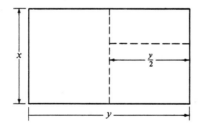

The total floor area is to be 820 square feet. Exterior walls cost 1.8 times as much as interior walls. What should the dimensions of the office be in order to minimize cost of materials?

14. A new discount department store is to have a floor area of 24,000 square feet. The store is to be rectangular in shape, having three cement block walls and a glass front. If glass costs 1.64 times as much as cement block, determine the dimensions of the store which will minimize the cost of materials?

15. An efficiency expert employed by the Tasty Food Company finds that far too much steel is being used in the manufacture of the cans in which the company packs its product. If the can used by Tasty has a volume of 96 cubic inches, and has the shape of a right circular cylinder, what dimensions (height and base radius) will result in the use of the least amount of steel? (*Hint*: Do not forget to account for the top and the bottom of the can as well as the lateral surface.)

*16. It is known that a nation is most subject to revolution when the discrepancy between public expectation of achievement by the current political regime and its actual achievement is as great as possible. There are various ways of measuring both achievement and expectation of achievement (the same unit of measure applies to both quantities). Let us call this unit the standard unit. Suppose that over the last 48 months the number of standard units of expected achievement was given by

$$E(t) = 34 - 4.7 \ln \left(\frac{1}{2} t^2 - 10t + 80 \right),$$

where t denotes time measured in months and $t = 48$ corresponds to the present time. Suppose further that the number of standard units of actual achievement over the same interval was given by

$$A(t) = -\frac{1}{6} t + 10.$$

During which of the preceding 48 months was revolution most likely?

17. Gas is being pumped into a balloon in such a way that the volume is increasing at a constant rate of 4 cubic feet per second. How rapidly is the radius of the balloon increasing at the time when the radius is 6?

18. Over the last 18 market days the price of one share of a certain stock was given by

$$c(t) = 37 + 8 \left[\frac{e^t - e^{-t}}{e^t + e^{-t}} \right],$$

where t denotes time in days with $t = 0$ corresponding to the first of the 18 days. At what rate was the price changing on the third of these 18 days? Did the price decrease at any time during this 18 day period? What was the average daily rate of change in price over the time interval $[0, 18]$?

19. An office manager observes that if there is just one typist in the typing pool, she will work approximately 38 hours per week. Each additional typist present, however, reduces the effectiveness of all. If there are t typists, then each one will work $38 - \frac{t^2}{4}$ hours per week. Find the number of typists that will produce the greatest amount of work in a week's time.

20. A conveyer belt is spewing gravel onto the top of a conical pile in such a fashion that the pile always has the form of a right circular cone whose height is equal to the radius of its base. If the gravel is being unloaded at the rate of 54 cubic feet per minute, at what rate is the height increasing when the height is 16 feet?

8 / *Integral Calculus*

The study of differential calculus, as treated in the two preceding chapters, arose from the need to determine the instantaneous rate of change of some quantity. By using the concept of limit, we were able to define the derivative of a function, and subsequently to solve not only problems involving instantaneous rates of change, but other problems as well.

The idea of instantaneous rate of change is an intuitive one. The concept of area is also an intuitive idea and even more fundamental. In the following pages we shall see that integral calculus arises from the need to determine areas of certain regions in the plane. Again, the concept of limit will be employed, though in a way which is different from its use in the differential calculus.

At first it will seem that the only connection between integral and differential calculus is the fact that both involve limits. For this reason, it should certainly come as no surprise that differential and integral calculus evolved separately over a period of several thousand years as independent branches of mathematics. Only within the last 400 years was the important and amazingly simple relationship between these topics discovered. In Sec. 8.5 we will see why we use the single word "calculus" to describe these two seemingly unrelated ideas.

8.1 THE Σ NOTATION

In the development of integral calculus it is necessary to deal with sums of numbers. Frequently the number of terms to be summed in a given discussion is far too great for an actual listing of the type

$$a_1 + a_2 + a_3 + \cdots + a_n$$

to be practical. For this reason, we adopt a concise method for expressing such sums which does not involve the listing of terms. Let $a_1, a_2, \ldots, a_n \in \mathcal{R}$. We write the symbol

$$\sum_{k=1}^{n} a_k,$$

read "the summation of a sub k from k equals one to k equals n," to mean

$$a_1 + a_2 + \cdots + a_n.$$

The symbol "$\sum$" is the capital letter "sigma" of the Greek alphabet, which corresponds to the Latin letter "S," the first letter of the word "sum." For example,

$$\sum_{k=1}^{6} k = 1 + 2 + 3 + 4 + 5 + 6,$$

$$\sum_{k=1}^{4} k^2 = 1^2 + 2^2 + 3^2 + 4^2 = 1 + 4 + 9 + 16,$$

and

$$\sum_{k=1}^{5} \frac{3}{k+1} = \frac{3}{1+1} + \frac{3}{2+1} + \frac{3}{3+1} + \frac{3}{4+1} + \frac{3}{5+1}.$$

The letter k appearing in these sums is called the *index of summation*. Since k is nothing more than a place-holder, we could just as well use any letter of the alphabet as the index of summation. Usually, however, we reserve the letters i, j, and k for this purpose. Thus we see that each of the following represents the same sum:

$$\sum_{i=1}^{n} a_i = \sum_{j=1}^{n} a_j = \sum_{k=1}^{n} a_k = a_1 + a_2 + \cdots + a_n.$$

With respect to the general summation,

$$\sum_{k=m}^{n} a_k,$$

the numbers m and n are called respectively the lower and upper limits of summation, and a_k is called the general term of summation. For example, the general term of the sum

$$\sum_{k=3}^{7} \frac{1}{k}$$

is $\frac{1}{k}$, the lower limit of summation is 3, and the upper limit of summation is 7. Here we have

$$\sum_{k=3}^{7} \frac{1}{k} = \frac{1}{3} + \frac{1}{4} + \frac{1}{5} + \frac{1}{6} + \frac{1}{7} = \frac{1377}{1260}.$$

Usually the index of summation appears explicitly in the general term of a given sum. It may happen, however, that the general term is independent of the index. That is, it may happen that $a_1 = a_2 = a_3 = \cdots = a_n = a$. In this case we agree that $\sum_{k=1}^{n} a$ shall denote

$$a + a + a + \cdots + a \ (n \text{ terms}).$$

For example,

$$\sum_{k=1}^{5} 2 = 2 + 2 + 2 + 2 + 2 = 5(2) = 10,$$

$$\sum_{i=1}^{7} 1 = 1 + 1 + 1 + 1 + 1 + 1 + 1 = 7(1) = 7,$$

and

$$\sum_{j=1}^{200} \frac{1}{2} = \underbrace{\frac{1}{2} + \frac{1}{2} + \cdots + \frac{1}{2}}_{200 \text{ terms}} = 200\left(\frac{1}{2}\right) = 100.$$

Suppose now that $a_1, a_2, \ldots, a_n, b_1, b_2, \ldots, b_n \in \mathcal{R}$. Then

$$\sum_{k=1}^{n} a_k + \sum_{k=1}^{n} b_k = (a_1 + a_2 + \cdots + a_n) + (b_1 + b_2 + \cdots + b_n)$$

$$= (a_1 + b_1) + (a_2 + b_2) + \cdots + (a_n + b_n)$$

$$= \sum_{k=1}^{n} (a_k + b_k).$$

Similarly, it can be shown that

$$\sum_{k=1}^{n} a_k - \sum_{k=1}^{n} b_k = \sum_{k=1}^{n} (a_k - b_k).$$

If $c \in \mathcal{R}$, then

$$c \sum_{k=1}^{n} a_k = c(a_1 + a_2 + \cdots + a_n)$$

$$= ca_1 + ca_2 + \cdots + ca_n$$

$$= \sum_{k=1}^{n} ca_k.$$

Summarizing these observations, we have the following result.

8.1 Theorem. *IF $c, a_k, b_k \in \Re$ for $k = 1, 2, \ldots, n$, THEN*

(i) $\displaystyle\sum_{k=1}^{n} a_k + \sum_{k=1}^{n} b_k = \sum_{k=1}^{n} (a_k + b_k),$

(ii) $\displaystyle\sum_{k=1}^{n} a_k - \sum_{k=1}^{n} b_k = \sum_{k=1}^{n} (a_k - b_k),$

and

(iii) $\displaystyle c \sum_{k=1}^{n} a_k = \sum_{k=1}^{n} ca_k.$

Example. Consider the sums $\displaystyle\sum_{k=1}^{4} k$ and $\displaystyle\sum_{k=1}^{4} \frac{1}{k}$. We have

$$\sum_{k=1}^{4} k = 1 + 2 + 3 + 4 = 10$$

and

$$\sum_{k=1}^{4} \frac{1}{k} = \frac{1}{1} + \frac{1}{2} + \frac{1}{3} + \frac{1}{4} = \frac{25}{12}.$$

Hence,

$$\sum_{k=1}^{4} k + \sum_{k=1}^{4} \frac{1}{k} = 10 + \frac{25}{12} = \frac{145}{12}.$$

Alternatively,

$$\sum_{k=1}^{4} k + \sum_{k=1}^{4} \frac{1}{k} = \sum_{k=1}^{4} \left(k + \frac{1}{k}\right)$$

$$= \sum_{k=1}^{4} \left(\frac{k^2 + 1}{k}\right)$$

$$= \frac{1^2 + 1}{1} + \frac{2^2 + 1}{2} + \frac{3^2 + 1}{3} + \frac{4^2 + 1}{4}$$

$$= 2 + \frac{5}{2} + \frac{10}{3} + \frac{17}{4} = \frac{145}{12}.$$

Likewise,

$$\sum_{k=1}^{4} 5k = 5 \sum_{k=1}^{4} k = 5(10) = 50.$$

Let us now consider some special sums. For example, suppose we want to find the sum of the first 100 positive integers. This is certainly no difficult task, although the process of actually adding up all these numbers is very time-consuming. Therefore, we consider the following idea. Let T denote this sum. Then

$$T = 1 + 2 + 3 + \cdots + 99 + 100.$$

But also,

$$T = 100 + 99 + 98 + \cdots + 2 + 1.$$

Hence,

$$2T = (1 + 100) + (2 + 99) + (3 + 98) + \cdots + (99 + 2) + (100 + 1)$$
$$= 101 + 101 + 101 + \cdots + 101 + 101 \ (100 \text{ terms})$$
$$= 100(101).$$

Thus, $T = \dfrac{100(101)}{2} = 5050.$

More generally, we have

$$\sum_{k=1}^{n} k = 1 + 2 + \cdots + n = n + (n - 1) + \cdots + 1.$$

Letting S denote this sum, we have

$$S = 1 + 2 + \cdots + n$$

and

$$S = n + (n - 1) + \cdots + 1$$

so that

$$2S = (n + 1) + [(n - 1) + 2] + \cdots + (n + 1)$$
$$= (n + 1) + (n + 1) + \cdots + (n + 1) \ (n \text{ terms})$$
$$= n(n + 1).$$

Hence,

$$S = \sum_{k=1}^{n} k = \frac{n(n + 1)}{2}.$$

Problem. Find the sum of the first 9 positive integers, first by using the formula
$\sum_{k=1}^{n} k = \frac{n(n+1)}{2}$, and then by actually summing these numbers.

Solution. Employing the indicated formula, we see that

$$\sum_{k=1}^{9} k = \frac{9(9+1)}{2} = \frac{9(10)}{2} = 45.$$

Actually adding, we have

$$1 + 2 + 3 + 4 + 5 + 6 + 7 + 8 + 9 = 45.$$

We shall now derive a formula, similar to the one just developed, for finding the sum of the squares of the first n positive integers,

$$1^2 + 2^2 + 3^2 + \cdots + n^2.$$

According to the binomial theorem, we have

$$(k+1)^3 - k^3 = (k^3 + 3k^2 + 3k + 1) - k^3 = 3k^2 + 3k + 1.$$

Hence,

$$\sum_{k=1}^{n} [(k+1)^3 - k^3] = \sum_{k=1}^{n} (3k^2 + 3k + 1)$$

$$= 3\sum_{k=1}^{n} k^2 + 3\sum_{k=1}^{n} k + \sum_{k=1}^{n} 1$$

$$= 3\sum_{k=1}^{n} k^2 + 3\left[\frac{n(n+1)}{2}\right] + n.$$

On the other hand, we may also write

$$\sum_{k=1}^{n} [(k+1)^3 - k^3] = [2^3 - 1^3] + [3^3 - 2^3] + \cdots + [(n+1)^3 - n^3]$$

$$= -1^3 + [2^3 - 2^3] + [3^3 - 3^3] + \cdots + [n^3 - n^3] + (n+1)^3$$

$$= (n+1)^3 - 1$$

$$= n^3 + 3n^2 + 3n.$$

Therefore, we must have

$$3\sum_{k=1}^{n} k^2 + \frac{3n(n+1)}{2} + n = n^3 + 3n^2 + 3n$$

so that

$$3 \sum_{k=1}^{n} k^2 = n^3 + 3n^2 + 2n - \frac{3n^2 + 3n}{2}$$

$$= \frac{1}{2}(2n^3 + 3n^2 + n)$$

$$= \frac{n}{2}(n + 1)(2n + 1).$$

Thus, $\sum_{k=1}^{n} k^2 = \frac{n}{6}(n + 1)(2n + 1).$

Problem. Find the sum of the squares of the first six positive integers.

Solution. We have $\sum_{k=1}^{6} k^2 = \frac{6}{6}[6 + 1][(2(6) + 1]$

$$= 7(13) = 91.$$

Notice that direct addition yields

$$1^2 + 2^2 + 3^2 + 4^2 + 5^2 + 6^2 = 1 + 4 + 9 + 16 + 25 + 36 = 91,$$

and the same result is obtained.

The two formulas developed in this section will be quite useful in our later work. For easy reference, we formalize these results in the following theorem.

8.2 Theorem. *IF n is a positive integer, THEN*

(i) $\sum_{k=1}^{n} k = \frac{n(n + 1)}{2}$,

and

(ii) $\sum_{k=1}^{n} k^2 = \frac{n(n + 1)(2n + 1)}{6}$.

In closing, we point out that formulas of the type presented here may be derived for $\sum_{k=1}^{n} k^3$, $\sum_{k=1}^{n} k^4, \ldots$, and in general for $\sum_{k=1}^{n} k^m$, where m is any positive integer. We shall have no need, however, for these results for $m > 2$.

Exercise Set 8.1

1. Expand each of the following:

    a. $\sum_{k=1}^{5} a_k$ b. $\sum_{i=2}^{5} a_i$ c. $\sum_{j=1}^{4} c_j$ d. $\sum_{i=1}^{6} (a_i + 2b_i)$

2. Expand each of the following and find the indicated sum:

   a. $\displaystyle\sum_{j=1}^{7} j$ b. $\displaystyle\sum_{j=1}^{4} (2j + 3)$ c. $\displaystyle\sum_{j=1}^{6} (-1)^j$ d. $\displaystyle\sum_{j=1}^{3} \frac{1}{2j}$

3. Expand each of the following and find the sum:

   a. $\displaystyle\sum_{k=1}^{4} (k - 1)$ b. $\displaystyle\sum_{k=1}^{4} \frac{1}{k + 1}$ c. $\displaystyle\sum_{k=1}^{4} (k - 1) + \sum_{k=1}^{4} \frac{1}{k + 1}$

   d. $\displaystyle\sum_{k=1}^{4} \left[(k - 1) + \frac{1}{k + 1} \right] = \sum_{k=1}^{4} \frac{k^2}{k + 1}$

4. a. Use Theorem 8.2 (i) to find $\displaystyle\sum_{k=1}^{25} k$.

   b. Use Theorem 8.2 (ii) to find $\displaystyle\sum_{i=1}^{30} k^2$.

5. Find the following sums:

   a. $\displaystyle\sum_{k=1}^{8} k^2$ b. $2\displaystyle\sum_{k=1}^{8} k^2$ c. $\displaystyle\sum_{k=1}^{8} 2k^2$

* * *

In problems 6–19, find the number represented by each summation.

6. $\displaystyle\sum_{k=0}^{5} (k - 3)$

7. $\displaystyle\sum_{k=2}^{6} \frac{k}{k - 1}$

8. $\displaystyle\sum_{i=4}^{7} \frac{3}{(i - 2)}$

9. $\displaystyle\sum_{i=4}^{7} (3i - 2)$

10. $\displaystyle\sum_{j=1}^{5} (-1)^j j^2$

11. $\displaystyle\sum_{j=1}^{5} (-1)^{j+1} j^2$

12. $\displaystyle\sum_{k=1}^{6} (-2)$

13. $\displaystyle\sum_{k=1}^{6} 38$

14. $\displaystyle\sum_{i=0}^{5} 2^i$

15. $\displaystyle\sum_{i=0}^{5} 2^{-i}$

16. $\displaystyle\sum_{j=1}^{62} j^2$

17. $\displaystyle\sum_{k=1}^{119} k$

18. $\displaystyle\sum_{j=1}^{12} 5j^2$

19. $\displaystyle\sum_{k=1}^{15} -3k$

In problems 20–27, write the first three and the last two terms of the indicated sum.

20. $\displaystyle\sum_{k=1}^{n} (x_k - x_{k-1})$

21. $\displaystyle\sum_{k=0}^{n-1} (x_{k+1} - x_k)$

22. $\displaystyle\sum_{k=1}^{n} m_k \, \Delta x_k$

23. $\displaystyle\sum_{k=1}^{n} M_k \, \Delta x_k$

24. $\displaystyle\sum_{i=1}^{n} (x_i - x_{i-1}) f(z_i)$

25. $\displaystyle\sum_{i=0}^{n-1} (x_{i+1} - x_i) f(z_{i+1})$

26. $\displaystyle\sum_{k=1}^{n} \frac{k}{n}$

27. $\displaystyle\sum_{k=1}^{n} \frac{1}{n}$

28. Show that $\displaystyle\sum_{i=1}^{n} (a_i - a_{i-1}) = a_n - a_0$.

29. Use the result of problem 28 to find each of the following sums:

a. $\displaystyle\sum_{i=1}^{8} [2^i - 2^{i-1}]$

b. $\displaystyle\sum_{i=1}^{4} [i^5 - (i-1)^5]$

c. $\displaystyle\sum_{k=1}^{12} \left[\frac{1}{k+1} - \frac{1}{k} \right]$

d. $\displaystyle\sum_{j=1}^{l} (c_j - c_{j-1})$

30. Show that $\displaystyle\sum_{i=1}^{n} a_i = \sum_{i=0}^{n-1} a_{i+1}$.

*31. Show that $\displaystyle\sum_{k=1}^{n} k^3 = \frac{n^2(n+1)^2}{4}$. (*Hint:* Begin by expanding $(k+1)^4 - k^4$ and apply the results of Theorem 8.2 to both sides of the resulting equation.)

8.2 SEQUENCES

As mentioned in the introduction to Chap. 8, the notion of a limit is basic to the development of integral calculus. The type of limit which we wish to consider is that of a sequence.

8.3 Definition. *A sequence is an infinite listing* $\{a_1, a_2, \ldots, a_n, \ldots\}$ *of objects. Elements of this set are called terms of the sequence. In particular,* a_1 *is the first term,* a_2 *the second term, and, in general,* a_n *is the n-th term.*

To be concise, we shall denote the sequence $\{a_1, a_2, \ldots, a_n, \ldots\}$ by $\{a_n\}$. A sequence may be specified either by writing the first few terms or by presenting an explicit formula for finding the n-th term of the sequence.

Example. Consider the sequence $\{a_n\}$ whose first few terms are $1, \dfrac{1}{2}, \dfrac{1}{3},$ $\dfrac{1}{4}, \ldots$. It is apparent that the terms of this sequence are obtained by taking the reciprocals of the positive integers in their usual ordering. An explicit formula for the n-th term of this sequence is

$$a_n = \frac{1}{n}.$$

Thus, $a_1 = \dfrac{1}{1}$, $a_2 = \dfrac{1}{2}$, $a_3 = \dfrac{1}{3}$, and, in general, $a_n = \dfrac{1}{n}$.

Example. Consider the sequence $\{b_n\}$ whose first few terms are $0, 2, 0, 18, 0,$ $162, 0, \ldots$. In this case it is not clear as to just what the next few terms of the sequence might be. The formula

$$b_n = \begin{cases} 2(3)^{n-2}, & \text{if } n = 2, 4, 6, \ldots \\ 0, & \text{if } n = 1, 3, 5, \ldots \end{cases}$$

gives the n-th term of a sequence whose first few terms are those listed. This example illustrates that the use of an explicit formula for the n-th term of a sequence is often preferable to a description of the sequence which simply lists the first few terms.

Most sequences considered in this chapter will be sequences of values of a function f. We shall be interested in knowing the behavior of a sequence $\{a_n\}$, where $a_n = f(n)$, as n increases without bound. In this regard we use the following definition.

8.4 Definition. *Let f be a function defined for all positive real numbers, and let $\{a_n\}$ be the sequence given by the rule*

$$a_n = f(n)$$

for each positive integer n. IF

$$L = \lim_{x \to \infty} f(x),$$

THEN L is called the limit of the sequence $\{a_n\}$, written

$$L = \lim_{n \to \infty} a_n.$$

Thus, we see that the problem of finding the limit of a given sequence is precisely that of finding the limit of a certain function associated with the sequence. Such limits may be found by using the methods of Sec. 7.6.

Example. Define sequences $\{a_n\}$, $\{b_n\}$, and $\{c_n\}$ by

$$a_n = \frac{3}{n},$$

$$b_n = -\frac{3}{2},$$

and $\qquad c_n = n + 1$

for each positive integer n. To find the limits of each of these sequences, we consider functions f, g, and h, defined by

$$f(x) = \frac{3}{x},$$

$$g(x) = -\frac{3}{2},$$

and $\qquad h(x) = x + 1.$

Clearly, if n is a positive integer, then

$$a_n = f(n),$$
$$b_n = g(n),$$

and $\qquad c_n = h(n).$

Now

$$\lim_{x \to \infty} f(x) = \lim_{x \to \infty} \frac{3}{x}$$

$$= 0,$$

$$\lim_{x \to \infty} g(x) = \lim_{x \to \infty} -\frac{3}{2}$$

$$= -\frac{3}{2},$$

and

$$\lim_{x \to \infty} h(x) = \lim_{x \to \infty} (x + 1)$$

$$= \infty.$$

Thus, by Definition 8.4,

$$\lim_{n \to \infty} a_n = 0,$$

$$\lim_{n\to\infty} b_n = -\frac{3}{2},$$

and

$$\lim_{n\to\infty} c_n = \infty.$$

Problem. Find $\lim\limits_{n\to\infty} \dfrac{5n-6}{10n^2+2n+1}$.

Solution. Let $a_n = \dfrac{5n-6}{10n^2+2n+1}$, and let f be given by

$$f(x) = \frac{5x-6}{10x^2+2x+1}.$$

Then $f(n) = a_n$ for each positive integer n, so that

$$\lim_{x\to\infty} f(x) = \lim_{x\to\infty} \frac{5x-6}{10x^2+2x+1}$$

$$= \lim_{x\to\infty} \frac{\dfrac{5}{x}-\dfrac{6}{x^2}}{10+\dfrac{2}{x}+\dfrac{1}{x^2}}$$

$$= \frac{0}{10} = 0.$$

Therefore, $\lim\limits_{n\to\infty} a_n = 0$.

Exercise Set 8.2

1. Write the first five terms of the sequence $\{a_n\}$ if

   a. $a_n = \dfrac{1}{n^2}$ b. $a_n = \dfrac{n-1}{n+1}$ c. $a_n = e^n$

2. Let $\{b_n\}$ be the sequence given by the rule $b_n = \dfrac{10}{n^3}$.

   a. Find b_1, b_{10}, b_{100}, and b_{1000}.
   b. Find a function $f(x)$, such that $f(n) = b_n$ for all positive integers n.
   c. Find $\lim\limits_{x\to\infty} f(x)$.
   d. Find $\lim\limits_{n\to\infty} b_n$.

3. Find the limit of the sequence $\{a_n\}$ if

   a. $a_n = \dfrac{1}{n^2}$ b. $a_n = \dfrac{n-1}{n+1}$ c. $a_n = e^n$

* * *

In problems 4–15, find the indicated limit.

4. $\lim\limits_{n\to\infty} \left(1 - \dfrac{1}{n}\right)$

5. $\lim\limits_{n\to\infty} (1 - n)$

6. $\lim\limits_{n\to\infty} \dfrac{5n - 6}{10n^2 + 2n + 1}$

7. $\lim\limits_{n\to\infty} \dfrac{6 - 5n^2}{10n^2 + 2n + 1}$

8. $\lim\limits_{n\to\infty} \dfrac{3 - 8n^2}{4 + 4n^2}$

9. $\lim\limits_{n\to\infty} \dfrac{(n + 1)(2n - 3)}{(3n + 4)(n - 3)}$

10. $\lim\limits_{n\to\infty} \dfrac{e^n}{e^{2n}}$

11. $\lim\limits_{n\to\infty} \dfrac{e^n}{2e^n}$

12. $\lim\limits_{n\to\infty} \left(\dfrac{2}{3} - \dfrac{5}{6n}\right)$

13. $\lim\limits_{n\to\infty} \left(\dfrac{2}{3} + \dfrac{5}{6n}\right)$

14. $\lim\limits_{n\to\infty} \left(6 - \dfrac{2}{n} + \dfrac{1}{n^2}\right)$

15. $\lim\limits_{n\to\infty} \left(-2 + \dfrac{1}{n} - \dfrac{1}{n^2}\right)$

In problems 16–25, for the given sequence $\{b_n\}$, find b_1, b_2, b_5, b_{100}, and b_{5000}. Find $\lim\limits_{n\to\infty} b_n$.

16. $b_n = 2^n$

17. $b_n = 2^{-n}$

18. $b_n = (-1)^n \left(\dfrac{1}{n}\right)$

19. $b_n = (-1)^{n+1} \left(\dfrac{1}{n^2}\right)$

20. $b_n = \dfrac{2{,}000{,}000}{n}$

21. $b_n = 5 - \dfrac{2{,}000{,}000}{n}$

22. $b_n = \dfrac{n^3 - 3n^2 + 5n - 18}{2n^2 + 3n - 15}$

23. $b_n = \dfrac{2n^2 + 3n - 15}{n^3 - 3n^2 + 5n - 18}$

24. $b_n = \dfrac{3}{4} + \dfrac{5}{n} + \dfrac{17}{n^2} + \dfrac{45}{n^3}$

25. $b_n = \dfrac{3}{4}n + \dfrac{5}{n} - \dfrac{17}{n^2} + \dfrac{45}{n^3}$

8.3 THE AREA OF A PLANE REGION

Everyone has, at one time or another, considered the concept of area. Indeed, elementary school children are confronted with the task of finding the area of a rectangle of base 4 and height 6, or the area of a circle of radius 5. It would seem that the idea of area is among the simplest in all of mathematics. Consider, however, the problem of finding the area of the region R appearing in Figure 8.1. This irregularly shaped region is certainly not one of the standard figures encountered in Euclidean geometry. There is no concise formula available for the computation of its area. How, then, do we approach the problem of finding the area of such a region? The goal of the present section is to suggest one way of attacking this problem.

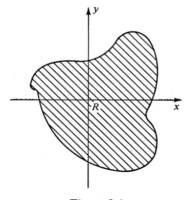

Figure 8.1

We shall restrict our attention initially to the computation of the area of a region R of the xy-plane which is bounded above by the graph of some positive continuous function $y = f(x)$, below by the x-axis, on the left by the vertical line $x = a$ and on the right by a second vertical line $x = b$ (see Figure 8.2). Our

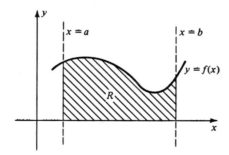

Figure 8.2

approach to determining the area of such a region will be to approximate the area by means of the areas of rectangles.

In Figure 8.3a, we have inscribed, within the region, several rectangles, each of which has a subinterval of $[a, b]$ as its base. In Figure 8.3b, we have super-scribed, about this region, several rectangles, each of which also has a subinterval of $[a, b]$ as its base. If we let $\underline{A}$ denote the sum of the areas of the inscribed rectangles, $\overline{A}$ the sum of the areas of the superscribed rectangles, and A the area of the region R, then both $\underline{A}$ and $\overline{A}$ approximate the actual area A. Clearly

$$\underline{A} \leqslant A \leqslant \overline{A}.$$

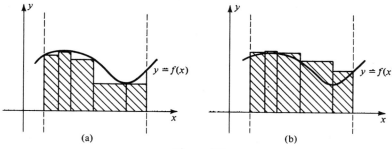

Figure 8.3

Moreover, we may improve the accuracy of each of these approximations by increasing the number of rectangles involved, doing so in such a way that the base of each component rectangle is decreased. Roughly speaking, *the more rectangles we use, and the smaller they are, the better they will "fit" our region.* It is precisely at this point that the concept of limit enters into the picture. For suppose that we have approximated the area of R by means of n inscribed rectangles, or by means of n superscribed rectangles, each of width $\dfrac{b - a}{n}$. If we denote by A_n the sum of the areas of the inscribed rectangles and by $\overline{A_n}$ the sum of the areas of the superscribed rectangles, then $A_n \leqslant A \leqslant \overline{A_n}$ for every positive integer n. If it happens that

$$\lim_{n \to \infty} A_n = \lim_{n \to \infty} \overline{A_n},$$

then the common value of these two limits must be the actual area of R.

We now illustrate this approach by finding the areas of a few particular regions.

Example. Let R be the region of the coordinate plane bounded above by the line $f(x) = \dfrac{1}{2}x + 1$, below by the x-axis, on the left by the y-axis, and on the right by the line $x = 1$ (see Figure 8.4). To obtain lower and upper rectangular approximations of the area of R, we divide the interval $[0, 1]$ into n subintervals of equal length:

$$I_1 = \left[0, \ \frac{1}{n} \right] \qquad I_k = \left[\frac{k - 1}{n}, \ \frac{k}{n} \right] \quad (k\text{th subinterval})$$

$$I_2 = \left[\frac{1}{n}, \ \frac{2}{n} \right]$$

$$\vdots \qquad \qquad \vdots$$

$$I_n = \left[\frac{n - 1}{n}, \ 1 \right],$$

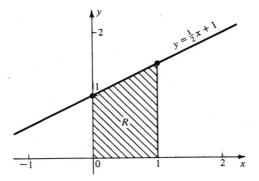

Figure 8.4

and construct rectangles r_k and $\overline{r}_k$, each based on the subinterval I_k, having respective heights $f\left(\dfrac{k-1}{n}\right)$ (the minimum value of $f(x)$ for $x \in I_k$) and $f\left(\dfrac{k}{n}\right)$ (the maximum value of $f(x)$ for $x \in I_k$). The rectangles $\underline{r}_k$ and $\overline{r}_k$ for $n = 4$ are illustrated in Figures 8.5a and 8.5b.

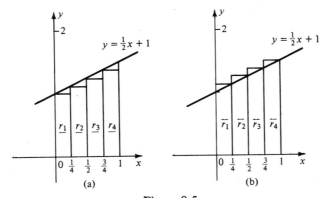

Figure 8.5

The area of the rectangle r_k is given by

$$\text{Area } (\underline{r}_k) = \frac{1}{n} f\left(\frac{k-1}{n}\right) = \frac{1}{n}\left[\frac{1}{2}\left(\frac{k-1}{n}\right) + 1\right]$$

$$= \frac{k}{2n^2} - \frac{1}{2n^2} + \frac{1}{n}$$

$$= \frac{k}{2n^2} + \frac{2n-1}{2n^2},$$

and the area of the rectangle $\overline{r_k}$ is given by

$$\text{Area } (\overline{r_k}) = \frac{1}{n} f\left(\frac{k}{n}\right) = \frac{1}{n}\left[\frac{1}{2}\left(\frac{k}{n}\right) + 1\right]$$

$$= \frac{k}{2n^2} + \frac{1}{n}.$$

Denoting the sum of the areas of the rectangles r_k by $\underline{A_n}$, we have

$$\underline{A_n} = \sum_{k=1}^{n}\left[\frac{k}{2n^2} + \frac{2n-1}{2n^2}\right] = \sum_{k=1}^{n}\frac{k}{2n^2} + \sum_{k=1}^{n}\frac{2n-1}{2n^2}$$

$$= \frac{1}{2n^2}\sum_{k=1}^{n} k + \frac{2n-1}{2n^2}\sum_{k=1}^{n} 1.$$

By Theorem 8.2, we know that $\sum_{k=1}^{n} k = \frac{n(n+1)}{2}$. Consequently,

$$\underline{A_n} = \frac{1}{2n^2}\left[\frac{n(n+1)}{2}\right] + \frac{2n-1}{2n^2}(n)$$

$$= \frac{1}{2n^2}\left(\frac{n^2}{2} + \frac{n}{2}\right) + \frac{2n-1}{2n}$$

$$= \frac{1}{4} + \frac{1}{4n} + 1 - \frac{1}{n}$$

$$= \frac{5}{4} - \frac{3}{4n}.$$

Denoting the sum of the areas of the rectangles $\overline{r_k}$ by $\overline{A_n}$, we have

$$\overline{A_n} = \sum_{k=1}^{n}\left[\frac{k}{2n^2} + \frac{1}{n}\right] = \sum_{k=1}^{n}\frac{k}{2n^2} + \sum_{k=1}^{n}\frac{1}{n} = \frac{1}{2n^2}\sum_{k=1}^{n} k + \frac{1}{n}\sum_{k=1}^{n} 1$$

$$= \frac{1}{2n^2}\left[\frac{n(n+1)}{2}\right] + \frac{1}{n}(n) = \frac{1}{2n^2}\left(\frac{n^2}{2} + \frac{n}{2}\right) + 1 = \frac{1}{4} + \frac{1}{4n} + 1 = \frac{5}{4} + \frac{1}{4n}.$$

Now $\underline{A_n} \leqslant \text{Area } (R) \leqslant \overline{A_n}$ for each positive integer n. Let us consider the limits of the sequences $\{\underline{A_n}\}$ and $\{\overline{A_n}\}$. We have

$$\lim_{n \to \infty} \underline{A_n} = \lim_{n \to \infty}\left(\frac{5}{4} - \frac{3}{4n}\right) = \frac{5}{4}$$

and

$$\lim_{n \to \infty} \overline{A_n} = \lim_{n \to \infty}\left(\frac{5}{4} + \frac{1}{4n}\right) = \frac{5}{4}.$$

Thus, we must have $\frac{5}{4} \leqslant$ Area $(R) \leqslant \frac{5}{4}$, or

$$\text{Area } (R) = \frac{5}{4}.$$

Example. Let D be the region of the plane bounded above by the line $G(x) = -\frac{1}{2}x + 4$, on the left by the line $x = 1$, on the right by the line $x = 5$, and below by the x-axis (see Figure 8.6). As in the preceding example, we begin by

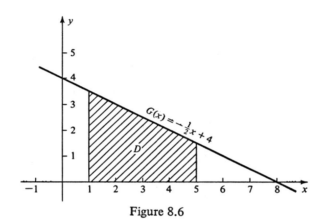

Figure 8.6

dividing the base interval $[1, 5]$ into n subintervals, each of length $\frac{1}{n} (5 - 1) = \frac{4}{n}$:

$$I_1 = \left[1, \ 1 + \frac{4}{n}\right]$$

$$I_2 = \left[1 + \frac{4}{n}, \ 1 + 2\left(\frac{4}{n}\right)\right]$$

.
.
.

$$I_k = \left[1 + (k - 1)\frac{4}{n}, \ 1 + k\left(\frac{4}{n}\right)\right] \quad (k\text{-th subinterval})$$

.
.
.

$$I_n = \left[1 + (n - 1)\frac{4}{n}, \ 5\right].$$

As before, we construct rectangles r_k and $\overline{r}_k$ with base I_k and respective heights $G\left(1 + k\left(\dfrac{4}{n}\right)\right)$ (the minimum value of $G(x)$ for $x \in I_k$), and $G\left(1 + (k - 1)\dfrac{4}{n}\right)$ (the maximum value of $G(x)$ for $x \in I_k$). Figures 8.7a and 8.7b illustrate the situation for $n = 4$.

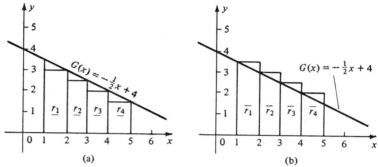

(a)   (b)

Figure 8.7

The area of the rectangle r_k is given by

$$\text{Area}\,(r_k) = \frac{4}{n}\left[-\frac{1}{2}\left(1 + \frac{4k}{n}\right) + 4\right]$$

$$= \frac{4}{n}\left[\frac{7}{2} - \frac{2k}{n}\right]$$

$$= \frac{14}{n} - \frac{8k}{n^2},$$

while the area of $\overline{r}_k$ is given by

$$\text{Area}\,(\overline{r}_k) = \frac{4}{n}\left[-\frac{1}{2}\left(1 + \frac{4(k - 1)}{n}\right) + 4\right]$$

$$= \frac{4}{n}\left[\frac{7}{2} - \frac{2k}{n} + \frac{2}{n}\right]$$

$$= \frac{14}{n} - \frac{8k}{n^2} + \frac{8}{n^2}.$$

Letting $A_n = \displaystyle\sum_{k=1}^{n} \text{Area}\,(r_k)$ and $\overline{A}_n = \displaystyle\sum_{k=1}^{n} \text{Area}\,(\overline{r}_k)$, we have

$$A_n = \sum_{k=1}^{n}\left(\frac{14}{n} - \frac{8k}{n^2}\right)$$

$$= \frac{14}{n} \sum_{k=1}^{n} 1 - \frac{8}{n^2} \sum_{k=1}^{n} k$$

$$= \frac{14}{n} (n) - \frac{8}{n^2} \cdot \frac{n(n+1)}{2}$$

$$= 14 - 4 - \frac{4}{n}$$

$$= 10 - \frac{4}{n}$$

and

$$\overline{A}_n = \sum_{k=1}^{n} \left(\frac{14}{n} - \frac{8k}{n^2} + \frac{8}{n^2} \right)$$

$$= \sum_{k=1}^{n} \left(\frac{14}{n} - \frac{8k}{n^2} \right) + \frac{8}{n^2} \sum_{k=1}^{n} 1$$

$$= 10 - \frac{4}{n} + \frac{8}{n^2} (n)$$

$$= 10 + \frac{4}{n}.$$

As in the preceding example, $\underline{A}_n \leqslant$ Area $(D) \leqslant \overline{A}_n$ for every positive integer n. Thus,

$$\lim_{n \to \infty} \underline{A}_n \leqslant \text{Area} (D) \leqslant \lim_{n \to \infty} \overline{A}_n.$$

Now

$$\lim_{n \to \infty} \underline{A}_n = \lim_{n \to \infty} \left(10 - \frac{4}{n} \right) = 10,$$

and

$$\lim_{n \to \infty} \overline{A}_n = \lim_{n \to \infty} \left(10 + \frac{4}{n} \right) = 10.$$

Hence $10 \leqslant$ Area $(D) \leqslant 10$ so that Area $(D) = 10$.

In each of the preceding examples, the region under consideration was a trapezoid (a plane quadrilateral, two of whose sides are parallel). There is a very simple formula for finding the area of a trapezoidal region R, namely, if b_1 and b_2 denote the lengths of the two parallel sides of R and if the perpendicular distance between these two sides is h, then

$$\text{Area} (R) = \frac{1}{2} (b_1 + b_2)h.$$

Using this formula in conjunction with the preceding example, we find that $b_1 = G(1) = -\frac{1}{2}(1) + 4 = \frac{7}{2}$, $b_2 = G(5) = -\frac{1}{2}(5) + 4 = \frac{3}{2}$, and $h = 5 - 1 = 4$ so that

$$\text{Area } (D) = \frac{1}{2}\left(\frac{7}{2} + \frac{3}{2}\right)(4) = 10.$$

Since the use of this formula in finding the area of D is considerably easier and less time consuming than the use of approximating rectangles, the question naturally arises as to why we should complicate matters by using rectangles and taking limits. The following example provides an answer to this question.

Example. Let E be the region of the plane bounded above on the interval $[0, 1]$ by the curve $s(x) = x^2$, below by the x-axis, and on the right by the line $x = 1$ (see Figure 8.8). Before finding the area of E, we observe that E is not a trapezoidal region. In fact, there is no simple formula from Euclidean geometry for finding the area of this type of region. Thus we have no choice but to use the rectangle approximation method to compute the area of E. To repeat what, by now, is a familiar construction, we divide the base interval $[0, 1]$ into n subintervals, each of length $\frac{1}{n}$:

$$I_1 = \left[0, \frac{1}{n}\right]$$

$$I_2 = \left[\frac{1}{n}, \frac{2}{n}\right]$$

.
.
.

$$I_k = \left[\frac{k-1}{n}, \frac{k}{n}\right] \quad (k\text{-th subinterval})$$

.
.
.

$$I_n = \left[\frac{n-1}{n}, 1\right].$$

We construct rectangles $\underline{r_k}$ and $\overline{r_k}$, having base I_k and respective heights $s\left(\frac{k-1}{n}\right) = \frac{(k-1)^2}{n^2}$ (the minimum value of $s(x)$ for $x \in I_k$) and $s\left(\frac{k}{n}\right) = \frac{k^2}{n^2}$ (the

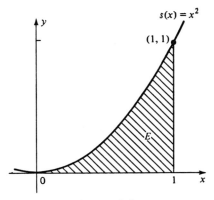

Figure 8.8

maximum value of $s(x)$ for $x \in I_k$). The rectangles $\underline{r_k}$ and $\overline{r_k}$ are illustrated in Figures 8.9a and 8.9b for $n = 5$.

Now

$$\text{Area}\,(\underline{r_k}) = \frac{1}{n} \left[\frac{(k - 1)^2}{n^2} \right] = \frac{1}{n^3}\,(k^2 - 2k + 1)$$

and

$$\text{Area}\,(\overline{r_k}) = \frac{1}{n} \left(\frac{k^2}{n^2} \right) = \frac{1}{n^3}\,(k^2).$$

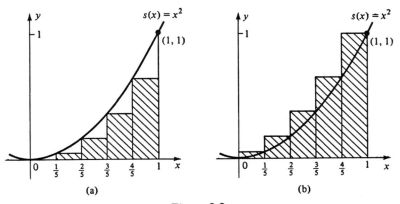

(a)   (b)

Figure 8.9

The sum of the areas of the inscribed rectangles $\underline{r_k}$ is

$$\underline{A_n} = \sum_{k=1}^{n} \text{Area}(\underline{r_k}) = \sum_{k=1}^{n} \frac{1}{n^3}(k^2 - 2k + 1)$$

$$= \frac{1}{n^3}\left[\sum_{k=1}^{n} k^2 - 2\sum_{k=1}^{n} k + \sum_{k=1}^{n} 1\right]$$

$$= \frac{1}{n^3}\left[\frac{n(n+1)(2n+1)}{6} - 2 \cdot \frac{n(n+1)}{2} + n\right]$$

$$= \frac{1}{n^3}\left[\frac{2n^3 + 3n^2 + n}{6} - n^2\right]$$

$$= \frac{1}{n^3}\left[\frac{2n^3 - 3n^2 + n}{6}\right]$$

$$= \frac{1}{3} - \frac{1}{2n} + \frac{1}{6n^2},$$

and the sum of the areas of the superscribed rectangles $\overline{r_k}$ is

$$\overline{A_n} = \sum_{k=1}^{n} \text{Area}(\overline{r_k}) = \sum_{k=1}^{n} \frac{1}{n^3}(k^2)$$

$$= \frac{1}{n^3}\sum_{k=1}^{n} k^2$$

$$= \frac{1}{n^3}\left[\frac{n(n+1)(2n+1)}{6}\right]$$

$$= \frac{1}{n^3}\left[\frac{2n^3 + 3n^2 + n}{6}\right]$$

$$= \frac{1}{3} + \frac{1}{2n} + \frac{1}{6n^2}.$$

Now

$$\lim_{n\to\infty} \underline{A_n} = \lim_{n\to\infty}\left(\frac{1}{3} - \frac{1}{2n} + \frac{1}{6n^2}\right) = \frac{1}{3}$$

and

$$\lim_{n\to\infty} \overline{A_n} = \lim_{n\to\infty}\left(\frac{1}{3} + \frac{1}{2n} + \frac{1}{6n^2}\right) = \frac{1}{3}.$$

Thus, from the inequality $\underline{A_n} \leqslant$ Area $(E) \leqslant \overline{A_n}$, we conclude that $\frac{1}{3} \leqslant$ Area $(E) \leqslant \frac{1}{3}$ so that Area $(E) = \frac{1}{3}$.

Exercise Set 8.3

1. Let R be the region bounded above by the curve $y = f(x)$, below by the x-axis, on the left by the line $x = 1$, and on the right by the line $x = 6$ as shown in the figure below.

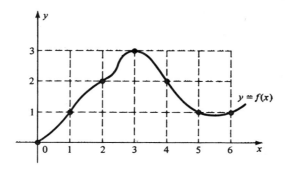

a. Inscribe, within the region R, a set of rectangles, each of which has a subinterval of $[0, 6]$ of length 1 as its base. Find the sum of the areas of the rectangles so chosen. Call this $\underline{A_5}$.

b. Superscribe, about the region R, a set of rectangles, each of which has a subinterval of $[0, 6]$ of length 1 as its base. Find the sum of the areas of the rectangles so chosen. Call this $\overline{A_5}$.

c. Repeat part a, choosing rectangles whose bases are of length $\frac{1}{2}$. Call the sum obtained $\underline{A_{10}}$.

d. Repeat part b, choosing rectangles whose bases are of length $\frac{1}{2}$. Call the sum obtained $\overline{A_{10}}$.

e. Compare $\underline{A_5}, \overline{A_5}, \underline{A_{10}}$, and $\overline{A_{10}}$.

f. Estimate the area of the region R.

2. Sketch the region R, bounded above by the line $y = x + 1$, on the left by the y-axis, on the right by the line $x = 2$, and below by the x-axis. Divide the interval $[0, 2]$ into subintervals of equal length.

a. What is the length of a subinterval I_k if there are 4 subintervals? 20 subintervals? n subintervals?

b. What are the coordinates of the endpoints of a subinterval I_k if there are 4 subintervals? 20 subintervals? n subintervals?

c. What is the height of an inscribed rectangle r_k whose base is the sub-interval I_k if there are 4 subintervals? 20 subintervals? n subintervals?

d. What is the area Area (r_k) of an inscribed rectangle r_k whose base is the subinterval I_k if there are 4 subintervals? 20 subintervals? n subintervals?

e. Repeat parts c and d for a superscribed rectangle $\overline{r}_k$ whose base is the subinterval I_k.

f. Find $\underline{A_n}$ if $\underline{A_n} = \displaystyle\sum_{k=1}^{n}$ Area (r_k), that is, $\underline{A_n}$ is the sum of the areas of the inscribed rectangles whose bases are the n subintervals I_k.

g. Find $\overline{A}_n$ if $\overline{A}_n = \displaystyle\sum_{k=1}^{n}$ Area $(\overline{r}_k)$, that is, $\overline{A}_k$ is the sum of the areas of the superscribed rectangles whose bases are the n subintervals I_k.

h. Find $\lim\limits_{n\to\infty} \underline{A_n}$ and $\lim\limits_{n\to\infty} \overline{A}_n$.

i. What is the area of the region R ?

3. Repeat problem 2 for the region R, bounded above by the portion of $f(x) = x^2$ between (0, 0) and (2, 4), below by the x-axis, and on the right by the line $x = 2$.

* * *

In problems 4–12, use the methods of this section to find the area of the given region R.

4. R is the region bounded above by the line $y = -2x + 4$, on the left by the y-axis, on the right by the line $x = 1$, and below by the x-axis.

5. R is the region bounded above by the line $y = -2x + 4$, on the left by the y-axis, and below by the segment [0, 2] of the x-axis.

6. R is the region bounded above by the graph of $f(x) = x^2 + 1$, on the left by y-axis, on the right by the line $x = 1$, and below by the x-axis.

7. R is the region bounded above by the graph of $f(x) = 2x^2 + 1$, on the left by the y-axis, on the right by the line $x = 3$, and below by the x-axis.

8. R is the region bounded above by the graph of $f(x) = 1 - x^2$, on the left by the y-axis, and below by the segment [0, 1] of the x-axis.

9. R is the region bounded above by the graph of $f(x) = 4 - x^2$, on the left by the y-axis, on the right by the line $x = 1$, and below by the x-axis.

10. Find the area of the region R bounded above by the segment $y = x$, $(0 \leqslant x \leqslant 1)$, and below by the curve $y = x^2$ $(0 \leqslant x \leqslant 1)$. (*Hint:* First find the area under the curve $y = x$ for $0 \leqslant x \leqslant 1$ and then find the area under the curve $y = x^2$ for $0 \leqslant x \leqslant 1$. Then subtract the latter from the former.)

11. Find the area of the region R bounded above by the portion of the curve $y = x^2$, $(1 \leqslant x \leqslant 2)$, below by the line $y = x$, $(1 \leqslant x \leqslant 2)$, and on the right by the line $x = 2$.

12. Find the area of the region R bounded above by the portion of the curve $y = 1 - x^2$, $\left(0 \leqslant x \leqslant \frac{\sqrt{2}}{2}\right)$, and below by the portion of the curve $y = x^2$, $\left(0 \leqslant x \leqslant \frac{\sqrt{2}}{2}\right)$.

8.4 THE DEFINITE INTEGRAL

In this section we consider integration, the second basic operation of the calculus. We recall that the process of differentiation involves the assignment of a number (the derivative) to a given function f at a point $a \in D_f$. Similarly, integration involves the assignment of a number (the *definite integral*) to a function f defined over a closed interval $I = [a, b]$. The problem of determining the definite integral of a given function is very similar to the method discussed in the preceding section for computing areas of irregularly shaped plane regions. Indeed, if f is a positive continuous function on $I = [a, b]$, then the integral of f over I is precisely the area of the region bounded above by the curve $y = f(x)$, below by the x-axis, and on the sides by the lines $x = a$ and $x = b$. We see, therefore, that we have already computed some integrals even before this concept has been presented!

In order to properly define the definite integral of a function, it is necessary to generalize a few of the ideas presented in Sec. 8.2. The first such generalization is as follows:

8.5 Definition. *A partition of the interval* $I = [a, b]$ *is a set of points* $P = \{x_0, x_1, x_2, \ldots, x_n\}$, *such that* $a = x_0 < x_1 < x_2 < \cdots < x_{n-1} < x_n = b$.
For a given partition P *of* I, *write*

$$\Delta x_k = x_k - x_{k-1} \ (k = 1, 2, \ldots, n).$$

The number

$$\text{Maximum } \{\Delta x_1, \Delta x_2, \ldots, \Delta x_n\}$$

is called the *norm* of the partition P, and is denoted by $\|P\|$.

We see that a partition is merely a subdivision of I which occurs as the result of the introduction of a finite number of intermediate points x_k into I. This partition need *not* be *even* in the sense that all subintervals are equal in length. The norm of the partition is simply the length of the largest subinterval $[x_{k-1}, x_k]$ determined by the partition.

Example. Let $I = [1, 5]$. The following are partitions of I:

$$P_1 = \left\{1, \frac{3}{2}, \frac{5}{2}, 4, 5\right\},$$

$$P_2 = \left\{ 1, \frac{4}{3}, \frac{3}{2}, \frac{9}{4}, \frac{9}{2}, 5 \right\},$$

and $$P_3 = \left\{ 1, 1 + \frac{4}{n}, 1 + \frac{8}{n}, \ldots, 1 + \frac{4k}{n}, \ldots, 1 + \frac{4(n-1)}{n}, 5 \right\}.$$

The norm of P_1 is given by

$$\|P_1\| = \text{Max} \left\{ \left(\frac{3}{2} - 1 \right), \left(\frac{5}{2} - \frac{3}{2} \right), \left(4 - \frac{5}{2} \right), (5 - 4) \right\}$$

$$= \text{Max} \left\{ \frac{1}{2}, 1, \frac{3}{2}, 1 \right\} = \frac{3}{2}.$$

The norm of P_2 is given by

$$\|P_2\| = \text{Max} \left\{ \left(\frac{4}{3} - 1 \right), \left(\frac{3}{2} - \frac{4}{3} \right), \left(\frac{9}{4} - \frac{3}{2} \right), \left(\frac{9}{2} - \frac{9}{4} \right), \left(5 - \frac{9}{2} \right) \right\}$$

$$= \text{Max} \left\{ \frac{1}{3}, \frac{1}{6}, \frac{3}{4}, \frac{9}{4}, \frac{1}{2} \right\} = \frac{9}{4}.$$

Since P_3 is an even partition $\left(\Delta x_k = \frac{4}{n} \text{ for } k = 1, 2, \ldots, n \right)$, we have $\|P_3\| = \frac{4}{n}$. The partitions P_1 and P_2 are illustrated in Figure 8.10.

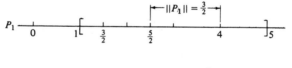

Figure 8.10

8.6 Definition. *Let f be a function defined on the interval $I = [a, b]$ and let $P = \{a = x_0, x_1, x_2, \ldots, x_n = b\}$ be a partition of I. For $k = 1, 2, \ldots, n$, choose a point $z_k \in [x_{k-1}, x_k]$ in any way whatever. The sum*

$$\sum_{k=1}^{n} f(z_k)(z_k - x_{k-1}) = \sum_{k=1}^{n} f(z_k) \Delta x_k$$

*is a **Riemann sum** for f over I.*

In case f is a positive continuous function over $[a, b]$, each Riemann sum for f provides an approximation to the area under the curve $y = f(x)$, from $x = a$ to $x = b$ (see Figure 8.11). This is because each term $f(z_k)\, \Delta x_k$ appearing in the Riemann sum is merely the area of the rectangle with base $[x_{k-1}, x_k]$ and height $f(z_k)$.

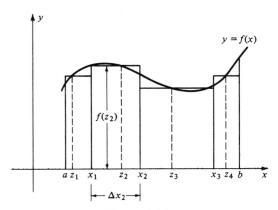

Figure 8.11

Note that the points z_k may be chosen in a purely arbitrary manner. In particular, if we choose z_k to be that value of $z \in [x_{k-1}, x_k]$ for which f achieves its *largest* value, the Riemann sum is called an *upper sum* for f. On the other hand, if z_k is chosen to be that value of $z \in [x_{k-1}, x_k]$ for which f achieves its *smallest* value, the Riemann sum is called a *lower sum* for f. If f is positive and continuous on $[a, b]$, then the upper sums and lower sums provide us with upper and lower estimates of the area under the curve $y = f(x)$ from $x = a$ to $x = b$, just as in Sec. 8.2. The only difference is that here we are not insisting that the partition of the interval $[a, b]$ be even.

Problem. Let P be the partition $P = \left\{0, \dfrac{1}{4}, \dfrac{1}{3}, \dfrac{1}{2}, \dfrac{4}{5}, 1\right\}$ of the interval $I = [0, 1]$. For each $k\,(k = 1, 2, 3, 4, 5)$, let z_k be the *midpoint* of the interval $[x_{k-1}, x_k]$, where x_k denotes the k-th point of subdivision. Let $f(x) = x^2$. Find the value of the Riemann sum S for f corresponding to this partition and the indicated choices of z_k.

Solution. We first determine the value of each z_k. We have

$$z_1 = \text{midpoint of } \left[0, \frac{1}{4}\right] = \frac{1}{8},$$

$$z_2 = \text{midpoint of } \left[\frac{1}{4}, \frac{1}{3}\right] = \frac{7}{24},$$

$$z_3 = \text{midpoint of } \left[\frac{1}{3}, \frac{1}{2}\right] = \frac{5}{12},$$

$$z_4 = \text{midpoint of } \left[\frac{1}{2}, \frac{4}{5}\right] = \frac{13}{20},$$

and
$$z_5 = \text{midpoint of } \left[\frac{4}{5}, 1\right] = \frac{9}{10}.$$

We now compute the value of $f(x)$ for $x = z_k$ ($k = 1, 2, 3, 4, 5$). In tabular form, we have

z_k	$\dfrac{1}{8}$	$\dfrac{7}{24}$	$\dfrac{5}{12}$	$\dfrac{13}{20}$	$\dfrac{9}{10}$
$f(z_k) = z_k^2$	$\dfrac{1}{64}$	$\dfrac{49}{576}$	$\dfrac{25}{144}$	$\dfrac{169}{400}$	$\dfrac{81}{100}$
Approximate value of $f(z_k)$	.02	.08	.17	.42	.81

In order to compute the Riemann sum, we need only one other piece of information, namely, the lengths Δx_k of the subintervals $[x_{k-1}, x_k]$ ($k = 1, 2, 3, 4, 5$) determined by the given partition. These are provided by the following table:

$[x_{k-1}, x_k]$	$\left[0, \frac{1}{4}\right]$	$\left[\frac{1}{4}, \frac{1}{3}\right]$	$\left[\frac{1}{3}, \frac{1}{2}\right]$	$\left[\frac{1}{2}, \frac{4}{5}\right]$	$\left[\frac{4}{5}, 1\right]$
Δx_k	$\dfrac{1}{4}$	$\dfrac{1}{12}$	$\dfrac{1}{6}$	$\dfrac{3}{10}$	$\dfrac{1}{5}$
Δx_k (approx.)	.25	.08	.17	.30	.20

In what follows, we shall use the symbol "$\approx$" to mean "approximately equals."
We are now ready to find the value of the sum $S = \sum\limits_{k=1}^{5} f(z_k) \, \Delta x_k$. We have

$$S = \sum_{k=1}^{5} f(z_k) \, \Delta x_k$$

$$\approx (.02)(.25) + (.08)(.08) + (.17)(.17) + (.42)(.30) + (.81)(.20)$$

$$\approx .005 + .006 + .029 + .126 + .162 = .328.$$

The value of this Riemann sum (.328) is the sum of the areas of the five rectangles appearing in Figure 8.12. This figure shows how the Riemann sum provides an approximation to the area of the region under the curve $y = x^2$ from $x = 0$ to $x = 1$. We recall that the actual area of the region under this curve is $\frac{1}{3}$. Thus, the Riemann sum computed in this example is quite accurate, even though the number of points of subdivision was relatively small.

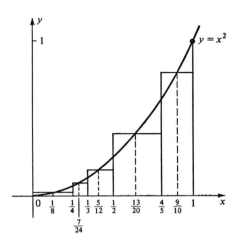

Figure 8.12

Problem. Find the upper and lower sums for $f(x) = x^2$ over $I = [0, 1]$ corresponding to the partition $P = \left\{0, \frac{1}{4}, \frac{1}{3}, \frac{1}{2}, \frac{4}{5}, 1\right\}$.

Solution. In order to compute the upper and lower sums for f we must know the maximum and minimum values of f on each subinterval $I_k = [x_{k-1}, x_k]$. We observe that $f(x) = x^2$ is increasing on I, so that f will achieve a maximum M_k on I_k for $x = x_k$ and a minimum m_k on I_k for $x = x_{k-1}$. The following table contains the values of M_k and m_k for $k = 1, 2, 3, 4, 5$.

$[x_{k-1}, x_k]$	$\left[0, \frac{1}{4}\right]$	$\left[\frac{1}{4}, \frac{1}{3}\right]$	$\left[\frac{1}{3}, \frac{1}{2}\right]$	$\left[\frac{1}{2}, \frac{4}{5}\right]$	$\left[\frac{4}{5}, 1\right]$
$M_k = f(x_k)$	$\frac{1}{16} \approx .06$	$\frac{1}{9} \approx .11$	$\frac{1}{4} = .25$	$\frac{16}{25} = .64$	1
$m_k = f(x_{k-1})$	0	$\frac{1}{16} \approx .06$	$\frac{1}{9} \approx .11$	$\frac{1}{4} = .25$	$\frac{16}{25} = .64$

Denoting the upper sum by $\bar{S}$ and the lower sum by $\underline{S}$, we have

$$\bar{S} = \sum_{k=1}^{5} M_k \, \Delta x_k \approx (.06)(.25) + (.11)(.08) + (.25)(.17) + (.64)(.30) + (1)(.20)$$

$$= .015 + .009 + .043 + .192 + .200 = .459,$$

and

$$\underline{S} = \sum_{k=1}^{5} m_k \, \Delta x_k$$

$$\approx 0(.25) + (.06)(.08) + (.11)(.17) + (.25)(.30) + (.64)(.20)$$

$$\approx 0 + .005 + .019 + 0.75 + .128 = .227.$$

$\bar{S}$ and $\underline{S}$ are illustrated in Figure 8.13. Compare these to Figure 8.12.

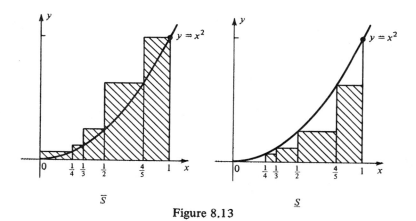

Figure 8.13

The upper and lower sums $\bar{S}$ and $\underline{S}$ are extremal for the partition P in the sense that each Riemann sum S for f corresponding to P satisfies the inequality

$$\underline{S} \leqslant S \leqslant \bar{S}.$$

With respect to the preceding examples, we found that $\underline{S} = .227$, and $\bar{S} = .459$. The Riemann sum obtained by choosing z_k to be the midpoint of $[x_{k-1}, x_k]$ was given by $S = .328$. Clearly

$$.227 < .328 < .459.$$

It should be understood that alteration of the partition P affects the values of all Riemann sums for the given function. Hence the above inequality is valid only if all sums are computed with respect to the same partition.

It would seem that the accuracy of the approximation to the area afforded by the Riemann sum should increase as we introduce additional points of subdivision, *provided these points are introduced in such a way that the norm of the partition becomes smaller.* This observation is the basic idea employed in the following definition.

8.7 Definition. *Let f be a function defined on the closed interval* $[a, b]$ *and let L be a real number. Let* $\{P_n\}$ *be any sequence of partitions of* $[a, b]$ *for which*

$$\lim_{n \to \infty} \|P_n\| = 0,$$

and for each n, let S_n *be a Riemann sum for f corresponding to the partition* P_n. *IF*

$$L = \lim_{n \to \infty} S_n,$$

*THEN f is **integrable** over* $[a, b]$, *and the number L is the **definite integral** of f over* $[a, b]$, *denoted by*

$$L = \int_a^b f(x)\, dx.$$

The notation used to represent the definite integral arises in a natural way from the definition of the integral. The symbol "$\int$" may be thought of as an elongated "*S*," suggesting that "sum" is involved. In particular, the Riemann sum may be represented symbolically as

$$\sum f(x)\, \Delta x,$$

where the subscripts have been omitted. Replacement of "$\sum$" by "$\int_a^b$" and of Δx by dx (indicating that the limit has been taken) gives

$$\int_a^b f(x)\, dx.$$

The letters a and b appearing on the integral sign tell us the interval over which the integral of f is computed. They are referred to as the lower and upper limits of integration, respectively. The following theorem, which we state without proof, provides us with a large class of integrable functions.

8.8 Theorem. *IF f is continuous on* $[a, b]$, *THEN f is integrable over* $[a, b]$.

Returning now to Definition 8.7, we see that if f is integrable over $[a, b]$, then any sequence of partitions whose norms approach zero and any family of Riemann sums corresponding to these partitions may be used in evaluating the integral of f. In particular, once we know that a function is integrable, its integral may be found by using either upper or lower sums corresponding to even partitions of the interval $[a, b]$. This is precisely the approach we used in Sec. 8.2 to find areas of certain regions of the plane. Since the functions $f(x) = x^2$ and $f(x) = -\frac{1}{2}x + 4$ are continuous, these functions must be integrable over any closed interval $[a, b]$. The results of Sec. 8.2 then show that

$$\int_1^5 \left(-\frac{1}{2}x + 4\right)dx = 10$$

and

$$\int_0^1 x^2\, dx = \frac{1}{3}.$$

Knowledge of the existence of the integral of a continuous function f over a closed interval $[a, b]$ is one thing but actual computation of the integral is quite another matter. We have seen that evaluation of integrals of the simplest of functions $\left(\text{e.g., } f(x) = -\frac{1}{2}x + 4 \text{ and } f(x) = x^2\right)$ is a formidable task indeed. It shall be our goal in Sec. 8.6 to develop a simple method for evaluating integrals which will free us from the rather cumbersome method provided by Definition 8.7.

We close by emphasizing once again that if f is a positive continuous function over the interval $[a, b]$, then $\int_a^b f(x)\, dx$ is simply the area of the region of the plane bounded above by the curve $y = f(x)$, below by the x-axis, and on the sides by the vertical lines $x = a$ and $x = b$.

Exercise Set 8.4

1. a. Let $P = \left\{0, \frac{1}{2}, \frac{7}{8}, \frac{4}{3}, 2\right\}$ be a partition of the interval $[0, 2]$. Find $x_0, x_1, x_2, x_3,$ and x_4. Find $\Delta x_1, \Delta x_2, \Delta x_3,$ and Δx_4. What is the norm, $\|P\|$, of this partition?

   b. Let $P = \{0, 0.1, 0.4, 1.4, 1.9, 2.0\}$ be a partition of the interval $[0, 2]$. Find $x_0, x_1, x_2, x_3, x_4,$ and x_5. Find $\Delta x_1, \Delta x_2, \Delta x_3, \Delta x_4,$ and Δx_5. Find $\|P\|$.

   c. Choose a partition P of the interval $[0, 2]$ such that $\|P\| = 0.3$.

2. Let R be the region bounded above by the curve $y = f(x)$, below by the x-axis, on the left by the line $x = 1$, and on the right by the line $x = 6$, as shown in the figure below.

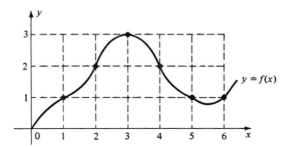

Let $P = \{1, 2, 3, 4, 5, 6\}$.
a. Find x_0, x_1, x_2, x_3, x_4, and x_5. Find Δx_k.
b. Choose $z_1 \in [1, 2], z_2 \in [2, 3], z_3 \in [3, 4], z_4 \in [4, 5]$, and $z_5 \in [5, 6]$.
c. Sketch the rectangle whose base is $[x_{k-1}, x_k]$ and whose height is $f(z_k)$ for $k = 1, 2, 3, 4, 5$.
d. Use the graph to estimate the Riemann sum,

$$\sum_{k=1}^{5} f(z_k)(x_k - x_{k-1}) = \sum_{k=1}^{n} f(z_k)\,\Delta x_k.$$

e. Repeat parts (a) through (d) for a different choice of z_1, z_2, z_3, z_4, and z_5.
f. Estimate the area of the region R.
g. Compare the estimated area obtained in this problem with the estimated area (for the same region) which was obtained in problem 1, Exercise Set 8.3.

3. Let $f(x) = \dfrac{1}{x}, I = [1, 2]$, and $P = \left\{1, \dfrac{5}{4}, \dfrac{4}{3}, \dfrac{3}{2}, 2\right\}$.
a. Determine x_0, x_1, x_2, x_3, and x_4.
b. Determine $\Delta x_1, \Delta x_2, \Delta x_3$, and Δx_4.
c. Find $\displaystyle\sum_{k=1}^{4} f(z_k)\,\Delta x_k$ if z_k is chosen to be that value of $z \in [x_{k-1}, x_k]$ for which f achieves the smallest value in the interval. Call this sum $\underline{S}$.
d. Find $\displaystyle\sum_{k=1}^{4} f(z_k)\,\Delta x_k$ if z_k is chosen to be that value of $z \in [x_{k-1}, x_k]$ for which f achieves the largest value in the interval. Call this sum $\overline{S}$.
e. Find $\displaystyle\sum_{k=1}^{4} f(z_k)\,\Delta x_k$ if z_k is chosen to be the midpoint of the interval $[x_{k-1}, x_k]$. Call this sum S_1.

f. Find $\sum\limits_{k=1}^{4} f(z_k)\,\Delta x_k$ if z_k is chosen in any way whatever, in the interval $[x_{k-1}, x_k]$. Call this sum S_2.

g. Compare $\underline{S}, \bar{S}, S_1$, and S_2.

h. Estimate the area of the region R which is bounded above by the curve $y = \dfrac{1}{x}$, below by the x-axis, on the left by the line $x = 1$, and on the right by the line $x = 2$.

4. a. If f is the function of problem 2, use the results of problem 2 to estimate
$$\int_1^6 f(x)\,dx.$$

b. If f is the function $f(x) = \dfrac{1}{x}$, use the results of problem 3 to estimate
$$\int_1^2 \frac{1}{x}\,dx.$$

$$* \; * \; *$$

In problems 5–10, find the lower sum $\underline{S}$ and the upper sum $\bar{S}$.

5. $f(x) = \dfrac{1}{x}$, $I = [1, 2]$, $P = \{1, 1.2, 1.4, 1.6, 1.8, 2\}$

6. $f(x) = \dfrac{1}{x}$, $I = [1, 2]$, $P = \left\{1, \dfrac{9}{8}, \dfrac{11}{8}, \dfrac{13}{8}, \dfrac{15}{8}, 2\right\}$

7. $f(x) = \dfrac{1}{x}$, $I = [1, 2]$, $P = \{1, 1.1, 1.2, 1.3, 1.4, 1.5, 1.6, 1.7, 1.8, 1.9, 2\}$

8. $f(x) = x^3$, $I = [0, 2]$, $P = \left\{0, \dfrac{1}{2}, 1, \dfrac{4}{3}, \dfrac{5}{3}, 2\right\}$

9. $f(x) = x^3$, $I = [0, 2]$, $P = \left\{0, \dfrac{1}{4}, \dfrac{1}{2}, \dfrac{3}{4}, 1, \dfrac{5}{4}, \dfrac{3}{2}, \dfrac{7}{4}, 2\right\}$

10. $f(x) = x^3$, $I = [0, 2]$, $P = \{0, .1, .2, .3, .5, .8, 1, 1.1, 1.5, 2\}$

In problems 11–16, construct a Riemann sum corresponding to the given partition and evaluate it.

11. $f(x) = e^x$, $I = [0, 1]$, $P = \{0, .1, .3, .8, 1\}$

12. $f(x) = e^{-x}$, $I = [0, 1]$, $P = \{0, .1, .3, .8, 1\}$

13. $f(x) = \ln x$, $I = [1, 4]$, $P = \{1, 1.5, 2, 2.5, 3, 3.5, 4\}$

14. $f(x) = \ln x$, $I = [1, 2]$, $P = \{1, 1.25, 1.3, 1.38, 1.72, 1.89, 2\}$

15. $f(x) = \sqrt{x}$, $I = \left[0, \dfrac{3}{2}\right]$, $P = \left\{0, \dfrac{1}{9}, \dfrac{1}{4}, \dfrac{3}{4}, 1, \dfrac{10}{9}, \dfrac{3}{2}\right\}$

16. $f(x) = \sqrt{x}$, $I = [0, 2]$, $P = \left\{ 0, \dfrac{9}{37}, \dfrac{18}{37}, \dfrac{27}{37}, \dfrac{36}{37}, \dfrac{45}{37}, \dfrac{54}{37}, \dfrac{63}{37}, 2 \right\}$

Use the results of problems 11 through 16 to estimate the following definite integrals:

17. $\displaystyle\int_0^1 e^x dx$

18. $\displaystyle\int_0^1 e^{-x} dx$

19. $\displaystyle\int_1^4 \ln x \, dx$

20. $\displaystyle\int_1^2 \ln x \, dx$

21. $\displaystyle\int_0^{\frac{3}{2}} \sqrt{x} \, dx$

22. $\displaystyle\int_0^2 \sqrt{x} \, dx$

23. Using the fact that (i) the integral of a positive continuous function over an interval $I = [a, b]$ is the area under the curve $y = f(x)$ from $x = a$ to $x = b$ and (ii) the area of a triangle with base b and height h is $\dfrac{1}{2} bh$, evaluate each of the following integrals:

a. $\displaystyle\int_0^3 x \, dx$

b. $\displaystyle\int_{-2}^0 (-x) \, dx$

c. $\displaystyle\int_0^9 \left(-\dfrac{1}{3} x + 3 \right) dx$

24. a. Sketch the function $g(x) = 2$ for $x \in [2, 5]$. Use your drawing to evaluate $\displaystyle\int_2^5 2 dx$.

b. Repeat part (a) for $g(x) = 3$ with $x \in [3, 7]$ to compute $\displaystyle\int_3^7 3 dx$.

c. Generalize parts (a) and (b) by evaluating $\displaystyle\int_a^b k \, dx$ where k is a positive constant and $a, b \in \mathcal{R}$ with $a < b$.

25. Construct and evaluate a Riemann sum for the function $f(x) = x^2 + x - 4$ over $[0, 2]$ corresponding to the partition $P = \{0, .4, .5, .8, 1, 1.1, 1.4, 1.5, 1.8, 2\}$. Use this sum to estimate $\displaystyle\int_0^2 (x^2 + x - 4) \, dx$.

8.5 PROPERTIES OF THE DEFINITE INTEGRAL

In this section we discuss certain properties enjoyed by the definite integral. Many of the results stated here seem obvious if the integral is interpreted as an area. We shall therefore attempt to illustrate the properties as they are presented, avoiding rigorous proofs. We first observe that if R_1 and R_2 are nonoverlapping regions of the plane and if $R = R_1 \cup R_2$, then the area of R is the sum of the areas of R_1 and R_2. In particular, let R be the region bounded above by the curve $y = f(x)$, below by the x-axis, and on the sides by the lines $x = a$ and $x = b$. If c is an intermediate point of the interval $[a, b]$, (i.e., $a \leqslant c \leqslant b$) let R_1 be the region bounded by $y = f(x)$, the x-axis, and the lines $x = a$ and $x = c$; and let R_2 be the region bounded by $y = f(x)$, the x-axis, and the lines $x = c$ and $x = b$. Then the area of R is the sum of the areas of R_1 and R_2 (see Figure 8.14).

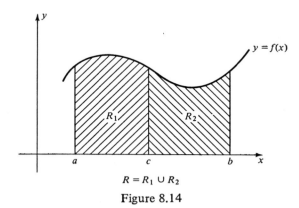

$$R = R_1 \cup R_2$$

Figure 8.14

We see, however, that the area of R is given by

$$\text{Area } (R) = \int_a^b f(x)dx,$$

while $\quad\quad \text{Area } (R_1) = \int_a^c f(x)dx \text{ and Area } (R_2) = \int_c^b f(x)dx.$

Hence, it must be the case that

$$\int_a^b f(x)dx = \int_a^c f(x)dx + \int_c^b f(x)dx.$$

Aside from providing an intuitive basis for the discussion, the restriction that f be a positive on $[a, b]$ is unnecessary. More generally, we have the following theorem:

8.9 Theorem. *IF f is continuous on the interval $[a, b]$ and $a \leqslant c \leqslant b$, THEN*

$$\int_a^b f(x)dx = \int_a^c f(x)dx + \int_c^b f(x)dx.$$

Example. It may be shown that

$$\int_1^2 (3x^2 + 1)dx = 8 \text{ and } \int_2^4 (3x^2 + 1)dx = 58.$$

Therefore, by Theorem 8.9, we have

$$\int_1^4 (3x^2 + 1)dx = \int_1^2 (3x^2 + 1)dx + \int_2^4 (3x^2 + 1)dx$$

$$= 8 + 58 = 64.$$

Suppose now that f is continuous on $[a, b]$, and let m and M be numbers such that

$$m \leqslant f(x) \leqslant M \text{ for all } x \in [a, b].$$

In particular, we may choose m to be the minimum value of f on $[a, b]$, and M to be the maximum value of f on $[a, b]$. If f is positive on $[a, b]$, then the region R under the curve $y = f(x)$ between $x = a$ and $x = b$ totally contains the rectangle $\underline{r}$ with base $[a, b]$ and height m, and is totally contained within the rectangle $\bar{r}$ with base $[a, b]$ and height M (see Figure 8.15).

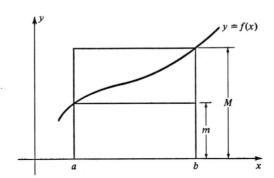

Figure 8.15

Hence, Area $(\underline{r}) \leqslant$ Area $(R) \leqslant$ Area $(\bar{r})$. But Area $(\underline{r}) = m(b - a)$ and Area $(\bar{r}) = M(b - a)$. Since Area $(R) = \int_a^b f(x)dx$, we have

$$m(b - a) \leqslant \int_a^b f(x)dx \leqslant M(b - a).$$

Again, the result remains true even if f is not positive over $[a, b]$. We formalize this observation as:

8.10 Theorem. *IF f is continuous over $[a, b]$ and if m and M are numbers such that*

$$m \leqslant f(x) \leqslant M \text{ for all } x \in [a, b],$$

THEN

$$m(b - a) \leqslant \int_a^b f(x)dx \leqslant M(b - a).$$

Example. Let $f(x) = 2x^2 + 3$ for $x \in [0, 1]$. Now

$$f'(x) = 4x,$$

so that $f'(x) > 0$ for all $x \in [0, 1]$. Hence, f is increasing on the interval $[0, 1]$, and must attain a minimum value m for $x = 0$ and a maximum value M for $x = 1$. We have $m = 2(0)^2 + 3 = 3$, $M = 2(1)^2 + 3 = 5$, and $b - a = 1 - 0 = 1$. Theorem 8.10 then implies that

$$3(1) \leqslant \int_0^1 (2x^2 + 3)dx \leqslant 5(1)$$

or

$$3 \leqslant \int_0^1 (2x^2 + 3)dx \leqslant 5.$$

In formulating the definition of the integral of a function f over an interval $[a, b]$, we assumed that $a < b$. It is convenient to extend our definition to the case when $a = b$ or $a > b$.

8.11 Definition. *IF f is integrable over the closed interval $[a, b]$, where $a < b$, THEN*

$$(i) \quad \int_a^a f(x)dx = 0$$

and

$$(ii) \quad \int_b^a f(x)dx = -\int_a^b f(x)dx.$$

We now observe that Theorem 8.9 remains valid even if c is not an intermediate point of the interval $[a, b]$. Indeed, suppose that $a < b < c$. Then

$$\int_a^c f(x)dx = \int_a^b f(x)dx + \int_b^c f(x)dx,$$

so that

$$\int_a^b f(x)dx = \int_a^c f(x)dx - \int_b^c f(x)dx$$

$$= \int_a^c f(x)dx + \int_c^b f(x)dx.$$

An analogous argument establishes the truth of the result for $c < a < b$.

Example. In Sec. 8.4 we saw that $\int_1^5 \left(-\frac{1}{2}x + 4\right)dx = 10$. According to Definition 8.11 we must have

$$\int_5^1 \left(-\frac{1}{2}x + 4\right)dx = -10.$$

We conclude this section with a result that will be quite useful when we develop a simple method for evaluating integrals.

8.12 Theorem. *IF f and g are functions, each continuous on* $[a,b]$*, and if* $k \in \mathcal{R}$*, THEN*

$$\text{i)} \quad \int_a^b [f(x) + g(x)] \, dx = \int_a^b f(x)dx + \int_a^b g(x)dx$$

and

$$\text{ii)} \quad \int_a^b k f(x)dx = k \int_a^b f(x)dx.$$

The content of this theorem is that the integral of a sum of functions is the sum of the integrals of the individual functions, and the integral of a constant times a function is the constant times the integral of the function. Since

$$\int_a^b [f(x) - g(x)] \, dx = \int_a^b [f(x) + (-1)g(x)] \, dx$$

$$= \int_a^b f(x)dx + \int_a^b (-1)g(x)dx$$

$$= \int_a^b f(x)dx + (-1) \int_a^b g(x)dx,$$

$$= \int_a^b f(x)dx - \int_a^b g(x)dx$$

we see that the integral of the difference of two functions is the integral of the first function minus the integral of the second.

Problem. Find $\int_0^1 \frac{7}{2}x^2 \, dx$.

Solution. We have $\int_0^1 \frac{7}{2}x^2 \, dx = \frac{7}{2} \int_0^1 x^2 \, dx$. We know that $\int_0^1 x^2 \, dx = \frac{1}{3}$ by results of the previous sections. Hence

$$\int_0^1 \frac{7}{2}x^2 \, dx = \frac{7}{2}\left(\frac{1}{3}\right) = \frac{7}{6}.$$

Problem. Find $\int_0^1 \left[\frac{4}{5}x^2 - \frac{5}{2}x\right] dx$, given that $\int_0^1 x \, dx = \frac{1}{2}$.

Solution. We have $\int_0^1 \left[\frac{4}{5}x^2 - \frac{5}{2}x\right] dx = \int_0^1 \frac{4}{5}x^2 \, dx - \int_0^1 \frac{5}{2}x \, dx =$

$$\frac{4}{5}\int_0^1 x^2 \, dx - \frac{5}{2}\int_0^1 x \, dx = \frac{4}{5}\left(\frac{1}{3}\right) - \frac{5}{2}\left(\frac{1}{2}\right) = \frac{4}{15} - \frac{5}{4} = -\frac{59}{60}.$$

Problem. Given that $\int_1^4 \sqrt{x} \, dx = \frac{14}{3}$ and $\int_1^4 \frac{1}{x^2} \, dx = \frac{3}{4}$, find

$$\int_1^4 \left[6\sqrt{x} + \frac{8}{3}\left(\frac{1}{x^2}\right)\right] dx.$$

Solution. $\int_1^4 \left[6\sqrt{x} + \frac{8}{3}\left(\frac{1}{x^2}\right)\right] dx = \int_1^4 6\sqrt{x} \, dx + \int_1^4 \frac{8}{3}\left(\frac{1}{x^2}\right) dx =$

$$6\int_1^4 \sqrt{x} \, dx + \frac{8}{3}\int_1^4 \frac{1}{x^2} \, dx = 6\left(\frac{14}{3}\right) + \frac{8}{3}\left(\frac{3}{4}\right) = 28 + 2 = 30.$$

Problem. Given that $\displaystyle\int_0^3 (x^2 + 1)dx = 12$ and $\displaystyle\int_3^4 (x^2 + 1)dx = \frac{40}{3}$, find $\displaystyle\int_0^4 (x^2 + 1)dx$.

Solution. By Theorem 8.6, we have

$$\int_0^4 (x^2 + 1)dx = \int_0^3 (x^2 + 1)dx + \int_3^4 (x^2 + 1)dx = 12 + \frac{40}{3} = \frac{76}{3}.$$

Exercise Set 8.5

1. Given

$$\int_0^1 x^2\, dx = \frac{1}{3}, \quad \int_1^2 x^2\, dx = \frac{7}{3}, \quad \int_0^1 1\,dx = 1, \text{ and} \quad \int_0^1 \sqrt{x}\,dx = \frac{2}{3},$$

find

a. $\displaystyle\int_0^2 x^2\, dx$ b. $\displaystyle\int_1^0 1\,dx$ c. $\displaystyle\int_0^1 (x^2 + 1)dx$ d. $\displaystyle\int_0^1 3\sqrt{x}\,dx$

e. $\displaystyle\int_1^2 -x^2\, dx$ f. $\displaystyle\int_0^1 \frac{9}{5}dx$ g. $\displaystyle\int_0^1 \sqrt{2x}\,dx$ h. $\displaystyle\int_0^1 (4\sqrt{x} - 7x^2)dx$

2. Let $f(x) = x^2 - 3x + 8$.
   a. Find the minimum value of $f(x)$ in $[0, 2]$.
   b. Find the maximum value of $f(x)$ in $[0, 2]$.
   c. Use Theorem 8.10 to bound $\displaystyle\int_0^2 (x^2 - 3x + 8)dx$.
   d. Interpret part c graphically.

* * *

In problems 3–12, evaluate the definite integrals using Theorem 8.12 and the following data:

$$\int_0^1 x^2\, dx = \frac{1}{3}, \quad \int_0^1 x^3\, dx = \frac{1}{4}, \quad \int_0^1 e^x\, dx = e - 1,$$

$$\int_0^1 1\,dx = 1, \text{ and} \quad \int_0^1 \sqrt{x}\,dx = \frac{2}{3}.$$

3. $\int_0^1 5x^3\,dx$

4. $\int_0^1 \sqrt{3}x^3\,dx$

5. $\int_0^1 (x^2 + 3x^3)\,dx$

6. $\int_0^1 (3x^2 - x^3)\,dx$

7. $\int_0^1 (e^x - 4)\,dx$

8. $\int_0^1 (2e^x - 5)\,dx$

9. $\int_0^1 \left(\frac{1}{2}x^3 - 3\sqrt{x}\right)dx$

10. $\int_0^1 \left(\frac{1}{4}x^2 - \frac{13}{2}\right)dx$

11. $\int_0^1 (10 + 2\sqrt{x})\,dx$

12. $\int_0^1 (x^3 + 2x^2 + 3e^x)\,dx$

In problems 13–18, use Theorem 8.10 to bound the given integral.

13. $\int_1^5 \frac{1}{x}\,dx$

14. $\int_1^5 -\frac{2}{x}\,dx$

15. $\int_{-2}^{-1} (x^2 + x - 5)\,dx$

16. $\int_{-3}^0 (x^2 + x - 5)\,dx$

17. $\int_{-2}^2 (x^3 + 1)\,dx$

18. $\int_{-2}^2 (x^3 - 3x + 5)\,dx$

19. Assuming that $\int_1^5 (x^2 + 1)\,dx = \frac{136}{3}$ and $\int_5^6 (x^2 + 1)\,dx = \frac{98}{3}$, find $\int_1^6 (x^2 + 1)\,dx$.

20. Assuming that $\int_1^2 \frac{1}{x}\,dx = \ln 2$ and $\int_2^3 \frac{1}{x}\,dx = \ln\frac{3}{2}$, find $\int_1^3 \frac{1}{x}\,dx$.

21. Assuming that $\int_{-2}^0 2x\,dx = -4$ and $\int_0^3 2x\,dx = 9$, find $\int_{-2}^3 2x\,dx$.

22. Assuming that $\int_{-2}^{0} 3x^2 \, dx = 8$ and $\int_{0}^{3} 3x^2 \, dx = 27$, find $\int_{-2}^{3} 3x^2 \, dx$.

8.6 ANTIDIFFERENTIATION AND THE FUNDAMENTAL THEOREM

In Chap. 6, consideration was given to the problem of finding the derivative of a function. Indeed, to each differentiable function F, it is possible to assign a second function F', the derivative of F, uniquely determined by F. The question, "Given a continuous function f, on $[a, b]$, is there a unique differentiable function F on $[a, b]$ such that $F'(x) = f(x)$ for each $x \in [a, b]$?", arises quite naturally. In essence, we are asking whether or not the operation of differentiation is reversible, and if so, does the reverse operation provide us with a *unique* function whose derivative is the original function.

If the functions F and f are related by

$$f(x) = F'(x) \quad \text{for } x \in [a, b],$$

we say that F is an *antiderivative* of f in $[a, b]$. For example, $F(x) = x^2 - 3x + 2$ is an antiderivative of $f(x) = 2x - 3$ because $F'(x) = 2x - 3 = f(x)$.

The question posed in the preceding paragraph may now be reformulated as "Does every continuous function on $[a, b]$ have a unique antiderivative on $[a, b]$?" The answer to this question lies in the following theorem, which really provides much more than we have anticipated:

8.13 Theorem. *IF f is continuous in* $[a, b]$ THEN *the function*

$$F(x) = \int_{a}^{x} f(t) \, dt$$

is an antiderivative of f for $x \in [a, b]$. That is,

$$\frac{d}{dx} \left[\int_{a}^{x} f(t) \, dt \right] = f(x) \text{ for } x \in [a, b].$$

This remarkable theorem not only tells us that continuous functions have antiderivatives, but also ties together the concepts of derivative and integral.

Problem. Find $F'(x)$ if:

(i) $F(x) = \int_{2}^{x} \frac{3t - 1}{t^2 + 4} \, dt$

(ii) $F(x) = \left[\displaystyle\int_{10}^{x} (u^3 + 2)du\right]\left[\displaystyle\int_{3}^{x} e^u du\right].$

Solution. (i) If $F(x) = \displaystyle\int_{2}^{x} \dfrac{3t - 1}{t^2 + 4} dt$, then according to Theorem 8.13,

$$F'(x) = \frac{3x - 1}{x^2 + 4}.$$

(ii) Here $F(x)$ is the product of two functions $G(x) = \displaystyle\int_{10}^{x} (u^3 + 2) \, du$ and $H(x) = \displaystyle\int_{3}^{x} e^u du.$

By the product rule we have

$$F'(x) = G(x) \, H'(x) + G'(x) \, H(x).$$

Now

$$H'(x) = \frac{d}{dx} \int_{3}^{x} e^u du = e^x$$

and

$$G'(x) = \frac{d}{dx} \int_{10}^{x} (u^3 + 2) \, du = x^3 + 2.$$

Thus

$$F'(x) = e^x \int_{10}^{x} (u^3 + 2) \, du + (x^3 + 2) \int_{3}^{x} e^u du.$$

Note that a continuous function f has many antiderivatives. Indeed, if F is any antiderivative of f, then $F + C$ is also an antiderivative of f for each $C \in \mathbb{R}$, since

$$\frac{d}{dx} [F(x) + C] = \frac{d}{dx} [F(x)] + \frac{d}{dx} [C] = F'(x) + 0 = F'(x).$$

It can be shown that the *only* way in which two antiderivatives of a continuous function can differ is by an additive constant. Thus if $F'(x) = f(x)$ and $G'(x) = f(x)$, then $F = G + C$ for some $C \in \mathbb{R}$.

The primary use of Theorem 8.13 is in evaluating integrals of continuous functions. If f is continuous on $[a, b]$ and if F is any antiderivative of f, then

$$F(x) = \int_a^x f(t)\, dt + C \text{ for all } x \in [a, b].$$

In particular,

$$F(a) = \int_a^a f(t)\, dt + C = 0 + C.$$

Hence, $C = F(a)$, and we have

$$F(x) = \int_a^x f(t)\, dt + F(a).$$

Letting $x = b$, we obtain

$$F(b) = \int_a^b f(t)\, dt + F(a)$$

or

$$\int_a^b f(t)\, dt = F(b) - F(a).$$

Observing that the letter "t" is merely a place-holder, we have established the following, known as the *Fundamental Theorem of Calculus*.

8.14 Theorem. *IF f is continuous on $[a, b]$ and F is any antiderivative of f, THEN*

$$\int_a^b f(x)\, dx = F(b) - F(a).$$

The difference $F(b) - F(a)$ is often denoted by $F(x)\Big|_a^b$.

Theorem 8.14 reduces the problem of finding the definite integral of a function f over an interval $[a, b]$ to that of finding a function whose derivative is f.

Example. In Sec. 8.3 we found that $\int_1^5 \left(-\frac{1}{2}x + 4\right) dx = 10$. Let us now evaluate this integral using the fundamental theorem. Our first task is that of finding an antiderivative for the function

$$f(x) = -\frac{1}{2}x + 4.$$

Looking first at the term $-\frac{1}{2}x$, we recall that the function $y = x^2$ has as its derivative

$$\frac{dy}{dx} = 2x.$$

If we modify this function by attaching the coefficient $-\frac{1}{4}$, we obtain

$$y = -\frac{1}{4}x^2$$

so that

$$\frac{dy}{dx} = -\frac{1}{4}(2x) = -\frac{1}{2}x.$$

Turning now to the constant term 4, we recall that the derivative of the function $y = x$ is $\frac{dy}{dx} = 1$. Letting $y = 4x$, we see that $\frac{dy}{dx} = 4$. Thus, the function

$$F(x) = -\frac{1}{4}x^2 + 4x$$

is an antiderivative of $f(x) = -\frac{1}{2}x + 4$. According to Theorem 8.14, we have

$$\int_1^5 \left(-\frac{1}{2}x + 4\right) dx = F(5) - F(1)$$

$$= \left[-\frac{1}{4}(5)^2 + 4(5)\right] - \left[-\frac{1}{4}(1)^2 + 4(1)\right]$$

$$= \frac{65}{4} - \frac{15}{4} = \frac{40}{4} = 10.$$

Example. We have previously seen that $\int_0^1 x^2 dx = \frac{1}{3}$. Let us again demonstrate this fact using Theorem 8.14. In attempting to find an antiderivative of $f(x) = x^2$, it is reasonable to consider the function $y = x^3$, since we know that differentiation reduces the exponent by one. For this function, we have $\frac{dy}{dx} = 3x^2$, which, except for the coefficient "3" arising in the differentiation process, fulfills our needs. If we let $F(x) = \frac{1}{3}x^3$, then $F'(x) = \frac{1}{3}(3)x^2 = x^2$, and F is an antiderivative of f. Invoking the fundamental theorem, we have

$$\int_0^1 x^2 dx = F(1) - F(0)$$

$$= \frac{1}{3}(1)^3 - \frac{1}{3}(0)^3 = \frac{1}{3}.$$

Since the process of integration is so closely linked to that of antidifferentia-tion, the latter operation has inherited notation similar to that employed in the former. If F is an antiderivative of f, we write

$$\int f(x)dx = F(x) + C$$

where C is an arbitrary constant. The symbol

$$\int f(x)dx$$

then represents the general antiderivative of f. Consistent with this notation, we often refer to the antiderivative of f as the *indefinite integral* of f.

We now list a number of formulas which will be useful in computing indefinite integrals:

8.15 Theorem.

(1) $\displaystyle\int [f(x) \pm g(x)] \, dx = \int f(x) \, dx \pm \int g(x) \, dx$

(2) $\displaystyle\int kf(x)dx = k \int f(x) \, dx \ (k \ a \ constant)$

(3) $\displaystyle\int 0 \, dx = C \ (C \ a \ constant)$

(4) $\displaystyle\int k \, dx = kx + C$

(5) $\displaystyle\int x \, dx = \frac{x^2}{2} + C$

(6) $\displaystyle\int x^n dx = \frac{x^{n+1}}{n+1} + C \ (n \ rational, n \neq -1)$

(7) $\displaystyle\int \frac{1}{x} dx = \ln x + C$

(8) $\displaystyle\int e^x dx = e^x + C.$

Problem. Evaluate the indefinite integral $\int (9x^6 - 3e^x + 4)\,dx$.

Solution. Parts (1) and (2) of Theorem 8.15 imply that

$$\int (9x^6 - 3e^x + 4)\,dx = 9\int x^6\,dx - 3\int e^x\,dx + \int 4\,dx.$$

According to part (6) of Theorem 8.15,

$$\int x^6\,dx = \frac{x^7}{7} + C_1,$$

by part (8) of the same theorem,

$$\int e^x\,dx = e^x + C_2,$$

and by part (4),

$$\int 4\,dx = 4x + C_3.$$

Combining these results, we obtain

$$\int (9x^6 - 3e^x + 4)\,dx = 9\left(\frac{x^7}{7} + C_1\right) - 3(e^x + C_2) + (4x + C_3)$$

$$= \frac{9}{7}x^7 - 3e^x + 4x + (9C_1 - 3C_2 + C_3)\cdot$$

$$= \frac{9}{7}x^7 - 3e^x + 4x + C,$$

where $C = 9C_1 - 3C_2 + C_3$.

Note that in the preceding example, since C_1, C_2, and C_3 are arbitrary, so is the constant C. Thus, instead of introducing a separate constant for each term arising in the antidifferentiation process, we need only add a single arbitrary constant at the end of the computation. For example, we will write

$$\int \left(\frac{1}{x} + e^x\right)dx = \int \frac{1}{x}\,dx + \int e^x\,dx$$

$$= \ln x + e^x + C$$

instead of

$$\int \left(\frac{1}{x} + e^x\right)dx = \int \frac{1}{x}\,dx + \int e^x\,dx$$

$$= \ln x + C_1 + e^x + C_2.$$

Problem. Evaluate the indefinite integral $\int \left(\frac{1}{x} - \frac{1}{\sqrt{x}} \right) dx$.

Solution. By part (1) of Theorem 8.15, we have

$$\int \left(\frac{1}{x} - \frac{1}{\sqrt{x}} \right) dx = \int \frac{1}{x} dx - \int \frac{1}{\sqrt{x}} dx$$

$$= \int \frac{1}{x} dx - \int x^{-\frac{1}{2}} dx$$

$$= \ln x - \frac{x^{-\frac{1}{2}+1}}{-\frac{1}{2}+1} + C$$

$$= \ln x - 2x^{\frac{1}{2}} + C$$

$$= \ln x - 2\sqrt{x} + C,$$

where C is an arbitrary constant.

Problem. Compute the definite integral $\int_0^2 (3x^2 - 2x + 7) dx$.

Solution. We first find an antiderivative of $f(x) = 3x^2 - 2x + 7$. Now

$$\int (3x^2 - 2x + 7) dx = 3 \int x^2 dx - 2 \int x \, dx + \int 7 \, dx$$

$$= 3 \left(\frac{x^3}{3} \right) - 2 \left(\frac{x^2}{2} \right) + 7x + C.$$

$$= x^3 - x^2 + 7x + C.$$

Letting $F(x) = x^3 - x^2 + 7x$, we have

$$\int_0^2 (3x^2 - 2x + 7) dx = F(x) \Big|_0^2 = F(2) - F(0)$$

$$= [2^3 - 2^2 + 7(2)] - [0^3 - 0^2 + 7(0)]$$

$$= 18 - 0 = 18.$$

Exercise Set 8.6

1. a. If $F(x) = 3x^2 - 5x + 2$, find $F'(x)$.
   b. Using the result of part (a), find an antiderivative of $f(x)$ if $f(x) = 6x - 5$.
   c. Find $\int (6x - 5) dx$.

2. a. If $F(x) = \sqrt{x}$, find $F'(x)$.

   b. Using the result of part (a), find an antiderivative of $f(x)$ if $f(x) = \frac{1}{2}x^{-\frac{1}{2}}$.

   c. Find $\int \frac{1}{2\sqrt{x}}\, dx$.

3. a. If $F(x) = e^{kx}$, k a constant, find $F'(x)$.
   b. Use the result of part (a) in order to find

   $$\int 8e^x dx, \int e^{-x} dx, \int e^{2x} dx, \text{ and } \int -3e^{2x} dx.$$

4. Use Theorem 8.15, part (6), to determine

   a. $\int x^3 dx$ b. $\int x^4 dx$ c. $\int x^{\frac{1}{3}} dx$ d. $\int x^{-2} dx$

   e. $\int \sqrt{x}\, dx$ f. $\int t^5 dt$ g. $\int y^{\frac{1}{5}} dy$ h. $\int z^{-4} dz$

5. Use the preceding results and the *Fundamental Theorem of Calculus* (Theorem 8.14) to evaluate

   a. $\int_1^4 (6x - 5) dx$ b. $\int_1^4 \frac{1}{2\sqrt{x}} dx$ c. $\int_0^1 e^{-x} dx$ d. $\int_{-1}^1 -3e^{2x} dx$

   e. $\int_{-1}^1 x^3 dx$ f. $\int_{-7}^{-2} x^{-2} dx$ g. $\int_{1/2}^{3/2} t^5 dt$ h. $\int_1^4 z^{-4} dz$

* * *

In problems 6–17, find the indicated antiderivative.

6. $\int 3x^5 dx$

7. $\int \frac{1}{3} x^3 dx$

8. $\int 7x^{\frac{1}{2}} dx$

9. $\int \frac{\sqrt{x}}{5} dx$

10. $\int \frac{1}{2} x^{-\frac{5}{3}} dx$

11. $\int \frac{6}{x^2} dx$

12. $\int \left(\frac{3}{4}x^3 - 2x^2 - \frac{1}{4} \right) dx$

13. $\int \left(\frac{2}{3}x^2 - \frac{2}{3}x + \frac{3}{2} \right) dx$

14. $\int \sqrt{5}\, dx$

15. $\int -2\, dx$

16. $\int x^{-1} dx$

17. $\int \frac{3}{7x} dx$

In problems 18–29, evaluate the indicated definite integral using the Fundamental Theorem of Calculus.

18. $\int_1^3 (3x^2 - x)dx$

19. $\int_{-2}^2 x^4\,dx$

20. $\int_{.1}^{.8} (-2x + 1)dx$

21. $\int_9^{16} \frac{1}{\sqrt{x}}dx$

22. $\int_1^e \frac{5}{x}dx$

23. $\int_0^4 3e^x dx$

24. $\int_2^8 x^{\frac{2}{3}}dx$

25. $\int_{-8}^{-1} x^{-\frac{2}{3}}dx$

26. $\int_1^e \left(\frac{x^3 - x}{x^2}\right) dx$

27. $\int_{1/2}^1 \frac{2x^4 - 7x^3}{x^8}dx$

28. $\int_0^{\ln 4} 2e^{-x}dx$

29. $\int_{-2}^{-1} \frac{e^x}{2}\,dx$

30. Sketch the graph of $f(x) = 2x^2 + 1$ for $x \in [-2, 2]$. Find the area under this curve from $x = -2$ to $x = 2$.

31. Without using Theorem 8.14, find $\int_{-3}^2 |x|\,dx$. (*Hint:* Sketch $y = |x|$ between $x = -3$ and $x = 2$, and interpret this integral as an area.)

32. (a) Show that the function $F(x) = e^{x^2} + 3$ is an antiderivative of $f(x) = 2xe^{x^2}$.

    (b) Find $\int_0^2 2xe^{x^2}\,dx$.

33. (a) Sketch the curves $f_1(x) = -x^2 + 1$ and $f_2(x) = x^2 - 1$ for $-1 \leqslant x \leqslant 1$.
    (b) Find the area of the region between the curves $y = f_1(x)$ and $y = f_2(x)$. (*Hint:* Consider the function $y = f_1(x) - f_2(x)$.)

In problems 34–41, use Theorem 8.13 to compute the indicated derivative.

34. $\frac{d}{dx} \int_4^x (t^2 - 3)dt$

35. $\frac{d}{dx} \int_{-10}^x t^3 e^t dt$

36. $\dfrac{d}{dx} \displaystyle\int_0^x (4t^3 - 3t^2 + t)\,dt$

37. $\dfrac{d}{dx} \displaystyle\int_5^x \dfrac{2w}{w^2 + 1}\,dw$

38. $\dfrac{d}{du} \displaystyle\int_1^u t \ln t\,dt$

39. $\dfrac{d}{dv} \displaystyle\int_0^v \dfrac{x^4 - 4}{x^4 + 4}\,dx$

40. $\dfrac{d}{dx} \left[x^3 \left(\displaystyle\int_1^x \ln t\,dt \right) \right]$

41. $\dfrac{d}{dw} \left[\left(\displaystyle\int_2^w \dfrac{1}{t}\,dt \right) \left(\displaystyle\int_3^w t^4\,dt \right) \right]$

42. Evaluate $\displaystyle\int_{-3}^2 F(x)\,dx$ where

$$F(x) = \begin{cases} x + 1, & \text{if } x \leqslant 0 \\ e^{-2x}, & \text{if } x > 0. \end{cases}$$

(*Hint:* $\displaystyle\int_{-3}^2 F(x)\,dx = \int_{-3}^0 F(x)\,dx + \int_0^2 F(x)\,dx.$)

43. Evaluate $\displaystyle\int_{-2}^6 g(t)\,dt$ if

$$g(t) = \begin{cases} -1, & \text{if } t \leqslant 0 \\ t - 1, & \text{if } 0 \leqslant t \leqslant 2 \\ \dfrac{t^2}{4}, & \text{if } t \leqslant 2. \end{cases}$$

8.7 INTEGRATION BY SUBSTITUTION

Having presented the fundamental theorem of calculus, we have effectively reduced the problem of integration to the problem of antidifferentiation. It is therefore essential that we develop skills in finding antiderivatives of commonly encountered functions. The formulas given in Theorem 8.15 provide us with a suitable foundation, in that they allow us to compute antiderivatives of some of the very basic functions encountered in our previous work. For example, using Theorem 8.15, we are able to integrate any polynomial function.

At this point in our study, however, there remain many functions which arise quite frequently, whose antiderivatives we are unable to compute. In this section, therefore, we shall introduce a general method of integration which will enable us to reduce the problem of integrating some seemingly complicated functions to the problem of integrating functions of the type considered in Theorem 8.15. This technique is known as the *change of variable method*, or the *substitution method*. Before we consider the details of this method, we introduce a new term closely related to the derivative of a function.

8.16 Definition. *Let the function $y = f(x)$ be differentiable at each point of the open interval (a, b), and let $x \in (a, b)$. Let the symbol dx denote any nonzero real number. The expression $f'(x)dx$ is the **differential** of y, and is denoted by the symbol dy, that is, $dy = f'(x)dx$.*

We observe that the differential of y is obtained by simply multiplying the derivative of $f(x)$ by dx. Formally, we have $\dfrac{dy}{dx} = u(x)$ if, and only if, $dy = u(x)dx$.

Example. Let $y = \sqrt{x^3 + 4}$. We may write this as

$$y = (x^3 + 4)^{\frac{1}{2}},$$

hence

$$\frac{dy}{dx} = \frac{3}{2}x^2(x^3 + 4)^{-\frac{1}{2}}.$$

The differential of y is, therefore,

$$dy = \frac{3}{2}x^2(x^3 + 4)^{-\frac{1}{2}}\,dx.$$

Turning now to the problem of integration, suppose we wish to determine

$$\int 2x\sqrt{x^2 + 4}\;dx.$$

The formulas of Theorem 8.15 cannot be applied directly to find this indefinite integral, but we will now show how the change of variable method can be used with Theorem 8.15 to find this integral. Our approach is to denote by some new letter, say u, the quantity $x^2 + 4$ appearing under the square-root sign. We have

$$u = x^2 + 4,$$

so that

$$\frac{du}{dx} = 2x.$$

We may rewrite the latter equation as

$$du = 2x\,dx.$$

We now substitute $u = x^2 + 4$ and $du = 2x\,dx$ into the original integral to obtain

$$\int 2x\sqrt{x^2 + 4}\;dx = \int \sqrt{u}\;du.$$

Note now that the latter integral can be determined using Theorem 8.15. In fact, we have

$$\int \sqrt{u}\, du = \int u^{1/2}\, du = \frac{u^{3/2}}{3/2} + C$$

$$= \frac{2}{3} u^{3/2} + C.$$

Recalling that $u = x^2 + 4$, we see that

$$\int 2x\sqrt{x^2 + 4}\, dx = \frac{2}{3}(x^2 + 4)^{3/2} + C.$$

In order to check our results, let us compute the derivative of $\frac{2}{3}(x^2 + 4)^{3/2} + C$ and be sure that it is $2x\sqrt{x^2 + 4}$. By the chain rule, we have

$$\frac{d}{dx}\left[\frac{2}{3}(x^2 + 4)^{3/2} + C\right]$$

$$= \frac{2}{3}\left(\frac{3}{2}\right)(x^2 + 4)^{1/2}(2x)$$

$$= 2x(x^2 + 4)^{1/2}$$

$$= 2x\sqrt{x^2 + 4}.$$

Problem. Find $\int x\sqrt{1 - x^2}\, dx$ by using substitution.

Solution. As in the preceding discussion, let $u = 1 - x^2$. Then $\frac{du}{dx} = -2x$, or $du = -2x\, dx$. We see that $x\, dx$ is present within the integral sign. We should, therefore, like to express this quantity in terms of u. Since $du = -2x\, dx$, it follows that $-\frac{1}{2} du = x dx$. Substituting, we obtain

$$\int x\sqrt{1 - x^2}\, dx = \int -\frac{1}{2}\sqrt{u}\, du$$

$$= -\frac{1}{2}\int u^{\frac{1}{2}}\, du = -\frac{1}{2}\frac{u^{\frac{3}{2}}}{3/2} + C = -\frac{1}{3}u^{\frac{3}{2}} + C.$$

Now $u = 1 - x^2$, so that

$$\int x\sqrt{1 - x^2}\, dx = -\frac{1}{3}(1 - x^2)^{3/2} + C.$$

Unfortunately, the method of substitution does not always allow us to compute the desired integral. For example, suppose we wish to compute $\int x \ln x \, dx$. We might try the substitution $u = \ln x$. This would result in $du = \frac{1}{x} dx$.

However, no factor of this type is present in the integral $\int x \ln x \, dx$. It appears, therefore, that the method of substitution might not be the best way to find this integral. In the next section we will develop another method of integration, different from the method of substitution, which will allow us to find integrals such as $\int x \ln x \, dx$.

Let us now attempt to extract the salient features of the substitution method in order that we may gain some insight into what is really involved. With regard to the integral

$$\int 2x \sqrt{x^2 + 4} \, dx$$

considered earlier, we made the substitution $u = u(x) = x^2 + 4$. The function

$$f(x) = \sqrt{x^2 + 4},$$

being a composite function, can be written as $f(u(x)) = \sqrt{u}$. The quantity $2x$ is recognized as being nothing more than $u'(x)$. Hence our integral takes the form

$$\int f(u(x)) \, u'(x) \, dx.$$

Note that $du = u'(x)dx$. Therefore, temporarily ignoring the variable x, the latter integral can be written as $\int f(u) \, du$, where, in this particular case, $f(u) = \sqrt{u}$.

In general, if an integral

$$\int f(x) \, dx$$

can be recognized as being of the form

$$\int f(u(x)) \, u'(x) \, dx,$$

then the method of substitution may be used to evaluate this integral. It should be noted that we are really using the chain rule in reverse fashion. The following theorem summarizes our observations.

8.17 Theorem. *IF* $\int f(u)\,du = F(u) + C$, *and* $u = u(x)$ *is a function differentiable on an interval I such that* $f(u(x))$ *is defined for* $x \in I$, *THEN*

$$\int_I f(u(x))u'(x)dx = F(u(x)) + C, \ (x \in I).$$

Proof. We need only show that the derivative of $F(u(x)) + C$ is $f(u(x))u'(x)$. By the chain rule,

$$\frac{d}{dx}\,[F(u(x)) + C] = \frac{d}{dx}\,[F(u(x))] + \frac{d}{dx}\,[C]$$

$$= F'(u)u'(x).$$

We are given that $\int f(u)\,du = F(u) + C$, so that

$$f(u) = F'(u).$$

Hence, $\dfrac{d}{dx}\,[F(u(x)) + C] = f(u(x))u'(x).$

Frequently, it is not apparent as to what substitution should be made in order to reduce a complicated integral to a less complicated one. Indeed, the task of recognizing a given integral as being of the type

$$\int f(u(x))u'(x)dx$$

is often the hardest part of the problem. The following problem illustrates certain integrals of frequent occurrence which are solvable by the change of variable method.

Problem. Determine the appropriate change of variable in each of the following integrals:

(i) $\displaystyle\int \frac{\ln x}{x}\,dx$

(ii) $\displaystyle\int x^2\, e^{x^3}\,dx$

(iii) $\displaystyle\int \frac{1}{x\sqrt{\ln x}}\,dx$

Solution. (i) If we write $\int \frac{\ln x}{x} dx$ as $\int \ln x \left(\frac{1}{x}\right) dx$, it is apparent that this integral is of the type $\int f(u(x))u'(x)dx$, where $f(u) = u, u(x) = \ln x$, and $du = \frac{1}{x} dx$. Hence, the substitution $u = \ln x$ is the one required.

(ii) Consider $\int x^2 e^{x^3} dx$. Here we observe that x^2 is "almost" the derivative. of x^3. It would seem, therefore, that the substitution $u = x^3$ is called for. In this case we find that $du = 3x^2 dx$, so that

$$x^2 dx = \frac{1}{3} du,$$

and

$$\int x^2 e^{x^3} dx = \frac{1}{3} \int e^{x^3} (3x^2)dx = \frac{1}{3} \int e^u du.$$

(iii) As in part (i), we write

$$\int \frac{1}{x\sqrt{\ln x}} dx$$

as

$$\int \frac{1}{\sqrt{\ln x}} \left(\frac{1}{x}\right) dx.$$

Letting $u(x) = \ln x$, we obtain

$$\int \frac{1}{\sqrt{\ln x}} \left(\frac{1}{x}\right) dx = \int \frac{1}{\sqrt{u}} du = \int u^{-\frac{1}{2}} du.$$

The reader should complete each of these examples by actually finding the indicated antiderivative.

In closing, we illustrate the manner in which substitution is used to evaluate definite integrals.

Problem. Evaluate the integral $\int_0^2 \frac{10x}{(5x^2 + 3)^2} dx$.

Solution. Writing $\int_0^2 \frac{10x}{(5x^2 + 3)^2} dx$ as $\int_0^2 \frac{1}{(5x^2 + 3)^2} (10x)dx$, it is clear that the substitution $u = 5x^2 + 3$ is required. From $u = 5x^2 + 3$, we obtain

$du = 10x \, dx$, so that

$$\int_0^2 \frac{10x}{(5x^2 + 3)^2} \, dx = \int_{x=0}^{x=2} \frac{1}{u^2} \, du.$$

Notice that the limits of integration are written as "$x = 0$" and "$x = 2$" rather than "0" and "2" as is our usual custom. The purpose of this notation is to emphasize that the numbers 0 and 2 are the correct limits of integration only with respect to the variable x and not with respect to the new variable u. Continuing, we have

$$\int_{x=0}^{x=2} \frac{1}{u^2} \, du = \int_{x=0}^{x=2} u^{-2} \, du = \frac{u^{-1}}{-1} \bigg|_{x=0}^{x=2} = -\frac{1}{u} \bigg|_{x=0}^{x=2}.$$

Recalling that $u = 5x^2 + 3$, we see that

$$-\frac{1}{u} \bigg|_{x=0}^{x=2} = -\frac{1}{5x^2 + 3} \bigg|_0^2 = \left[-\frac{1}{5(2)^2 + 3} \right] - \left[-\frac{1}{5(0)^2 + 3} \right]$$

$$= -\frac{1}{23} + \frac{1}{3} = \frac{20}{69}.$$

Hence, $\displaystyle\int_0^2 \frac{10x}{(5x^2 + 3)^2} \, dx = \frac{20}{69}$. This integral may be evaluated directly without converting from the variable u back to the variable x, provided we determine the appropriate limits of integration relative to u, which correspond to 0 and 2 as the limits of integration relative to x. Indeed, since $u = 5x^2 + 3$, we have

$$u = 3 \text{ when } x = 0$$

and

$$u = 23 \text{ when } x = 2.$$

Hence,

$$\int_0^2 \frac{10x}{(5x^2 + 3)^2} \, dx = \int_3^{23} u^{-2} \, du = -\frac{1}{u} \bigg|_3^{23} = -\frac{1}{23} + \frac{1}{3} = \frac{20}{69}.$$

Either of these methods is acceptable.

Exercise Set 8.7

1. Find the differential du if:
   a. $u = 3x - 2$ b. $u = x^2 + 5$ c. $u = \ln(x + 1)$

2. Use the substitution $u = 3x - 2$ and $du = 3dx$ to determine the following indefinite integrals:

a. $\displaystyle\int 3(3x - 2)^5 \, dx$ b. $\displaystyle\int 3\sqrt{3x - 2} \, dx$ c. $\displaystyle\int \frac{3}{3x - 2} \, dx$

d. $\displaystyle\int (3x - 2)^5 \, dx$ e. $\displaystyle\int \sqrt{3x - 2} \, dx$ f. $\displaystyle\int \frac{1}{3x - 2} \, dx$

3. Use the substitution $u = x^2 + 5$ and $du = 2x \, dx$ to determine the following indefinite integrals:

a. $\displaystyle\int 2x(x^2 + 5)^{-3} \, dx$ b. $\displaystyle\int 2x\sqrt{x^2 + 5} \, dx$ c. $\displaystyle\int \frac{2x}{x^2 + 5} \, dx$

d. $\displaystyle\int 3x(x^2 + 5)^{-3} \, dx$ e. $\displaystyle\int -\frac{x}{2} \sqrt{x^2 + 5} \, dx$ f. $\displaystyle\int \frac{x}{x^2 + 5} \, dx$

4. Use the substitution $u = \ln(x + 1)$ and $du = \dfrac{1}{x + 1} \, dx$ to determine the following indefinite integrals:

a. $\displaystyle\int \frac{\ln(x + 1)}{x + 1} \, dx$ b. $\displaystyle\int \frac{\sqrt{\ln(x + 1)}}{x + 1} \, dx$ c. $\displaystyle\int \frac{1}{(x + 1)[\ln(x + 1)]} \, dx$

5. Complete the evaluation of the following definite integrals:

a. $\displaystyle\int_0^1 3(3x - 2)^5 \, dx = \int_{x=0}^{x=1} u^5 \, du = \frac{u^6}{6} \Big|_{x=0}^{x=1} = \frac{(3x - 2)^6}{6} \Big|_0^1 =$

b. $\displaystyle\int_0^1 3(3x - 2)^5 \, dx = \int_{-2}^1 u^5 \, du = \frac{u^6}{6} \Big|_{-2}^1 =$

c. $\displaystyle\int_1^6 \sqrt{3x - 2} \, dx =$

d. $\displaystyle\int_0^5 \frac{x}{x^2 + 5} \, dx =$

* * *

In problems 6–17, use the method of substitution to determine the specified indefinite integrals.

6. $\displaystyle\int \frac{2x}{x^2 + 1} \, dx$ 7. $\displaystyle\int \frac{2x}{\sqrt{x^2 + 1}} \, dx$

8. $\displaystyle\int x^4 \sqrt{x^5 + 1} \, dx$ 9. $\displaystyle\int 3x^4 (x^5 + 1)^7 \, dx$

10. $\displaystyle\int \frac{3x^2}{\sqrt{x^3 + 10}}\, dx$

11. $\displaystyle\int x\,(2x^2 + 3)^{\frac{1}{3}}\, dx$

12. $\displaystyle\int xe^{-x^2}\, dx$

13. $\displaystyle\int xe^{x^2+3}\, dx$

14. $\displaystyle\int \frac{1}{x\,(\ln x)^2}\, dx$

15. $\displaystyle\int \frac{e^{2x}}{2 + e^{2x}}\, dx$

16. $\displaystyle\int (2x + 1)\,\sqrt{x^2 + x + 4}\, dx$

17. $\displaystyle\int \frac{6x^2 + 2x}{\sqrt{2x^3 + x^2 + 5}}\, dx$

In problems 18–29, evaluate the indicated definite integral.

18. $\displaystyle\int_{1}^{2} 3x^2\,(x^3 + 4)^2\, dx$

19. $\displaystyle\int_{0}^{1} (2x - 3)\,\sqrt{x^2 - 3x + 4}\, dx$

20. $\displaystyle\int_{e}^{2e} \frac{1}{x\,\ln x}\, dx$

21. $\displaystyle\int_{1}^{e} \frac{\ln x}{x}\, dx$

22. $\displaystyle\int_{0}^{2} 6x^2\,e^{x^3}\, dx$

23. $\displaystyle\int_{0}^{2} 6x^2\,e^{x^3+1}\, dx$

24. $\displaystyle\int_{0}^{1/2} \frac{2x}{\sqrt{1 - x^2}}\, dx$

25. $\displaystyle\int_{0}^{1} \frac{x}{\sqrt{4 - x^2}}\, dx$

26. $\displaystyle\int_{1}^{2} (1 - 4x)\,\sqrt{-2x^2 + x + 10}\, dx$

27. $\displaystyle\int_{-2}^{-1} (1 - 4x)\,\sqrt{-2x^2 + x + 10}\, dx$

28. $\displaystyle\int_{1}^{\sqrt{e}} \frac{\ln x^2}{x}\, dx$

29. $\displaystyle\int_{-1}^{1} \frac{x^2}{2x^3 + 8}\, dx$

8.8 INTEGRATION BY PARTS

The change of variable method provides us with a way of evaluating certain integrals involving composite functions. This method was seen to be a consequence of the chain rule for differentiation. We now develop another technique of integration which is based on the rule for finding the derivative of the product of two functions.

Recall that if $u = u(x)$ and $v = v(x)$ are differentiable functions, then

$$\frac{d}{dx}\,[uv] = u\frac{dv}{dx} + v\frac{du}{dx}$$

or in differential notation,

$$d(uv) = u\,dv + v\,du.$$

We may rewrite this equation as

$$u\,dv = d(uv) - v\,du.$$

We now integrate both sides to obtain

$$\int u\,dv = \int d(uv) - \int v\,du.$$

However, $d(uv) = \dfrac{d}{dx}[uv]\,dx$. Therefore,

$$\int d(uv) = \int \frac{d}{dx}[uv]\,dx$$

$$= uv.$$

Hence, $\displaystyle\int u\,dv = uv - \int v\,du.$

The constant C is not explicitly present in the final equation. Since each of the integrals remaining in this equation will also involve a constant of integration, we merely incorporate C into these constants.

The rather simple equation

$$\int u\,dv = uv - \int v\,du$$

provides us with a powerful tool for computing antiderivatives.

Consider, for example, the problem of finding

$$\int x \ln x \, dx.$$

If we let $u = \ln x$ and $dv = x\,dx$, then $du = \dfrac{1}{x}dx$, and $v = \dfrac{x^2}{2}$. Hence,

$$\int u\,dv = \int (\ln x)\,x\,dx,$$

and

$$uv - \int v\,du = \frac{x^2}{2}\ln x - \int \frac{x^2}{2}\left(\frac{1}{x}\,dx\right)$$

$$= \frac{x^2}{2}\ln x - \frac{1}{2}\int x\,dx$$

$$= \frac{x^2}{2} \ln x - \frac{1}{2} \left(\frac{x^2}{2} \right)$$

$$= \frac{x^2}{2} \left(\ln x - \frac{1}{2} \right).$$

Thus,

$$\int x \ln x \, dx = \frac{x^2}{2} \left(\ln x - \frac{1}{2} \right) + C.$$

Letting $F(x) = \frac{x^2}{2} \left(\ln x - \frac{1}{2} \right)$, we see that

$$F'(x) = \frac{x^2}{2} \left(\frac{1}{x} \right) + x \left(\ln x - \frac{1}{2} \right)$$

$$= \frac{x}{2} + x \ln x - \frac{x}{2} = x \ln x,$$

so the integration is correct.

The method provided by the equation $\int u \, dv = uv - \int v \, du$ is known as *integration by parts*. We reemphasize that this method comes about quite naturally as the reverse application of the product rule for differentiation. As we might expect, the difficult aspect of effective use of this tool lies in choosing u and dv.

Problem. Evaluate $\int x e^x dx$.

Solution. Here there are several possible ways of choosing u and dv. Indeed, we might let (i) $u = e^x$ and $dv = x \, dx$; or (ii) $u = x e^x$ and $dv = dx$; or (iii) $u = x$ and $dv = e^x \, dx$. Let us consider these possibilities in order.

(i) If $u = e^x$ and $dv = x \, dx$, then $du = e^x \, dx$ and $v = \int x \, dx = \frac{x^2}{2}$. Thus, $\int u \, dv = \int x e^x \, dx$ and $uv - \int v \, du = \frac{x^2}{2} e^x - \int \frac{x^2}{2} e^x \, dx$. Now the integral $\int \frac{x^2}{2} e^2 \, dx$ appears to be even more difficult to evaluate than the original integral $\int x e^x \, dx$. For this reason, we abandon this first possibility in favor of one of the others.

(ii) If $u = x e^x$ and $dv = dx$, then $du = (x e^x + e^x) \, dx$ and $v = x$. Hence,

$$\int u \, dv = \int x e^x \, dx$$

$$= uv - \int v \, du$$

$$= x^2 \, e^x - \int x(xe^x + e^x) \, dx.$$

Obviously, this approach is even less tractable than the first. We therefore move on to the final possibility.

(iii) If $u = x$ and $dv = e^x \, dx$, we obtain $du = dx$ and $v = e^x$. Thus,

$$\int u \, dv = \int xe^x \, dx$$

$$= uv - \int v \, du$$

$$= xe^x - \int e^x \, dx$$

$$= xe^x - e^x + C.$$

Apparently this is the best choice of u and dv.

The preceding problem suggests that we should be prepared to make many in-correct decisions in attempting to master the technique of integration by parts.

Problem. Evaluate the integral $\int_0^1 x^2 \, e^x \, dx$.

Solution. Let $u = x^2$ and $dv = e^x \, dx$. Thus, $du = 2x \, dx$ and $v = e^x$. Hence,

$$\int x^2 \, e^x \, dx = x^2 \, e^x - 2 \int xe^x \, dx.$$

We must perform a second integration by parts to evaluate $\int xe^x \, dx$. This has already been done, however, in the preceding problem. Hence, we have

$$\int_0^1 x^2 \, e^x \, dx = x^2 \, e^x \, \bigg|_0^1 - 2 \, [xe^x - e^x] \, \bigg|_0^1$$

$$= 1^2 \cdot e^1 - 0^2 \cdot e^0 - 2 \, [(1e^1 - e^1) - (0e^0 - e^0)]$$

$$= -e - 2.$$

Problem. Evaluate $\int_1^e \ln x \, dx$.

Solution. We first compute the antiderivative $\int \ln x\, dx$ using integration by parts. To do this, we observe that $\ln x\, dx = (\ln x)(1\, dx)$. Letting $u = \ln x$ and $dv = 1\, dx$, we have $du = \dfrac{1}{x}\, dx$, and $v = x$. Thus,

$$\int \ln x\, dx = \int u\, dv$$

$$= uv - \int v\, du$$

$$= x \ln x - \int x\left(\frac{1}{x}\, dx\right)$$

$$= x \ln x - \int dx$$

$$= x \ln x - x,$$

and

$$\int_{1}^{e} \ln x\, dx = (x \ln x - x)\,\Big|_{1}^{e}$$

$$= (e \ln e - e) - (1 \ln 1 - 1) = 1.$$

Exercise Set 8.8

1. Use integration by parts to find $\int xe^{-x}\, dx$.

   a. If $u = x$, find du.

   b. If $dv = e^{-x}\, dx$, find v.

   c. Find uv.

   d. Find $\int v\, du$.

   e. Find $uv - \int v\, du = \int xe^{-x}\, dx$.

   f. Check the answer by differentiation.

2. Use integration by parts to find $\int x \ln 3x\, dx$.

   a. If $u = \ln 3x$, find du.

   b. If $dv = x\, dx$, find v.

   c. Find uv.

  d. Find $\int v\,du$.

  e. Find $uv - \int v\,du = \int x \ln 3x\,dx$.

  f. Check the answer by differentiation.

3. Evaluate the definite integral, $\displaystyle\int_0^5 x\sqrt{x+4}\,dx$.

  a. If $u = x$, find du.

  b. If $dv = (x+4)^{\frac{1}{2}}\,dx$, find v.

  c. Find uv.

  d. Find $\int v\,du$.

  e. Find $uv\,\Big|_0^5 - \displaystyle\int_0^5 v\,du = \int_0^5 x\sqrt{x+4}\,dx$.

4. Use integration by parts to find $\int (\ln t)^2\,dt$.

  a. If $u = (\ln t)^2$, find du.
  b. If $dv = dt$, find v.
  c. Find uv.

  d. Use a second integration by parts to find $\int v\,du$.

  e. Find $uv - \int v\,du = \int (\ln t)^2\,dt$.

  f. Check the answer by differentiation.

* * *

In problems 5–14, compute the indicated indefinite integrals.

5. $\displaystyle\int xe^{3x}\,dx$

6. $\displaystyle\int 2x\,e^{-x}\,dx$

7. $\displaystyle\int x^3 \ln x\,dx$

8. $\displaystyle\int x^3 (\ln x)^2\,dx$

9. $\displaystyle\int (x\,e^x)^2\,dx$

10. $\displaystyle\int \sqrt{x}\ln x\,dx$

11. $\displaystyle\int x^2 \sqrt{x}\ln x\,dx$

12. $\displaystyle\int \frac{\ln x}{x^2}\,dx$

13. $\int x\sqrt{x+1}\,dx$ 14. $\int x^2\sqrt{x-4}\,dx$

In problems 15-20, evaluate the given definite integral.

15. $\int_{1}^{8} \sqrt[3]{x}\,\ln x\,dx$ 16. $\int_{1}^{2} x^5 \ln x\,dx$

17. $\int_{1}^{3} \frac{\ln x}{x^3}\,dx$ 18. $\int_{0}^{4} 2x\,e^{4x}\,dx$

19. $\int_{0}^{13} -xe^{-3x}\,dx$ 20. $\int_{0}^{3} x^2\,e^x\,dx$

8.9 IMPROPER INTEGRALS

For an integrable function f, we have provided meaning for the symbol $\int_{a}^{b} f(x)\,dx$. Moreover, if f is positive and continuous on $[a, b]$, then $\int_{a}^{b} f(x)\,dx$ represents the area between the curve $y = f(x)$ and the x-axis from $x = a$ to $x = b$. In applications of the integral to statistics and to other related disciplines, it often happens that one or both of the limits of integration is infinite. It is our purpose in this section to adequately define such integrals and to show how they may be evaluated.

8.18 Definition. *Let $a, b \in \mathbb{R}$.*

(i) *IF* $\lim\limits_{t\to\infty}\left[\int_{a}^{t} f(x)\,dx\right]$ *exists, THEN*

$$\int_{a}^{\infty} f(x)\,dx = \lim_{t\to\infty}\left[\int_{a}^{t} f(x)\,dx\right].$$

(ii) *IF* $\lim\limits_{t\to-\infty}\left[\int_{t}^{b} f(x)\,dx\right]$ *exists, THEN*

$$\int_{-\infty}^{b} f(x)\,dx = \lim_{t\to-\infty}\left[\int_{t}^{b} f(x)\,dx\right].$$

Both of the integrals $\int_{a}^{\infty} f(x)\,dx$ and $\int_{-\infty}^{b} f(x)\,dx$ are called *improper inte-*

grals because one of the limits of integration is infinite. If one of the limits of Definition 8.18 is a real number, then the improper integral defined by that limit is said to *converge*. Otherwise, the improper integral is said to *diverge*. These concepts are now illustrated by means of several solved problems.

Problem. Find $\int_0^\infty e^{-2x}\, dx$. Does $\int_0^\infty e^{-2x}\, dx$ converge?

Solution. We have

$$\int_0^t e^{-2x}\, dx = -\frac{1}{2} e^{-2x}\Big|_0^t = \frac{1}{2} - \frac{1}{2e^{2t}}.$$

Since $\lim_{t\to\infty} \dfrac{1}{2e^{2t}} = 0$, this yields $\lim_{t\to\infty} \left(\dfrac{1}{2} - \dfrac{1}{2e^{2t}}\right) = \dfrac{1}{2}$. Therefore $\int_0^\infty e^{-2x}dx$ converges and

$$\int_0^\infty e^{-2x}\, dx = \lim_{t\to\infty}\left[\int_0^t e^{-2x}\, dx\right]$$

$$= \lim_{t\to\infty}\left(\frac{1}{2} - \frac{1}{2e^{2t}}\right)$$

$$= \frac{1}{2}.$$

Figure 8.16 illustrates the geometric nature of the improper integral just evaluated. Notice that the integral

$$\int_0^\infty e^{-2x}\, dx$$

may be interpreted as an area just as in the case of the ordinary definite integral of a positive function. In this case, however, the region whose area is represented is of infinite extent. Namely, it is the area of the region bounded the curve $y = e^{-2x}$, the nonnegative x-axis, and the y-axis.

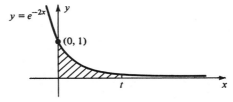

Figure 8.16

Problem. Evaluate the improper integral $\int_{-\infty}^{-1} \frac{1}{x^2} dx$.

Solution. Proceeding according to Definition 8.18, we have

$$\int_{-\infty}^{-1} x^{-2} dx = \lim_{t \to -\infty} \left[\int_t^{-1} x^{-2} dx \right]$$

$$= \lim_{t \to -\infty} \left[\left(-\frac{1}{x} \right) \Big|_t^{-1} \right]$$

$$= \lim_{t \to -\infty} \left[-\frac{1}{(-1)} - \left(-\frac{1}{(t)} \right) \right]$$

$$= \lim_{t \to -\infty} \left(1 + \frac{1}{t} \right).$$

As $|t|$ becomes large without bound, the fraction $\frac{1}{t}$ approaches zero. Thus,

$\lim_{t \to -\infty} \left(1 + \frac{1}{t} \right) = 1$, so that

$$\int_{-\infty}^{-1} \frac{1}{x^2} dx = \lim_{t \to -\infty} \left[\int_t^{-1} x^{-2} dx \right] = 1.$$

Problem. Show that the improper integral $\int_1^{\infty} \frac{1}{\sqrt{x}} dx$ diverges.

Solution. By definition,

$$\int_1^{\infty} \frac{1}{\sqrt{x}} dx = \lim_{t \to \infty} \left[\int_1^t \frac{1}{\sqrt{x}} dx \right] = \lim_{t \to \infty} \left[\int_1^t x^{-\frac{1}{2}} dx \right]$$

$$= \lim_{t \to \infty} \left[2x^{\frac{1}{2}} \Big|_1^t \right]$$

$$= \lim_{t \to \infty} (2\sqrt{t} - 2)$$

$$= \infty.$$

The improper integral $\int_1^{\infty} \frac{1}{\sqrt{x}} dx$ diverges because it is infinite.

We close this section by considering an improper integral in which both limits of integration are infinite.

8.19 Definition. *Let* $a \in \mathbb{R}$. *IF* $\displaystyle\int_{-\infty}^{a} f(x)\,dx$ *and* $\displaystyle\int_{a}^{\infty} f(x)\,dx$ *both converge*

THEN

$$\int_{-\infty}^{\infty} f(x)\,dx = \int_{-\infty}^{a} f(x)\,dx + \int_{a}^{\infty} f(x)\,dx.$$

Example. Consider the function $f(x) = e^{-|x|}$ whose graph is shown in Figure 8.17.

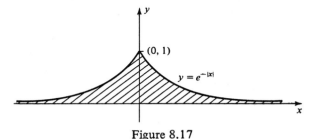

Figure 8.17

Suppose we wish to determine the area of the region bounded by this graph and the x-axis. A natural way to do this is to find the area of that portion of the region lying to the right of the y-axis and add this to the area of that portion of the region lying to the left of the y-axis. These areas are respectively

$$\int_{0}^{\infty} e^{-|x|}\,dx = \lim_{t \to \infty}\left[\int_{0}^{t} e^{-|x|}\,dx\right] = \lim_{t \to \infty}\left[\int_{0}^{t} e^{-x}\,dx\right]$$

$$= \lim_{t \to \infty}\left[-e^{-x}\ \Big|_{0}^{t}\ \right]$$

$$= \lim_{t \to \infty}(-e^{-t} + 1)$$

$$= 1$$

and

$$\int_{-\infty}^{0} e^{-|x|}\,dx = \lim_{t \to -\infty}\left[\int_{t}^{0} e^{-|x|}\,dx\right] = \lim_{t \to -\infty}\left[\int_{t}^{0} e^{x}\,dx\right]$$

$$= \lim_{t \to -\infty}\left[e^{x}\ \Big|_{t}^{0}\ \right]$$

$$= \lim_{t \to -\infty}(1 - e^{t})$$

$$= 1.$$

Thus, the area of the region bounded by the graph and the x-axis is

$$\int_{-\infty}^{\infty} e^{-|x|}\, dx = \int_{-\infty}^{0} e^{-|x|}\, dx + \int_{0}^{\infty} e^{-|x|}\, dx = 1 + 1 = 2.$$

If $\int_{-\infty}^{a} f(x)\, dx$ and $\int_{a}^{\infty} f(x)\, dx$ both converge, then, according to Definition

8.19, $\int_{-\infty}^{\infty} f(x)\, dx = \int_{-\infty}^{a} f(x)\, dx + \int_{a}^{\infty} f(x)\, dx$. Moreover, it can also be shown

that, if b is any other real number, then $\int_{-\infty}^{b} f(x)\, dx$ and $\int_{b}^{\infty} f(x)\, dx$ also con-

verge and $\int_{-\infty}^{\infty} f(x)\, dx = \int_{-\infty}^{b} f(x)\, dx + \int_{b}^{\infty} f(x)\, dx$. This means that the

number $\int_{-\infty}^{\infty} f(x)\, dx$ does not depend on the choice of the number a.

Exercise Set 8.9

1. a. Use the method of substitution to find $\int 2x\, e^{-x^2}\, dx$.

   b. Find $\int_{0}^{t} 2x\, e^{-x^2}\, dx$.

   c. Find $\lim\limits_{t \to \infty} \int_{0}^{t} 2x\, e^{-x^2}\, dx$.

   d. Find $\int_{0}^{\infty} 2x\, e^{-x^2}\, dx = \lim\limits_{t \to \infty} \int_{0}^{t} 2x\, e^{-x^2}\, dx$.

   e. Does $\int_{0}^{\infty} 2x\, e^{-x^2}\, dx$ converge?

2. a. Find $\int \dfrac{dx}{x^2}$.

   b. Find $\int_{1}^{t} \dfrac{dx}{x^2}$.

c. Find $\displaystyle\lim_{t\to\infty} \int_1^t \frac{dx}{x^2}$.

d. Find $\displaystyle\int_1^\infty \frac{dx}{x^2} = \lim_{t\to\infty} \int_1^t \frac{dx}{x^2}$.

e. Does $\displaystyle\int_1^\infty \frac{dx}{x^2}$ converge?

3. a. Use the method of substitution to find $\displaystyle\int \frac{2x}{x^2+1}\, dx$.

b. Find $\displaystyle\int_0^t \frac{2x}{x^2+1}\, dx$.

c. Find $\displaystyle\lim_{t\to\infty} \int_0^t \frac{2x}{x^2+1}\, dx$.

d. Find $\displaystyle\int_0^\infty \frac{2x}{x^2+1}\, dx = \lim_{t\to\infty} \int_0^t \frac{2x}{x^2+1}\, dx$.

e. Does $\displaystyle\int_0^\infty \frac{2x}{x^2+1}\, dx$ converge?

4. a. Find $\displaystyle\int \frac{e^x}{(e^x+1)^2}\, dx$.

b. Find $\displaystyle\int_0^t \frac{e^x}{(e^x+1)^2}\, dx$.

c. Find $\displaystyle\lim_{t\to\infty} \int_0^t \frac{e^x}{(e^x+1)^2}\, dx$.

d. Find $\displaystyle\int_t^0 \frac{e^x}{(e^x+1)^2}\, dx$.

e. Find $\displaystyle\lim_{t\to-\infty} \int_t^0 \frac{e^x}{(e^x+1)^2}\, dx$.

f. Find $\displaystyle\int_{-\infty}^\infty \frac{e^x}{(e^x+1)^2}\, dx$, if it exists.

* * *

In problems 5–21, evaluate the improper integral provided it converges.

Note: (i) $\lim\limits_{t\to\infty} \dfrac{t^n}{e^t} = 0$ for each positive integer n.

(ii) $\lim\limits_{t\to\infty} \dfrac{\ln t}{t^n} = 0$ for each positive integer n.

5. $\displaystyle\int_2^\infty e^{-3x}\, dx$

6. $\displaystyle\int_{-\infty}^{-1} e^{2x}\, dx$

7. $\displaystyle\int_{-\infty}^0 e^{5x}\, dx$

8. $\displaystyle\int_0^\infty e^{2x+1}\, dx$

9. $\displaystyle\int_1^\infty \frac{1}{x\sqrt{x}}\, dx$

10. $\displaystyle\int_{16}^\infty \frac{4}{\sqrt{x}}\, dx$

11. $\displaystyle\int_1^\infty \frac{\ln x}{x}\, dx$

12. $\displaystyle\int_2^\infty \frac{1}{x(\ln x)^2}\, dx$

13. $\displaystyle\int_0^\infty xe^{-x}\, dx$

14. $\displaystyle\int_{-\infty}^0 2x\, e^{x^2}\, dx$

15. $\displaystyle\int_{-\infty}^\infty \frac{x}{x^2+1}\, dx$

16. $\displaystyle\int_{-\infty}^\infty \frac{x}{(x^2+1)^2}\, dx$

17. $\displaystyle\int_1^\infty \frac{3x^2}{x^3+1}\, dx$

18. $\displaystyle\int_{-\infty}^{-10} \frac{2x+1}{x^2+x-9}\, dx$

19. $\displaystyle\int_1^\infty x^2\, e^{-x}\, dx$

20. $\displaystyle\int_1^\infty \frac{\ln x}{x^3}\, dx$

21. $\displaystyle\int_{-\infty}^\infty f(x)\, dx$ where $f(x) = \begin{cases} 0, & \text{if } x < 0 \\ 2e^{-2x}, & \text{if } x \geqslant 0 \end{cases}$

8.10 APPLICATIONS OF THE DEFINITE INTEGRAL

We are well aware of the fact that the definite integral of a positive continuous function f over an interval $[a, b]$ gives the area of the region under the curve $y = f(x)$ from a to b. What is the geometric situation, however, if f is negative over $[a, b]$? We first observe that each Riemann sum for f over $[a, b]$ must be nonpositive. Indeed, given a partition $P = \{a = x_0, x_1, x_2, \ldots, x_n = b\}$ of $[a, b]$, the quantity $\Delta x_k = x_k - x_{k-1}$ is positive for each k. Regardless of our choice of $z_k \in [x_{k-1}, x_k]$, we have $f(z_k) \leqslant 0$. Hence,

$$f(z_k) \Delta x_k \leqslant 0 \text{ for } k = 1, 2, \ldots, n,$$

so that

$$\sum_{k=1}^{n} f(z_k) \Delta x_k \leqslant 0.$$

It must then happen that $\displaystyle\int_a^b f(x)\,dx \leqslant 0$ because this definite integral is the limit of a sequence of nonpositive Riemann sums. Therefore, the latter integral does not represent an area. Observe, however, that if $f(x) \leqslant 0$ for all $x \in [a, b]$, then $-f(x) \geqslant 0$ for all $x \in [a, b]$. Hence

$$\int_a^b -f(x)\,dx = -\int_a^b f(x)\,dx \geqslant 0$$

does represent an area, the area between the curve $y = -f(x)$ and the x-axis from a to b.

Since the graph of f is symmetric to the graph of $-f$ with respect to the x-axis (see Figure 8.18), the area of the region R_1 *below* the curve $y = -f(x)$ (i.e., between the curve $y = -f(x)$ and the x-axis) from a to b and the area of the region R_2 *above* the curve $y = f(x)$ (i.e., between the x-axis and the curve $y = f(x)$) from a to b must be the same.

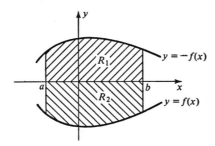

Figure 8.18

Since the area of R_1 is given by $-\displaystyle\int_a^b f(x)\,dx$, the area of R_2 is $-\displaystyle\int_a^b f(x)\,dx$.

Problem. Find the area of the region of the plane between the curve $y = -x\sqrt{x^2 + 5}$ and the x-axis from $x = 0$ to $x = 2$.

Solution. We begin by sketching the graph of the function $y = -x\sqrt{x^2 + 5}$ over the interval $[0, 2]$. This graph appears in Figure 8.19.

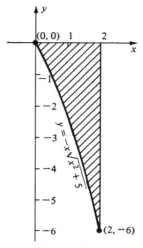

Figure 8.19

We see that $-x\sqrt{x^2 + 5} \leqslant 0$ for all $x \in [0, 2]$. Thus, the area of the region between the curve $y = -x\sqrt{x^2 + 5}$ and the x-axis from $x = 0$ to $x = 2$ is given by

$$A = -\int_0^2 -x\sqrt{x^2 + 5}\, dx = \int_0^2 x\sqrt{x^2 + 5}\, dx.$$

We evaluate this integral using substitution. If $u = x^2 + 5$, then $du = 2x\, dx$, so that $\frac{1}{2}\, du = x\, dx$. For $x = 0$, we have $u = 5$; and for $x = 2$, we have $u = 9$. Consequently,

$$A = \int_0^2 x\sqrt{x^2 + 5}\, dx = \int_5^9 \frac{1}{2}\sqrt{u}\, du$$

$$= \frac{1}{2}\int_5^9 u^{\frac{1}{2}}\, du$$

$$= \frac{1}{2}\left.\frac{u^{\frac{3}{2}}}{3/2}\right|_5^9$$

$$= 9 - \frac{5}{3}\sqrt{5}.$$

Suppose now that f is continuous on $[a, b]$, but that f is neither positive nor negative over the entire interval $[a, b]$. To determine the area of the region between the curve $y = f(x)$ and the x-axis from $x = a$ to $x = b$, we must first determine those subintervals of $[a, b]$ over which f is positive and those subintervals of $[a, b]$ over which f is negative. Having done this, we may then find the area of the region in question by adding up the areas of the regions bounded below or above by those subintervals of $[a, b]$ over which f is positive or negative respectively. The following problem illustrates this procedure.

Problem. Find the area of the region between the curve $y = x^2 - 4$ and the x-axis from $x = -3$ to $x = 4$.

Solution. As in the preceding problem, we begin by sketching the graph of the function $y = x^2 - 4$ over the interval $[-3, 4]$ (see Figure 8.20).

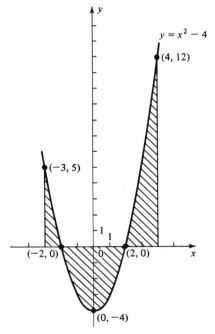

Figure 8.20

From this graph we see that f is positive on the intervals $[-3, -2]$ and $[2, 4]$; and that f is negative on the interval $[-2, 2]$. Therefore, the area A of the shaded portion of Figure 8.20 is given by

$$A = \int_{-3}^{-2} (x^2 - 4) \, dx - \int_{-2}^{2} (x^2 - 4) \, dx + \int_{2}^{4} (x^2 - 4) \, dx$$

$$= \left(\frac{x^3}{3} - 4x \right) \bigg|_{-3}^{-2} - \left(\frac{x^3}{3} - 4x \right) \bigg|_{-2}^{2} + \left(\frac{x^3}{3} - 4x \right) \bigg|_{2}^{4}$$

$$= 29.$$

Suppose now that f and g are nonnegative continuous functions on the closed interval $[a, b]$ and that $g(x) \leqslant f(x)$ for all $x \in [a, b]$. Let R_1 denote the region under the curve $y = f(x)$, let R_2 denote the region under the curve $y = g(x)$, and let R_3 denote the region between the curves $y = f(x)$ and $y = g(x)$, all from $x = a$ to $x = b$ (see Figure 8.21).

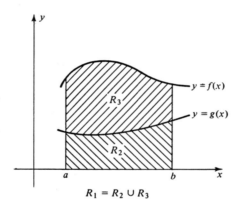

$$R_1 = R_2 \cup R_3$$

Figure 8.21

We take for granted the intuitively evident fact that

$$\text{Area}\,(R_3) = \text{Area}\,(R_1) - \text{Area}\,(R_2).$$

Since Area $(R_1) = \int_a^b f(x) \, dx$ and Area $(R_2) = \int_a^b g(x) \, dx$, we have

$$\text{Area}\,(R_3) = \int_a^b f(x) \, dx - \int_a^b g(x) \, dx$$

$$= \int_a^b [f(x) - g(x)] \, dx.$$

Problem. Find the points of intersection of the graphs of $f(x) = \frac{1}{2}x + 3$ and $g(x) = x^2$, and compute the area of the region between these two curves from the first point of intersection to the second.

Solution. The curves $y = f(x)$ and $y = g(x)$ will meet if and only if $x^2 = \frac{1}{2}x + 3$. This is equivalent to the quadratic equation $x^2 - \frac{1}{2}x - 3 = 0$, which has solutions $x = -\frac{3}{2}$ and $x = 2$. The graphs of the functions f and g over the interval $\left[-\frac{3}{2}, 2\right]$ appear in Figure 8.22.

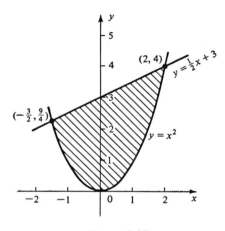

Figure 8.22

For $x \in \left[-\frac{3}{2}, 2\right]$ we have $g(x) \leqslant f(x)$, so that the area of the shaded region in Figure 8.22 is given by

$$A = \int_{-3/2}^{2} \left(\frac{1}{2}x + 3 - x^2\right) dx$$

$$= \left(\frac{1}{4}x^2 + 3x - \frac{1}{3}x^3\right) \Bigg|_{-3/2}^{2}$$

$$= \left[\frac{1}{4}(2)^2 + 3(2) - \frac{1}{3}(2)^3 \right] - \left[\frac{1}{4}\left(-\frac{3}{2}\right)^2 + 3\left(-\frac{3}{2}\right) - \frac{1}{3}\left(-\frac{3}{2}\right)^3 \right]$$

$$= \frac{13}{3} - \left(-\frac{45}{16}\right) = \frac{343}{48}.$$

We recall from Sec. 7.9 that the derivative of a function F at a point x represents the instantaneous rate of change of F at x. Suppose that, in a particular situation, we are given a function which represents the rate of change of some quantity F. That is, suppose we are given F'. Then the total change in F between $x = a$ and $x = b$ can be computed using the fundamental theorem of calculus. Namely, we have

$$F(b) - F(a) = \int_a^b F'(x)\, dx.$$

The usefulness of this observation is shown in the following solved problems.

Problem. R. Q. King, production engineer for the Allied Flange Corporation, found that the rate at which his company's product came off the assembly line over the first 216 man-hours of January was approximated by

$$R(t) = \frac{13}{2}\sqrt[3]{t} + \frac{5}{4},$$

where t represents time measured in man-hours. Assuming that production rate continues to be represented by this function, estimate the number of units that will be manufactured over the next 513 man-hours.

Solution. We denote by $N(t)$ the number of units produced up to time t. Thus $\dfrac{dN}{dt} = R(t) = \dfrac{13}{2}\sqrt[3]{t} + \dfrac{5}{4}$. Beginning our count of N at $t = 216$, and continuing over the next 513 man-hours, we see that the number of units manufactured in this time interval is

$$N(729) - N(216) = \int_{216}^{729} N'(t)\, dt$$

$$= \int_{216}^{729} \left[\frac{13}{2} t^{1/3} + \frac{5}{4} \right] dt$$

$$= \left[\frac{13}{2}\left(\frac{3}{4}\right) t^{4/3} + \frac{5}{4} t \right] \Bigg|_{216}^{729}$$

$$= \left[\frac{39}{8}(729)^{4/3} + \frac{5}{4}(729) \right] - \left[\frac{39}{8}(216)^{4/3} + \frac{5}{4}(216) \right]$$

$$\approx 32{,}896 - 6588 = 26{,}308 \text{ units.}$$

Problem. J. S. Finlay, consultant to the Las Vegas Chamber of Commerce, discovers that H. Ewes is buying up the city at a rate given by the function

$$f(t) = \frac{100}{3} e^{-\frac{1}{3}t},$$

where t represents time measured in months and r denotes the rate of change of the percentage P of the city owned at time t by Mr. Ewes. Making what seems to be a justified assumption, that Ewes' resources are unlimited, what percentage of Las Vegas will he own 3 years after his initial investment (i.e., $t = 0$) in the city?

Solution. The percentage of Las Vegas owned by Ewes at time t_0 is determined by the fact that

$$P(t_0) - P(0) = \int_0^{t_0} P'(t)\, dt = \int_0^{t_0} \frac{100}{3} e^{-\frac{1}{3}t}\, dt$$

$$= -100 e^{-\frac{1}{3}t} \Big|_0^{t_0}$$

$$= 100 - 100 e^{-\frac{1}{3}t_0}.$$

Since $P(0) = 0$, we have

$$P(t_0) = 100 - 100 e^{-\frac{1}{3}t_0}.$$

In particular, three years after his initial investment, Ewes will own $P(36)\%$ of the city (3 years = 36 months). Now

$$P(36) = 100 - 100 e^{-\frac{1}{3}(36)} = 100 - 100 e^{-12} = 100 - \frac{100}{e^{12}}.$$

Now $e > 2$. Hence $e^{12} > 2^{12} = 4096$, so that $\dfrac{100}{e^{12}} < \dfrac{100}{4096} \approx .02\%$.

We see that Mr. Ewes will own in excess of 98% of the city.

In this section we have introduced a few applications of the integral. The principal use was in finding areas of plane regions. It should be emphasized that we have barely scratched the surface of the wealth of possible applications of integration. The uses of the integral in the physical sciences are limitless. But, more surprisingly, the integral can be just as basic a tool to the psychologist or the management scientist as it is to the physicist. Indeed, in business, the integral is widely used in solving inventory problems by means of marginal analysis. In sociology, the theoretical aspects of social change are analyzed through the use of integral calculus. In literally all quantitative fields the methods of probability and statistics are used, and the integral plays a central role in these two disciplines. Most importantly, however, the integral is indispensible as a tool for developing more sophisticated mathematical techniques.

Exercise Set 8.10

1. a. Sketch the graph of the function $f(x) = \dfrac{2x}{x^2 + 4}$ over the interval $\left[0, \dfrac{1}{2}\right]$.

   b. Write the definite integral which represents the area of the region between the curve $y = f(x)$ and the x-axis from $x = 0$ to $x = \dfrac{1}{2}$.

   c. Find the area of this region by evaluating the definite integral obtained in part b.

2. a. Sketch the graph of the function $f(x) = x^2 - x - 6$ over the interval $[0, 3]$.

   b. Write the definite integral which represents the area of the region between the curve $y = f(x)$ and the x-axis from $x = 0$ to $x = 3$.

   c. Find the area of this region by evaluating the definite integral obtained in part b.

3. a. Sketch the graph of the functions $f(x) = x$ and $g(x) = x^3$ over the interval $[-1, 1]$.

   b. Write the definite integral which represents the area of the region bounded by the two curves from $x = 0$ to $x = 1$. Evaluate this integral.

   c. Write the definite integral which represents the area of the region bounded by the two curves from $x = -1$ to $x = 0$. Evaluate this integral.

   d. Find the total area of the region between the two curves from $x = -1$ to $x = 1$.

4. a. Find the points of intersection of the graphs of $f(x) = x - x^2$ and $g(x) = -x$.

   b. Sketch the graphs of f and g between the points of intersection.

   c. Write the definite integral which represents the area of the region bounded by the two curves between the points of intersection.

   d. Find the area of this region.

5. Let t measure the time elapsed, in a swimming event, since the starting gun. If Mark Martin can swim at the rate of $8 - \dfrac{\sqrt{t}}{4}$ feet per second at time t, how far can he swim in 25 seconds?

* * *

In problems 6–14, sketch the graph of $y = f(x)$. Then find the area of the region between the x-axis and the curve $y = f(x)$ from $x = a$ to $x = b$.

6. $f(x) = x^2 - 1$ $a = -2$ $b = 3$

7. $f(x) = x^3$ $a = -2$ $b = 1$

8. $f(x) = 3 - x^2$ $a = 1$ $b = 3$

9. $f(x) = -x^2 - x - 2$ $a = -1$ $b = 1$

10. $f(x) = \dfrac{x}{x^2 + 1}$ $a = -2$ $b = 1$

11. $f(x) = \dfrac{\ln x}{x}$ $a = 1$ $b = 6$

12. $f(x) = x e^{-x^2}$ $a = 0$ $b = 2$

13. $f(x) = x e^{-2x}$ $a = 0$ $b = 2$

14. $f(x) = \begin{cases} x - 1, & \text{if } x \leqslant 0 \\ -1 - x, & \text{if } x > 0 \end{cases}$ $a = -2$ $b = 3$

In problems 15–20, find the points of intersection of the two curves $y = f(x)$ and $y = g(x)$. Then find the area of the region which is enclosed by the two curves.

15. $f(x) = x^2 - 1$ $g(x) = -x^2 + 1$

16. $f(x) = x^2$ $g(x) = \dfrac{1}{2}x + 2$

17. $f(x) = x^3$ $g(x) = x^2 \ (x \geqslant 0)$

18. $f(x) = |x|$ $g(x) = -x^2 + 2$

19. $f(x) = \dfrac{1}{x}$ $g(x) = -x + 3 \ (x > 0)$

20. $f(x) = |x| - 1$ $g(x) = \left| \dfrac{1}{2}x \right|$

21. The rate at which the percentage of a light sensitive chemical, in an experimental type of film, darkens when the film is exposed is given by $r(t) = t^{\frac{1}{2}}$, where t denotes time in seconds $(0 \leqslant t)$. Determine the percentage of the chemical which has darkened 16 seconds after exposure. When will all of the chemical have darkened?

22. The rate of production of Wendell buggy whips over the first 64 man-hours of 1910 was given by $r(t) = 6t^{\frac{1}{2}}$. How many buggy whips were produced over the next 36 man-hours (assume the same production rate)?

23. The average value of a continuous function f over the interval $[a, b]$ is defined by the rule

$$A(f) = \frac{1}{b-a} \int_a^b f(x)\, dx.$$

Find the average value of the function $f(x) = x - \ln x$ over the interval $[1, e]$.

24. Repeat problem 23 for the function $f(x) = \dfrac{x}{\sqrt[3]{x^2 - 3}}$ over the interval $[2, 6]$.

25. Find an approximate value for the sum

$$S = 1 + \frac{1}{\sqrt{2}} + \frac{1}{\sqrt{3}} + \cdots + \frac{1}{\sqrt{30}}.$$

$\left(\text{Hint: Let } f(x) = \dfrac{1}{\sqrt{x}} \text{ and then } S \approx \displaystyle\int_1^{30} f(x)\, dx\ .\right)$

26. Find an approximate value for the sum $T = 1^2 + 2^2 + \cdots + 10^2$ using the method suggested in problem 25. How does your approximation compare with the actual value of T?

27. A psychologist discovers that the length of time required for a laboratory animal to perform a certain task, having already performed this task $k - 1$ times, is given by

$$f(k) = \frac{A}{\sqrt[3]{k}},$$

where A is the amount of time required for the animal to perform the task the first time. If the animal requires 8 minutes to perform the task the first time, approximately how long will it take him to perform the task 27 times? (*Hint:* Apply the method of the preceding two problems.)

28. Approximate the average amount of time required for the laboratory animal in problem 27 to perform the task 64 times.

29. It is possible to show that the length of a curve C given by a continuously differentiable function $y = f(x)$ for $a \leqslant x \leqslant b$ may be found using the formula

$$L(C) = \int_a^b \sqrt{1 + (f'(x))^2}\, dx.$$

If $f(x) = 3x - 5$ then find the length of that portion of the graph of f for $2 \leqslant x \leqslant 4$. Since the graph of f in this case is a straight line, check your answer by using the distance formula.

8.11 THE TRAPEZOIDAL RULE

It sometimes happens that we are required to evaluate the definite integral of a function whose antiderivative is not known. In such a situation, the fundamental theorem of calculus is of little use, since its practical application requires an explicitly written antiderivative. We recall, however, that the integral of a continuous function f over a closed interval $[a, b]$ always exists, and is *defined* not in terms of antiderivatives of f, but rather, as the limit of a sequence of Riemann sums. This would indicate that the problem of finding the definite integral of such a function might best be solved by reverting back to the basic definition of the integral.

Let f be continuous on $[a, b]$ and let P_n be any sequence of partitions of $[a, b]$ whose norms converge to zero. Suppose that S_n and T_n are any Riemann sums corresponding to P_n. Thus, according to Definition 8.7,

$$\int_a^b f(x)dx = \lim_{n \to \infty} S_n = \lim_{n \to \infty} T_n.$$

It can then be shown that

$$\int_a^b f(x)dx = \lim_{n \to \infty} \frac{1}{2}(S_n + T_n).$$

This means that, for a fixed n,

$$\int_a^b f(x)dx \approx \frac{1}{2}(S_n + T_n).$$

In particular, let P_n be the even partition given by

$$P_n = \left\{ a, a + \frac{b-a}{n}, a + \frac{2(b-a)}{n}, \ldots, a + \frac{(n-1)(b-a)}{n}, b \right\}.$$

In this case $\Delta x_k = \dfrac{b-a}{n}, k = 1, \ldots, n.$ Consider the Riemann sums

$$S_n = \sum_{k=1}^n f(x_{k-1}) \Delta x_k = \sum_{k=1}^n f(x_{k-1}) \left(\frac{b-a}{n} \right)$$

and

$$T_n = \sum_{k=1}^n f(x_k) \Delta x_k = \sum_{k=1}^n f(x_k) \left(\frac{b-a}{n} \right).$$

We have

$$\frac{S_n + T_n}{2} = \frac{1}{2}\left[\sum_{k=1}^{n} f(x_{k-1})\left(\frac{b-a}{n}\right) + \sum_{k=1}^{n} f(x_k)\left(\frac{b-a}{n}\right)\right]$$

$$= \frac{b-a}{2n}\left[\sum_{k=1}^{n} f(x_{k-1}) + \sum_{k=1}^{n} f(x_k)\right]$$

$$= \frac{b-a}{2n}\left[f(x_0) + \sum_{k=2}^{n} f(x_{k-1}) + \sum_{k=1}^{n-1} f(x_k) + f(x_n)\right]$$

$$= \frac{b-a}{2n}\left[f(x_0) + \sum_{k=1}^{n-1} f(x_k) + \sum_{k=1}^{n-1} f(x_k) + f(x_n)\right]$$

$$= \frac{b-a}{2n}\left[f(x_0) + \sum_{k=1}^{n-1} 2f(x_k) + f(x_n)\right].$$

Thus

$$\int_a^b f(x)dx \approx \frac{b-a}{2n}\left[f(x_0) + 2f(x_1) + 2f(x_2) + \cdots + 2f(x_{n-1}) + f(x_n)\right].$$

The latter formula, known as the *trapezoidal rule*, provides us with an effective technique for approximating integrals whose exact values cannot be obtained.

The trapezoidal rule derives its name from the fact that the quantity

$$\frac{b-a}{2n}\left[f(x_0) + \sum_{k=1}^{n-1} 2f(x_k) + f(x_n)\right]$$

$$= \frac{f(x_0) + f(x_1)}{2}\,\Delta x + \frac{f(x_1) + f(x_2)}{2}\,\Delta x + \cdots + \frac{f(x_{n-1}) + f(x_n)}{2}\,\Delta x$$

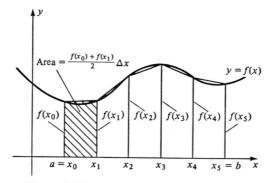

Figure 8.23. The trapezoidal rule for $n = 5$.

represents the sum of the areas of n trapezoids, each of thickness $\Delta x = \dfrac{b-a}{n}$ (see Figure 8.23).

Problem. Use the trapezoidal rule with $n = 4$ to approximate

$$\int_0^1 \frac{1}{1+x^2}\, dx.$$

Solution. The graph of $f(x) = \dfrac{1}{1+x^2}$ appears in Figure 8.24. In this example,

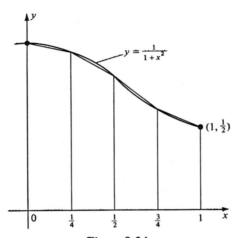

Figure 8.24

$a = 0$, $b = 1$, and $n = 4$. Hence $\dfrac{b-a}{2n} = \dfrac{1-0}{2(4)} = \dfrac{1}{8}$. The partition of $[0, 1]$ corre-

sponding to $n = 4$ is given by

$$P = \left\{0, \frac{1}{4}, \frac{1}{2}, \frac{3}{4}, 1\right\}.$$

We now gather in tabular form the information needed for application of the trapezoidal rule:

k	0	1	2	3	4
x_k	0	$\frac{1}{4}$	$\frac{1}{2}$	$\frac{3}{4}$	1
$f(x_k)$	1	$\frac{16}{17}$	$\frac{4}{5}$	$\frac{16}{25}$	$\frac{1}{2}$

We have

$$\int_0^1 \frac{1}{1+x^2}\, dx \approx \frac{1}{8}\left[1 + 2\left(\frac{16}{17}\right) + 2\left(\frac{4}{5}\right) + 2\left(\frac{16}{25}\right) + \frac{1}{2}\right]$$

$$= \frac{1}{8}\left[1 + \frac{32}{17} + \frac{8}{5} + \frac{32}{25} + \frac{1}{2}\right]$$

$$\approx \frac{1}{8}\left[1.00 + 1.88 + 1.60 + 1.28 + .50\right]$$

$$= \frac{1}{8}(6.26)$$

$$\approx 0.78.$$

Example. The function $f(x) = e^{-x^2}$ arises quite frequently in the study of statistics. Though this function is defined and continuous over the entire set $\mathcal{R}$, it is impossible to find an antiderivative for f among the functions we have studied. In order to integrate this function it is, therefore, necessary to use an approximative method. Let us estimate the integral

$$\int_0^2 e^{-x^2}\, dx$$

using the trapezoidal rule with $n = 8$. Here we have $a = 0$, and $b = 2$, so that $\frac{b-a}{2n} = \frac{2-0}{16} = \frac{1}{8}$. The partition of $[0, 2]$ for this example is

$$P = \left\{0, \frac{1}{4}, \frac{1}{2}, \frac{3}{4}, 1, \frac{5}{4}, \frac{3}{2}, \frac{7}{4}, 2\right\}.$$

In constructing the following table, the values of e^{-x^2} are obtained from the tables in Appendix I:

k	0	1	2	3	4	5	6	7	8
x_k	0	$\frac{1}{4}$	$\frac{1}{2}$	$\frac{3}{4}$	1	$\frac{5}{4}$	$\frac{3}{2}$	$\frac{7}{4}$	2
$f(x_k)$	1	.865	.780	.573	.367	.213	.107	.048	.019

We now have

$$\int_0^2 e^{-x^2}\, dx \approx \frac{1}{8}\left[1 + 2(.865) + 2(.780) + 2(.573)\right.$$

$$\left. + 2(.367) + 2(.213) + 2(.107) + 2(.048) + .019\right]$$

$$= \frac{1}{8} [1 + 1.730 + 1.560 + 1.140 + .734 + .426 + .214 + .096 + .019]$$

$$= \frac{1}{8} [6.919]$$

$$\approx .865.$$

Exercise Set 8.11

1. Let $f(x) = x^2$ from $x = 0$ to $x = 2$.
   a. If P is an even partition of $[0, 2]$ with $n = 4$, find x_0, x_1, x_2, x_3, and x_4.
      Find $\dfrac{b - a}{2n}$.
   b. Find $f(x_0), f(x_1), f(x_2), f(x_3)$, and $f(x_4)$.
   c. Find $\dfrac{b - a}{2n} [f(x_0) + 2f(x_1) + 2f(x_2) + 2f(x_3) + f(x_4)]$.

   d. The result of part (c) is the approximation of $\displaystyle\int_0^2 x^2 \, dx$ found by using

      the trapezoidal rule with $n = 4$. Find $\displaystyle\int_0^2 x^2 \, dx$ using the Fundamental

      Theorem of Calculus. Compare the results.

2. By the Fundamental Theorem of Calculus $\displaystyle\int_1^2 \frac{1}{x} \, dx = \ln 2$.

   a. Use the trapezoidal rule, with $n = 8$, to approximate $\displaystyle\int_1^2 \frac{1}{x} \, dx$.

   b. Use the result of part (a) to approximate $\ln 2$.
   c. Compare the value for $\ln 2$ obtained in part (b) with the value for $\ln 2$ listed in Appendix II.

3. Let $f(x) = \sqrt{1 - x^2}$ from $x = 0$ to $x = 0.8$
   a. If P is an even partition of $[0, 0.8]$ with $n = 4$, find x_0, x_1, x_2, x_3, and
      x_4. Find $\dfrac{b - a}{2n}$.
   b. Find $f(x_0), f(x_1), f(x_2), f(x_3)$, and $f(x_4)$.
   c. Find $\dfrac{b - a}{2n} [f(x_0) + 2f(x_1) + 2f(x_2) + 2f(x_3) + f(x_4)]$.
   d. If P is an even partition of $[0, 0.8]$ with $n = 8$, find x_k for $k = 0, 1, \ldots 8$.
      Find $\dfrac{b - a}{2n}$.
   e. Find $f(x_k)$ for $k = 0, 1, \ldots 8$.

f. Find $\dfrac{b-a}{2n}[f(x_0) + \displaystyle\sum_{k=1}^{7} 2f(x_k) + f(x_8)]$.

g. Approximate $\displaystyle\int_{0}^{0.8} \sqrt{1 - x^2}\, dx$.

* * *

In problems 4–9, use the trapezoidal rule for the given n to approximate each of the given integrals. Then evaluate the integral using the Fundamental Theorem of Calculus and compare your results.

4. $\displaystyle\int_{-1}^{2} 3x^3\, dx \qquad\qquad n = 12$

5. $\displaystyle\int_{1}^{4} (4x + 5)\, dx \qquad n = 4$

6. $\displaystyle\int_{0}^{2} \dfrac{2x}{x^2 + 1}\, dx \qquad n = 8$

7. $\displaystyle\int_{4}^{9} \dfrac{1}{\sqrt{x}}\, dx \qquad\qquad n = 10$

8. $\displaystyle\int_{0}^{1} x\sqrt{x^2 + 1}\, dx \qquad n = 4$

9. $\displaystyle\int_{-1}^{1} 2xe^{-x}\, dx \qquad\qquad n = 4$

10. Use the trapezoidal rule to approximate $\displaystyle\int_{0}^{1} \dfrac{1}{1 + x^2}\, dx$ with $n = 4$, $n = 6$, $n = 9$. Compare the three results.

11. Use the result of problem 10 to approximate π, given that $\displaystyle\int_{0}^{1} \dfrac{1}{1 + x^2}\, dx = \dfrac{\pi}{4}$.

12. Explain why the trapezoidal rule gives the *exact* value of the integral $\displaystyle\int_{a}^{b} (kx + 1)\, dx$, where $a, b, k \in \mathcal{R}$.

13. Approximate $\displaystyle\int_1^3 e^{-x^2}\, dx$.

14. Approximate $\displaystyle\int_1^{1.6} \sqrt{1 + x^3}\, dx$.

8.12 DIFFERENTIAL EQUATIONS

In this section we show how the integral may be used to solve some first-order differential equations. We then demonstrate a few relevant applications of differential equations.

By a *differential equation* we mean an equation involving a function and its derivatives. For notational simplicity, we agree to denote the derivative of a function $y = f(x)$ by y' as well as by the more customary $f'(x)$ or $\dfrac{dy}{dx}$.

Example. The equations

     (i) $y'' - 3xy' + x^2 y = e^x$,

     (ii) $(y'')^2 - 144\, y = 0$,

     (iii) $\dfrac{dy}{dx} + 2\, xy = 0$,

and   (iv) $f'(x) + 3x^2 f(x) - 9x = 0$

are examples of differential equations.

A function s is said to be a *solution* of a differential equation in an interval I, provided substitution of $s(x)$ for y, $s'(x)$ for y', $s''(x)$ for y'', etc., makes the equation true for all $x \in I$.

Example. The function of $s(x) = x^4$ is a solution of equation (ii) in the preceding example, for if $y = x^4$, then

$$(y'')^2 - 144y = (12x^2)^2 - 144x^4 = 0.$$

Similarly, the function of $g(x) = e^{-x^2}$ is a solution of equation (iii), for if $y = e^{-x^2}$, then

$$\frac{dy}{dx} + 2xy = -2xe^{-x^2} + 2xe^{-x^2} = 0.$$

We recall that the degree of a polynomial function is the highest power to which the variable is raised in the expression defining the function. An idea very similar to this gives us the order of a differential equation. A differential equation is of *order n* if *n* is the highest order derivative appearing in the equation. In the preceding examples, equations (i) and (ii) are of order 2, while equations (iii) and (iv) are of the first order. We shall restrict our attention to *first-order differential equations.*

The simplest type of first-order differential equations is one which can be solved by direct integration. Namely, the solution of the equation

$$y' = f(x)$$

is given by

$$y = \int f(x)\, dx.$$

Problem. Find the solution of the first order differential equation

$$y' = x^2 + e^x.$$

Solution. As we have observed, the solution can be found by "integrating both sides of the equation." Namely,

$$y = \int (x^2 + e^x)\, dx$$

$$= \int x^2\, dx + \int e^x\, dx$$

$$= \frac{1}{3}x^3 + e^x + C,$$

where C is an arbitrary constant.

Consider the preceding equation once again. For every constant C the function $y = \frac{1}{3}x^3 + e^x + C$ is a solution of the given differential equation. Moreover, if s is any solution of the given differential equation, then there is a constant C such that $s(x) = \frac{1}{3}x^3 + e^x + C$. In other words, as C varies, the functions $y = \frac{1}{3}x^3 + e^x + C$ yield all solutions of the given differential equation $y' = x^2 + e^x$. For this reason, $y = \frac{1}{3}x^3 + e^x + C$, where C is an arbitrary constant, is called the *general solution* of the differential equation $y' = x^2 + e^x$. If a particular value of C is chosen, the resulting solution is called a *particular solution* of the given differential equation. For example, if $C = -5$, then $y = \frac{1}{3}x^3 + e^x - 5$ is a particular solution of the differential equation $y' = x^2 + e^x$.

Problem. Find the particular solution of the differential equation

$$y' = x^2 + e^x$$

which has the property that $y = 2$ when $x = 0$.

Solution. The general solution is $y = \frac{1}{3}x^3 + e^x + C$. In addition, it is required that $y = 2$ when $x = 0$. This leads to the equation

$$2 = \frac{1}{3}0^3 + e^0 + C = 1 + C,$$

so $C = 1$. In other words, the particular solution

$$y = \frac{1}{3}x^3 + e^x + 1$$

satisfies the condition that $y = 2$ when $x = 0$.

Let us now consider a more general kind of differential equation.

8.20 Definition. *A first-order differential equation of the form*

$$f(x) + g(y)y' = 0$$

is a separable equation.

This equation is separable since it is possible to "separate" the variables x and y. If we rewrite y' as $\frac{dy}{dx}$, we obtain

$$f(x) + g(y)\frac{dy}{dx} = 0.$$

Multiplying both sides by dx gives

$$f(x)\, dx + g(y)dy = 0.$$

Formally integrating, we have

$$\int f(x)dx + \int g(y)dy = C.$$

If we write $F(x) = \int f(x)\, dx$ and $G(y) = \int g(y)\, dy$, then y satisfies the equation

$$F(x) + G(y) = C,$$

or

$$G(y) = C - F(x).$$

In many cases the latter equation can be solved to yield y as an explicit function of x. The method of solution of a separable differential equation is now illustrated by means of several solved problems.

Problem. Solve the differential equation

$$2y = 3xy'.$$

Solution. Writing $2y - 3xy' = 0$ and dividing by xy, we have

$$\frac{2}{x} - \frac{3}{y}y' = 0.$$

If we let $f(x) = \dfrac{2}{x}$ and $g(y) = -\dfrac{3}{y}$, then it is seen that the preceding equation is separable in the sense of Definition 8.20. The solution is

$$\int \frac{2}{x} dx + \int -\frac{3}{y} dy = C,$$

or

$$2 \ln x - 3 \ln y = C.$$

In order to put our answer in a more concise form, we write $C = -\ln K$, where K is a constant. Then

$$3 \ln y = \ln K + 2 \ln x$$
$$= \ln K + \ln x^2$$
$$= \ln (Kx^2).$$

Since $3 \ln y = \ln y^3$, we have

$$\ln y^3 = \ln (Kx^2)$$

so that

$$y^3 = Kx^2.$$

Therefore $y = (Kx^2)^{\frac{1}{3}} = (K)^{\frac{1}{3}} x^{\frac{2}{3}}$. Denoting $(K)^{\frac{1}{3}}$ by C_1 our solution is

$$y = C_1 x^{\frac{2}{3}}.$$

Problem. Find the particular solution of the equation

$$4x + \frac{2}{y}y' = 0$$

which has the property that $y = 3$ when $x = 0$.

Solution. If we let $f(x) = 4x$ and $g(y) = \dfrac{2}{y}$ in Definition 8.20, then the given equation is seen to be separable. The solution is

$$\int 4x \, dx + \int \frac{2}{y} \, dy = C,$$

or

$$2x^2 + 2 \ln y = C.$$

Therefore,

$$\ln y^2 = C - 2x^2$$

which yields

$$y^2 = e^{C - 2x^2} = e^C e^{-2x^2}.$$

Letting $K = \sqrt{e^C}$, we have

$$y = \pm K e^{-x^2}.$$

Since K itself is an arbitrary positive constant, it is unnecessary to carry along the $\pm$ signs. The general solution is $y = K e^{-x^2}$ where K is now completely arbitrary (for example, K could be negative). In addition, we want $y = 3$ when $x = 0$, or $3 = K e^0 = K$. Therefore, the particular solution is

$$y = 3 e^{-x^2}.$$

We now consider a second type of differential equation, the so-called linear equation.

8.21 Definition. *A first order differential equation of the form*

$$y' + f(x)y = g(x)$$

is a linear differential equation.

Example. The equations

    (i) $y' - 3xy = 2x^2$,

    (ii) $y' + x^2 y = 10$,

    (iii) $\dfrac{dy}{dx} + y = 6x$,

and (iv) $\dfrac{dy}{dx} - 3x^2 y = 2x + 1$

are examples of linear differential equations.

The linear differential equation is solved by means of an *integrating factor*. An integrating factor is a function by which we multiply the equation in order to reduce the problem of finding its solution to a problem in integration. The method is now illustrated:

Given $\dfrac{dy}{dx} + f(x)y = g(x)$, we multiply through by $e^{\int f(x)dx}$, obtaining

$$e^{\int f(x)dx} \cdot \frac{dy}{dx} + e^{\int f(x)dx} \cdot f(x) \cdot y = g(x)e^{\int f(x)dx}.$$

But the left-hand side of this equation is nothing more than

$$\frac{d}{dx}[y\, e^{\int f(x)dx}],$$

because, by the product rule for differentiation, we have

$$\frac{d}{dx}[y\, e^{\int f(x)dx}]$$

$$= y\frac{d}{dx}[e^{\int f(x)dx}] + e^{\int f(x)dx} \cdot \frac{dy}{dx}$$

$$= y\, e^{\int f(x)dx} \cdot \frac{d}{dx}\left[\int f(x)dx\right] + e^{\int f(x)dx}\frac{dy}{dx}$$

$$= y \cdot f(x) \cdot e^{\int f(x)dx} + e^{\int f(x)dx} \cdot \frac{dy}{dx},$$

which is the left-hand side of our equation. Thus, in place of

$$e^{\int f(x)dx} \cdot \frac{dy}{dx} + e^{\int f(x)dx} \cdot f(x) \cdot y = g(x)e^{\int f(x)dx},$$

we may write

$$\frac{d}{dx}[y\, e^{\int f(x)dx}] = g(x)\, e^{\int f(x)dx}.$$

Integrating, we obtain

$$y\, e^{\int f(x)dx} = \int [g(x)e^{\int f(x)dx}]\, dx + C,$$

or

$$y = e^{-\int f(x)dx}\left[\int [g(x)e^{\int f(x)dx}]\, dx + C\right].$$

Problem. Solve the differential equation

$$\frac{dy}{dx} + \left(\frac{2x+1}{x}\right) y = e^{-2x}.$$

Solution. We recognize this equation as being linear, with $f(x) = \dfrac{2x+1}{x}$ and $g(x) = e^{-2x}$. We must multiply the equation by

$$e^{\int f(x)dx} = e^{\int \frac{2x+1}{x}dx} = e^{2x + \ln x} = e^{2x} \cdot e^{\ln x} = e^{2x} \cdot x.$$

So we obtain

$$xe^{2x} \frac{dy}{dx} + (2x+1) e^{2x} y = x.$$

The left side of this equation is simply

$$\frac{d}{dx} [xe^{2x}y].$$

Thus, we have

$$\frac{d}{dx} [xe^{2x}y] = x,$$

so that

$$xe^{2x}y = \frac{x^2}{2} + C,$$

or

$$y = \frac{1}{2} x e^{-2x} + \frac{C}{x} e^{-2x}.$$

Problem. Find the particular solution of the equation

$$(x^2 + 2) \frac{dy}{dx} + 4xy = x$$

which has the property that $y = 1$ when $x = 0$.

Solution. At first glance, the differential equation

$$(x^2 + 2) \frac{dy}{dx} + 4xy = x$$

does not appear to be linear. However, division of both sides of the equation by $x^2 + 2$ gives

$$\frac{dy}{dx} + \frac{4x}{x^2 + 2} y = \frac{x}{x^2 + 2},$$

which is linear, with $f(x) = \dfrac{4x}{x^2 + 2}$ and $g(x) = \dfrac{x}{x^2 + 2}$. We see that

$$\int f(x)dx = \int \frac{4x}{x^2 + 2} dx$$

$$= 2 \int \frac{2x}{x^2 + 2} dx$$

$$= 2 \ln (x^2 + 2).$$

Hence, the integrating factor for this equation is

$$e^{\int f(x)dx} = e^{2 \ln (x^2 + 2)} = (x^2 + 2)^2.$$

Multiplying both sides of the latter differential equation by $(x^2 + 2)^2$, we have

$$(x^2 + 2)^2 \frac{dy}{dx} + 4x(x^2 + 2)y = x(x^2 + 2),$$

or

$$\frac{d}{dx} [(x^2 + 2)^2 y] = x^3 + 2x.$$

Integrating, we obtain

$$(x^2 + 2)^2 y = \frac{x^4}{4} + x^2 + C,$$

so that

$$y = (x^2 + 2)^{-2} \left(\frac{x^4}{4} + x^2 + C \right)$$

as the general solution. To find the particular solution in question, we substitute $x = 0$ and $y = 1$ into the general solution to obtain

$$1 = (0^2 + 2)^{-2} \left(\frac{0^2}{4} + 0^2 + C \right).$$

Hence, $C = 4$ and

$$y = (x^2 + 2)^{-2} \left(\frac{x^4}{4} + x^2 + 4 \right).$$

Let us now see how first-order differential equations are applied. Often the rate at which some quantity changes with respect to time is known or easily determined.

Example. Suppose the rate at which the population of the United States changes is known to be proportional to the number of people who presently comprise this population. We may indicate this fact mathematically in the form of the equation

$$\frac{dP}{dt} = kP,$$

where P represents the population at time t, measured in years, and k is a constant of proportionality. An initial condition is obtained by observing that the population of the U.S. in 1950 (time $t = 0$) was about 180,000,000. Thus, we have derived the equations

$$\frac{dP}{dt} = kP$$

$$P(0) = 180,000,000$$

to describe the behavior of our population. Now the equation

$$\frac{dP}{dt} = kP$$

is separable, and is solved by writing

$$\frac{dP}{P} = k \, dt.$$

Integrating, we have $\ln P = kt + C$ hence

$$P = e^{kt+C} = e^C e^{kt} = Ke^{kt},$$

where $K = e^C$. Since $P(0) = 180,000,000$, we have

$$180,000,000 = P(0) = Ke^0 = K.$$

Hence,

$$P = (180,000,000) \, e^{kt}$$

is the equation describing the population of the U.S. at time t. There is, however, the undetermined constant k present in this equation. In order that we may assign a value to this constant, it is necessary that we have one additional piece of information. A quick check of Census Bureau records shows that the 1970 population (time $t = 20$) was 200,000,000. Hence

$$200,000,000 = (180,000,000) \, e^{20k}$$

so that

$$e^{20k} = \frac{2}{1.8} = 1.11.$$

Thus $20k = \ln(1.11)$ which gives $k = \dfrac{\ln(1.11)}{20}$. Therefore,

$$P = (180,000,000)\, e^{\frac{\ln(1.11)}{20}t}.$$

It is interesting to use the preceding equation to predict the population of the United States in the year 2000. Here we have $t = 50$, so that

$$P = (180,000,000)\, e^{\frac{\ln(1.11)50}{20}}$$

$$= (180,000,000)(1.11)^{\frac{5}{2}}.$$

Using logarithms, we see that

$$(1.11)^{\frac{5}{2}} = 1.3,$$

hence $P = 234$ million.

Problem. It is known that the rate at which the number of bacteria in a given culture increases is proportional to the number of bacteria present in the culture. The particular bacterium, Chilium-PC, is such that its number triples in 2 hours. If a culture of Chilium-PC is taken and initially contains 4000 bacteria, how many bacteria will be present 24 hours later?

Solution. Letting $x = x(t)$ denote the number of bacteria present at time t, the rate at which this population will increase is given by

$$\frac{dx}{dt} = kx$$

where k is a constant of proportionality. This is the same equation that we solved in the preceding example. So the solution is

$$x = Ke^{kt}.$$

At time $t = 0$, 4000 bacteria are present. Hence, $4000 = Ke^0 = K$, so that

$$x = 4000\, e^{kt}.$$

We use the fact that the number of Chilium-PC triples every 2 hours to evaluate the constant k. We see that

$$x(2) = 3(4000) = 12,000,$$

so that

$$12,000 = 4000\ e^{2k},$$

and $e^{2k} = 3$. Thus, $k = \dfrac{1}{2} \ln 3$, and the solution of our differential equation is

$$x = 4000\ e^{\frac{\ln 3}{2} t}$$

$$= 4000\ (3)^{\frac{t}{2}}.$$

At time $t = 24$, the number of bacterial present is $4000\ (3)^{12} = 2,125,764,000$.

Problem. Sociologist H. E. Silver, an authority on racial compatibility, found in a study of the Island of Wu that this island's population (100,000 people) in 1940 was comprised solely of Caucasians. During that year, people of Mongoloid descent began to come to the island at a rate of 3000 per year. At the same time, however, others began to move off the island. Moreover, Mr. Silver found that a Mongoloid was just as likely to move off of Wu as was a Caucasian, and that the net effect of the input and output of people was that the island's population remained constant.

    (i) How many Mongoloids were present on Wu in 1970?
    (ii) Will the Mongoloids ever be sole inhabitants of the island?

Solution. Let x denote the number of Mongoloids on Wu at time t. The rate at which this number is changing is given by

$$\frac{dx}{dt} = (\text{rate of entry}) - (\text{rate of exit}).$$

Now the rate at which Mongoloids are entering Wu is 3000 per year. The rate at which they leave at time t is somewhat more difficult to ascertain. Since the total population remains constant, an input of 3000 Mongoloids will force an output of 3000 persons. At time t, the likelihood that a Mongoloid will leave depends on the proportion of Mongoloids on the island at this time. This proportion is given by

$$\frac{x}{100,000}.$$

Consequently, the number of Mongoloids that leave at time t is given by $\dfrac{x}{100,000}$

$(3000) = \dfrac{3x}{100}$. Thus

$$\frac{dx}{dt} = 3000 - \frac{3}{100}x.$$

We also have the initial condition $x(0) = 0$, since there were no Mongoloids on Wu in 1940 ($t = 0$). The equation

$$\frac{dx}{dt} + \frac{3}{100} x = 3000$$

is seen to be linear, with $f(t) = \frac{3}{100}$.

An integrating factor is given by

$$e^{\int f(t) dt} = e^{\int \frac{3}{100} dt} = e^{\frac{3t}{100}}.$$

Multiplying through by $e^{\frac{3t}{100}}$, we have

$$e^{\frac{3t}{100}} \frac{dx}{dt} + \frac{3}{100} e^{\frac{3t}{100}} x = 3000 \, e^{\frac{3t}{100}}.$$

The left side of this equation is

$$\frac{d}{dt} \left[x \, e^{\frac{3t}{100}} \right].$$

Hence, we have

$$\frac{d}{dt} \left[x \, e^{\frac{3t}{100}} \right] = 3000 \, e^{\frac{3t}{100}},$$

so that

$$x e^{\frac{3t}{100}} = 3000 \int e^{\frac{3t}{100}} \, dt$$

$$= 3000 \left(\frac{100}{3} \right) \left(e^{\frac{3t}{100}} \right) + C$$

$$= 100,000 \, e^{\frac{3t}{100}} + C.$$

Solving for x, we obtain

$$x = 100,000 + C e^{-\frac{3t}{100}}.$$

We now use the initial condition $x(0) = 0$ to find that $0 = 100,000 + C$, or $C = -100,000$. The solution for x is given by

$$x = 100,000 - 100,000 \, e^{-\frac{3t}{100}}.$$

At time $t = 30$ (the year 1970), the number of Mongoloids on Wu was

$$100{,}000 - 100{,}000\, e^{-\frac{90}{100}}$$

$$= 100{,}000 \left(1 - e^{-\frac{9}{10}}\right)$$

$$\approx 100{,}000\,(1 - .41) = 59{,}000.$$

As time increases without bound, $e^{-\frac{3t}{100}}$ approaches 0. Hence x approaches 100,000 as t increases without bound so that eventually there will be few if any Caucasians left on Wu.

Exercise Set 8.12

1. a. Use direct integration to find the general solution of the differential equation $y' = x^{\frac{1}{2}}$.

   b. Find the particular solution of the differential equation $y' = x^{\frac{1}{2}}$ which has the property that $y = 1$ when $x = 1$.

2. a. Find the general solution of the separable differential equation $y\,dx + x\,dy = 2dy$.

   b. Find the particular solution of the differential equation $y\,dx + x\,dy = 2dy$, which has the property that $y = 1$ when $x = 0$.

3. a. Rewrite the differential equation $xy' - y = x^3\sqrt{1 - x^2}$ in the form $y' + f(x)y = g(x)$.

   b. Find the integrating factor $e^{\int f(x)dx}$.

   c. Multiply the differential equation, in its rewritten form, by the integrating factor.

   d. Find the general solution of the differential equation obtained in part (c).

   e. Find the particular solution of the differential equation $xy' - y = x^3\sqrt{1 - x^2}$, which has the property that $y = -1$ when $x = 1$.

* * *

In problems 4–9, find the general solution of the indicated differential equation.

4. $2x^2 - 4y^3 y' = 0$

5. $\dfrac{3}{x} + \dfrac{2y'}{y^2} = 0$

6. $(y')^2 = 81\,y$

7. $y' - 3x^2 y = 3x^2$

8. $y' + \dfrac{2x}{x^2 + 9} y = 2x$

9. $4xy' - 48x^3 y = 64x^3$

In problems 10–13, find the general solution of the given differential equation. Then find the particular solution which satisfies the stated condition.

10. $\begin{cases} x^3 - \sqrt{y}\, y' = 0 \\ y = 1 \text{ when } x = 0 \end{cases}$

11. $\begin{cases} y' - 2x^2\, y = x^2 \\ y = 10 \text{ when } x = 0 \end{cases}$

12. $\begin{cases} 3x^{-1} + 5y'y^{-3} = 0 \\ y = 4 \text{ when } x = 1 \end{cases}$

13. $\begin{cases} x^6 \dfrac{dy}{dx} + x^5 y = x + 1 \\ y = 0 \text{ when } x = 1 \end{cases}$

14. At time $t = 0$, a certain culture of bacteria numbers 2400. This strain of bacteria grows at a rate proportional to the number of bacteria present and quadruples its population every 9 hours. How many bacteria are present at time $t = 21$?

15. In 1950 the pheasant population in a certain state park was approximately 3200. By 1952 the population had increased to 3600. Without external interference the population increases at a rate proportional to the number present at any given time. What was the population in the year 1958?

16. The rate at which radioactive nuclei decay is proportional to the number of nuclei present in a given sample. In a certain sample 25% of the original number of nuclei have decayed over a period of 80 years. What percentage of the original number of nuclei will remain after 320 years?

17 In a chemical reaction a certain chemical (call it A) is converted to another chemical (say B). The rate at which A is converted to B is proportional to the amount of A present at any instant. Twenty percent of the original amount of A is converted in four minutes.
    i) What percent of A will have been converted in twelve minutes?
    ii) When will ninety percent of A have been converted to B?

18. It is known that the rate at which an object cools is proportional to the difference in the temperature of the object and the medium in which the object is situated. At time $t = 0$, a magnum of Asti Spumante whose temperature is $70°$ is placed in a refrigerator whose temperature is held constant at $0°$. After 10 minutes, the champagne has cooled to $64°$.

a) What is the temperature of the champagne 18 minutes after it was put into the refrigerator?

b) When will it be ready for drinking?

(*Hint:* The authors agree that sparkling wines are best appreciated at a temperature of $42°$.)

*19. A test shows that at time $t = 0$ the air in a room of volume 8000 cubic feet contains 0.2 percent carbon dioxide. At this time fresh air with a carbon dioxide content of 0.05 percent is pumped into the room at a rate of 2000 cubic feet per minute. Four minutes later the air in the room is re-tested. What is the carbon dioxide content at that time?

*20. A tropical fish enthusiast is attempting to regulate the salinity of his 50 gallon salt water tank. Initially, there are two pounds of salt dissolved in the water. A solution composed of .01 pounds of salt per gallon of water is introduced into the tank at a rate of 5 gallons per minute. The resulting mixture is continuously stirred and is allowed to flow out of the tank at the rate of 5 gallons per minute. How much salt is dissolved in the tank 8 minutes after this procedure begins?

9 | Matrices and Linear Equations

It happens that many problems arising in the real world can be mathematically formulated as a system of two linear equations in two unknowns. Some indication of this fact was given in Chap. 2. There we saw that the solution of such problems involved finding all pairs of numbers which simultaneously satisfied the two equations.

It is the purpose of this chapter to show how certain problems may be formulated as systems of three or more linear equations in three or more unknowns, and to extend the methods of Chap. 2 so that we may solve such systems. The following problems, which indicate the diverse ways in which systems of linear equations arise, serve to establish the tone of the subsequent discussion.

Problem. Acme Corporation manufactures three items: gears, pulleys, and chains. On June 10 the total number of items produced was 300. Twice the number of gears produced equaled five times the number of chains, while the number of gears produced was equal to the number of all other items produced. How many chains, gears, and pulleys were manufactured by Acme Corporation that day?

Solution. Let x denote the number of gears, y the number of pulleys, and z the number of chains produced on June 10. Since the total number of items produced was 300, we have

$$x + y + z = 300.$$

We also observe that

$$2x = 5z,$$

or

$$2x - 5z = 0.$$

Since the number of gears produced equaled the number of other items produced, we also have

$$x = y + z$$

or

$$x - y - z = 0.$$

Thus, the mathematical formulation of this problem is a system of three linear equations in three unknowns given by

$$\begin{cases} x + y + z = 300 \\ 2x \qquad - 5z = 0 \\ x - y - z = 0. \end{cases}$$

We will postpone, temporarily, the solution of this system.

Problem. Mr. Jones, a chemist for a certain pharmaceutical company, is required to mix a compound composed of three different substances, A, B, and C. He plans to prepare exactly 300 ounces of the mixture in which the number of ounces of substance C is two-fifths the number of ounces of substance A, and the number of ounces of substance A equals the combined number of ounces of B and C. How many ounces of each chemical should he use?

Solution. Let x, y, and z denote the number of ounces of substances A, B, and C, respectively. The total weight of the mixture is 300 ounces, so that

$$x + y + z = 300.$$

Since the amount of substance C is two-fifths the amount of substance A, we obtain

$$z = \frac{2}{5}x,$$

or

$$2x - 5z = 0.$$

We also observe that

$$x = y + z,$$

hence

$$x - y - z = 0.$$

Thus, the mathematical formulation of this problem is the system

$$\begin{cases} x + y + \ z = 300 \\ 2x \quad\ \ - 5z = 0 \\ x - y - \ z = 0, \end{cases}$$

which happens to be the same set of equations obtained in the first problem. Again, we temporarily postpone the solution of this system.

Problem. In a small community there are 300 registered voters. The number of independents is two and one-half times the number of Republicans. The number of Republicans and Democrats combined equals the number of independents. How many of the voters are Republicans, how many are Democrats, and how many are independents?

Solution. Since the total number of voters is 300, we have

$$x + y + z = 300,$$

where x denotes the number of independents, y the number of Democrats, and z the number of Republicans. The problem also states that $x = \dfrac{5}{2} z$, or

$$2x - 5z = 0.$$

Finally, we have

$$x = y + z,$$

or

$$x - y - z = 0.$$

Again, the system

$$\begin{cases} x + y + \ z = 300 \\ 2x \quad\ \ - 5z = 0 \\ x - y - \ z = 0 \end{cases}$$

arises as the mathematical formulation of the problem at hand. It happens that the solution for this particular system of equations is

$$x = 150,$$
$$y = 90,$$
$$z = 60.$$

Thus, in the first problem, 150 gears, 90 pulleys, and 60 chains were produced. In the second problem, our chemist should mix 150 ounces of substance A, 90 ounces of substance B, and 60 ounces of substance C. In the third problem, there are 150 independents, 90 Democrats, and 60 Republicans in our small community.

It is important to observe that the same system of equations arose in the formulation of each of the preceding problems. Indeed, there are many word problems having this set of equations as its mathematical description. We see, then, that the solving of the "abstract" system of linear equations will lead to the solution of a large collection of "practical" problems. Herein lies the real power of abstract mathematics. Namely, by approaching a problem in general terms, we solve not just one problem, but a host of mathematically related problems. The rest of this chapter is devoted to the solution of a general system of linear equations.

9.1 LINEAR EQUATIONS IN *n* UNKNOWNS

In Chap. 2 we defined a linear equation in two unknowns to be an equation of the type $ax + by = c$, where $a, b, c \in \mathcal{R}$. We could just as well have designated the two unknowns by x_1 and x_2 and the coefficients by a_1 and a_2. Then our equation would have had the form $a_1 x_1 + a_2 x_2 = c$. Using this subscripted notation, the following definition is a natural extension of the idea of a linear equation.

9.1 Definition. *Let n be a positive integer, and let $a_1, a_2, \ldots a_n, b \in \mathcal{R}$. An equation of the type*

$$a_1 x_1 + a_2 x_2 + \cdots + a_n x_n = b$$

is a linear equation in the n unknowns $x_1, x_2, \ldots x_n$.

Example. The equation

$$3w - 2x + 4y - 7z = 12$$

is a linear equation in four unknowns. Using subscript notation, this same equation may be written as

$$3x_1 - 2x_2 + 4x_3 - 7x_4 = 12.$$

Example. The equation

$$-\frac{5}{2}j + 820k - \frac{\sqrt{5}}{2}l + 6m - 13n = 9$$

is a linear equation in five unknowns. Using subscript notation, this same equation may be written as

$$-\frac{5}{2}t_1 + 820t_2 - \frac{\sqrt{5}}{2}t_3 + 6t_4 - 13t_5 = 9.$$

Having defined a single linear equation, we now want to consider a set of m linear equations in n unknowns. For purposes of formally defining a system of linear equations it is convenient to use a double subscript to denote the coefficients of the unknowns. The symbol a_{ij} will be used to designate the coefficient of the j-th unknown in the i-th equation.

9.2 Definition. *Let m and n be positive integers and let a_{ij}, $b_i \in \mathcal{R}$ for $i = 1, 2, \ldots m$ and $j = 1, 2, \ldots n$. The set*

$$\begin{cases} a_{11}x_1 + a_{12}x_2 + \cdots + a_{1n}x_n = b_1 \\ a_{21}x_1 + a_{22}x_2 + \cdots + a_{2n}x_n = b_2 \\ \quad \cdot \qquad \cdot \qquad\qquad \cdot \qquad \cdot \\ \quad \cdot \qquad \cdot \qquad\qquad\quad \cdot \quad\ \cdot \\ \quad \cdot \qquad \cdot \qquad\qquad\qquad \cdot \ \ \cdot \\ a_{m1}x_1 + a_{m2}x_2 + \cdots + a_{mn}x_n = b_m \end{cases}$$

is a system of m linear equations in n unknowns.

Example. The set

$$\begin{cases} 3w + 2x \quad\ - 4z = 8 \\ w + \ x - \ y + \frac{1}{2}z = -4 \\ w \quad\ + 3y + \ z = \frac{1}{2} \\ -13w - \frac{1}{2}x + \frac{1}{4}y + 6z = \sqrt{5} \end{cases}$$

is a system of 4 equations in 4 unknowns. Using subscript notation, this same system may be written as

$$\begin{cases} 3x_1 + 2x_2 + 0x_3 - 4x_4 = 8 \\ x_1 + \ x_2 - \ x_3 + \frac{1}{2}x_4 = -4 \\ x_1 + 0x_2 + 3x_3 + \ x_4 = \frac{1}{2} \\ -13x_1 - \frac{1}{2}x_2 + \frac{1}{4}x_3 + 6x_4 = \sqrt{5}. \end{cases}$$

9.3 Definition. *An ordered set of n real numbers, denoted by $(s_1, s_2, \ldots, s_n)$, is a solution of the system*

$$\begin{cases} a_{11}x_1 + a_{12}x_2 + \cdots + a_{1n}x_n = b_1 \\ a_{21}x_1 + a_{22}x_2 + \cdots + a_{2n}x_n = b_2 \\ \cdot \qquad \cdot \qquad \qquad \cdot \qquad \cdot \\ \cdot \qquad \cdot \qquad \qquad \cdot \qquad \cdot \\ \cdot \qquad \cdot \qquad \qquad \cdot \qquad \cdot \\ a_{m1}x_1 + a_{m2}x_2 + \cdots + a_{mn}x_n = b_m \end{cases}$$

if and only if the numbers $s_1, s_2, \ldots, s_n$, when substituted for $x_1, x_2, \ldots, x_n$ respectively, satisfy each equation of the system.

Example. The ordered set $(150, 90, 60)$ is a solution of the system

$$\begin{cases} x + y + z = 300 \\ 2x - 5z = 0 \\ x - y - z = 0 \end{cases}$$

since

$$\begin{cases} 150 + 90 + 60 = 300 \\ 2(150) - 5(60) = 0 \\ 150 - 90 - 60 = 0. \end{cases}$$

In the case of a system of two equations in two unknowns we recall that there could be one solution, no solution, or many solutions. Likewise, in the case of a system of m equations in n unknowns, there can be one solution, no solution, or many solutions.

9.4 Definition. *Two systems of equations are **equivalent** if they have exactly the same solutions.*

Example. The systems

$$\begin{cases} x_1 + x_2 + x_3 = 3 \\ x_1 - 2x_2 - x_3 = 4 \\ 2x_1 + x_2 + x_3 = 5 \end{cases} \quad \text{and} \quad \begin{cases} 2x_1 - 4x_2 - 2x_3 = 8 \\ 2x_1 + x_2 + x_3 = 5 \\ 3x_2 + 2x_3 = -1 \end{cases}$$

are equivalent, since it can be shown that the only solution of each of these systems is the ordered triple $(2, -3, 4)$.

We want to develop a method for solving systems of m equations in n unknowns by successively replacing a given system by an equivalent system of a more elementary nature. We shall employ a technique which closely resembles the method of elimination discussed in Chap. 2 for systems of two equations in two unknowns. But before proceeding to the general case, we examine a special type of system, whose solution is particularly easy to obtain.

Example. Consider the system of four equations in four unknowns

$$\left\{\begin{array}{rcl} 2x_1 + x_2 - 3x_3 - x_4 &=& 2 \\ 3x_2 + x_3 + 2x_4 &=& 6 \\ -2x_3 + x_4 &=& 0 \\ 4x_4 &=& -8. \end{array}\right.$$

The solution to this system of equations can be found by successive substitution. Starting with the fourth equation, we see that $x_4 = -2$. Substituting -2 for x_4 in the third equation, we have

$$-2x_3 + (-2) = 0$$

or

$$x_3 = -1.$$

Using these values of x_3 and x_4 in the second equation, we have

$$3x_2 + (-1) + 2(-2) = 6$$

or

$$x_3 = \frac{11}{3}.$$

Finally, from the first equation, we obtain

$$2x_1 + \frac{11}{3} - 3(-1) - (-2) = 2$$

or

$$x_1 = -\frac{10}{3}.$$

Therefore, the solution of the given system is the ordered set of numbers $\left(-\frac{10}{3}, \frac{11}{3}, -1, -2\right).$

A system of equations, such as the system of the preceding example, is said to be in triangular form.

9.5 Definition. *A system of m linear equations in n unknowns, n $\geqslant$ m, of the type*

$$\begin{cases} a_{11}x_1 + a_{12}x_2 + a_{13}x_3 + \cdots + a_{1n}x_n = b_1 \\ \qquad\quad a_{22}x_2 + a_{23}x_3 + \cdots + a_{2n}x_n = b_2 \\ \qquad\qquad\qquad\qquad \vdots \qquad\qquad\qquad \vdots \\ \qquad\qquad\qquad\qquad a_{mm}x_m + \cdots + a_{mn}x_n = b_m, \end{cases}$$

where $a_{kj} = 0$ if $j < k$ (k = 1, 2, . . . m), is in triangular form.

If a system of equations is in triangular form and $a_{11}, \ldots, a_{mm}$ are all non-zero numbers, then this system can always be solved using substitution, although the procedure may not be quite as simple as the previous example seems to indicate. Let us consider another problem, but in this case we will consider three equations in five unknowns.

Problem. Given the triangular system

$$\begin{cases} 2x_1 - 3x_2 + x_3 - x_4 + x_5 = 2 \\ \qquad\quad x_2 - 2x_3 + x_4 - 2x_5 = 0 \\ \qquad\qquad\quad 3x_3 + 3x_4 - x_5 = 0, \end{cases}$$

find the solutions.

Solution. Here we have fewer equations than unknowns, but we may proceed in the same manner as in the previous example. From the third equation we conclude that

$$x_3 = \frac{1}{3}(x_5 - 3x_4),$$

where x_4 and x_5 can be any real numbers. We now substitute this expression into the second equation, obtaining

$$x_2 - 2\left[\frac{1}{3}(x_5 - 3x_4)\right] + x_4 - 2x_5 = 0.$$

This equation simplifies to

$$x_2 = \frac{8}{3}x_5 - 3x_4.$$

Substituting the expressions for x_2 and x_3 into the first equation, we have

$$2x_1 - 3\left[\frac{8}{3}x_5 - 3x_4\right] + \frac{1}{3}(x_5 - 3x_4) - x_4 + x_5 = 2,$$

or

$$2x_1 - 8x_5 + 9x_4 + \frac{1}{3}x_5 - x_4 - x_4 + x_5 = 2.$$

This simplifies to

$$2x_1 - \frac{20}{3}x_5 + 7x_4 = 2,$$

so that

$$x_1 = 1 + \frac{10}{3}x_5 - \frac{7}{2}x_4.$$

The solution set of the given system consists of all ordered quintuples of real numbers of the form

$$\left(1 + \frac{10}{3}x_5 - \frac{7}{2}x_4, \frac{8}{3}x_5 - 3x_4, \frac{1}{3}x_5 - x_4, x_4, x_5\right),$$

where x_4 and x_5 may be any real numbers. If, for example, we let $x_4 = 1$ and $x_5 = 0$, then the ordered quintuple $\left(-\frac{5}{2}, -3, -1, 1, 0\right)$ is obtained. If we let $x_4 = 2$ and $x_5 = -3$, then the quintuple $(-16, -14, -3, 2, -3)$ is obtained. We see, in general, that each particular solution to the given system depends upon our choice of values for x_4 and x_5. For this reason, the system of equations studied in this example is called *dependent*.

Given a system of linear equations, the last two examples show that it would be very useful to be able to find an equivalent triangular system. In this case, the latter system can be solved if the coefficients $a_{11}, \ldots, a_{mm}$ are all nonzero numbers. The next section is devoted to that task.

Before closing, however, it should be noted that not every system of linear equations has a solution. For example, it follows from the work of Chap. 2 that the system

$$x_1 - 2x_2 = -4$$

$$-3x_1 + 6x_2 = 1$$

has no solution. In fact, the next section shows that an attempt to put this system in triangular form results in the equivalent system

$$x_1 - 2x_2 = -4$$

$$0x_1 + 0x_2 = 11$$

which clearly does not have a solution. In general, a system of m linear equations in n unknowns which does not have a solution is called *inconsistent*.

Exercise Set 9.1

1. Rewrite the following linear equations in subscripted form, using x_1, x_2, x_3, and x_4.

   a. $2w - 3x + 7y - z = 3$

   b. $\dfrac{4}{3}r - \dfrac{1}{3}s + \dfrac{7}{5}t + \dfrac{31}{2}u = 0$

   c. $\sqrt{2}m - 64n - \dfrac{p}{2} + (\sqrt{2} + 5)q = -\dfrac{1}{2}$

2. Consider the following system of linear equations:

$$\begin{cases} x_1 - x_2 + 2x_3 + \sqrt{5}x_4 = -4 \\ -\dfrac{x_1}{3} + \dfrac{x_2}{4} + \dfrac{2x_3}{3} - \quad x_4 = \sqrt{7} \\ x_1 \qquad\qquad + \quad x_4 = \dfrac{3}{8} \end{cases}$$

   a. Determine the coefficients $a_{11}, a_{34}, a_{23}, a_{32}$, and a_{21}.

   b. Determine the constants b_1, b_2, and b_3.

   c. Rewrite the system of linear equations in subscripted form, using t_1, t_2, t_3, and t_4.

3. Consider the following system of linear equations:

$$\begin{cases} 2x_1 - 2x_2 + \dfrac{1}{2}x_3 = 5 \\ -x_1 + 3x_2 - 2x_3 = 0 \\ \dfrac{3}{2}x_1 + \dfrac{2}{3}x_2 - \dfrac{17}{4}x_3 = 1 \end{cases}$$

   a. Is the ordered triple $(5, 3, 2)$ a solution of the system?

   b. Is the ordered triple $(3, 5, 2)$ a solution of the system?

   c. Is the ordered set $(0, 5, 3, 2)$ a solution of the system?

   d. Is the ordered pair $(5, 3)$ a solution of the system?

   e. Is the ordered triple $(3, 3, 10)$ a solution of the system?

4. Solve the following system of linear equations:

$$\begin{cases} x_1 + x_2 - x_3 = 15 \\ x_2 - x_3 = -1 \\ x_3 = 2 \end{cases}$$

* * *

In problems 5–14, solve the given system of equations.

5. $\begin{cases} t_1 + 5t_2 - 3t_3 = -9 \\ 5t_2 - t_3 = 8 \\ 2t_3 = 4 \end{cases}$

6. $\begin{cases} 3z_1 + z_2 + 5z_3 = 3 \\ 3z_2 - 2z_3 = -11 \\ 2z_3 = -4 \end{cases}$

7. $\begin{cases} -l_1 + 3l_2 - 5l_3 = -1 \\ \qquad 2l_2 + \ l_3 = 8 \\ \qquad\qquad l_3 = 7 \end{cases}$

8. $\begin{cases} 3x_1 + 3x_2 + 3x_3 = 1 \\ \qquad 3x_2 + 3x_3 = 1 \\ \qquad\qquad 3x_3 = 1 \end{cases}$

9. $\begin{cases} y_1 + y_2 - y_3 = \dfrac{1}{4} \\ \qquad y_2 - y_3 = \dfrac{1}{3} \\ \qquad\qquad y_3 = \dfrac{1}{2} \end{cases}$

10. $\begin{cases} t_1 - 3t_2 + \qquad t_3 = -\sqrt{2} \\ \qquad 2t_2 - \qquad t_3 = \sqrt{2} \\ \qquad\qquad \sqrt{2}\,t_3 = 1 \end{cases}$

11. $\begin{cases} -5x_1 - 2x_2 + \ x_3 = -100 \\ \qquad 8x_2 - \ 3x_3 = -1 \\ \qquad\qquad \dfrac{1}{2}x_3 = 4 \end{cases}$

12. $\begin{cases} \dfrac{4}{3}x + \quad y - \quad z = \dfrac{5}{8} \\ \quad -\dfrac{1}{2}y + \dfrac{3}{2}z = -\dfrac{1}{2} \\ \qquad\qquad \dfrac{2}{3}z = 1 \end{cases}$

13. $\begin{cases} 2w + 3x - 4y + \ z = 8 \\ \quad -2x + \ y - \ z = 2 \\ \qquad\qquad 3y + 5z = 0 \\ \qquad\qquad\qquad 5z = -2 \end{cases}$

14. $\begin{cases} -x_1 - \ x_2 - x_3 + 9x_4 = \dfrac{1}{2} \\ \qquad 4x_2 - x_3 + 5x_4 = \dfrac{1}{2} \\ \qquad\qquad -x_3 - \ x_4 = \dfrac{1}{2} \\ \qquad\qquad\qquad 3x_4 = \dfrac{1}{2} \end{cases}$

In problems 15–22, find the general solution of the given system of equations. Find a particular solution of the system of equations.

15. $\begin{cases} 3x_1 + 2x_2 - x_3 = 4 \\ \qquad x_2 + x_3 = 5 \end{cases}$

16. $\begin{cases} -2x_1 - 3x_2 + 5x_3 = -1 \\ \qquad -2x_2 + \ x_3 = 2 \end{cases}$

17. $\begin{cases} -t_1 - \ t_2 + t_3 = -8 \\ \qquad 3t_2 + t_3 = -6 \end{cases}$

18. $\begin{cases} 4t_1 + 6t_2 + \ t_3 = 0 \\ \qquad 4t_2 + 2t_3 = 3 \end{cases}$

19. $\begin{cases} m_1 + m_2 - m_3 - m_4 = -1 \\ \quad -m_2 + m_3 - m_4 = 2 \\ \qquad -m_3 + m_4 = 0 \end{cases}$

20. $\begin{cases} m_1 + 2m_2 - 2m_3 - \ m_4 = 2 \\ \quad -m_2 + 2m_3 + 2m_4 = 1 \\ \qquad -m_3 - \ m_4 = 1 \end{cases}$

21. $\begin{cases} t_1 + \ t_2 - t_3 + t_4 + 2t_5 = 6 \\ \qquad 2t_2 + t_3 + t_4 + \ t_5 = -5 \\ \qquad\qquad t_4 - \ t_5 = 0 \end{cases}$

22. $\begin{cases} -\dfrac{1}{2}x_1 + \dfrac{3}{2}x_2 + \dfrac{5}{2}x_3 - x_4 = 4 \\ \qquad \dfrac{2}{3}x_2 + \dfrac{5}{3}x_3 - x_4 = 1 \\ \qquad\qquad \dfrac{1}{4}x_3 - x_4 = 0 \end{cases}$

9.2 SOLUTIONS OF SYSTEMS OF LINEAR EQUATIONS

Let

$$\begin{cases} a_{11}x_1 + a_{12}x_2 + \cdots + a_{1n}x_n = b_1 \\ a_{21}x_1 + a_{22}x_2 + \cdots + a_{2n}x_n = b_2 \\ \qquad \vdots \qquad \qquad \vdots \qquad \qquad \vdots \\ a_{m1}x_1 + a_{m2}x_2 + \cdots + a_{mn}x_n = b_m \end{cases}$$

be a system of m linear equations in n unknowns. By an elementary operation on this system, we mean any one of the following:

(i) interchanging the positions of two equations within the system,
(ii) multiplying both sides of an equation in the system by a nonzero constant,

or

(iii) replacing the equation $a_{k1}x_1 + a_{k2}x_2 + \cdots a_{kn}x_n = b_k$ by the sum of itself and any other equation $a_{j1}x_1 + a_{j2} + \cdots + a_{jn} = b_j$ in the system.

Steps (ii) and (iii) are frequently combined in a single operation.

The importance of elementary operations lies in the fact that performance of an elementary operation on a system of linear equations yields an equivalent system. In view of this fact, we may successively perform several elementary operations on a system of equations, being assured that the resulting system will be equivalent to the original one. In particular, by successive applications of elementary operations, it is always possible to reduce a system of linear equations to triangular form. If the resulting system has a solution it may then be solved, as was done in Sec. 9.1, by using substitution.

Let us now examine some problems which will demonstrate this procedure. Keep in mind that the elementary operations will be applied in such a way as to obtain an equivalent system in triangular form.

Problem. Solve the system of linear equations

$$\begin{cases} 3x - 2y + z = 6 \\ x + 2y - 2z = 4 \\ 2x - 3y - z = 0. \end{cases}$$

Solution. To solve this system of equations, we shall apply elementary operations successively until we obtain an equivalent system which is in triangular form and then apply the method of substitution. Equivalent systems will be written on the left hand side of the page, the operation used to obtain the system on the right hand side of the page.

$$\begin{cases} 3x - 2y + z = 6 \\ x + 2y - 2z = 4 \\ 2x - 3y - z = 0 \end{cases}$$

This is the original system.

$$\begin{cases} x + 2y - 2z = 4 \\ 3x - 2y + z = 6 \\ 2x - 3y - z = 0 \end{cases}$$

The first and second equations were interchanged.

$$\begin{cases} x + 2y - 2z = 4 \\ - 8y + 7z = -6 \\ 2x - 3y - z = 0 \end{cases}$$

The second equation was replaced by the sum of itself and -3 times the first equation.

$$\begin{cases} x + 2y - 2z = 4 \\ - 8y + 7z = -6 \\ - 7y + 3z = -8 \end{cases}$$

The third equation was replaced by the sum of itself and -2 times the first equation.

$$\begin{cases} x + 2y - 2z = 4 \\ y - \dfrac{7}{8}z = \dfrac{3}{4} \\ - 7y + 3z = -8 \end{cases}$$

Both sides of the second equation were multiplied by $-\dfrac{1}{8}$.

$$\begin{cases} x + 2y - 2z = 4 \\ y - \dfrac{7}{8}z = \dfrac{3}{4} \\ - \dfrac{25}{8}z = -\dfrac{11}{4} \end{cases}$$

The third equation was replaced by the sum of itself and 7 times the second equation.

From the third equation, we find that

$$z = \frac{22}{25}.$$

Substituting this value into the second equation yields

$$y - \frac{7}{8}\left(\frac{22}{25}\right) = \frac{3}{4}$$

or

$$y = \frac{38}{25}.$$

Substituting into the first equation, we have

$$x + 2\left(\frac{38}{25}\right) - 2\left(\frac{22}{25}\right) = 4$$

or

$$x = \frac{68}{25}.$$

Therefore, the ordered triple $\left(\dfrac{68}{25}, \dfrac{38}{25}, \dfrac{22}{25}\right)$ is the solution of our system.

Example. Let us now consider the system of equations

$$\begin{cases} x + y + z = 300 \\ 2x \quad\;\; - 5z = 0 \\ x - y - z = 0 \end{cases}$$

which arose in the examples in the introduction to this chapter and show that the solution $(150, 90, 60)$ can be obtained by the method just illustrated.

$$\begin{cases} x + y + z = 300 \\ 2x \quad\;\; - 5z = 0 \\ x - y - z = 0 \end{cases}$$ This is the original system.

$$\begin{cases} x + y + z = 300 \\ x - y - z = 0 \\ 2x \quad\;\; - 5z = 0 \end{cases}$$ The second and third equations were inter-changed.

$$\begin{cases} x + y + z = 300 \\ \quad - 2y - 2z = -300 \\ 2x \quad\;\; - 5z = 0 \end{cases}$$ The second equation was replaced by the sum of itself and -1 times the first equation.

$$\begin{cases} x + y + z = 300 \\ \quad - 2y - 2z = -300 \\ \quad - 2y - 7z = -600 \end{cases}$$ The third equation was replaced by the sum of itself and -2 times the first equation.

$$\begin{cases} x + y + z = 300 \\ \quad - 2y - 2z = -300 \\ \quad\quad\;\; - 5z = -300 \end{cases}$$ The third equation was replaced by the sum of itself and -1 times the second equation.

From $-5z = -300$, we obtain $z = 60$. Substituting into the second equation, we have $-2y - 2(60) = -300$, or $y = 90$. Finally, from the first equation, we see that $x + 90 + 60 = 300$ or $x = 150$. Therefore the solution to the system of equations is the ordered triple $(150, 90, 60)$.

We now solve a system of three linear equations in four unknowns by using the same general procedure.

Problem. Solve the system

$$\begin{cases} x_1 - x_2 - 2x_3 + 2x_4 = 5 \\ 2x_1 - x_2 - 7x_3 + 5x_4 = 10 \\ 3x_1 - 3x_2 - 5x_3 + 8x_4 = 16. \end{cases}$$

Solution.

$$\begin{cases} x_1 - x_2 - 2x_3 + 2x_4 = 5 \\ 2x_1 - x_2 - 7x_3 + 5x_4 = 10 \\ 3x_1 - 3x_2 - 5x_3 + 8x_4 = 16 \end{cases}$$ This is the original system.

$$\begin{cases} x_1 - x_2 - 2x_3 + 2x_4 = 5 \\ x_2 - 3x_3 + x_4 = 0 \\ 3x_1 - 3x_2 - 5x_3 + 8x_4 = 16 \end{cases}$$

The second equation was replaced by the sum of itself and -2 times the first equation.

$$\begin{cases} x_1 - x_2 - 2x_3 + 2x_4 = 5 \\ x_2 - 3x_3 + x_4 = 0 \\ x_3 + 2x_4 = 1 \end{cases}$$

The third equation was replaced by the sum of itself and -3 times the first equation.

The system is now in triangular form and may be solved by substitution as in the preceding problems. From the third equation, we see that

$$x_3 = 1 - 2x_4 \quad \text{(where } x_4 \text{ is any real number)}.$$

We replace x_3 in the second equation by $1 - 2x_4$ and obtain

$$x_2 - 3(1 - 2x_4) + x_4 = 0$$

from which it follows that

$$x_2 = 3 - 7x_4.$$

Substituting $1 - 2x_4$ for x_3 and $3 - 7x_4$ for x_2 in the first equation, we have

$$x_1 - (3 - 7x_4) - 2(1 - 2x_4) + 2x_4 = 5$$

which reduces to

$$x_1 = 10 - 13x_4.$$

We conclude that the solution of this system of equations consists of all quadruples of the form

$$(10 - 13x_4, \quad 3 - 7x_4, \quad 1 - 2x_4, \quad x_4),$$

where x_4 is any real number. If, for example, $x_4 = 0$, then the quadruple $(10, 3, 1, 0)$ results. This is one particular solution of the system. If $x_4 = -2$, then $(36, 17, 5, -2)$ is the resulting solution. This is another particular solution of the system. Thus we see that this system of equations has many solutions.

We conclude this section by investigating a system of three linear equations in three unknowns which has no solution. This will become apparent as we attempt to solve this using elementary operations to obtain triangular form.

Problem. Attempt to solve the system of equations

$$\begin{cases} 4x - 2y + 5z = 3 \\ 2x + 3y - z = 2 \\ 2x - 5y + 6z = 4. \end{cases}$$

Solution. We follow the methods already discussed using the elementary operations to obtain an equivalent triangular system.

$$\begin{cases} 4x - 2y + 5z = 3 \\ 2x + 3y - z = 2 \\ 2x - 5y + 6z = 4 \end{cases}$$ This is the original system.

$$\begin{cases} 4x - 2y + 5z = 3 \\ -8y + 7z = -1 \\ 2x - 5y + 6z = 4 \end{cases}$$ The second equation was replaced by the sum of -2 times itself and the first equation.

$$\begin{cases} 4x - 2y + 5z = 3 \\ -8y + 7z = -1 \\ 8y - 7z = -5 \end{cases}$$ The third equation was replaced by the sum of -2 times itself and the first equation.

$$\begin{cases} 4x - 2y + 5z = 3 \\ -8y + 7z = -1 \\ 0 = -6 \end{cases}$$ The third equation was replaced by the sum of the second and third equations.

We see that this resulting system of equations has no solution since $0 \neq -6$. Therefore, the original system of equations has no solution, that is, the system of equations is inconsistent.

Exercise Set 9.2

1. Consider the equation $3x_1 - 2x_2 + \frac{5}{2}x_3 - \frac{3}{17}x_4 = 41$.

    a. Multiply the equation by: $2, -3, \frac{1}{3}, -\frac{5}{2}$, and 34.

    b. Are the resulting equations equivalent to the original equation?

    c. Multiply the given equation by 0.

    d. Is the resulting equation equivalent to the original equation?

2. Let E_1 be the equation $-\frac{1}{2}w + \frac{5}{3}x + 2y - \frac{9}{2}z = 1$ and E_2 be the equation

    $12w + 4x - 3y - \frac{1}{2}z = -5$.

    a. Add E_1 and E_2. Call this equation E_3.

    b. Is every solution of E_1 a solution of E_3?

    c. Is every solution of E_3 a solution of E_1?

    d. Is every solution of the system of equations $\begin{cases} E_1 \\ E_3 \end{cases}$ a solution of the system

    of equations $\begin{cases} E_1 \\ E_2 \end{cases}$?

    e. Find the sum of E_2 and 24 times E_1. Call this equation E_4.

    f. Is every solution of the system of equations $\begin{cases} E_1 \\ E_4 \end{cases}$ a solution of the system

    of equations $\begin{cases} E_1 \\ E_2 \end{cases}$?

3. Consider the following system of equations:
$$\begin{cases} x + 2y - z = 9 \\ 2x - y + z = -3 \\ -3x + 5y + 8z = -4 \end{cases}$$

a. Replace the second equation by the sum of itself and -2 times the first equation.

b. Replace the third equation by the sum of itself and 3 times the first equation.

c. Multiply the second equation by $-\dfrac{1}{5}$.

d. Replace the third equation by the sum of itself and -11 times the second equation.

e. The resulting system of linear equations is in triangular form. Solve this system by substitution.

f. What is the solution of the original system of equations?

* * *

In problems 4–19 solve (if possible) the given system of linear equations. To do this, apply successively elementary operations until an equivalent system in triangular form is obtained. If the system has no solution indicate this with the statement "the system of equations is inconsistent."

4. $\begin{cases} 2x + 3y - z = 6 \\ -x - y - z = -1 \\ 5x - 5y - 3z = 3 \end{cases}$

5. $\begin{cases} 2x + 3y - z = 9 \\ -x - y - z = -1 \\ 5x - 5y - 3z = 11 \end{cases}$

6. $\begin{cases} 2x_1 - 3x_2 + x_3 = 2 \\ x_1 + 5x_2 - x_3 = 7 \\ -6x_1 + 8x_2 - 4x_3 = -2 \end{cases}$

7. $\begin{cases} 2x_1 + x_2 - x_3 = -2 \\ 6x_1 - x_2 + x_3 = 6 \\ -x_1 + \dfrac{1}{2}x_2 + 3x_3 = 5 \end{cases}$

8. $\begin{cases} 3x - 2y + 7z = 4 \\ 5x + y - 2z = 1 \\ 2x + 3y - 9z = 5 \end{cases}$

9. $\begin{cases} 2x + 3y + z = 6 \\ 3x + 2y - 2z = 2 \\ x + 4y = 7 \end{cases}$

10. $\begin{cases} 4t_1 + t_2 + 3t_3 = -1 \\ 3t_1 - 2t_2 = -2 \\ 4t_1 + 2t_3 = 0 \end{cases}$

11. $\begin{cases} 3t_1 + t_3 = 3 \\ 6t_1 + 6t_2 - 10t_3 = -12 \\ -2t_2 + 4t_3 = 6 \end{cases}$

12. $\begin{cases} 3r - s + 2t = 11 \\ -4r + 6s - 2t = 1 \\ r + s - t = -3 \end{cases}$

13. $\begin{cases} 4u + \dfrac{8}{5}w = 8 \\ 6u - 2v = -10 \\ 8v + 24w = 16 \end{cases}$

14. $\begin{cases} 4x_1 + 5x_2 - 3x_3 = 0 \\ 7x_1 - 2x_2 + 4x_3 = -2 \\ 5x_1 + 17x_2 - 13x_2 = -5 \end{cases}$

15. $\begin{cases} t_1 + t_2 + t_3 = 1 \\ 2t_1 + 2t_3 = -1 \\ t_1 - t_2 + t_3 = 3 \end{cases}$

16.
$$\begin{cases} -z_1 + \dfrac{2}{3}z_2 - 2z_3 = 4 \\ 2z_1 + z_2 + 3z_3 = 0 \\ 3z_1 - 2z_2 + 6z_3 = -10 \end{cases}$$

17.
$$\begin{cases} 4x - 4y = 11 + z \\ 7x + 2z = -4y - 14 \\ y + z = -5 + 37x \end{cases}$$

18.
$$\begin{cases} 2x_1 + 3x_2 + 4x_3 + 3x_4 = -5 \\ -x_1 - 4x_2 + 5x_3 + x_4 = 0 \\ 3x_1 - x_2 - x_3 + 2x_4 = 3 \\ 2x_1 - 3x_3 - 2x_4 = 8 \end{cases}$$

19.
$$\begin{cases} 3w - 4x - 5y + 8z = 2 \\ 9w + x - 7z = 3 \\ w - 4x - 6y + 2z = -7 \\ w - \dfrac{1}{2}x + y - \dfrac{3}{2}z = 0 \end{cases}$$

In problems 20–25, find the general solution of the given system of equations. Find a particular solution of the system of equations.

20.
$$\begin{cases} -x_1 + 4x_2 - 5x_3 = 8 \\ -3x_1 + x_2 + x_3 = 24 \end{cases}$$

21.
$$\begin{cases} 2x_1 + 5x_2 + 6x_3 = -3 \\ 3x_1 - x_2 - x_3 = 1 \end{cases}$$

22.
$$\begin{cases} x_1 + x_2 + x_3 - x_4 = -4 \\ x_1 + 2x_2 - x_3 + 2x_4 = 1 \\ x_1 - 2x_2 + 2x_3 - x_4 = 3 \end{cases}$$

23.
$$\begin{cases} x_1 - 2x_2 - x_3 - x_4 = 7 \\ x_1 - 2x_2 + 3x_3 - 2x_4 = -5 \\ 2x_1 + x_2 - 2x_3 + x_4 = 0 \end{cases}$$

24.
$$\begin{cases} 2x_1 - x_2 + 3x_3 + 4x_4 = 0 \\ x_1 + 5x_2 - 4x_3 + 2x_4 = 8 \end{cases}$$

25.
$$\begin{cases} x_1 + 4x_2 + 8x_3 - x_4 = 7 \\ 2x_1 + 8x_2 + 16x_3 - x_4 = 12 \end{cases}$$

26. Find three numbers whose sum is 50 if twice the second is 3 more than the first and if the second plus the third is 4 more than the first.

27. Mr. Cox, Census Bureau representative, discovered in a survey of 550 heads of households who were interviewed, that there were twice as many who lived in rented apartments as there were those who live in houses which they owned, while the number of those who rented houses was only one third as great as the number who rented apartments. How many of those interviewed lived in a house which they owned?

28. A man has $3.80 worth of change in his pocket, consisting solely of nickels, dimes, and quarters. The number of nickels is one less than twice the number of dimes, and the value of the nickels and quarters combined is $2.50. How many nickels, how many dimes, and how many quarters does he have?

29. The Kern Corporation produces three products: abrasive discs, polishers, and cutting tools. One of Kern's customers has placed an order for a total of 208 items, 100 of which are abrasive discs and polishers. If the number of cutting tools exceeds by 12, three times the number of polishers, how many of each item was ordered?

30. A tank may be filled by three pipes. If all three pipes are used the tank can be filled at the rate of 240 gallons per minute. The second pipe, used alone, fills the tank 20 gallons per minute faster than the first and third combined. The first pipe fills the tank 30 gallons per minute slower than the third. At what rate would the tank fill if only the second pipe were used?

31. Mr. White plans to invest $2000 in the stock of three different companies. Coco is selling at $130/share, Echo at $12/share, and Whirlwind, a highly speculative stock, at $17/share. He prefers to purchase twice as many shares of the blue chip stocks as the speculative stock. On the basis of the dividends paid previously, he expects a yearly return of $5.60/share on the Coco stock, $.80/share on the Echo stock, and $1.00/share on the Whirlwind stock. If he wants a total yearly return of $100 on his investment, how many shares of each stock should he buy?

32. Mr. Graham, the manager of a bookstore, has allocated $604 for the purchase of paperbacks. He has 90 linear feet of shelving available on which to display the books. On the average, an Arro book occupies 1 inch of shelf space, a Boro book $1\frac{3}{4}$ inches, and a Corrow book $2\frac{1}{8}$ inches. Mr. Graham estimates that he will sell twice as many Arro books as Boro and Corrow books combined. If Arro books cost $.50 each, Boro books cost $.90 each, and Corrow books cost $1.80 each, how many of each kind should he purchase?

9.3 MATRICES

Consider the system of equations

$$\begin{cases} x + y + z = 300 \\ 2x \quad - 5z = 0 \\ x - y - z = 0 \end{cases}$$

which was solved in the preceding section. If we examine the method used to solve this system, or in fact, any system of linear equations, we see that the letters representing the unknowns play a very minor role in obtaining the solution. Actually, the symbols used to denote the unknowns can be changed at will without altering the system. These symbols merely serve to establish the position of the coefficients. It is the coefficients and constant terms which really determine the solution set of the system.

Let us support this statement by solving the aforementioned system, just as we did in Sec. 9.2, but with some omissions. If we omit the symbols for the unknowns and the equality signs, the coefficients and constants of the original system take the form

$$\begin{cases} 1 & 1 & 1 & 300 \\ 2 & 0 & -5 & 0 \\ 1 & -1 & -1 & 0. \end{cases}$$

Interchanging the second and third rows, we have

$$\begin{cases} 1 & 1 & 1 & 300 \\ 1 & -1 & -1 & 0 \\ 2 & 0 & -5 & 0. \end{cases}$$

Replacing the second row by the sum of itself and -1 times the first row, we obtain

$$\begin{bmatrix} 1 & 1 & 1 & 300 \\ 0 & -2 & -2 & -300 \\ 2 & 0 & -5 & 0. \end{bmatrix}$$

Replacing the third row by the sum of itself and -2 times the first row, we have

$$\begin{bmatrix} 1 & 1 & 1 & 300 \\ 0 & -2 & -2 & -300 \\ 0 & -2 & -7 & -600. \end{bmatrix}$$

Finally, replacing the third row by the sum of itself and -1 times the second row, we obtain

$$\begin{bmatrix} 1 & 1 & 1 & 300 \\ 0 & -2 & -2 & -300 \\ 0 & 0 & -5 & -300. \end{bmatrix}$$

This array of numbers can be interpreted as the triangular system

$$\begin{cases} x + y + z = 300 \\ -2y - 2z = -300 \\ -5z = -300 \end{cases}$$

which may be solved by substitution as before.

In the preceding discussion, each array of numbers had a special significance, each being derived from a system of linear equations. Such arrays of numbers, which are encountered quite frequently not only in the context of systems of linear equations but in many diverse areas of mathematics, are called *matrices*. It is useful to adopt the following definition of a matrix.

9.6 Definition. *Let m and n be positive integers. An $m \times n$ **matrix** is a rectangular array of real numbers of the form*

$$\begin{bmatrix} a_{11} & a_{12} & a_{13} & \cdots & a_{1n} \\ a_{12} & a_{22} & a_{23} & \cdots & a_{2n} \\ \cdot & \cdot & \cdot & & \cdot \\ \cdot & \cdot & \cdot & & \cdot \\ \cdot & \cdot & \cdot & & \cdot \\ a_{m1} & a_{m2} & a_{m3} & \cdots & a_{mn} \end{bmatrix},$$

where a_{ij} represents the number in the i-th row and j-th column.

The numbers m and n, called the *dimensions* of the matrix, indicate the number of rows (a row is horizontal) and the number of columns (a column is vertical), respectively, appearing in the matrix.

Example.

$$\begin{bmatrix} 3 & 1 & -\dfrac{2}{3} & 5 \\ 9 & 0 & 8 & -12 \\ 2 & \dfrac{1}{2} & \dfrac{1}{2} & 13 \end{bmatrix} \quad \text{is a 3} \times \text{4 matrix.}$$

and

$$\begin{bmatrix} 1 & 2 \\ -2 & -2 \\ 3 & -6 \\ 8 & 5 \end{bmatrix} \quad \text{is a 4} \times \text{2 matrix.}$$

If the number of rows and the number of columns in a matrix are equal, say both are equal to n, then the matrix is a *square matrix of order n.*

Example.

$$\begin{bmatrix} 1 & 0 & -2 \\ 2 & 6 & \dfrac{1}{2} \\ \dfrac{3}{13} & -5 & 1 \end{bmatrix} \quad \text{is a square matrix of order 3.}$$

A matrix having only one row is a *row matrix.* A matrix having only one column is a *column matrix.*

Example. $[2 \ \ -3 \ \ 8 \ \ 2]$ is a row matrix,

while $\begin{bmatrix} 0 \\ 1 \\ 1 \end{bmatrix}$ is a column matrix.

Frequently, matrices are denoted by capital letters. Thus, we may write

$$A = \begin{bmatrix} a_{11} & a_{12} & \cdots & a_{1n} \\ a_{21} & a_{22} & \cdots & a_{2n} \\ \vdots & \vdots & & \vdots \\ a_{m1} & a_{m2} & \cdots & a_{mn} \end{bmatrix} \quad \text{and} \quad B = \begin{bmatrix} b_{11} & b_{12} & \cdots & b_{1n} \\ b_{21} & b_{22} & \cdots & b_{2n} \\ \vdots & \vdots & & \vdots \\ b_{m1} & b_{m2} & \cdots & b_{mn} \end{bmatrix}.$$

9.7 Definition. *The matrices A and B are equal, written A = B, provided*
   (i) *A and B have the same dimensions,*
and (ii) *elements in corresponding positions of A and B are equal.*

Example.

If $A = \begin{bmatrix} 2 & 1 & 0 \\ 1 & 3 & -5 \end{bmatrix}$, $\qquad B = \begin{bmatrix} 1 & 3 & -5 \\ 2 & 1 & 0 \end{bmatrix}$, $\qquad C = \begin{bmatrix} 2 & 1 & 0 \\ 1 & 3 & -5 \end{bmatrix}$, $\qquad$ and

$D = \begin{bmatrix} 1 & 3 & -5 & 2 \\ 2 & 1 & 0 & 2 \end{bmatrix}$, then $A = C, A \neq B, A \neq D, B \neq D$.

Let us now consider further the concept of a matrix in relation to our dis‑ cussion of linear equations. Given the system of m equations in n unknowns

$$\begin{cases} a_{11} x_1 + a_{12} x_2 + \cdots + a_{1n} x_n = b_1 \\ a_{21} x_1 + a_{22} x_2 + \cdots + a_{2n} x_n = b_2 \\ \quad \cdot \qquad \cdot \qquad \qquad \cdot \qquad \cdot \\ \quad \cdot \qquad \cdot \qquad \qquad \cdot \qquad \cdot \\ \quad \cdot \qquad \cdot \qquad \qquad \cdot \qquad \cdot \\ a_{m1} x_1 + a_{m2} x_2 + \cdots + a_{mn} x_n = b_m \end{cases}$$

the matrix

$$A = \begin{bmatrix} a_{11} & a_{12} & \cdots & a_{1n} \\ a_{21} & a_{22} & \cdots & a_{2n} \\ \cdot & \cdot & & \cdot \\ \cdot & \cdot & & \cdot \\ \cdot & \cdot & & \cdot \\ a_{m1} & a_{m2} & \cdots & a_{mn} \end{bmatrix}$$

is the *coefficient matrix* of the system and the matrix

$$\overline{A} = \begin{bmatrix} a_{11} & a_{12} & \cdots & a_{1n} & b_1 \\ a_{21} & a_{22} & \cdots & a_{2n} & b_2 \\ \cdot & \cdot & & \cdot & \cdot \\ \cdot & \cdot & & \cdot & \cdot \\ \cdot & \cdot & & \cdot & \cdot \\ a_{m1} & a_{m2} & \cdots & a_{mn} & b_m \end{bmatrix}$$

is the *augmented matrix* of the system.

Example. Given the system of equations

$$\begin{cases} 3x - 2y + z = 3 \\ 4x + 4y + 5z = 0 \\ -2x + 2y - 2z = 1 \end{cases}$$

the coefficient matrix is

$$\begin{bmatrix} 3 & -2 & 1 \\ 4 & 4 & 5 \\ -2 & 2 & -2 \end{bmatrix}$$

and the augmented matrix is

$$\begin{bmatrix} 3 & -2 & 1 & 3 \\ 4 & 4 & 5 & 0 \\ -2 & 2 & -2 & 1 \end{bmatrix}.$$

Corresponding to elementary operations on a system of linear equations we have elementary row operations on a matrix. By an elementary row operation on a matrix A, we mean one of the following:

(i) interchanging any two rows of A,
(ii) multiplying any row of A by a nonzero constant,

or (iii) replacing the k-th row of A by the sum of itself and any other row of A.

Steps (ii) and (iii) are frequently combined into a single row operation.

Two $m \times n$ matrices A and B are said to be equivalent if they are augmented matrices of equivalent systems of equations. If A is the augmented matrix of a system of linear equations and if the matrix B is obtained from A by a sequence of elementary row operations, then B is the augmented matrix of a system of linear equations which is equivalent to the system represented by A. Therefore A and B are equivalent matrices.

A matrix A is a triangular matrix if the first $j - 1$ entries of the j-th row are zero, for $j = 1, 2, \ldots m$.

Example.

$$A = \begin{bmatrix} 3 & 0 & 4 & 1 \\ 0 & 2 & 2 & 5 \\ 0 & 0 & -1 & \frac{1}{2} \end{bmatrix} \text{ and } B = \begin{bmatrix} 1 & 2 & -1 & 5 & 8 \\ 0 & 3 & 6 & -2 & 2 \\ 0 & 0 & 0 & 0 & 0 \end{bmatrix} \text{ are triangular.}$$

By successively applying elementary row operations to any matrix A, we may reduce it to an equivalent triangular matrix. Let us consider a problem which illustrates the process of reducing a matrix to triangular form.

Problem. Reduce the matrix A to triangular form if

$$A = \begin{bmatrix} 3 & 6 & 0 & 15 \\ 2 & 2 & 8 & -4 \\ 3 & 0 & -9 & 12 \end{bmatrix}.$$

Solution. We write equivalent matrices in succession, noting at the right of the page the row operation which was applied at each step in the process of this reduction.

$$\begin{bmatrix} 3 & 6 & 0 & 15 \\ 2 & 2 & 8 & -4 \\ 3 & 0 & -9 & 12 \end{bmatrix}$$ This is the original matrix.

$$\begin{bmatrix} 1 & 2 & 0 & 5 \\ 2 & 2 & 8 & -4 \\ 3 & 0 & -9 & 12 \end{bmatrix}$$ The entries in row 1 were multiplied by $\frac{1}{3}$.

$$\begin{bmatrix} 1 & 2 & 0 & 5 \\ 0 & -2 & 8 & -14 \\ 3 & 0 & -9 & 12 \end{bmatrix}$$ Row 2 was replaced by the sum of itself and -2 times row 1.

$$\begin{bmatrix} 1 & 2 & 0 & 5 \\ 0 & -2 & 8 & -14 \\ 0 & -6 & -9 & -3 \end{bmatrix}$$ Row 3 was replaced by the sum of itself and -3 times row 1.

$$\begin{bmatrix} 1 & 2 & 0 & 5 \\ 0 & -2 & 8 & -14 \\ 0 & 0 & -33 & 39 \end{bmatrix}$$ Row 3 was replaced by the sum of itself and -3 times row 2.

Let us now consider a problem which illustrates the process of finding the solution to a system of linear equations using a matrix representation.

Problem. Solve the system

$$\begin{cases} 2x + 3y - z = 4 \\ -x + y + 2z = -3 \\ 3x - y - 2z = 1 \end{cases}$$

by reducing the augmented matrix of this system to triangular form.

Solution. The augmented matrix of this system is

$$A = \begin{bmatrix} 2 & 3 & -1 & 4 \\ -1 & 1 & 2 & -3 \\ 3 & -1 & -2 & 1 \end{bmatrix}.$$

Interchanging the positions of row 1 and row 2, we have

$$\begin{bmatrix} -1 & 1 & 2 & -3 \\ 2 & 3 & -1 & 4 \\ 3 & -1 & -2 & 1 \end{bmatrix}.$$

Multiplying row 1 by 2 and adding to row 2 yields

$$\begin{bmatrix} -1 & 1 & 2 & -3 \\ 0 & 5 & 3 & -2 \\ 3 & -1 & -2 & 1 \end{bmatrix}.$$

We now multiply row 1 by 3 and add to row 3, obtaining

$$\begin{bmatrix} -1 & 1 & 2 & -3 \\ 0 & 5 & 3 & -2 \\ 0 & 2 & 4 & -8 \end{bmatrix}.$$

At this point it is convenient to multiply row 3 by $\frac{1}{2}$. This gives

$$\begin{bmatrix} -1 & 1 & 2 & -3 \\ 0 & 5 & 3 & -2 \\ 0 & 1 & 2 & -4 \end{bmatrix}.$$

Multiplying row 3 by -5 and adding to row 2, we have

$$\begin{bmatrix} -1 & 1 & 2 & -3 \\ 0 & 0 & -7 & 18 \\ 0 & 1 & 2 & -4 \end{bmatrix}.$$

We now switch rows 2 and 3 to obtain the triangular matrix

$$B = \begin{bmatrix} -1 & 1 & 2 & -3 \\ 0 & 1 & 2 & -4 \\ 0 & 0 & -7 & 18 \end{bmatrix}.$$

This is precisely the augmented matrix of the system

$$\begin{cases} -x + y + 2z = -3 \\ \quad\; y + 2z = -4 \\ \quad\quad\; -7z = 18. \end{cases}$$

By substitution, we see that $z = -\dfrac{18}{7}$,

$$y = -4 - 2\left(-\frac{18}{7}\right) = -\frac{28}{7} + \frac{36}{7} = \frac{8}{7},$$

and

$$x = \frac{8}{7} + 2\left(-\frac{18}{7}\right) + 3 = \frac{8}{7} - \frac{36}{7} + \frac{21}{7} = -1.$$

So $\left(-1, \dfrac{8}{7}, -\dfrac{18}{7}\right)$ is the solution to our system of equations.

Exercise Set 9.3

1. Consider the matrix

$$A = \begin{bmatrix} 3 & 0 & 1 & 5 \\ 2 & 2 & -3 & 0 \\ 1 & -4 & 2 & 3 \end{bmatrix}.$$

a. Determine a_{22}, a_{31}, a_{34}, and a_{24}.
b. What are the dimensions of the matrix A?
c. If the matrix A is the augmented matrix of a system of linear equations, write the system of equations which is represented, using x_1, x_2, x_3, and x_4.

2. a. Give an example of a 3 X 5 matrix.
b. Give an example of a 5 X 3 matrix.
c. Give an example of a square matrix of order 4.
d. Give an example of a row matrix.
e. Give an example of a column matrix.

3. In parts a and b replace the variables by numbers so that the matrices A and B are equal.

a. $A = \begin{bmatrix} 1 & 4 \\ x & 0 \end{bmatrix}$ and $B = \begin{bmatrix} 1 & 4 \\ 3 & y \end{bmatrix}$

b. $A = \begin{bmatrix} x-1 & 7 \\ -3 & 6 \end{bmatrix}$ and $B = \begin{bmatrix} 8 & 7 \\ -3 & 3y \end{bmatrix}$

4. Write the coefficient matrix and the augmented matrix of the following system of linear equations:

$$\begin{cases} 3x & - & z = 4 \\ 2x + 4y + 3z = 6 \\ x - 8y & = 0 \end{cases}$$

5. Given the matrix

$$A = \begin{bmatrix} 2 & 4 & -1 & -5 \\ 1 & -1 & 3 & 8 \\ -3 & 3 & 3 & 12 \end{bmatrix},$$

perform successively the indicated row operations:
a. Interchange row 1 and row 2.
b. Replace row 2 by the sum of itself and -2 times row 1.
c. Replace row 3 by the sum of itself and 3 times row 1.
d. Has the matrix been reduced to triangular form?

* * *

In problems 6–9, write the coefficient matrix and the augmented matrix of the given system of linear equations.

6. $\begin{cases} \dfrac{1}{2}x_1 - \dfrac{2}{3}x_2 + 6x_3 = \dfrac{3}{4} \\ -x_1 + x_2 = 0 \\ \dfrac{1}{5}x_1 - \dfrac{2}{3}x_2 + 9x_3 = 12 \end{cases}$

7. $\begin{cases} 3r - 6s + 19t = 144 \\ -r - s + 35t = 2 \\ -2r + 5s - 8t = -39 \end{cases}$

8. $\begin{cases} w_1 - w_2 - w_3 = 1 \\ w_3 - w_1 + 5w_2 = 10 \\ 100 + 4w_1 - 6w_3 = 8 \end{cases}$

9. $\begin{cases} 3l - 2m + n - 5p = 0 \\ -\dfrac{1}{2}l - 3m + n = 8 \\ 14m - 3p = 1 \end{cases}$

In problems 10–17, reduce the given matrix to triangular form.

10. $\begin{bmatrix} 2 & 4 & -1 & 8 \\ 3 & -2 & 1 & 5 \end{bmatrix}$

11. $\begin{bmatrix} \dfrac{1}{2} & \dfrac{1}{3} & -1 & 7 \\ 3 & 0 & 0 & -6 \end{bmatrix}$

12. $\begin{bmatrix} 1 & -2 & -2 & 3 \\ 0 & -9 & 4 & 6 \\ -8 & 0 & 5 & 1 \end{bmatrix}$

13. $\begin{bmatrix} 1 & -1 & 1 & -1 \\ 1 & 3 & -2 & 7 \\ -2 & -6 & 4 & -14 \end{bmatrix}$

14. $\begin{bmatrix} 0 & 6 & 3 & 1 \\ 4 & 2 & 2 & 9 \\ 1 & 1 & 1 & 4 \\ 6 & -6 & 6 & -3 \end{bmatrix}$

15. $\begin{bmatrix} 0 & 1 & 0 & 1 \\ 1 & 0 & -1 & 1 \\ 2 & 0 & 0 & -1 \\ -1 & 2 & -1 & -2 \end{bmatrix}$

16. $\begin{bmatrix} 2 & -6 \\ 3 & -5 \\ 4 & -4 \\ 5 & 13 \end{bmatrix}$

17. $\begin{bmatrix} \dfrac{2}{3} & 0 & \dfrac{1}{2} \\ -1 & 0 & 2 \\ 3 & \dfrac{1}{4} & \dfrac{9}{4} \\ -\dfrac{3}{4} & 1 & -3 \end{bmatrix}$

In problems 18–25, find the solution to the given system of linear equations by using matrix representation.

18. $\begin{cases} 6x + 3y - 2z = 8 \\ -x - 15y + z = -12 \\ 2x + 6y + 4z = -1 \end{cases}$

19. $\begin{cases} x + y + z = \dfrac{6}{5} \\ 2x - 2y + \dfrac{1}{5}z = -1 \\ -\dfrac{2}{5}x + \dfrac{2}{5}y + \dfrac{4}{25}z = \dfrac{2}{5} \end{cases}$

20. $\begin{cases} 2x + y + 8z = 4 \\ 3x - y - 2z = -7 \\ x + 2y - 3z = -21 \end{cases}$

21. $\begin{cases} 6x_1 + 2x_2 + 4x_3 = -8 \\ -3x_1 - x_2 + x_3 = -20 \\ 8x_1 - 2x_2 - 3x_3 = 0 \end{cases}$

22. $\begin{cases} 8r - s + 12t = -24 \\ -2r + 2s + 3t = 1 \\ r + \dfrac{1}{2}s + \dfrac{1}{2}t = \dfrac{3}{2} \end{cases}$

23. $\begin{cases} 3u_1 - 2u_2 + u_3 = 4 \\ -6u_1 + 4u_2 - 2u_3 = 3 \\ u_1 + 3u_2 + 2u_3 = -1 \end{cases}$

24. $\begin{cases} 4x + 2y - z = -1 \\ -3x + y + 3z = 6 \\ 2x + 3y - \dfrac{1}{4}z = 3 \end{cases}$

25. $\begin{cases} 2x_1 - x_2 + 3x_3 = 2\sqrt{2} - 6 \\ 4x_2 + 7x_3 = 5 \\ -3x_1 + 2x_2 = 6 - 3\sqrt{2} \end{cases}$

9.4 ADDITION AND MULTIPLICATION OF MATRICES

As stated in the preceding section, the usefulness of matrices is by no means restricted to the solution of systems of linear equations. Indeed, applications of matrix theory abound in many branches of mathematics, as well as in other seemingly unrelated fields. Wherever there is a need for concise storage of information in a rectangular array, matrices can play a vital role. In this and the following section, we consider addition and multiplication of matrices and examine some of the algebraic properties which are useful in the applications of matrix theory.

We begin by defining addition for $m \times n$ matrices.

9.8 Definition. *Given two $m \times n$ matrices*

$$A = \begin{bmatrix} a_{11} & a_{12} & \cdots & a_{1n} \\ a_{21} & a_{22} & \cdots & a_{2n} \\ \vdots & \vdots & & \vdots \\ a_{m1} & a_{m2} & \cdots & a_{mn} \end{bmatrix} \quad and \quad B = \begin{bmatrix} b_{11} & b_{12} & \cdots & b_{1n} \\ b_{21} & b_{22} & \cdots & b_{2n} \\ \vdots & \vdots & & \vdots \\ b_{m1} & b_{m2} & \cdots & b_{mn} \end{bmatrix},$$

their sum is the $m \times n$ matrix

$$A + B = \begin{bmatrix} a_{11} + b_{11} & a_{12} + b_{12} & \cdots & a_{1n} + b_{1n} \\ a_{21} + b_{21} & a_{22} + b_{22} & \cdots & a_{2n} + b_{2n} \\ \vdots & \vdots & \vdots & \vdots \\ a_{m1} + b_{m1} & a_{m2} + b_{m2} & \cdots & a_{mn} + b_{mn} \end{bmatrix}.$$

In other words, the matrix $A + B$ is obtained by adding the elements in corresponding positions in the matrices A and B. Addition is defined, therefore, only for matrices of like dimensions.

Example. If

$$A = \begin{bmatrix} 3 & 1 & 0 & 9 \\ -2 & 1 & 4 & 6 \\ 7 & 0 & 5 & -5 \end{bmatrix} \text{ and } B = \begin{bmatrix} -14 & 8 & 6 & 6 \\ 2 & 7 & -4 & -3 \\ 9 & 12 & -5 & 6 \end{bmatrix},$$

then

$$A + B = \begin{bmatrix} 3 + (-14) & 1 + 8 & 0 + 6 & 9 + 6 \\ -2 + 2 & 1 + 7 & 4 + (-4) & 6 + (-3) \\ 7 + 9 & 0 + 12 & 5 + (-5) & -5 + 6 \end{bmatrix} = \begin{bmatrix} -11 & 9 & 6 & 15 \\ 0 & 8 & 0 & 3 \\ 16 & 12 & 0 & 1 \end{bmatrix}.$$

The $m \times n$ matrix having a zero in each position is called the *zero matrix* and will be denoted by the symbol 0. The matrix 0 has the property that $A + 0 = 0 + A = A$ where A and 0 have the same dimensions.

Just as there corresponds to each real number a, a real number $-a$, having the property that $a + (-a) = 0$, there corresponds to each matrix A, a matrix $-A$, such that $A + (-A) = 0$. If

$$A = \begin{bmatrix} a_{11} & a_{12} & \cdots & a_{1n} \\ a_{21} & a_{22} & \cdots & a_{2n} \\ \cdot & \cdot & & \cdot \\ \cdot & \cdot & & \cdot \\ \cdot & \cdot & & \cdot \\ a_{m1} & a_{m2} & \cdots & a_{mn} \end{bmatrix} \text{ then } -A = \begin{bmatrix} -a_{11} & -a_{12} & \cdots & -a_{1n} \\ -a_{21} & -a_{22} & \cdots & -a_{2n} \\ \cdot & \cdot & & \cdot \\ \cdot & \cdot & & \cdot \\ \cdot & \cdot & & \cdot \\ -a_{m1} & -a_{m2} & \cdots & -a_{mn} \end{bmatrix}.$$

Having defined addition for matrices, we now define subtraction for matrices.

9.9 Definition. *IF A and B are $m \times n$ matrices, THEN $A - B = A + (-B)$.*

Example. If

$$A = \begin{bmatrix} 3 & 1 & 2 \\ -4 & 0 & -5 \\ 6 & 7 & 5 \end{bmatrix} \text{ and } B = \begin{bmatrix} 2 & -2 & 4 \\ 8 & 10 & -5 \\ 3 & 1 & 6 \end{bmatrix},$$

then

$$A - B = \begin{bmatrix} 1 & 3 & -2 \\ -12 & -10 & 0 \\ 3 & 6 & -1 \end{bmatrix}.$$

Having defined addition and subtraction, we now turn our attention to defining multiplication for matrices. While the definition of addition is quite straightforward, the definition of multiplication is somewhat more complicated.

9.10 Definition. *IF the m × n matrix A and the n × p matrix B are given by*

$$A = \begin{bmatrix} a_{11} & a_{12} & \cdots & a_{1n} \\ a_{21} & a_{22} & \cdots & a_{2n} \\ \cdot & \cdot & & \cdot \\ \cdot & \cdot & & \cdot \\ \cdot & \cdot & & \cdot \\ a_{m1} & a_{m2} & \cdots & a_{mn} \end{bmatrix} \quad and \quad B = \begin{bmatrix} b_{11} & b_{12} & \cdots & b_{1p} \\ b_{21} & b_{22} & \cdots & b_{2p} \\ \cdot & \cdot & & \cdot \\ \cdot & \cdot & & \cdot \\ \cdot & \cdot & & \cdot \\ b_{n1} & b_{n2} & \cdots & b_{np} \end{bmatrix}$$

*THEN the **product** AB is the m × p matrix*

$$AB = \begin{bmatrix} c_{11} & c_{12} & \cdots & c_{1p} \\ c_{21} & c_{22} & \cdots & c_{2p} \\ \cdot & \cdot & & \cdot \\ \cdot & \cdot & & \cdot \\ \cdot & \cdot & & \cdot \\ c_{m1} & c_{m2} & \cdots & c_{mp} \end{bmatrix},$$

where $c_{ij} = \sum_{k=1}^{n} a_{ik} b_{kj} = a_{i1} b_{1j} + a_{i2} b_{2j} + \cdots + a_{in} b_{nj}.$

Example. (i) If

$$A = \begin{bmatrix} 2 & 1 \\ -3 & -4 \end{bmatrix} \text{ and } B = \begin{bmatrix} 2 & 5 \\ -1 & 6 \end{bmatrix},$$

then

$$AB = \begin{bmatrix} (2)(2) + (1)(-1) & (2)(5) + (1)(6) \\ (-3)(2) + (-4)(-1) & (-3)(5) + (-4)(6) \end{bmatrix} = \begin{bmatrix} 3 & 16 \\ -2 & -39 \end{bmatrix}$$

and

$$BA = \begin{bmatrix} (2)(2) + (5)(-3) & (2)(1) + (5)(-4) \\ (-1)(2) + (6)(-3) & (-1)(1) + (6)(-4) \end{bmatrix} = \begin{bmatrix} -11 & -18 \\ -20 & -25 \end{bmatrix}.$$

(ii) If

$$C = \begin{bmatrix} 2 & 0 & -3 \\ 4 & -5 & 2 \end{bmatrix} \text{ and } D = \begin{bmatrix} -8 \\ 3 \\ 7 \end{bmatrix},$$

then

$$CD = \begin{bmatrix} 2(-8) + (0)(3) + (3)(7) \\ 4(-8) + (-5)(3) + (2)(7) \end{bmatrix} = \begin{bmatrix} -37 \\ -33 \end{bmatrix}$$

and *DC* is not defined.

We observe that, in order for the product AB to be defined, it is necessary that the number of columns in A be equal to the number of rows in B. In particular, if both A and B are square matrices of order n, then the product AB and the product BA are both defined and are also square matrices of order n. The product AB is not necessarily equal to the product BA as was illustrated in the preceding example.

Exercise Set 9.4

If $A = \begin{bmatrix} a & b \\ c & d \end{bmatrix}$ and $B = \begin{bmatrix} e & f \\ g & h \end{bmatrix}$, then $A + B = \begin{bmatrix} a+e & b+f \\ c+g & d+h \end{bmatrix}$ and

$$AB = \begin{bmatrix} ae+bg & af+bh \\ ce+dg & cf+dh \end{bmatrix}.$$

1. Use the above pattern to find $A + B$ and AB if

$$A = \begin{bmatrix} 2 & -1 \\ -5 & 4 \end{bmatrix} \quad \text{and} \quad B = \begin{bmatrix} -3 & 0 \\ 4 & 6 \end{bmatrix}.$$

2. Use the same pattern to find $A + B$ and AB if

$$A = \begin{bmatrix} -3 & 0 \\ 4 & 6 \end{bmatrix} \quad \text{and} \quad B = \begin{bmatrix} 2 & -1 \\ -5 & 4 \end{bmatrix}.$$

3. Find $A - B$ and $B - A$ for the matrices A and B of problem 1.

4. An extension of the above pattern for multiplication suggests that to find the number belonging in the i-th row and j-th column of the matrix AB, we "multiply" the i-th row of A by the j-th column of B. Suppose

$$\begin{bmatrix} 5 & 2 & 1 \\ -3 & 1 & 7 \\ 0 & 4 & 2 \end{bmatrix} \begin{bmatrix} 1 & 2 & -3 \\ 0 & -2 & 1 \\ 1 & -3 & 2 \end{bmatrix} = \begin{bmatrix} c_{11} & c_{12} & c_{13} \\ c_{21} & c_{22} & c_{23} \\ c_{31} & c_{32} & c_{33} \end{bmatrix},$$

where the first matrix is denoted by A, the second by B, and the resulting matrix by AB.

a. Find the element c_{11} of the matrix AB by "multiplying" the first row of A by the first column of B. That is, find $(5)(1) + (2)(0) + (1)(1)$.

b. Find the element c_{12} of the matrix AB by "multiplying" the first row of A by the second column of B. That is, find $(5)(2) + (2)(-2) + (1)(-3)$.

c. Find the element c_{13} of the matrix AB by "multiplying" the first row of A by the third column of B. That is, find $(5)(-3) + (2)(1) + (1)(2)$.

d. Find the element c_{21} of the matrix AB by "multiplying" the second row of A by the first column of B. That is, find $(-3)(1) + (1)(0) + (7)(1)$.

e. Find the elements $c_{22}, c_{23}, c_{31}, c_{32},$ and c_{33} of the matrix AB.

f. Find the product BA.

5. Let

$$A = \begin{bmatrix} 1 & 3 & -2 & 2 \\ 4 & 0 & 3 & -3 \end{bmatrix} \quad \text{and} \quad B = \begin{bmatrix} 2 & -1 & 4 \\ 0 & 5 & -3 \\ 3 & 2 & 2 \\ 1 & 0 & -3 \end{bmatrix}.$$

a. What are the dimensions of A?
b. What are the dimensions of B?
c. What are the dimensions of AB?
d. Find the product AB.

* * *

In problems 6–13, find $A + B, A - B$, and $B - A$ for the given matrices A and B.

6. $A = \begin{bmatrix} 2 & -6 \\ -8 & -7 \end{bmatrix}$ $\qquad B = \begin{bmatrix} -13 & 3 \\ 8 & 9 \end{bmatrix}$

7. $A = \begin{bmatrix} \dfrac{1}{4} & -\dfrac{1}{3} \\ \dfrac{3}{2} & 6 \end{bmatrix}$ $\qquad B = \begin{bmatrix} \dfrac{1}{2} & \dfrac{5}{6} \\ -3 & -12 \end{bmatrix}$

8. $A = \begin{bmatrix} -6 \\ -\dfrac{1}{2} \\ 8 \\ 4 \end{bmatrix}$ $\qquad B = \begin{bmatrix} \dfrac{5}{2} \\ 2 \\ 3 \\ -4 \end{bmatrix}$

9. $A = [1 \quad 3 \quad 9 \quad 15]$ $\qquad B = [-12 \quad 7 \quad -2 \quad 34]$

10. $A = \begin{bmatrix} -13 & 2 & 9 & 4 & -6 \\ 8 & -7 & -4 & 0 & 2 \end{bmatrix}$ $\qquad B = \begin{bmatrix} 3 & 6 & 14 & 7 & 8 \\ -5 & 5 & 2 & 0 & -17 \end{bmatrix}$

11. $A = \begin{bmatrix} 9 & 2 \\ -4 & \dfrac{1}{2} \\ 0 & \dfrac{1}{7} \end{bmatrix}$ $\qquad B = \begin{bmatrix} -13 & 3 \\ 7 & \dfrac{1}{2} \\ 2 & \dfrac{2}{3} \end{bmatrix}$

12. $A = \begin{bmatrix} 3 & 1 & 2 & 0 \\ -4 & 1 & 9 & 6 \\ 3 & 8 & 8 & 10 \end{bmatrix}$ $\qquad B = \begin{bmatrix} -5 & 4 & 3 & 2 \\ 2 & -7 & 8 & 3 \\ 0 & -9 & 12 & 20 \end{bmatrix}$

13. $A = \begin{bmatrix} 3 & 1 & 2 \\ 4 & 0 & 9 \\ 0 & 5 & -4 \end{bmatrix}$ $\qquad B = \begin{bmatrix} -2 & 2 & 3 \\ 0 & 1 & 2 \\ -3 & 5 & 0 \end{bmatrix}$

14. Find B if $A = \begin{bmatrix} 3 & -2 & 4 \\ 5 & 8 & 7 \\ -1 & 6 & 0 \end{bmatrix}$ and $A - B = \begin{bmatrix} 13 & 9 & 2 \\ 3 & 7 & 16 \\ 5 & 1 & 8 \end{bmatrix}$.

In problems 15–20, find the dimensions of A, the dimensions of B, and the dimensions of AB. Find the product AB.

15. $A = [3 \quad 9 \quad 2 \quad 4]$ $\qquad B = \begin{bmatrix} 4 \\ -2 \\ 5 \\ -2 \end{bmatrix}$

16. $A = \begin{bmatrix} 3 \\ 9 \\ 2 \\ 4 \end{bmatrix}$ $\qquad$ $B = [4 \ -2 \ 5 \ -2]$

17. $A = \begin{bmatrix} 1 & 3 \\ -2 & 7 \end{bmatrix}$ $\qquad$ $B = \begin{bmatrix} 2 & 4 \\ -1 & -2 \end{bmatrix}$

18. $A = \begin{bmatrix} 1 & -3 \\ 3 & 2 \\ 7 & -2 \end{bmatrix}$ $\qquad$ $B = \begin{bmatrix} 4 & 2 & 1 \\ -5 & 6 & -3 \end{bmatrix}$

19. $A = \begin{bmatrix} 4 & 2 & 1 \\ -5 & 6 & -3 \end{bmatrix}$ $\qquad$ $B = \begin{bmatrix} 1 & -3 \\ 3 & 2 \\ 7 & -2 \end{bmatrix}$

20. $A = \begin{bmatrix} 3 & 0 & 0 \\ 0 & -2 & 0 \\ 0 & 0 & 4 \end{bmatrix}$ $\qquad$ $B = \begin{bmatrix} -2 & 0 & 0 \\ 0 & 5 & 0 \\ 0 & 0 & -3 \end{bmatrix}$

In problems 21–32, find $A + A$, $A + B$, $A - B$, AA, AB, and BA for the given matrices A and B.

21. $A = \begin{bmatrix} 3 & 4 \\ -1 & 2 \end{bmatrix}$ $\qquad$ $B = \begin{bmatrix} 1 & 0 \\ 0 & 1 \end{bmatrix}$

22. $A = \begin{bmatrix} 3 & 4 \\ -1 & 2 \end{bmatrix}$ $\qquad$ $B = \begin{bmatrix} \dfrac{1}{5} & -\dfrac{2}{5} \\ \dfrac{1}{10} & \dfrac{3}{10} \end{bmatrix}$

23. $A = \begin{bmatrix} 0 & 1 \\ -1 & 0 \end{bmatrix}$ $\qquad$ $B = \begin{bmatrix} 5 & -4 \\ 3 & 7 \end{bmatrix}$

24. $A = \begin{bmatrix} 1 & -3 \\ -8 & 5 \end{bmatrix}$ $\qquad$ $B = \begin{bmatrix} 1 & 1 \\ 1 & 1 \end{bmatrix}$

25. $A = \begin{bmatrix} -3 & 2 \\ 5 & 1 \end{bmatrix}$ $\qquad$ $B = \begin{bmatrix} -\dfrac{1}{13} & \dfrac{2}{13} \\ \dfrac{5}{13} & \dfrac{3}{13} \end{bmatrix}$

26. $A = \begin{bmatrix} 3 & 1 & 2 \\ 4 & 2 & 9 \\ -10 & 5 & -4 \end{bmatrix}$ $\qquad$ $B = \begin{bmatrix} 1 & 0 & 0 \\ 0 & 1 & 0 \\ 0 & 0 & 1 \end{bmatrix}$

27. $A = \begin{bmatrix} 6 & -15 & 8 \\ -1 & -1 & 39 \\ \sqrt{2} & \sqrt{3} & -\sqrt{5} \end{bmatrix}$ $\qquad$ $B = \begin{bmatrix} 0 & 0 & 0 \\ 0 & 0 & 0 \\ 0 & 0 & 0 \end{bmatrix}$

28. $A = \begin{bmatrix} 3 & 1 & 2 \\ 4 & 0 & 9 \\ 0 & 5 & -4 \end{bmatrix}$ $\qquad$ $B = \begin{bmatrix} -2 & 2 & 3 \\ 0 & 1 & 2 \\ -3 & 5 & 0 \end{bmatrix}$

29. $A = \begin{bmatrix} 2 & 1 & -2 \\ 0 & 1 & 1 \\ 3 & 0 & -2 \end{bmatrix}$ $B = \begin{bmatrix} -\dfrac{2}{5} & \dfrac{2}{5} & \dfrac{3}{5} \\[2mm] \dfrac{3}{5} & \dfrac{2}{5} & -\dfrac{2}{5} \\[2mm] -\dfrac{3}{5} & \dfrac{3}{5} & \dfrac{2}{5} \end{bmatrix}$

30. $A = \begin{bmatrix} 3 & -2 & 1 \\ 0 & 4 & 9 \\ 0 & 0 & -7 \end{bmatrix}$ $B = \begin{bmatrix} -1 & -2 & 5 \\ 0 & 3 & -4 \\ 0 & 0 & 5 \end{bmatrix}$

31. $A = \begin{bmatrix} 2 & 2 & -2 \\ 0 & 3 & -1 \\ 0 & 0 & 1 \end{bmatrix}$ $B = \begin{bmatrix} 4 & -3 & 3 \\ 0 & 3 & 8 \\ 0 & 0 & -9 \end{bmatrix}$

32. $A = \begin{bmatrix} 2 & 0 & 2 \\ 1 & 3 & 5 \\ 5 & 3 & 1 \end{bmatrix}$ $B = \begin{bmatrix} \dfrac{1}{4} & -\dfrac{1}{8} & \dfrac{1}{8} \\[2mm] -\dfrac{1}{2} & \dfrac{1}{6} & \dfrac{1}{6} \\[2mm] \dfrac{1}{4} & \dfrac{1}{8} & -\dfrac{1}{8} \end{bmatrix}$

9.5 THE ALGEBRA OF SQUARE MATRICES

Let us investigate more fully the algebraic properties of sets of square matrices. We will restrict our attention at this point to the set M_3 consisting of all square matrices of order 3. All that is done here for square matrices of order 3 remains valid for square matrices of order n.

If A and B are square matrices of order 3, then the sums $A + B, B + A$ and the products AB, BA, are all defined. We will consider first some of the properties relating to addition. If

$$A = \begin{bmatrix} a_{11} & a_{12} & a_{13} \\ a_{21} & a_{22} & a_{23} \\ a_{31} & a_{32} & a_{33} \end{bmatrix} \quad \text{and} \quad B = \begin{bmatrix} b_{11} & b_{12} & b_{13} \\ b_{21} & b_{22} & b_{23} \\ b_{31} & b_{32} & b_{33} \end{bmatrix},$$

then

$$A + B = \begin{bmatrix} a_{11} + b_{11} & a_{12} + b_{12} & a_{13} + b_{13} \\ a_{21} + b_{21} & a_{22} + b_{22} & a_{23} + b_{23} \\ a_{31} + b_{31} & a_{32} + b_{32} & a_{33} + b_{33} \end{bmatrix}$$

and

$$B + A = \begin{bmatrix} b_{11} + a_{11} & b_{12} + a_{12} & b_{13} + a_{13} \\ b_{21} + a_{21} & b_{22} + a_{22} & b_{23} + a_{23} \\ b_{31} + a_{31} & b_{32} + a_{32} & b_{33} + a_{33} \end{bmatrix}.$$

Using the fact that addition of real numbers is commutative, it is evident from the definition of equality of matrices that $A + B = B + A$. Therefore, addition of matrices satisfies a commutative property. It is also relatively easy to show that addition of matrices satisfies an associative property. This is left for the exercises.

The 3×3 matrix 0, whose entries are all 0, has the property that $A + 0 = 0 + A = A$. Thus, sometimes we refer to the matrix 0 as the *additive identity matrix* for M_3.

To each matrix A in M_3 there corresponds another matrix $-A$ in M_3, such that $A + (-A) = 0$. Therefore we refer to the matrix $-A$ as the *additive inverse* of A.

Example. If

$$A = \begin{bmatrix} 2 & 7 & -5 \\ 6 & -1 & 0 \\ -9 & 4 & -12 \end{bmatrix} \text{ then } -A = \begin{bmatrix} -2 & -7 & 5 \\ -6 & 1 & 0 \\ 9 & -4 & 12 \end{bmatrix}.$$

Thus, in many ways, the collection of all 3×3 matrices under the operation of addition is reminiscent of the set of all real numbers under the operation of addition of real numbers. But what about multiplication? Are the algebraic properties of multiplication of matrices in M_3 similar to the algebraic properties of the multiplication of real numbers?

A straightforward, though somewhat lengthy, argument can be used to show that if A, B, and C are 3×3 matrices, then $A(BC) = (AB)C$, that is, multiplication of square matrices of order 3 is an associative operation (see problem 4, Exercise Set 9.5).

The question as to whether multiplication is commutative must be answered in the negative, since for

$$A = \begin{bmatrix} 1 & -1 & 2 \\ 3 & 0 & 2 \\ -2 & 1 & -1 \end{bmatrix} \text{ and } B = \begin{bmatrix} 3 & 0 & 4 \\ -1 & -2 & 1 \\ 2 & 3 & -4 \end{bmatrix}$$

we have

$$AB = \begin{bmatrix} 8 & 8 & -5 \\ 13 & 6 & 4 \\ -9 & -5 & -3 \end{bmatrix}$$

while

$$BA = \begin{bmatrix} -5 & 1 & 2 \\ -9 & 2 & -7 \\ 19 & -6 & 14 \end{bmatrix}.$$

It is natural to ask whether there is there a matrix in M_3 which has a property for matrices similar to the property possessed by the number 1 for multiplication of real numbers. The following answers this question in the affirmative. If

$$A = \begin{bmatrix} a_{11} & a_{12} & a_{13} \\ a_{21} & a_{22} & a_{23} \\ a_{31} & a_{32} & a_{33} \end{bmatrix} \text{ and } I = \begin{bmatrix} 1 & 0 & 0 \\ 0 & 1 & 0 \\ 0 & 0 & 1 \end{bmatrix},$$

then

$$AI = IA = \begin{bmatrix} a_{11} & a_{12} & a_{13} \\ a_{21} & a_{22} & a_{23} \\ a_{31} & a_{32} & a_{33} \end{bmatrix}.$$

Hence, the matrix I is a *multiplicative identity* for the set M_3.

One might now inquire as to whether each member of M_3, other than the matrix 0, has a multiplicative inverse. This is not the case as the following example illustrates.

Example.

$$\text{Let } A = \begin{bmatrix} 1 & 0 & 0 \\ 0 & 1 & 0 \\ 0 & 0 & 0 \end{bmatrix}. \text{ If } B = \begin{bmatrix} b_{11} & b_{12} & b_{13} \\ b_{21} & b_{22} & b_{23} \\ b_{31} & b_{32} & b_{33} \end{bmatrix}$$

is a multiplicative inverse of A, then we must have $AB = I$. But

$$AB = \begin{bmatrix} 1 & 0 & 0 \\ 0 & 1 & 0 \\ 0 & 0 & 0 \end{bmatrix} \begin{bmatrix} b_{11} & b_{12} & b_{13} \\ b_{21} & b_{22} & b_{23} \\ b_{31} & b_{32} & b_{33} \end{bmatrix} = \begin{bmatrix} b_{11} & b_{12} & b_{13} \\ b_{21} & b_{22} & b_{23} \\ 0 & 0 & 0 \end{bmatrix}$$

and

$$\begin{bmatrix} b_{11} & b_{12} & b_{13} \\ b_{21} & b_{22} & b_{23} \\ 0 & 0 & 0 \end{bmatrix} \neq \begin{bmatrix} 1 & 0 & 0 \\ 0 & 1 & 0 \\ 0 & 0 & 1 \end{bmatrix} = I$$

regardless of how the entries b_{ij} are chosen.

Some matrices in M_3 do have *multiplicative inverses* (i.e., A has an inverse C, provided $AC = I$ and $CA = I$). Those which do are called *nonsingular*, while those which do not are called *singular*. It can be shown that a nonsingular matrix has exactly one multiplicative inverse (see problem *23, Exercise Set 9.5). If the matrix A is nonsingular, we denote its multiplicative inverse by the symbol A^{-1}. We shall tackle the problem of determining the multiplicative inverses of nonsingular matrices later in this chapter.

We close this section by demonstrating how knowledge of the inverse of the matrix of coefficients of a system of linear equations can be used to solve the system. Consider the system

$$\begin{cases} a_{11}x_1 + a_{12}x_2 + a_{13}x_3 = b_1 \\ a_{21}x_1 + a_{22}x_2 + a_{23}x_3 = b_2 \\ a_{31}x_1 + a_{32}x_2 + a_{33}x_3 = b_3. \end{cases}$$

The matrix of coefficients of this system is

$$A = \begin{bmatrix} a_{11} & a_{12} & a_{13} \\ a_{21} & a_{22} & a_{23} \\ a_{31} & a_{32} & a_{33} \end{bmatrix}.$$

If we let $X = \begin{bmatrix} x_1 \\ x_2 \\ x_3 \end{bmatrix}$ and $B = \begin{bmatrix} b_1 \\ b_2 \\ b_3 \end{bmatrix}$, then

$$AX = \begin{bmatrix} a_{11} & a_{12} & a_{13} \\ a_{21} & a_{22} & a_{23} \\ a_{31} & a_{32} & a_{33} \end{bmatrix} \begin{bmatrix} x_1 \\ x_2 \\ x_3 \end{bmatrix} = \begin{bmatrix} a_{11}x_1 + a_{12}x_2 + a_{13}x_3 \\ a_{21}x_1 + a_{22}x_2 + a_{23}x_3 \\ a_{31}x_1 + a_{32}x_2 + a_{33}x_3 \end{bmatrix}$$

$$= \begin{bmatrix} b_1 \\ b_2 \\ b_3 \end{bmatrix}$$

$$= B.$$

Thus, the given system of equations may be represented in matrix form as $AX = B$.

Suppose now that A is nonsingular and that $C = A^{-1}$. Then $CA = I$. Consequently, we have

$$C(AX) = (CA)X = IX = X.$$

Since $AX = B$ it follows, by substitution, that

$$X = CB.$$

But this equation expresses the matrix X in terms of the matrices C and B. Thus the original system of equations is now solved.

Problem. Show that the matrix $C = \begin{bmatrix} -3 & 2 & -2 \\ -1 & 0 & 1 \\ 2 & -1 & 1 \end{bmatrix}$ is the multiplicative inverse of the matrix

$$A = \begin{bmatrix} 1 & 0 & 2 \\ 3 & 1 & 5 \\ 1 & 1 & 2 \end{bmatrix}$$

and use this fact to solve the system of equations

$$\begin{cases} x_1 \qquad + 2x_3 = -4 \\ 3x_1 + x_2 + 5x_3 = 3 \\ x_1 + x_2 + 2x_3 = 2. \end{cases}$$

Solution. We have

$$CA = \begin{bmatrix} -3 & 2 & -2 \\ -1 & 0 & 1 \\ 2 & -1 & 1 \end{bmatrix} \begin{bmatrix} 1 & 0 & 2 \\ 3 & 1 & 5 \\ 1 & 1 & 2 \end{bmatrix} = \begin{bmatrix} 1 & 0 & 0 \\ 0 & 1 & 0 \\ 0 & 0 & 1 \end{bmatrix} = I.$$

Similarly, $AC = I$. Thus $C = A^{-1}$. Now the matrix A is the coefficient matrix of the given system. In matrix notation, this system may be represented as $AX = B$, where

$$X = \begin{bmatrix} x_1 \\ x_2 \\ x_3 \end{bmatrix} \quad \text{and} \quad B = \begin{bmatrix} -4 \\ 3 \\ 2 \end{bmatrix}.$$

Hence $(CA)X = CB$ or $X = CB$ and

$$X = \begin{bmatrix} x_1 \\ x_2 \\ x_3 \end{bmatrix} = CB = \begin{bmatrix} -3 & 2 & -2 \\ -1 & 0 & 1 \\ 2 & -1 & 1 \end{bmatrix} \begin{bmatrix} -4 \\ 3 \\ 2 \end{bmatrix} = \begin{bmatrix} 14 \\ 6 \\ -9 \end{bmatrix}.$$

Therefore, by the definition of equality for matrices, $x_1 = 14$, $x_2 = 6$, and $x_3 = -9$ and the triple $(14, 6, -9)$ is the solution of the system.

Exercise Set 9.5

Consider M_2, the set of all 2×2 matrices.
1. a. Show that $A + B = B + A$ for $A, B \in M_2$.
   b. Show that $A + (B + C) = A + (B + C)$ for $A, B, C \in M_2$.
   c. What is the zero matrix of M_2?
   d. Show that $A + 0 = 0 + A = A$, for $A \in M_2$.
   e. If $A = \begin{bmatrix} a_{11} & a_{12} \\ a_{21} & a_{22} \end{bmatrix}$, find $-A$.
   f. Show that $A + (-A) = 0$ for $A \in M_2$.
2. a. Give an example to show that $AB \neq BA$ for some $A, B \in M_2$.
   b. Show that $A(BC) = (AB)C$ for $A, B, C \in M_2$.
   c. What is the identity matrix for multiplication in M_2?

d. Show that $AC = CA = I$ if

$$A = \begin{bmatrix} 2 & 3 \\ 1 & 2 \end{bmatrix} \quad \text{and} \quad C = \begin{bmatrix} 2 & -3 \\ -1 & 2 \end{bmatrix}.$$

Is A nonsingular? Is C nonsingular?

3. Consider the system of linear equations

$$\begin{cases} 3x_1 + 4x_2 = -20 \\ -x_1 + 2x_2 = 5. \end{cases}$$

a. Write the coefficient matrix A of the given system of equations.

b. Let $X = \begin{bmatrix} x_1 \\ x_2 \end{bmatrix}$ and $B = \begin{bmatrix} -20 \\ 5 \end{bmatrix}.$ Show that the matrix equation $AX = B$

represents the given system of equations.

c. Show that $A^{-1} = \begin{bmatrix} \dfrac{1}{5} & -\dfrac{2}{5} \\ \dfrac{1}{10} & \dfrac{3}{10} \end{bmatrix}$

d. Find $A^{-1}B$.

e. Since $X = \begin{bmatrix} x_1 \\ x_2 \end{bmatrix} = A^{-1}B$, find x_1 and x_2.

f. What is the solution of the given system of linear equations?

* * *

4. Show that $A(BC) = (AB)C$ for $A, B, C \in M_3$.

In problems 5–12, show that $C = A^{-1}$ for the given matrices A and C.

5. $A = \begin{bmatrix} 1 & 0 \\ 0 & 1 \end{bmatrix}$ $\qquad\qquad C = \begin{bmatrix} 1 & 0 \\ 0 & 1 \end{bmatrix}$

6. $A = \begin{bmatrix} 4 & -3 \\ 2 & -1 \end{bmatrix}$ $\qquad\qquad C = \begin{bmatrix} -\dfrac{1}{2} & \dfrac{3}{2} \\ -1 & 2 \end{bmatrix}$

7. $A = \begin{bmatrix} -\dfrac{1}{7} & \dfrac{3}{7} \\ \dfrac{1}{7} & \dfrac{1}{14} \end{bmatrix}$ $\qquad\qquad C = \begin{bmatrix} -1 & 6 \\ 2 & 2 \end{bmatrix}$

8. $A = \begin{bmatrix} 2 & -1 \\ 7 & 4 \end{bmatrix}$ $\qquad\qquad C = \begin{bmatrix} -4 & -1 \\ 7 & 2 \end{bmatrix}$

9. $A = \begin{bmatrix} -1 & 2 & 1 \\ 5 & -8 & -6 \\ -3 & 5 & 4 \end{bmatrix}$ $C = \begin{bmatrix} 2 & 3 & 4 \\ 2 & 1 & 1 \\ -1 & 1 & 2 \end{bmatrix}$

10. $A = \begin{bmatrix} \dfrac{3}{7} & \dfrac{2}{7} & \dfrac{1}{7} \\[6pt] -\dfrac{1}{7} & \dfrac{4}{7} & \dfrac{2}{7} \\[6pt] \dfrac{1}{7} & \dfrac{3}{7} & \dfrac{5}{7} \end{bmatrix}$ $C = \begin{bmatrix} 2 & -1 & 0 \\ 1 & 2 & -1 \\ -1 & -1 & 2 \end{bmatrix}$

11. $A = \begin{bmatrix} 2 & 0 & 1 \\ 3 & 1 & 2 \\ 1 & 0 & 1 \end{bmatrix}$ $C = \begin{bmatrix} 1 & 0 & -1 \\ -1 & 1 & -1 \\ -1 & 0 & 2 \end{bmatrix}$

12. $A = \begin{bmatrix} 1 & -2 & 5 \\ 0 & 1 & -4 \\ -1 & 2 & -4 \end{bmatrix}$ $C = \begin{bmatrix} 4 & 2 & 3 \\ 4 & 1 & 4 \\ 1 & 0 & 1 \end{bmatrix}$

In problems 13–22, solve the system of linear equations using the fact that the matrix C is the multiplicative inverse of the coefficient matrix of the system.

13. $\begin{cases} 4x - 3y = 9 \\ 2x - y = 2 \end{cases}$ $C = \begin{bmatrix} -\dfrac{1}{2} & \dfrac{3}{2} \\[6pt] -1 & 2 \end{bmatrix}$

14. $\begin{cases} 4x - 3y = -1 \\ 2x - y = 0 \end{cases}$ $C = \begin{bmatrix} -\dfrac{1}{2} & \dfrac{3}{2} \\[6pt] -1 & 2 \end{bmatrix}$

15. $\begin{cases} -x_1 + 6x_2 = -3 \\ 2x_1 + 2x_2 = 7 \end{cases}$ $C = \begin{bmatrix} -\dfrac{1}{7} & \dfrac{3}{7} \\[6pt] \dfrac{1}{7} & \dfrac{1}{14} \end{bmatrix}$

16. $\begin{cases} 2z_1 - z_2 = 55 \\ 7z_1 + 4z_2 = -21 \end{cases}$ $C = \begin{bmatrix} -4 & -1 \\ 7 & 2 \end{bmatrix}$

17. $\begin{cases} -4t_1 - t_2 = \dfrac{1}{2} \\[6pt] 7t_1 + 2t_2 = -\dfrac{3}{2} \end{cases}$ $C = \begin{bmatrix} 2 & -1 \\ 7 & 4 \end{bmatrix}$

18. $\begin{cases} 2z_1 + 3z_2 + 4z_3 = 2 \\ 2z_1 + z_2 + z_3 = -1 \\ z_1 + z_2 + 2z_3 = 5 \end{cases}$ $C = \begin{bmatrix} -1 & 2 & 1 \\ 5 & -8 & -6 \\ -3 & 5 & 4 \end{bmatrix}$

19. $\begin{cases} -\ x_1 + 2x_2 +\ x_3 = 12 \\ \ \ \ 5x_1 - 8x_2 - 6x_3 = -8 \\ -3x_1 + 5x_2 + 4x_3 = 1 \end{cases}$ $\qquad C = \begin{bmatrix} 2 & 3 & 4 \\ 2 & 1 & 1 \\ -1 & 1 & 2 \end{bmatrix}$

20. $\begin{cases} \ \ \ x - 2y + 5z = -13 \\ \qquad\ \ y - 4z = -17 \\ -x + 2y - 4z = 19 \end{cases}$ $\qquad C = \begin{bmatrix} 4 & 2 & 3 \\ 4 & 1 & 4 \\ 1 & 0 & 1 \end{bmatrix}$

21. $\begin{cases} -2x + 2y + 3z = 2 \\[4pt] \ \ \ x + 4y -\ z = 4 \\[4pt] \ \ 6x \qquad\ + 2z = -8 \end{cases}$ $\qquad C = \begin{bmatrix} -\dfrac{1}{13} & \dfrac{1}{26} & -\dfrac{7}{52} \\[8pt] \dfrac{1}{13} & \dfrac{11}{52} & -\dfrac{1}{104} \\[8pt] \dfrac{3}{13} & -\dfrac{3}{26} & \dfrac{5}{52} \end{bmatrix}$

22. $\begin{cases} \ \ \ 3w_1 - 2w_2 + 3w_3 = 5 \\[4pt] \ \ \ 4w_1 + 2w_2 - 2w_3 = 10 \\[4pt] -2w_1 + 6w_2 + 3w_3 = 8 \end{cases}$ $\qquad C = \begin{bmatrix} \dfrac{9}{77} & \dfrac{12}{77} & -\dfrac{1}{77} \\[8pt] -\dfrac{4}{77} & \dfrac{15}{154} & \dfrac{9}{77} \\[8pt] \dfrac{2}{11} & -\dfrac{1}{11} & \dfrac{1}{11} \end{bmatrix}$

*23. Let A, B, and C be square nonsingular matrices of the same order. If $AB = BA = I$ and $AC = CA = I$, show that $B = C$.

9.6 DETERMINANTS

In this section we consider a function which assigns to each square matrix A of order n, a real number, called the *determinant* of A. After establishing a few properties of determinants we shall be able to devise a procedure for determining the inverse of any nonsingular square matrix. As indicated in the preceding section, this will provide us with a particularly elegant method of solving some systems of linear equations.

9.11 Definition. *IF $A = \begin{bmatrix} a & b \\ c & d \end{bmatrix}$ is a 2 $\times$ 2 matrix, THEN the determinant of A is the number*

$$ad - bc.$$

We denote the determinant of A by $|\,A\,|$ or $\begin{vmatrix} a & b \\ c & d \end{vmatrix}$, that is

$$|\,A\,| = \begin{vmatrix} a & b \\ c & d \end{vmatrix} = ad - bc.$$

Example. If

$$A = \begin{bmatrix} 3 & -\dfrac{1}{3} \\ 2 & -5 \end{bmatrix} \quad \text{and} \quad B = \begin{bmatrix} 4 & 3 \\ -3 & -2 \end{bmatrix},$$

then

$$|A| = \begin{vmatrix} 3 & -\dfrac{1}{3} \\ 2 & -5 \end{vmatrix} = 3(-5) - 2\left(-\dfrac{1}{3}\right) = -15 + \dfrac{2}{3} = -\dfrac{43}{3},$$

and

$$|B| = \begin{vmatrix} 4 & 3 \\ -3 & -2 \end{vmatrix} = 4(-2) - (-3)(3) = -8 - (-9) = 1.$$

The product of the matrices A and B is

$$AB = \begin{bmatrix} 3 & -\dfrac{1}{3} \\ 2 & -5 \end{bmatrix} \begin{bmatrix} 4 & 3 \\ -3 & -2 \end{bmatrix} = \begin{bmatrix} 13 & \dfrac{29}{3} \\ 23 & 16 \end{bmatrix}.$$

Thus,

$$|AB| = \begin{vmatrix} 13 & \dfrac{29}{3} \\ 23 & 16 \end{vmatrix} = 13(16) - 23\left(\dfrac{29}{3}\right) = 208 - \dfrac{667}{3}$$

$$= -\dfrac{43}{3}.$$

On the other hand, $|A||B| = \left(-\dfrac{43}{3}\right)(1) = -\dfrac{43}{3}$. Thus, we see that $|AB| = |A||B|$ for these particular matrices. In fact, it is readily proved that this result is true for all square matrices A and B of order 2 (see problem *29, Exercise Set 9.6).

Now let

$$A = \begin{bmatrix} a_{11} & a_{12} & a_{13} \\ a_{21} & a_{22} & a_{23} \\ a_{31} & a_{32} & a_{33} \end{bmatrix}$$

be a square matrix of order 3. If we delete one row and one column of A, we are left with a 2×2 matrix. For example, deletion of the first row and second column of A yields

$$\begin{bmatrix} a_{11} & a_{12} & a_{13} \\ a_{21} & a_{22} & a_{23} \\ a_{31} & a_{32} & a_{33} \end{bmatrix} \rightarrow \begin{bmatrix} a_{21} & a_{23} \\ a_{31} & a_{33} \end{bmatrix}.$$

The determinant of the 2×2 matrix obtained by crossing out the row (row i) and the column (column j) containing the element a_{ij} is called the *minor* of a_{ij}. For example, the minor of a_{12} in the matrix A is

$$\begin{vmatrix} a_{21} & a_{23} \\ a_{31} & a_{33} \end{vmatrix} = a_{21}a_{33} - a_{31}a_{23}.$$

Let us denote the minor of a_{ij} by A_{ij}. Keep in mind that A_{ij} is a number (the determinant of the matrix obtained by deleting row i and column j), and not a matrix.

9.12 Definition. *IF*

$$A = \begin{bmatrix} a_{11} & a_{12} & a_{13} \\ a_{21} & a_{22} & a_{23} \\ a_{31} & a_{32} & a_{33} \end{bmatrix}$$

*is a 3×3 matrix, THEN the number $(-1)^{i+j} A_{ij}$ is the **cofactor** of the element a_{ij} $(i = 1, 2, 3; j = 1, 2, 3)$.*

This definition shows that the cofactor of a_{ij} is either plus or minus the minor of a_{ij}. Following is an example in which we compute cofactors of two elements of a 3×3 matrix.

Example. If

$$A = \begin{bmatrix} -2 & 3 & -3 \\ 1 & 5 & 4 \\ 6 & 10 & -7 \end{bmatrix}$$

then the cofactor of $a_{23}(a_{23} = 4)$ is given by

$$(-1)^{2+3}A_{23} = (-1)^{2+3} \begin{vmatrix} -2 & 3 \\ 6 & 10 \end{vmatrix} = - \begin{vmatrix} -2 & 3 \\ 6 & 10 \end{vmatrix}$$

$$= -(-20 - 18)$$

$$= 38.$$

The cofactor of $a_{11}(a_{11} = -2)$ is

$$(-1)^{1+1}A_{11} = (-1)^{1+1} \begin{vmatrix} 5 & 4 \\ 10 & -7 \end{vmatrix} = \begin{vmatrix} 5 & 4 \\ 10 & -7 \end{vmatrix}$$
$$= -35 - 40$$
$$= -75.$$

Using the idea of cofactors, we now define the determinant of a 3 X 3 matrix.

9.13 Definition. *Let*

$$A = \begin{bmatrix} a_{11} & a_{12} & a_{13} \\ a_{21} & a_{22} & a_{23} \\ a_{31} & a_{32} & a_{33} \end{bmatrix}$$

*be a 3 X 3 matrix. The **determinant** of A, written*

$$|A| \ or \ \begin{vmatrix} a_{11} & a_{12} & a_{13} \\ a_{21} & a_{22} & a_{23} \\ a_{31} & a_{32} & a_{33} \end{vmatrix} ,$$

is

$$|A| = (-1)^{1+1}a_{11}A_{11} + (-1)^{1+2}a_{12}A_{12} + (-1)^{1+3}a_{13}A_{13}$$

$$= a_{11} \begin{vmatrix} a_{22} & a_{23} \\ a_{32} & a_{33} \end{vmatrix} - a_{12} \begin{vmatrix} a_{21} & a_{23} \\ a_{31} & a_{33} \end{vmatrix} + a_{13} \begin{vmatrix} a_{21} & a_{22} \\ a_{31} & a_{32} \end{vmatrix} .$$

We see that evaluation of the determinant of a 3 X 3 matrix involves multiplying each element of the first row by its cofactor, and then summing these products.

Example. According to the preceding definition, we see that

$$\begin{vmatrix} 3 & 1 & -2 \\ 2 & 4 & -1 \\ 3 & -3 & 5 \end{vmatrix} = (-1)^{1+1}a_{11}A_{11} + (-1)^{1+2}a_{12}A_{12} + (-1)^{1+3}a_{13}A_{13}$$

$$= (-1)^2 (3) \begin{vmatrix} 4 & -1 \\ -3 & 5 \end{vmatrix} + (-1)^3(1) \begin{vmatrix} 2 & -1 \\ 3 & 5 \end{vmatrix} + (-1)^4(-2) \begin{vmatrix} 2 & 4 \\ 3 & -3 \end{vmatrix}$$

$$= 3(20 - 3) - (10 + 3) - 2(-6 - 12)$$

$$= 74.$$

Since the determinant of a 3 × 3 matrix is defined in terms of the elements of the first row, we sometimes refer to this method of evaluation as *expansion relative to the first row*. It is possible to show that a determinant can be evaluated by expansion relative to any row or any column.

Problem. Evaluate the determinant of the preceding example by expansion relative to the second row, and then by expansion relative to the third column.

Solution. Expanding relative to the second row, we have

$$\begin{vmatrix} 3 & 1 & -2 \\ 2 & 4 & -1 \\ 3 & -3 & 5 \end{vmatrix} = (-1)^{2+1} a_{21} A_{21} + (-1)^{2+2} a_{22} A_{22} + (-1)^{2+3} a_{23} A_{23}$$

$$= (-1)^3 (2) \begin{vmatrix} 1 & -2 \\ -3 & 5 \end{vmatrix} + (-1)^4 (4) \begin{vmatrix} 3 & -2 \\ 3 & 5 \end{vmatrix} + (-1)^5 (-1) \begin{vmatrix} 3 & 1 \\ 3 & -3 \end{vmatrix}$$

$$= -2(5 - 6) + 4(15 + 6) - (-9 - 3)$$

$$= 74.$$

Expanding relative to the third column, we have

$$\begin{vmatrix} 3 & 1 & -2 \\ 2 & 4 & -1 \\ 3 & -3 & 5 \end{vmatrix} = (-1)^{1+3} a_{13} A_{13} + (-1)^{2+3} a_{23} A_{23} + (-1)^{3+3} a_{33} A_{33}$$

$$= (-1)^4 (-2) \begin{vmatrix} 2 & 4 \\ 3 & -3 \end{vmatrix} + (-1)^5 (-1) \begin{vmatrix} 3 & 1 \\ 3 & -3 \end{vmatrix} + (-1)^6 (5) \begin{vmatrix} 3 & 1 \\ 2 & 4 \end{vmatrix}$$

$$= -2(-6 - 12) + (-9 - 3) + 5(12 - 2)$$

$$= 74.$$

In order to define the determinant of an arbitrary $n \times n$ matrix (for $n \geqslant 3$), we again appeal to the concept of cofactor. If

$$A = \begin{bmatrix} a_{11} & a_{12} & \cdots & a_{1n} \\ a_{21} & a_{22} & \cdots & a_{2n} \\ \vdots & \vdots & & \vdots \\ a_{n1} & a_{n2} & \cdots & a_{nn} \end{bmatrix},$$

then the cofactor of the element a_{ij} is the number $(-1)^{i+j} A_{ij}$, where A_{ij} is the minor of a_{ij} (i.e., A_{ij} is the determinant obtained by deletion of row i and column j from A).

9.14 Definition. *IF*

$$A = \begin{bmatrix} a_{11} & a_{12} & \cdots & a_{1n} \\ a_{21} & a_{22} & \cdots & a_{2n} \\ \cdot & \cdot & & \cdot \\ \cdot & \cdot & & \cdot \\ \cdot & \cdot & & \cdot \\ a_{n1} & a_{n2} & \cdots & a_{nn} \end{bmatrix}$$

is an n × n matrix (for n ⩾ 3), THEN

$$|A| = \begin{vmatrix} a_{11} & a_{12} & \cdots & a_{1n} \\ a_{21} & a_{22} & \cdots & a_{2n} \\ \cdot & \cdot & & \cdot \\ \cdot & \cdot & & \cdot \\ \cdot & \cdot & & \cdot \\ a_{n1} & a_{n2} & \cdots & a_{nn} \end{vmatrix}$$

$$= (-1)^{1+1} a_{11} A_{11} + (-1)^{1+2} a_{12} A_{12} + \cdots + (-1)^{1+n} a_{1n} A_{1n}.$$

This definition assigns meaning for the determinant of a matrix of order n in terms of minors which are determinants of order $n - 1$. Determinants of order $n - 1$ are defined, however, in terms of determinants of order $n - 2$, etc. This will eventually reduce to a number of determinants of order 3 which we can evaluate. Thus, the evaluation of a particular determinant may require many successive applications of the definition. We now illustrate the method for finding determinants of matrices of order 4.

Problem. Evaluate $|A|$ if

$$A = \begin{bmatrix} 2 & 1 & 0 & -3 \\ 1 & 3 & 2 & 2 \\ -1 & 0 & 1 & 4 \\ 2 & 2 & 4 & -3 \end{bmatrix}.$$

Solution. Applying Definition 9.14, we have

$$|A| = \begin{vmatrix} 2 & 1 & 0 & -3 \\ 1 & 3 & 2 & 2 \\ -1 & 0 & 1 & 4 \\ 2 & 2 & 4 & -3 \end{vmatrix}$$

$$= (-1)^{1+1} a_{11} A_{11} + (-1)^{1+2} a_{12} A_{12} + (-1)^{1+3} a_{13} A_{13} + (-1)^{1+4} a_{14} A_{14}$$

$$= (-1)^2(2) \begin{vmatrix} 3 & 2 & 2 \\ 0 & 1 & 4 \\ 2 & 4 & -3 \end{vmatrix} + (-1)^3(1) \begin{vmatrix} 1 & 2 & 2 \\ -1 & 1 & 4 \\ 2 & 4 & -3 \end{vmatrix} + (-1)^4(0) \begin{vmatrix} 1 & 3 & 2 \\ -1 & 0 & 4 \\ 2 & 2 & -3 \end{vmatrix}$$

$$+ (-1)^5(-3) \begin{vmatrix} 1 & 3 & 2 \\ -1 & 0 & 1 \\ 2 & 2 & 4 \end{vmatrix}$$

$$= 2(-45) - (-21) + 3(12)$$

$$= -33.$$

The values of the determinants of the 3 × 3 matrices should be verified.

As was the case for determinants of 3 × 3 matrices, we may expand a determinant of an $n \times n$ matrix relative to any row or any column. The choice of the row or column to be used in the expansion is dictated by the ease with which the computations can be performed relative to the chosen row or column. A time-saving observation is that the row or column containing the most zeros should be used.

Problem. Evaluate the determinant

$$\begin{vmatrix} 3 & 2 & \frac{1}{2} \\ 0 & 5 & 0 \\ -1 & 2 & 4 \end{vmatrix}.$$

Solution. We expand relative to the second row, since the second row has two zero entries, obtaining

$$\begin{vmatrix} 3 & 2 & \frac{1}{2} \\ 0 & 5 & 0 \\ -1 & 2 & 4 \end{vmatrix} = (-1)^{2+1}(0) \begin{vmatrix} 2 & \frac{1}{2} \\ 2 & 4 \end{vmatrix} + (-1)^{2+2}(5) \begin{vmatrix} 3 & \frac{1}{2} \\ -1 & 4 \end{vmatrix} + (-1)^{3+2}(0) \begin{vmatrix} 3 & 2 \\ -1 & 2 \end{vmatrix}$$

$$= 5 \begin{vmatrix} 3 & \frac{1}{2} \\ -1 & 4 \end{vmatrix}$$

$$= \frac{125}{2}.$$

Note that, in the preceding computations, it is not necessary to actually evaluate the first and third determinants because they are both multiplied by zero. On the other hand, if the original determinant is computed by an expansion relative to the first row, third row, or second column, then it will be necessary to evaluate the determinants resulting in those expansions because none of the entries in those rows or columns are zero.

Problem. Evaluate the determinant

$$\begin{vmatrix} 2 & 0 & 0 & 0 & 0 \\ 0 & -1 & 0 & 0 & 0 \\ 0 & 0 & 3 & 0 & 0 \\ 0 & 0 & 0 & \frac{1}{2} & 0 \\ 0 & 0 & 0 & 0 & 4 \end{vmatrix}.$$

Solution. Expanding relative to the first row, we have

$$\begin{vmatrix} 2 & 0 & 0 & 0 & 0 \\ 0 & -1 & 0 & 0 & 0 \\ 0 & 0 & 3 & 0 & 0 \\ 0 & 0 & 0 & \frac{1}{2} & 0 \\ 0 & 0 & 0 & 0 & 4 \end{vmatrix} = 2(-1)^{1+1} \begin{vmatrix} -1 & 0 & 0 & 0 \\ 0 & 3 & 0 & 0 \\ 0 & 0 & \frac{1}{2} & 0 \\ 0 & 0 & 0 & 4 \end{vmatrix}$$

$$= 2 \begin{vmatrix} -1 & 0 & 0 & 0 \\ 0 & 3 & 0 & 0 \\ 0 & 0 & \frac{1}{2} & 0 \\ 0 & 0 & 0 & 4 \end{vmatrix} \qquad = (2)(-1)^{1+1}(-1) \begin{vmatrix} 3 & 0 & 0 \\ 0 & \frac{1}{2} & 0 \\ 0 & 0 & 4 \end{vmatrix}$$

$$= -2 \begin{vmatrix} 3 & 0 & 0 \\ 0 & \frac{1}{2} & 0 \\ 0 & 0 & 4 \end{vmatrix} \qquad = -2(-1)^{1+1}(3) \begin{vmatrix} \frac{1}{2} & 0 \\ 0 & 4 \end{vmatrix} = -6(2 - 0) = -12.$$

Note that up to this point we have defined determinants only for matrices of order n, where $n \geqslant 2$. For technical convenience, we define the determinant of a 1×1 matrix to be the single entry of the matrix, that is, if $A = [a]$ then $|A| = a$.

Exercise Set 9.6

1. Evaluate the determinant of each of the given 2×2 matrices.

a. $\begin{bmatrix} 3 & -1 \\ 2 & 4 \end{bmatrix}$ b. $\begin{bmatrix} 3 & 7 \\ -3 & -7 \end{bmatrix}$ c. $\begin{bmatrix} 0 & 8 \\ -2 & 5 \end{bmatrix}$ d. $\begin{bmatrix} \sqrt{2} & \frac{1}{2} \\ \sqrt{2} & 2\sqrt{2} \end{bmatrix}$

2. Let

$$A = \begin{bmatrix} 5 & 8 \\ 3 & -6 \end{bmatrix} \quad \text{and} \quad B = \begin{bmatrix} -1 & -11 \\ 2 & 10 \end{bmatrix}.$$

   a. Find $|A|$ and $|B|$.
   b. Find AB and $|AB|$.
   c. Compare $|AB|$ and $|A||B|$.

3. Let

$$A = \begin{bmatrix} 3 & -1 & 5 \\ 8 & 0 & 2 \\ -2 & -9 & 1 \end{bmatrix}.$$

   a. Delete the *first row* and the *third column* of A.
   b. The element a_{13} is common to the *first row* and the *third column* of A. What number is a_{13} for the given matrix A?
   c. The determinant of the matrix which remains after deletion of the first row and the third column of A is the minor A_{13} of a_{13}. What is A_{13} in this case?
   d. The cofactor of a_{13} is the number $(-1)^{1+3} A_{13}$. What is the value of the cofactor of a_{13}?
   e. In a similar fashion, determine the cofactor of a_{11} and the cofactor of a_{12}.

4. Let

$$A = \begin{bmatrix} 3 & -1 & 5 \\ 8 & 0 & 2 \\ -2 & -9 & 1 \end{bmatrix}.$$

   a. Using the results of problem 3, find a_{11} times its cofactor.
   b. Using the results of problem 3, find a_{12} times its cofactor.
   c. Using the results of problem 3, find a_{13} times its cofactor.
   d. Find the sum of these three products. This is the determinant of A.

5. a. For the matrix A of problem 4, determine the cofactors of each of the elements of the second column of A.
   b. Use the elements of the second column of A and the cofactors of these elements to find the determinant of A.
   c. Use the elements of the third row of A and the cofactors of these elements to find $|A|$.

* * *

In problems 6–9, evaluate $|A|$, $|B|$, $|A||B|$, AB, and $|AB|$.

6. $A = \begin{bmatrix} 2 & 1 \\ 3 & -5 \end{bmatrix}$, $B = \begin{bmatrix} 1 & -1 \\ 1 & 0 \end{bmatrix}$
       7. $A = \begin{bmatrix} -6 & 14 \\ 7 & -3 \end{bmatrix}$, $B = \begin{bmatrix} 1 & 0 \\ 0 & 1 \end{bmatrix}$

8. $A = \begin{bmatrix} \frac{1}{2} & \frac{1}{3} \\ \frac{2}{3} & -\frac{1}{2} \end{bmatrix}$, $B = \begin{bmatrix} 2 & 3 \\ -2 & 3 \end{bmatrix}$

9. $A = \begin{bmatrix} 1 & 1 \\ 1 & 1 \end{bmatrix}$, $B = \begin{bmatrix} 3 & 8 \\ -21 & 5 \end{bmatrix}$

In problems 10–23, find $|A|$ by expansion relative to the first row. Find $|A|$ by expansion relative to the second column.

10. $A = \begin{bmatrix} 3 & 2 & 2 \\ -1 & -3 & -2 \\ 4 & 5 & 0 \end{bmatrix}$

11. $A = \begin{bmatrix} \frac{1}{2} & 2 & -3 \\ 4 & 6 & -1 \\ 2 & \frac{1}{4} & 3 \end{bmatrix}$

12. $A = \begin{bmatrix} 1 & 0 & 0 \\ 0 & 1 & 0 \\ 0 & 0 & 1 \end{bmatrix}$

13. $A = \begin{bmatrix} 2 & 0 & 0 \\ 0 & -1 & 0 \\ 0 & 0 & 3 \end{bmatrix}$

14. $A = \begin{bmatrix} 0 & 5 & 2 \\ 9 & 4 & 0 \\ 0 & 3 & -18 \end{bmatrix}$

15. $A = \begin{bmatrix} -7 & 2 & 4 \\ 2 & 3 & \frac{1}{2} \\ 4 & 8 & -5 \end{bmatrix}$

16. $A = \begin{bmatrix} 3 & 0 & 0 \\ 5 & 7 & 0 \\ -2 & 1 & 8 \end{bmatrix}$

17. $A = \begin{bmatrix} -2 & -4 & 6 \\ 0 & 3 & 7 \\ 0 & 0 & 5 \end{bmatrix}$

18. $A = \begin{bmatrix} -2 & 1 & 4 & 2 \\ 3 & 3 & -2 & 6 \\ -1 & -1 & 2 & -2 \\ 2 & -4 & -4 & -8 \end{bmatrix}$

19. $A = \begin{bmatrix} 0 & 6 & 9 & -3 \\ 7 & 0 & -4 & 8 \\ 13 & 6 & 1 & 5 \\ 0 & -2 & -3 & 1 \end{bmatrix}$

20. $A = \begin{bmatrix} 4 & 3 & 0 & 0 \\ 0 & -2 & 0 & 1 \\ 0 & 0 & \frac{1}{2} & 0 \\ 0 & 2 & 0 & \frac{1}{3} \end{bmatrix}$

21. $A = \begin{bmatrix} 3 & 2 & 1 & 8 \\ 0 & \frac{1}{2} & .4 & 1 \\ 0 & 0 & -5 & 3 \\ 0 & 0 & 0 & 2 \end{bmatrix}$

22. $A = \begin{bmatrix} 1 & 0 & -1 & 2 \\ 0 & 2 & -3 & -2 \\ 1 & 1 & 1 & -1 \\ 4 & 0 & 2 & 1 \end{bmatrix}$

23. $A = \begin{bmatrix} 1 & 1 & 1 & 1 \\ -2 & 2 & -2 & 2 \\ -1 & 1 & 3 & 1 \\ -1 & 2 & 2 & 3 \end{bmatrix}$

24. Evaluate the determinant of each of the matrices A, B, and C.

$$A = \begin{bmatrix} 3 & -1 & 2 \\ 4 & 0 & 5 \\ 3 & -1 & 2 \end{bmatrix} \qquad B = \begin{bmatrix} 0 & 1 & 3 & 4 \\ -7 & -1 & 0 & 5 \\ 0 & 1 & 3 & 4 \\ 2 & 3 & -6 & 6 \end{bmatrix} \qquad C = \begin{bmatrix} -1 & 5 \\ -1 & 5 \end{bmatrix}$$

Note that each of the matrices has two identical rows.

25. Compare the values of the determinants

$$\begin{vmatrix} 1 & -3 \\ 2 & 4 \end{vmatrix} \quad \text{and} \quad \begin{vmatrix} 2 & 4 \\ 1 & -3 \end{vmatrix}.$$

26. Compare the values of the determinants

$$\begin{vmatrix} 2 & 4 \\ 3 & 5 \end{vmatrix} \quad \text{and} \quad \begin{vmatrix} 2k & 4k \\ 3 & 5 \end{vmatrix} \quad (k \in \mathcal{R}, k \neq 0).$$

27. Compare the values of the determinants

$$\begin{vmatrix} 5 & -2 \\ 3 & 4 \end{vmatrix} \quad \text{and} \quad \begin{vmatrix} (5+3k) & (-2+4k) \\ 3 & 4 \end{vmatrix} \quad (k \in \mathcal{R}).$$

28. Compare the values of the determinants

$$\begin{vmatrix} -3 & 3 \\ 2 & -1 \end{vmatrix} \quad \text{and} \quad \begin{vmatrix} -3 & (3-3k) \\ 2 & (-1+2k) \end{vmatrix} \quad (k \in \mathcal{R}).$$

*29. Let A and B be matrices each of order 2. Prove that $|AB| = |A||B|$.

9.7 PROPERTIES OF DETERMINANTS

Although it is certainly not difficult to evaluate determinants, the process of expansion relative to a given row or column can be rather lengthy and time-consuming if the matrix under consideration is of large order. It is the purpose of the present section to establish a few facts concerning determinants which will, in some cases, make the task of their evaluation easier.

We begin by considering the effects of elementary row operations on a determinant. Suppose that

$$A = \begin{bmatrix} 3 & 1 & 2 \\ -2 & 0 & 4 \\ 1 & 4 & 3 \end{bmatrix}.$$

The determinant of A is $|A| = -54$. Let $c \in \mathcal{R}$ with $c \neq 0$, and multiply the first row of A by c, to obtain

$$\begin{bmatrix} 3c & c & 2c \\ -2 & 0 & 4 \\ 1 & 4 & 3 \end{bmatrix}.$$

The determinant of this matrix is given by

$$\begin{vmatrix} 3c & c & 2c \\ -2 & 0 & 4 \\ 1 & 4 & 3 \end{vmatrix} = 3c \begin{vmatrix} 0 & 4 \\ 4 & 3 \end{vmatrix} - c \begin{vmatrix} -2 & 4 \\ 1 & 3 \end{vmatrix} + 2c \begin{vmatrix} -2 & 0 \\ 1 & 4 \end{vmatrix}$$

$$= 3c(-16) - c(-10) + 2c(-8) = -48c + 10c - 16c = -54c.$$

Thus, we see that

$$\begin{vmatrix} 3c & c & 2c \\ -2 & 0 & 4 \\ 1 & 4 & 3 \end{vmatrix} = c \begin{vmatrix} 3 & 1 & 2 \\ -2 & 0 & 4 \\ 1 & 4 & 3 \end{vmatrix}.$$

If we multiply any other row (or column) of A by c, then by expanding relative to that row (or column), we find that the determinant of the resulting matrix is c times the determinant of the original matrix.

Example.

$$\begin{vmatrix} \frac{1}{2} & -\frac{3}{2} & -2 \\ -\frac{4}{3} & \frac{1}{3} & \frac{7}{3} \\ 0 & -\frac{2}{5} & -\frac{1}{10} \end{vmatrix} = -\frac{1}{2} \begin{vmatrix} -1 & 3 & 4 \\ -\frac{4}{3} & \frac{1}{3} & \frac{7}{3} \\ 0 & -\frac{2}{5} & -\frac{1}{10} \end{vmatrix}$$

$$= -\frac{1}{2}\left(-\frac{1}{3}\right) \begin{vmatrix} -1 & 3 & 4 \\ 4 & 1 & -7 \\ 0 & -\frac{2}{5} & -\frac{1}{10} \end{vmatrix} = -\frac{1}{2}\left(-\frac{1}{3}\right)\left(-\frac{1}{10}\right) \begin{vmatrix} -1 & 3 & 4 \\ 4 & 1 & -7 \\ 0 & 4 & 1 \end{vmatrix}$$

$$= -\frac{1}{60}\left((-1)^{3+2}(4) \begin{vmatrix} -1 & 4 \\ 4 & -7 \end{vmatrix} + (-1)^{3+3}(1) \begin{vmatrix} -1 & 3 \\ 4 & 1 \end{vmatrix}\right)$$

$$= -\frac{1}{60}[-4(7-16) + 1(-1-12)] = -\frac{23}{60}.$$

Returning to the matrix $A = \begin{bmatrix} 3 & 1 & 2 \\ -2 & 0 & 4 \\ 1 & 4 & 3 \end{bmatrix}$, whose determinant equals -54, let us now see what effect switching the positions of two rows has on the value of the determinant. Interchanging the second and third rows, we have

$$\begin{vmatrix} 3 & 1 & 2 \\ 1 & 4 & 3 \\ -2 & 0 & 4 \end{vmatrix} = 3 \begin{vmatrix} 4 & 3 \\ 0 & 4 \end{vmatrix} - \begin{vmatrix} 1 & 3 \\ -2 & 4 \end{vmatrix} + 2 \begin{vmatrix} 1 & 4 \\ -2 & 0 \end{vmatrix}$$

$$= 3(16) - (4+6) + 2(8) = 54.$$

Thus,

$$\begin{vmatrix} 3 & 1 & 2 \\ 1 & 4 & 3 \\ -2 & 0 & 4 \end{vmatrix} = - \begin{vmatrix} 3 & 1 & 2 \\ -2 & 0 & 4 \\ 1 & 4 & 3 \end{vmatrix}.$$

Finally, let us add two rows of A together and evaluate the resulting determinant. Replacing row 1 by the sum of rows 2 and 1, we have

$$\begin{vmatrix} 1 & 1 & 6 \\ -2 & 0 & 4 \\ 1 & 4 & 3 \end{vmatrix} = 1\begin{vmatrix} 0 & 4 \\ 4 & 3 \end{vmatrix} - 1\begin{vmatrix} -2 & 4 \\ 1 & 3 \end{vmatrix} + 6\begin{vmatrix} -2 & 0 \\ 1 & 4 \end{vmatrix}$$

$$= -16 - (-6 - 4) + 6(8)$$

$$= -54.$$

Thus, the determinant of the resulting matrix is equal to the determinant of the original matrix.

The properties illustrated in the foregoing discussion are true for determinants of arbitrary matrices of order n. These properties are summarized in the following theorem.

9.15 Theorem. *Let*

$$A = \begin{bmatrix} a_{11} & a_{12} & \cdots & a_{1n} \\ a_{21} & a_{22} & \cdots & a_{2n} \\ \cdot & \cdot & & \cdot \\ \cdot & \cdot & & \cdot \\ \cdot & \cdot & & \cdot \\ a_{n1} & a_{n2} & \cdots & a_{nn} \end{bmatrix}$$

be an arbitrary square matrix of order n.

(i) *IF a row or column of A is multiplied by a constant c, THEN the determinant of the resulting matrix is $c|A|$.*

(ii) *IF the positions of two rows or of two columns of A are interchanged, THEN the determinant of the resulting matrix is $-|A|$.*

(iii) *IF a row of A is replaced by the sum of itself and any other row of A, THEN the determinant of the resulting matrix is $|A|$.*

(iv) *IF a column of A is replaced by the sum of itself and any other column of A, THEN the determinant of the resulting matrix is $|A|$.*

Consider now the effect of replacing the k-th row of A by the sum of itself and c times the j-th row of A $(c \neq 0)$. Let us illustrate the case for a 3×3 matrix. Let

$$A = \begin{bmatrix} a_{11} & a_{12} & a_{13} \\ a_{21} & a_{22} & a_{23} \\ a_{31} & a_{32} & a_{33} \end{bmatrix}$$

and suppose we replace the second row of A by the sum of itself and c times the first row. The resulting determinant is

$$\begin{vmatrix} a_{11} & a_{12} & a_{13} \\ ca_{11}+a_{21} & ca_{12}+a_{22} & ca_{13}+a_{23} \\ a_{31} & a_{32} & a_{33} \end{vmatrix} = c \begin{vmatrix} a_{11} & a_{12} & a_{13} \\ a_{11}+\dfrac{a_{21}}{c} & a_{12}+\dfrac{a_{22}}{c} & a_{13}+\dfrac{a_{23}}{c} \\ a_{31} & a_{32} & a_{33} \end{vmatrix}$$

$$= -c \begin{vmatrix} -a_{11} & -a_{12} & -a_{13} \\ a_{11}+\dfrac{a_{21}}{c} & a_{12}+\dfrac{a_{22}}{c} & a_{13}+\dfrac{a_{23}}{c} \\ a_{31} & a_{32} & a_{33} \end{vmatrix} = -c \begin{vmatrix} -a_{11} & -a_{12} & -a_{13} \\ \dfrac{a_{21}}{c} & \dfrac{a_{22}}{c} & \dfrac{a_{23}}{c} \\ a_{31} & a_{32} & a_{33} \end{vmatrix}$$

$$= c \begin{vmatrix} a_{11} & a_{12} & a_{13} \\ \dfrac{a_{21}}{c} & \dfrac{a_{22}}{c} & \dfrac{a_{23}}{c} \\ a_{31} & a_{32} & a_{33} \end{vmatrix} = \begin{vmatrix} a_{11} & a_{12} & a_{13} \\ a_{21} & a_{22} & a_{23} \\ a_{31} & a_{32} & a_{33} \end{vmatrix} = |A|.$$

We see that the determinant obtained through this process is precisely $|A|$. As before, this result holds for determinants of arbitrary order n.

Therefore, parts (iii) and (iv) of Theorem 9.15 can be stated more generally as

(iii)′ IF a row of A is replaced by the sum of itself and a constant multiple of any other row of A, THEN the determinant of the resulting matrix is $|A|$.

(iv)′ IF a column of A is replaced by the sum of itself and a constant multiple of any other column of A, THEN the determinant of the resulting matrix is $|A|$.

It is clear that a determinant having a row or a column, all of whose entries are zero, must have the value zero, since expansion relative to such a row or column results in a sum of 0's. We may deduce from this fact that if one row of a given determinant is equal to a constant times some other row, then the determinant must be 0. To verify this for the determinant of a matrix A of order 3, suppose that $a_{21} = ka_{11}, a_{22} = ka_{12}, a_{23} = ka_{13}$ for some constant k.

Thus $|A| = \begin{vmatrix} a_{11} & a_{12} & a_{13} \\ a_{21} & a_{22} & a_{23} \\ a_{31} & a_{32} & a_{33} \end{vmatrix} = \begin{vmatrix} a_{11} & a_{12} & a_{13} \\ ka_{11} & ka_{12} & ka_{13} \\ a_{31} & a_{32} & a_{33} \end{vmatrix} = k \begin{vmatrix} a_{11} & a_{12} & a_{13} \\ a_{11} & a_{12} & a_{13} \\ a_{31} & a_{32} & a_{33} \end{vmatrix}$

$= k \begin{vmatrix} a_{11} & a_{12} & a_{13} \\ 0 & 0 & 0 \\ a_{31} & a_{32} & a_{33} \end{vmatrix} = 0,$

where the row of 0's is obtained by replacing row 2 by the sum of itself and -1 times row 1.

Problem. Evaluate the determinants

$$\text{(i)} \begin{vmatrix} 3 & 2 & 0 \\ -4 & 1 & 0 \\ 5 & 3 & 0 \end{vmatrix}, \quad \text{(ii)} \begin{vmatrix} 2 & 1 & 4 & 5 \\ -3 & 6 & 7 & 2 \\ 6 & 3 & 12 & 15 \\ 0 & 2 & 0 & 9 \end{vmatrix}, \quad \text{(iii)} \begin{vmatrix} 2 & 1 & 2 \\ -3 & 4 & -3 \\ 0 & 18 & 0 \end{vmatrix}$$

Solution. (i) $\begin{vmatrix} 3 & 2 & 0 \\ -4 & 1 & 0 \\ 5 & 3 & 0 \end{vmatrix} = 0$, since the third column contains 0 in each position.

(ii) $\begin{vmatrix} 2 & 1 & 4 & 5 \\ -3 & 6 & 7 & 2 \\ 6 & 3 & 12 & 15 \\ 0 & 2 & 0 & 9 \end{vmatrix} = 0$, since the third row is 3 times the first row.

(iii) $\begin{vmatrix} 2 & 1 & 2 \\ -3 & 4 & -3 \\ 0 & 18 & 0 \end{vmatrix} = 0$, since the first and third columns are identical.

Recall that a square triangular matrix is one of the form

$$A = \begin{bmatrix} a_{11} & a_{12} & a_{13} & \cdots & a_{1n} \\ 0 & a_{22} & a_{23} & \cdots & a_{2n} \\ 0 & 0 & a_{33} & \cdots & a_{3n} \\ \vdots & \vdots & \vdots & & \vdots \\ 0 & 0 & 0 & \cdots & a_{nn} \end{bmatrix}.$$

9.16 Theorem. *IF*

$$A = \begin{bmatrix} a_{11} & a_{12} & a_{13} & \cdots & a_{1n} \\ 0 & a_{22} & a_{23} & \cdots & a_{2n} \\ 0 & 0 & a_{33} & \cdots & a_{3n} \\ \vdots & \vdots & \vdots & & \vdots \\ 0 & 0 & 0 & \cdots & a_{nn} \end{bmatrix}$$

THEN

$$|A| = a_{11} a_{22} a_{33} \cdots a_{nn}.$$

That is, the determinant of a triangular matrix A is equal to the product of the entries along the *principal diagonal* (the diagonal passing from the upper left corner of A to the lower right corner of A).

We now illustrate how the results of this section may be used to simplify the evaluation of determinants of large order.

Problem. Given

$$A = \begin{bmatrix} 6 & 2 & 4 & 0 \\ 3 & 0 & 5 & 4 \\ 1 & 1 & 8 & 2 \\ 3 & -4 & 2 & 2 \end{bmatrix},$$

find $|A|$.

Solution. Instead of simply expanding $|A|$ relative to a particular row or column, we shall attempt to express $|A|$ as the product of some number and a triangular determinant by use of elementary row and column operations.

$$|A| = \begin{vmatrix} 6 & 2 & 4 & 0 \\ 3 & 0 & 5 & 4 \\ 1 & 1 & 8 & 2 \\ 3 & -4 & 2 & 2 \end{vmatrix}$$

$$= - \begin{vmatrix} 1 & 1 & 8 & 2 \\ 3 & 0 & 5 & 4 \\ 6 & 2 & 4 & 0 \\ 3 & -4 & 2 & 2 \end{vmatrix} \quad \text{Switch rows 1 and 3.}$$

$$= - \begin{vmatrix} 1 & 1 & 8 & 2 \\ 0 & -3 & -19 & -2 \\ 6 & 2 & 4 & 0 \\ 3 & -4 & 2 & 2 \end{vmatrix} \quad \begin{array}{l}\text{Row 2 was replaced by the sum of} \\ \text{itself and } -3 \text{ times row 1.}\end{array}$$

$$= - \begin{vmatrix} 1 & 1 & 8 & 2 \\ 0 & -3 & -19 & -2 \\ 0 & -4 & -44 & -12 \\ 3 & -4 & 2 & 2 \end{vmatrix} \quad \begin{array}{l}\text{Row 3 was replaced by the sum of} \\ \text{itself and } -6 \text{ times row 1.}\end{array}$$

$$= - \begin{vmatrix} 1 & 1 & 8 & 2 \\ 0 & -3 & -19 & -2 \\ 0 & -4 & -44 & -12 \\ 0 & -7 & -22 & -4 \end{vmatrix} \quad \begin{array}{l}\text{Row 4 was replaced by the sum of} \\ \text{itself and } -3 \text{ times row 1.}\end{array}$$

$$= \begin{vmatrix} 1 & 1 & 8 & 2 \\ 0 & -4 & -44 & -12 \\ 0 & -3 & -19 & -2 \\ 0 & -7 & -22 & -4 \end{vmatrix} \quad \text{Row 2 and 3 were interchanged.}$$

$$= -4 \begin{vmatrix} 1 & 1 & 8 & 2 \\ 0 & 1 & 11 & 3 \\ 0 & -3 & -19 & -2 \\ 0 & -7 & -22 & -4 \end{vmatrix} \quad \text{Row 2 was multiplied by } -\frac{1}{4}.$$

$$= -4 \begin{vmatrix} 1 & 1 & 8 & 2 \\ 0 & 1 & 11 & 3 \\ 0 & 0 & 14 & 7 \\ 0 & -7 & -22 & -4 \end{vmatrix}$$ Row 3 was replaced by the sum of itself and 3 times row 2.

$$= -4 \begin{vmatrix} 1 & 1 & 8 & 2 \\ 0 & 1 & 11 & 3 \\ 0 & 0 & 14 & 7 \\ 0 & 0 & 55 & 17 \end{vmatrix}$$ Row 4 was replaced by the sum of itself and 7 times row 2.

$$= -4(14) \begin{vmatrix} 1 & 1 & 8 & 2 \\ 0 & 1 & 11 & 3 \\ 0 & 0 & 1 & \frac{1}{2} \\ 0 & 0 & 55 & 17 \end{vmatrix}$$ Row 3 was multiplied by $\frac{1}{14}$.

$$= -4(14) \begin{vmatrix} 1 & 1 & 8 & 2 \\ 0 & 1 & 11 & 3 \\ 0 & 0 & 1 & \frac{1}{2} \\ 0 & 0 & 0 & -\frac{21}{2} \end{vmatrix}$$ Row 4 was replaced by the sum of itself and -55 times row 3.

$$= -4(14)(1)(1)(1)\left(-\frac{21}{2}\right)$$

$$= 588.$$

Exercise Set 9.7

1. Use Theorem 9.15 i, ii, iii, or iv to justify each of the following statements:

a. $\begin{vmatrix} 2 & -3 \\ 1 & 2 \end{vmatrix} = - \begin{vmatrix} 1 & 2 \\ 2 & -3 \end{vmatrix}$ b. $\begin{vmatrix} 2 & -3 \\ 1 & 2 \end{vmatrix} = - \begin{vmatrix} -2 & 3 \\ 1 & 2 \end{vmatrix}$ c. $\begin{vmatrix} 2 & -3 \\ 1 & 2 \end{vmatrix} = \begin{vmatrix} 2 & -3 \\ 3 & -1 \end{vmatrix}$

d. $-\begin{vmatrix} 2 & -3 \\ 1 & 2 \end{vmatrix} = \begin{vmatrix} -3 & 2 \\ 2 & 1 \end{vmatrix}$ e. $\begin{vmatrix} 2 & -3 \\ 1 & 2 \end{vmatrix} = \begin{vmatrix} -1 & -3 \\ 3 & 2 \end{vmatrix}$ f. $\begin{vmatrix} 2 & -3 \\ -2 & -4 \end{vmatrix} = -2\begin{vmatrix} 2 & -3 \\ 1 & 2 \end{vmatrix}$

2. Each of the following determinants has the value zero. State the reason in each case.

a. $\begin{vmatrix} 3 & -9 & 0 \\ 2 & 4 & 7 \\ -1 & 3 & 0 \end{vmatrix}$ b. $\begin{vmatrix} 5 & 4 & -2 \\ 3 & -2 & 1 \\ 7 & 8 & -4 \end{vmatrix}$ c. $\begin{vmatrix} -3 & 2 & 4 \\ -3 & 2 & 7 \\ -3 & 2 & -3 \end{vmatrix}$

3. Find the value of k which makes each of the following statements true:

a. $\begin{vmatrix} -4 & -3 & 3 \\ 5 & 0 & 7 \\ -9 & 6 & 4 \end{vmatrix} = k \begin{vmatrix} -4 & 1 & 3 \\ 5 & 0 & 7 \\ -9 & -2 & 4 \end{vmatrix}$

b. $\begin{vmatrix} -2 & 4 & 5 \\ 0 & -3 & 3 \\ 6 & 8 & -7 \end{vmatrix} = k \begin{vmatrix} -2 & 4 & 5 \\ 0 & 1 & -1 \\ 6 & 8 & -7 \end{vmatrix}$

c. $\begin{vmatrix} \frac{3}{2} & 2 & -2 \\ \frac{8}{3} & -1 & 5 \\ -\frac{7}{6} & 4 & \frac{2}{3} \end{vmatrix} = k \begin{vmatrix} 9 & 2 & -2 \\ 16 & -1 & 5 \\ -7 & 4 & \frac{2}{3} \end{vmatrix}$

4. Let

$$|A| = \begin{vmatrix} 31 & -19 & 28 \\ 0 & -20 & 63 \\ 0 & 0 & 107 \end{vmatrix}.$$

a. Expand the given determinant relative to the first column.
b. Evaluate the determinant using the expansion of part a.
c. Evaluate the determinant using Theorem 9.16.

5. Let

$$|A| = \begin{vmatrix} 2 & -2 & 4 & 8 \\ 1 & 0 & 3 & 9 \\ 4 & 2 & -2 & 6 \\ 3 & 0 & -3 & 8 \end{vmatrix}.$$

a. Express the given determinant as the product of some number and a triangular determinant.
b. Evaluate the given determinant.

* * *

In problems 6–23, evaluate the given determinant.

6. $\begin{vmatrix} 1 & 3 & 7 & -9 \\ 0 & -1 & 2 & \frac{1}{2} \\ 0 & 0 & 1 & 14 \\ 0 & 0 & 0 & -1 \end{vmatrix}$

7. $\begin{vmatrix} 2 & 3 & 7 & -9 \\ 0 & \frac{1}{2} & 2 & -1 \\ 0 & 0 & -1 & 14 \\ 0 & 0 & 0 & 2 \end{vmatrix}$

8. $\begin{vmatrix} 1 & 3 & 1 \\ 2 & 3 & 2 \\ 0 & -5 & 4 \end{vmatrix}$

9. $\begin{vmatrix} 3 & 6 & 3 \\ 2 & 3 & 2 \\ 0 & -5 & 4 \end{vmatrix}$

10. $\begin{vmatrix} 3 & 1 & 5 \\ -2 & 0 & 2 \\ 6 & 0 & -1 \end{vmatrix}$

11. $\begin{vmatrix} 4 & -5 & 1 \\ -7 & 1 & 0 \\ -2 & -3 & 0 \end{vmatrix}$

12. $\begin{vmatrix} \frac{1}{2} & \frac{2}{3} & -1 \\ \frac{5}{12} & -\frac{2}{3} & 2 \\ 1 & 0 & 0 \end{vmatrix}$

13. $\begin{vmatrix} \frac{2}{5} & -\frac{2}{5} & \frac{4}{5} \\ 1 & 0 & 0 \\ 5 & -5 & 5 \end{vmatrix}$

14. $\begin{vmatrix} 12 & 2 & -15 \\ 0 & 1 & 0 \\ -11 & 0 & 6 \end{vmatrix}$

15. $\begin{vmatrix} 1 & 4 & 0 \\ -1 & 7 & 3 \\ 4 & 1 & 0 \end{vmatrix}$

16. $\begin{vmatrix} 4 & -2 & 6 \\ 0 & 3 & -9 \\ 2 & 4 & -2 \end{vmatrix}$

17. $\begin{vmatrix} -3 & 2 & 4 \\ 2 & 0 & -6 \\ 1 & 8 & -4 \end{vmatrix}$

18. $\begin{vmatrix} 0 & -4 & 3 \\ 2 & 1 & 7 \\ 8 & 0 & -6 \end{vmatrix}$

19. $\begin{vmatrix} 5 & 7 & 9 \\ -3 & 2 & 2 \\ 1 & 4 & 1 \end{vmatrix}$

20. $\begin{vmatrix} 9 & 0 & 0 & -4 \\ 2 & 1 & 0 & -2 \\ -3 & 0 & 3 & 0 \\ 1 & -2 & 3 & 2 \end{vmatrix}$

21. $\begin{vmatrix} 0 & 1 & 2 & -1 \\ 2 & 0 & -1 & -1 \\ -1 & 3 & 0 & 2 \\ -2 & 1 & 1 & 0 \end{vmatrix}$

22. $\begin{vmatrix} 1 & 2 & -2 & 5 \\ 2 & -3 & -5 & 1 \\ -2 & 0 & 2 & 4 \\ -1 & 1 & 5 & -3 \end{vmatrix}$

23. $\begin{vmatrix} -4 & 0 & 4 & 1 \\ 0 & 6 & -12 & -3 \\ 3 & 6 & -12 & -3 \\ 0 & 1 & -2 & 1 \end{vmatrix}$

24. Let l be a line in the plane passing through the point (x_0, y_0) and having slope m. Show that the equation of l is given by

$$\begin{vmatrix} 1 & m & 0 \\ x & y & 1 \\ x_0 & y_0 & 1 \end{vmatrix} = 0.$$

25. Let l be the line determined by the points (x_1, y_1) and (x_2, y_2). Show that the equation of l is given by

$$\begin{vmatrix} 1 & x & y \\ 1 & x_1 & y_1 \\ 1 & x_2 & y_2 \end{vmatrix} = 0.$$

9.8 MATRIX INVERSION AND CRAMER'S RULE

In this, the concluding section of the chapter, we develop two methods for solving systems of linear equations, both of which use determinants. The first method uses matrix representation which was presented in Sec. 9.5, using a technique for determining the multiplicative inverse of the coefficient matrix, if such an inverse exists. The second method is known as *Cramer's rule* and relies directly upon the evaluation of determinants which are related to the system.

We begin by considering two ways in which new matrices can be constructed from a given matrix.

9.17 Definition. *IF*

$$A = \begin{bmatrix} a_{11} & a_{12} & \cdots & a_{1n} \\ a_{21} & a_{22} & \cdots & a_{2n} \\ \cdot & \cdot & & \cdot \\ \cdot & \cdot & & \cdot \\ \cdot & \cdot & & \cdot \\ a_{m1} & a_{m2} & \cdots & a_{mn} \end{bmatrix},$$

THEN the matrix

$$A^T = \begin{bmatrix} a_{11} & a_{21} & \cdots & a_{m1} \\ a_{12} & a_{22} & \cdots & a_{m2} \\ \cdot & \cdot & & \cdot \\ \cdot & \cdot & & \cdot \\ \cdot & \cdot & & \cdot \\ a_{1n} & a_{2n} & \cdots & a_{mn} \end{bmatrix},$$

obtained by interchanging the rows and columns of A, is the **transpose** *of A.*

Example. (i) If $A = \begin{bmatrix} 3 & 1 & 2 \\ 0 & -4 & 3 \\ 7 & -9 & 6 \end{bmatrix}$, then $A^T = \begin{bmatrix} 3 & 0 & 7 \\ 1 & -4 & -9 \\ 2 & 3 & 6 \end{bmatrix}$.

(ii) If $B = \begin{bmatrix} 5 & 9 \\ -3 & 3 \\ 2 & \frac{1}{2} \end{bmatrix}$, then $B^T = \begin{bmatrix} 5 & -3 & 2 \\ 9 & 3 & \frac{1}{2} \end{bmatrix}$.

9.18 Definition. *Given a square matrix*

$$A = \begin{bmatrix} a_{11} & a_{12} & \cdots & a_{1n} \\ a_{21} & a_{22} & \cdots & a_{2n} \\ \cdot & \cdot & & \cdot \\ \cdot & \cdot & & \cdot \\ \cdot & \cdot & & \cdot \\ a_{n1} & a_{n2} & \cdots & a_{nn} \end{bmatrix},$$

let C_{ij} denote the cofactor of a_{ij} $(i, j = 1, 2, \ldots, n)$. The matrix

$$C = \begin{bmatrix} C_{11} & C_{21} & \cdots & C_{n1} \\ C_{12} & C_{22} & \cdots & C_{n2} \\ \cdot & \cdot & & \cdot \\ \cdot & \cdot & & \cdot \\ \cdot & \cdot & & \cdot \\ C_{1n} & C_{2n} & \cdots & C_{nn} \end{bmatrix}$$

is the **adjoint** *of the matrix A.*

We see that the adjoint of A is the transpose of the matrix of the cofactors of the elements of A.

Example. If $A = \begin{bmatrix} 4 & -1 \\ 5 & 6 \end{bmatrix}$ then $C_{11} = 6, C_{12} = -5, C_{21} = -(-1) = 1,$ and $C_{22} = 4.$

Thus, the adjoint of A is $C = \begin{bmatrix} 6 & -5 \\ 1 & 4 \end{bmatrix}^T = \begin{bmatrix} 6 & 1 \\ -5 & 4 \end{bmatrix}.$

Example. If

$$A = \begin{bmatrix} 1 & 3 & 2 \\ -2 & 3 & 3 \\ 4 & 0 & 1 \end{bmatrix},$$

then

$$C_{11} = \begin{vmatrix} 3 & 3 \\ 0 & 1 \end{vmatrix} = 3, \qquad C_{12} = - \begin{vmatrix} -2 & 3 \\ 4 & 1 \end{vmatrix} = 14, \qquad C_{13} = \begin{vmatrix} -2 & 3 \\ 4 & 0 \end{vmatrix} = -12,$$

$$C_{21} = - \begin{vmatrix} 3 & 2 \\ 0 & 1 \end{vmatrix} = -3, \qquad C_{22} = \begin{vmatrix} 1 & 2 \\ 4 & 1 \end{vmatrix} = -7, \qquad C_{23} = - \begin{vmatrix} 1 & 3 \\ 4 & 0 \end{vmatrix} = 12,$$

$$C_{31} = \begin{vmatrix} 3 & 2 \\ 3 & 3 \end{vmatrix} = 3, \qquad C_{32} = - \begin{vmatrix} 1 & 2 \\ -2 & 3 \end{vmatrix} = -7, \text{ and } \quad C_{33} = \begin{vmatrix} 1 & 3 \\ -1 & 2 \end{vmatrix} = 9.$$

Hence, the adjoint of A is the matrix

$$C = \begin{bmatrix} 3 & -3 & 3 \\ 14 & -7 & -7 \\ -12 & 12 & 9 \end{bmatrix}.$$

Consider now the matrix C^*, obtained by dividing each entry of C by $|A|$. For the given matrix A, $|A| = 21$ and

$$C^* = \begin{bmatrix} \frac{1}{7} & -\frac{1}{7} & \frac{1}{7} \\ \frac{2}{3} & -\frac{1}{3} & -\frac{1}{3} \\ -\frac{4}{7} & \frac{4}{7} & \frac{3}{7} \end{bmatrix}.$$

If we now multiply the matrices C^* and A, we obtain

$$C*A = \begin{bmatrix} \frac{1}{7} & -\frac{1}{7} & \frac{1}{7} \\ \frac{2}{3} & -\frac{1}{3} & -\frac{1}{3} \\ -\frac{4}{7} & \frac{4}{7} & \frac{3}{7} \end{bmatrix} \begin{bmatrix} 1 & 3 & 2 \\ -2 & 3 & 3 \\ 4 & 0 & 1 \end{bmatrix} = \begin{bmatrix} 1 & 0 & 0 \\ 0 & 1 & 0 \\ 0 & 0 & 1 \end{bmatrix} = I = AC*.$$

Thus $C* = A^{-1}$.

The technique illustrated in the preceding example can be applied in general to find the inverse of any nonsingular square matrix. We state this as a theorem which we will not prove.

9.19 Theorem. *Let*

$$A = \begin{bmatrix} a_{11} & a_{12} & \cdots & a_{1n} \\ a_{21} & a_{22} & \cdots & a_{2n} \\ \cdot & \cdot & & \cdot \\ \cdot & \cdot & & \cdot \\ \cdot & \cdot & & \cdot \\ a_{n1} & a_{n2} & \cdots & a_{nn} \end{bmatrix}$$

and let

$$C = \begin{bmatrix} C_{11} & C_{21} & \cdots & C_{n1} \\ C_{12} & C_{22} & \cdots & C_{n2} \\ \cdot & \cdot & & \cdot \\ \cdot & \cdot & & \cdot \\ \cdot & \cdot & & \cdot \\ C_{1n} & C_{2n} & \cdots & C_{nn} \end{bmatrix}$$

be the adjoint of A. IF $|A| \neq 0$, THEN A is nonsingular and $A^{-1} = \dfrac{C}{|A|}$, where $\dfrac{C}{|A|}$ is the matrix obtained by dividing each entry of C by the number $|A|$.

Example. Consider the system of equations

$$\begin{cases} -2x + 3y + z = 14 \\ x - 2y + 5z = -3 \\ -x + 4y - 7z = 7. \end{cases}$$

The matrix of coefficients of this system is

$$A = \begin{bmatrix} -2 & 3 & 1 \\ 1 & -2 & 5 \\ -1 & 4 & -7 \end{bmatrix}.$$

Letting $X = \begin{bmatrix} x \\ y \\ z \end{bmatrix}$ and $B = \begin{bmatrix} 14 \\ -3 \\ 7 \end{bmatrix}$, this system may be written in matrix form as

$$AX = B.$$

Let us solve this matrix equation by finding A^{-1}. We first determine the cofactors C_{ij} of elements of A. We have

$$C_{11} = \begin{vmatrix} -2 & 5 \\ 4 & -7 \end{vmatrix} = -6; \qquad C_{12} = -\begin{vmatrix} 1 & 5 \\ -1 & -7 \end{vmatrix} = 2; \qquad C_{13} = \begin{vmatrix} 1 & -2 \\ -1 & 4 \end{vmatrix} = 2;$$

$$C_{21} = -\begin{vmatrix} 3 & 1 \\ 4 & -7 \end{vmatrix} = 25; \qquad C_{22} = \begin{vmatrix} -2 & 1 \\ -1 & -7 \end{vmatrix} = 15; \qquad C_{23} = -\begin{vmatrix} -2 & 3 \\ -1 & 4 \end{vmatrix} = 5;$$

$$C_{31} = \begin{vmatrix} 3 & 1 \\ -2 & 5 \end{vmatrix} = 17; \qquad C_{32} = -\begin{vmatrix} -2 & 1 \\ 1 & 5 \end{vmatrix} = 11; \qquad C_{33} = \begin{vmatrix} -2 & 3 \\ 1 & -2 \end{vmatrix} = 1.$$

Hence, the adjoint of A is given by

$$C = \begin{bmatrix} -6 & 25 & 17 \\ 2 & 15 & 11 \\ 2 & 5 & 1 \end{bmatrix}.$$

Now

$$|A| = \begin{vmatrix} -2 & 3 & 1 \\ 1 & -2 & 5 \\ -1 & 4 & -7 \end{vmatrix} = -2\begin{vmatrix} -2 & 5 \\ 4 & -7 \end{vmatrix} - 3\begin{vmatrix} 1 & 5 \\ -1 & -7 \end{vmatrix} + \begin{vmatrix} 1 & -2 \\ -1 & 4 \end{vmatrix}$$

$$= -2(-6) - 3(-2) + 2$$
$$= 20.$$

Hence,

$$A^{-1} = \begin{bmatrix} -\dfrac{3}{10} & \dfrac{5}{4} & \dfrac{17}{20} \\[6pt] \dfrac{1}{10} & \dfrac{3}{4} & \dfrac{11}{20} \\[6pt] \dfrac{1}{10} & \dfrac{1}{4} & \dfrac{1}{20} \end{bmatrix}$$

Multiplying both sides of the equation $AX = B$ by A^{-1}, we have

$$(A^{-1}A)X = A^{-1}B,$$

or

$$X = A^{-1}B.$$

Now

$$A^{-1}B = \begin{bmatrix} -\dfrac{3}{10} & \dfrac{5}{4} & \dfrac{17}{20} \\[2mm] \dfrac{1}{10} & \dfrac{3}{4} & \dfrac{11}{20} \\[2mm] \dfrac{1}{10} & \dfrac{1}{4} & \dfrac{1}{20} \end{bmatrix} \begin{bmatrix} 14 \\ -3 \\ 7 \end{bmatrix} = \begin{bmatrix} -2 \\ 3 \\ 1 \end{bmatrix}.$$

Thus,

$$\begin{bmatrix} x \\ y \\ z \end{bmatrix} = \begin{bmatrix} -2 \\ 3 \\ 1 \end{bmatrix}.$$

so that $x = -2$, $y = 3$, $z = 1$.

Let us now consider *Cramer's rule*, a slightly different approach to the problem of solving linear equations. For the sake of conciseness, we shall work with systems of three equations in three unknowns. Given the system

$$\begin{cases} a_{11}x_1 + a_{12}x_2 + a_{13}x_3 = b_1 \\ a_{21}x_1 + a_{22}x_2 + a_{23}x_3 = b_2 \\ a_{31}x_1 + a_{32}x_2 + a_{33}x_3 = b_3, \end{cases}$$

let

$$A = \begin{bmatrix} a_{11} & a_{12} & a_{13} \\ a_{21} & a_{22} & a_{23} \\ a_{31} & a_{32} & a_{33} \end{bmatrix}$$

be the matrix of coefficients. Thus,

$$x_1|A| = x_1 \begin{vmatrix} a_{11} & a_{12} & a_{13} \\ a_{21} & a_{22} & a_{23} \\ a_{31} & a_{32} & a_{33} \end{vmatrix}$$

$$= \begin{vmatrix} a_{11}x_1 & a_{12} & a_{13} \\ a_{21}x_1 & a_{22} & a_{23} \\ a_{31}x_1 & a_{32} & a_{33} \end{vmatrix} \qquad \text{By part (i) of Theorem 9.15.}$$

$$= \begin{vmatrix} a_{11}x_1 + a_{12}x_2 & a_{12} & a_{13} \\ a_{21}x_1 + a_{22}x_2 & a_{22} & a_{23} \\ a_{31}x_1 + a_{32}x_2 & a_{32} & a_{33} \end{vmatrix} \qquad \text{Replace column 1 by the sum of itself and } x_2 \text{ times column 2.}$$

$$= \begin{vmatrix} a_{11}x_1 + a_{12}x_2 + a_{13}x_3 & a_{12} & a_{13} \\ a_{21}x_1 + a_{22}x_2 + a_{23}x_3 & a_{22} & a_{23} \\ a_{31}x_1 + a_{32}x_2 + a_{33}x_3 & a_{32} & a_{33} \end{vmatrix} \qquad \text{Replace column 1 by the sum of itself and } x_3 \text{ times column 3.}$$

If the triple (x_1, x_2, x_3) is to be a solution of our system, then we must have

$$\begin{vmatrix} a_{11}x_1 + a_{12}x_2 + a_{13}x_3 & a_{12} & a_{13} \\ a_{21}x_1 + a_{22}x_2 + a_{23}x_3 & a_{22} & a_{23} \\ a_{31}x_1 + a_{32}x_2 + a_{33}x_3 & a_{32} & a_{33} \end{vmatrix} = \begin{vmatrix} b_1 & a_{12} & a_{13} \\ b_2 & a_{22} & a_{23} \\ b_3 & a_{32} & a_{33} \end{vmatrix},$$

that is,

$$x_1 |A| = \begin{vmatrix} b_1 & a_{12} & a_{13} \\ b_2 & a_{22} & a_{23} \\ b_3 & a_{32} & a_{33} \end{vmatrix}.$$

Hence,

$$x_1 = \frac{\begin{vmatrix} b_1 & a_{12} & a_{13} \\ b_2 & a_{22} & a_{23} \\ b_3 & a_{32} & a_{33} \end{vmatrix}}{|A|} = \frac{\begin{vmatrix} b_1 & a_{12} & a_{13} \\ b_2 & a_{22} & a_{23} \\ b_3 & a_{32} & a_{33} \end{vmatrix}}{\begin{vmatrix} a_{11} & a_{12} & a_{13} \\ a_{21} & a_{22} & a_{23} \\ a_{31} & a_{32} & a_{33} \end{vmatrix}}.$$

An analogous argument shows that

$$x_2 = \frac{\begin{vmatrix} a_{11} & b_1 & a_{13} \\ a_{21} & b_2 & a_{23} \\ a_{31} & b_3 & a_{33} \end{vmatrix}}{|A|} = \frac{\begin{vmatrix} a_{11} & b_1 & a_{13} \\ a_{21} & b_2 & a_{23} \\ a_{31} & b_3 & a_{33} \end{vmatrix}}{\begin{vmatrix} a_{11} & a_{12} & a_{13} \\ a_{21} & a_{22} & a_{23} \\ a_{31} & a_{32} & a_{33} \end{vmatrix}}$$

and

$$x_3 = \frac{\begin{vmatrix} a_{11} & a_{12} & b_1 \\ a_{21} & a_{22} & b_2 \\ a_{31} & a_{32} & b_3 \end{vmatrix}}{|A|} = \frac{\begin{vmatrix} a_{11} & a_{12} & b_1 \\ a_{21} & a_{22} & b_2 \\ a_{31} & a_{32} & b_3 \end{vmatrix}}{\begin{vmatrix} a_{11} & a_{12} & a_{13} \\ a_{21} & a_{22} & a_{23} \\ a_{31} & a_{32} & a_{33} \end{vmatrix}}.$$

The conclusions of our discussion are presented in the following theorem stated in its general form.

9.20 Theorem. (*Cramer's rule*): *Let the linear system of n equations in n unknowns*

$$\begin{cases} a_{11}x_1 + a_{12}x_2 + \cdots + a_{nn}x_n = b_1 \\ a_{21}x_1 + a_{22}x_2 + \cdots + a_{2n}x_n = b_2 \\ \quad \cdot \qquad \cdot \qquad \qquad \cdot \qquad \cdot \\ \quad \cdot \qquad \cdot \qquad \qquad \cdot \qquad \cdot \\ \quad \cdot \qquad \cdot \qquad \qquad \cdot \qquad \cdot \\ a_{n1}x_1 + a_{n2}x_2 + \cdots + a_{nn}x_n = b_n, \end{cases}$$

be given. IF A denotes the matrix of coefficients, and B(j) denotes the matrix obtained by replacing the j-th column of A by the column whose entries are $b_1, b_2, \ldots b_n$, THEN the solution of the system is given by

$$x_1 = \frac{|B(1)|}{|A|}, x_2 = \frac{|B(2)|}{|A|}, \ldots, x_n = \frac{|B(n)|}{|A|},$$

provided $|A| \neq 0$. IF $|A| = 0$, THEN the system does not have a unique solution.

Problem. Solve the system

$$\begin{cases} -2x + 3y + z = 14 \\ x - 2y + 5z = -3 \\ -x + 4y - 7z = 7 \end{cases}$$

using Cramer's rule.

Solution. The matrix of coefficients is

$$A = \begin{bmatrix} -2 & 3 & 1 \\ 1 & -2 & 5 \\ -1 & 4 & -7 \end{bmatrix}.$$

Thus,

$$|A| = \begin{vmatrix} -2 & 3 & 1 \\ 1 & -2 & 5 \\ -1 & 4 & -7 \end{vmatrix} = 20.$$

Now

$$|B(1)| = \begin{vmatrix} 14 & 3 & 1 \\ -3 & -2 & 5 \\ 7 & 4 & -7 \end{vmatrix} = 14 \begin{vmatrix} -2 & 5 \\ 4 & -7 \end{vmatrix} - 3 \begin{vmatrix} -3 & 5 \\ 7 & -7 \end{vmatrix} + \begin{vmatrix} -3 & -2 \\ 7 & 4 \end{vmatrix}$$

$$= 14(-6) - 3(-14) + 2 = -40,$$

$$|B(2)| = \begin{vmatrix} -2 & 14 & 1 \\ 1 & -3 & 5 \\ -1 & 7 & -7 \end{vmatrix} = -2\begin{vmatrix} -3 & 5 \\ 7 & -7 \end{vmatrix} - 14\begin{vmatrix} 1 & 5 \\ -1 & -7 \end{vmatrix} + \begin{vmatrix} 1 & -3 \\ -1 & 7 \end{vmatrix}$$

$$= -2(-14) - 14(-2) + 4 = 60,$$

and

$$|B(3)| = \begin{vmatrix} -2 & 3 & 14 \\ 1 & -2 & -3 \\ -1 & 4 & 7 \end{vmatrix} = -2\begin{vmatrix} -2 & -3 \\ 4 & 7 \end{vmatrix} - 3\begin{vmatrix} 1 & -3 \\ -1 & 7 \end{vmatrix} + 14\begin{vmatrix} 1 & -2 \\ -1 & 4 \end{vmatrix}$$

$$= -2(-2) - 3(4) + 14(2) = 20.$$

Thus,

$$x = \frac{|B(1)|}{|A|} = \frac{-40}{20} = -2,$$

$$y = \frac{|B(2)|}{|A|} = \frac{60}{20} = 3,$$

and

$$z = \frac{|B(3)|}{|A|} = \frac{20}{20} = 1.$$

In this chapter we have developed three matrix methods for solving systems of linear equations, namely, reduction of the augmented matrix to triangular form, inversion of the matrix of coefficients, and Cramer's rule. The particular method used in a given situation depends primarily on the personal preference of the user. It is well to mention that any of these methods may be used in conjunction with a digital computer to solve systems of large dimension in a relatively short length of time. We again emphasize that the value of matrices is in no way confined to the study of systems of equations, though the presentation of matrices in connection with systems of equations serves to promote the basic skills required for effective application of matrix theory to other topics.

Exercise Set 9.8

1. For each given matrix A, find the transpose, A^T (see Definition 9.17).

a. $A = \begin{bmatrix} 2 & 4 & -1 \\ 3 & 0 & 6 \\ 1 & -1 & -4 \end{bmatrix}$ b. $A = \begin{bmatrix} \dfrac{5}{3} & \dfrac{2}{5} & -\dfrac{1}{8} \\ \dfrac{21}{6} & -7 & \dfrac{3}{4} \\ -\dfrac{6}{7} & \dfrac{5}{3} & \dfrac{2}{7} \end{bmatrix}$ c. $A = \begin{bmatrix} a & b & c \\ d & e & f \\ g & h & i \end{bmatrix}$

2. Let

$$A = \begin{bmatrix} 2 & -1 & 3 \\ -1 & 4 & 1 \\ 1 & 2 & -2 \end{bmatrix}.$$

a. Let C_{ij} be the cofactor of a_{ij} for each element a_{ij} in the matrix A. Find $C_{11}, C_{12}, C_{13}, C_{21}, C_{22}, C_{23}$, and C_{31}, C_{32}, and C_{33}.

b. Find the matrix C, the adjoint of the matrix A (see Definition 9.18).

3. Let

$$A = \begin{bmatrix} 2 & -1 & 3 \\ -1 & 4 & 1 \\ 1 & 2 & -2 \end{bmatrix}.$$

a. Find $|A|$.

b. Use Theorem 9.19 and the results of problem 2(b) to find A^{-1}.

4. Given the system of linear equations

$$\begin{cases} 2x - y + 3z = -2 \\ -x + 4y + z = 7 \\ x + 2y - 2z = -2 \end{cases}$$

the coefficient matrix is

$$A = \begin{bmatrix} 2 & -1 & 3 \\ -1 & 4 & 1 \\ 1 & 2 & -2 \end{bmatrix}.$$

Use the results of problem 3(b) to solve the system of equations.

5. Given the system of linear equations

$$\begin{cases} 2x - y + 3z = -2 \\ -x + 4y + z = 7 \\ x + 2y - 2z = -2 \end{cases}$$

use Cramer's rule, Theorem 9.20, to solve the system.

* * *

In problems 6–11 determine whether A is singular or nonsingular. If A is nonsingular then find A^{-1}.

6. $A = \begin{bmatrix} 2 & -3 \\ 5 & 2 \end{bmatrix}$

7. $A = \begin{bmatrix} 14 & -12 \\ 6 & -5 \end{bmatrix}$

8. $A = \begin{bmatrix} 2 & -4 & 5 \\ 1 & 3 & 6 \\ -2 & 0 & 4 \end{bmatrix}$

9. $A = \begin{bmatrix} 2 & 3 & -1 \\ 1 & -2 & 2 \\ 4 & 4 & -2 \end{bmatrix}$

10. $A = \begin{bmatrix} 3 & 8 & 4 \\ 2 & -1 & 3 \\ -1 & \dfrac{1}{2} & -\dfrac{3}{2} \end{bmatrix}$

11. $A = \begin{bmatrix} 1 & 3 & 0 & 3 \\ 2 & 0 & 4 & -2 \\ 4 & -8 & 2 & 0 \\ 0 & 0 & 1 & 5 \end{bmatrix}$

In problems 12–17, use the results of problems 6–11 to solve the given system of equations. Also, solve the given system using Cramer's rule.

12. $\begin{cases} 2x - 3y = 4 \\ 5x + 2y = -3 \end{cases}$

13. $\begin{cases} 14r - 12s = 8 \\ 6r - 5s = 6 \end{cases}$

14. $\begin{cases} 2x - 4y + 5z = 0 \\ x + 3y + 6z = 12 \\ -2x + 4z = -2 \end{cases}$

15. $\begin{cases} 2x_1 + 3x_2 - x_3 = -4 \\ x_1 - 2x_2 + 2x_3 = 8 \\ 4x_1 + 4x_2 - 2x_3 = 10 \end{cases}$

16. $\begin{cases} 3u + 8v + 4w = 0 \\ 2u - v + 3w = -1 \\ -u + \dfrac{1}{2}v - \dfrac{3}{2}w = -4 \end{cases}$

17. $\begin{cases} y_1 + 3y_2 + 3y_4 = 6 \\ 2y_1 + 4y_3 - 2y_4 = -4 \\ 4y_1 - 8y_2 + 2y_3 = 4 \\ y_3 + 5y_4 = 24 \end{cases}$

Appendix I | Exponential Functions

x	e^x	e^{-x}	x	e^x	e^{-x}
0.00	1.0000	1.0000	2.0	7.3891	0.1353
0.05	1.0513	0.9512	2.1	8.1662	0.1225
0.10	1.1052	0.9048	2.2	9.0250	0.1108
0.15	1.1618	0.8607	2.3	9.9742	0.1003
0.20	1.2214	0.8187	2.4	11.0232	0.0907
0.25	1.2840	0.7788	2.5	12.1825	0.0821
0.30	1.3499	0.7408	2.6	13.4637	0.0743
0.35	1.4191	0.7047	2.7	14.8797	0.0672
0.40	1.4918	0.6703	2.8	16.4446	0.0608
0.45	1.5683	0.6376	2.9	18.1741	0.0550
0.50	1.6487	0.6065	3.0	20.0855	0.0498
0.55	1.7333	0.5769	3.1	22.1980	0.0450
0.60	1.8221	0.5488	3.2	24.5325	0.0408
0.65	1.9155	0.5220	3.3	27.1126	0.0369
0.70	2.0138	0.4966	3.4	29.9641	0.0334
0.75	2.1170	0.4724	3.5	33.1155	0.0302
0.80	2.2255	0.4493	3.6	36.5982	0.0273
0.85	2.3396	0.4274	3.7	40.4473	0.0247
0.90	2.4596	0.4066	3.8	44.7012	0.0224
0.95	2.5857	0.3867	3.9	49.4024	0.0202
1.0	2.7183	0.3679	4.0	54.5982	0.0183
1.1	3.0042	0.3329	4.1	60.3403	0.0166
1.2	3.3201	0.3012	4.2	66.6863	0.0150
1.3	3.6693	0.2725	4.3	73.6998	0.0136
1.4	4.0552	0.2466	4.4	81.4509	0.0123
1.5	4.4817	0.2231	4.5	90.0171	0.0111
1.6	4.9530	0.2019	4.6	99.4843	0.0101
1.7	5.4739	0.1827	4.7	109.9471	0.0091
1.8	6.0496	0.1653	4.8	121.5104	0.0082
1.9	6.6859	0.1496	4.9	134.2898	0.0074
			5.0	148.4132	0.0067
			6.0	403.4288	0.0025
			7.0	1096.6332	0.0009
			8.0	2980.9580	0.0003
			9.0	8103.0840	0.0001
			10.0	22026.4658	0.00005

Appendix II | *Natural Logarithm Function*

x	ln x	x	ln x	x	ln x
0.0	......	3.5	1.2528	7.0	1.9459
0.1	−2.3026	3.6	1.2809	7.1	1.9601
0.2	−1.6094	3.7	1.3083	7.2	1.9741
0.3	−1.2040	3.8	1.3350	7.3	1.9879
0.4	−0.9163	3.9	1.3610	7.4	2.0015
0.5	−0.6931	4.0	1.3863	7.5	2.0149
0.6	−0.5108	4.1	1.4110	7.6	2.0281
0.7	−0.3567	4.2	1.4351	7.7	2.0412
0.8	−0.2231	4.3	1.4586	7.8	2.0541
0.9	−0.1054	4.4	1.4816	7.9	2.0669
1.0	0.0000	4.5	1.5041	8.0	2.0794
1.1	0.0953	4.6	1.5261	8.1	2.0919
1.2	0.1823	4.7	1.5476	8.2	2.1041
1.3	0.2624	4.8	1.5686	8.3	2.1163
1.4	0.3365	4.9	1.5892	8.4	2.1282
1.5	0.4055	5.0	1.6094	8.5	2.1401
1.6	0.4700	5.1	1.6292	8.6	2.1518
1.7	0.5306	5.2	1.6487	8.7	2.1633
1.8	0.5878	5.3	1.6677	8.8	2.1748
1.9	0.6419	5.4	1.6864	8.9	2.1861
2.0	0.6931	5.5	1.7047	9.0	2.1972
2.1	0.7419	5.6	1.7228	9.1	2.2083
2.2	0.7885	5.7	1.7405	9.2	2.2192
2.3	0.8329	5.8	1.7579	9.3	2.2300
2.4	0.8755	5.9	1.7750	9.4	2.2407
2.5	0.9163	6.0	1.7918	9.5	2.2513
2.6	0.9555	6.1	1.8083	9.6	2.2618
2.7	0.9933	6.2	1.8245	9.7	2.2721
2.8	1.0296	6.3	1.8406	9.8	2.2824
2.9	1.0647	6.4	1.8563	9.9	2.2925
3.0	1.0986	6.5	1.8718	10.0	2.3026
3.1	1.1314	6.6	1.8871		
3.2	1.1632	6.7	1.9021		
3.3	1.1939	6.8	1.9169		
3.4	1.2238	6.9	1.9315		

Solutions to Exercises

1. a) T   b) F   c) F   d) T   e) F   f) T   g) F   h) T
2. a) T   b) F   c) T   d) T   e) T   f) T
3. a) {Bob, Carol, Ted, Alice}
   b) {Sunday, Monday, Tuesday, Wednesday, Thursday, Friday, Saturday}
   c) {7}   d) { }   e) {3,cat,?,T}
   f) {(1,1), (1,2), (1,3), (1,4), (1,5), (1,6), (2,1), (2,2), (2,3), (2,4), (2,5), (2,6), (3,1), (3,2), (3,3), (3,4), (3,5), (3,6), (4,1), (4,2), (4,3), (4,4), (4,5), (4,6), (5,1), (5,2), (5,3), (5,4), (5,5), (5,6), (6,1), (6,2), (6,3), (6,4), (6,5), (6,6)}
4. a) {x: x is a living man}   b) {y: y is a doctor and a surgeon}
   c) {x: x is a vowel}   d) {e: e is an even integer}
   e) {s: s is a student who does not hold a scholarship}
   f) {x: x is a man and x weighs 3 tons}
5. a) T   c) F   e) F
6. a) $5 \in$ {4,5,John Doe}   c) $3 \notin$ {\$,a,Hollywood}   e) $12 \in$ {12}
   g) {a,g,h} $\notin$ {□: □ is a letter of the alphabet}

1. For example: a) B = {2,3}, C = {1,2,3,4}   b) B = {t}, C = {1,b}
   c) B = {9}, C = {△}   d) B = {x: x is a college student}, C = {j,v}
   e) B = {Florida, Ohio}, C = {x: x is a city in the U.S.}
   f) B = { }, C = {*,△}   g) B = {Atlantic Ocean}, C = {Hudson River}
   h) B = {p: p is a selling price greater than \$15}, C = {a,b,\$2}
2. For example: a) {1}, {3,1}, {1,2,3}   b) {a}, {1,q}, {a,q,m}
   c) {9}, {4}, { }   d) {Los Angeles}, {Baltimore}, {y: y is a city in Iowa}
   e) {z: z is a person under 17 years of age}, {b: b is a baby boy}, {t: t is a girl under 18 years of age}   f) {Jean}, {Jill}, {Jane}
   g) {q: q is a jet pilot}, {q: q is a licensed pilot}, {q: q is a pilot}
   h) {Arnold Palmer}, {Fred Flintstone}, {Margaret Chase Smith}

3. a) T   c) T   e) F   g) F   4. a) T   c) T   e) F
5. a) $\subseteq$ c) $\in$ e) $\subseteq$ g) $\subseteq$ i) $\in$

Exercise Set 1.3

1. $A \subseteq B, B \subseteq A, A = B$ 2. $B \subseteq A$ 3. $A \subseteq B, B \subseteq A, A = B$
4. $A \subseteq B, B \subseteq A, A = B$ 5. Neither set is a subset of the other
6. Neither set is a subset of the other
7. $A \subseteq B, B \subseteq A, A = B$ 9. $A \subseteq B, B \subseteq A, A = B = \{\ \}$ 11. $A \subseteq B$
13. F   15. F   17. F   19. F   21. F   23. F

Exercise Set 1.4

1. $3, -3, (2)(3), 2 + 3, 2 - 3, (3) \cdot (-3), 1,067,582, 0, -5280$

2.
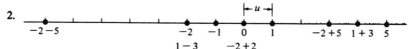

3. a) $3 \cdot 4$ b) $(-3) \cdot (-4)$ c) $(-1) \cdot (-10)$ d) $(4) \cdot (-3)$
5. a) $1,2; 1,2,3,4,6,12; 1,2,3,4,5,6,10,12,15,20,30,60; 1,3,7,21; 1,7; 1$
   c) $\{x: x \text{ is an integer}\}$
6. a) $\{5n: n \text{ is a positive integer}\} = \{5,10,15,\dots\}$
   c) $\{5n: n \text{ is an integer}\} = \{0,5,-5,10,-10,15,-15,\dots\}$
7. $1100, 902, 1100, -902$ 9. $-28, -28, -28, 28$
11. $-15, 15, -8, 3000$ 13. $90, 90, -90, -90$ 15. $m \cdot n, -(m \cdot n)$
17. a) 7   b) 1   c) 0   d) -3 e) -17

Exercise Set 1.5

1. $15 = \dfrac{15}{1}$, $-2 = \dfrac{-2}{1}, \dfrac{15}{-2}$, $1.5 = \dfrac{15}{10}$, $\dfrac{3}{7+9} = \dfrac{3}{16}$, $\dfrac{3}{7-9} = \dfrac{3}{-2}$, $4 - 15 = \dfrac{-11}{1}$, $0 = \dfrac{0}{7}$,
$\dfrac{251}{75} + \dfrac{681}{796} = \dfrac{250,871}{59,700}$, $\dfrac{251}{75} \cdot \dfrac{681}{796} = \dfrac{170,931}{59,700}$

2.
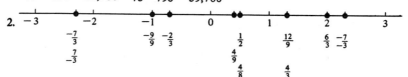

3. $\dfrac{3}{12} = \dfrac{7}{28}$, $\dfrac{-3}{8} = \dfrac{3}{-8}$, $\dfrac{19}{21} \neq \dfrac{23}{25}$, $\dfrac{-12}{51} = \dfrac{-16}{68}$
4. $r + s = \dfrac{31}{15}$, $r \cdot s = \dfrac{2}{3}$, $r/s = \dfrac{6}{25}$
   $r + s = \dfrac{-103}{72}$, $r \cdot s = \dfrac{-5}{6}$, $r/s = \dfrac{-135}{32}$
   $r + s = \dfrac{49}{10}$, $r \cdot s = \dfrac{18}{5}$, $r/s = \dfrac{40}{9}$
   $r + s = \dfrac{69}{-56}$, $r \cdot s = \dfrac{9}{28}$, $r/s = \dfrac{7}{16}$
5. a) $A \subseteq B$ b) $A \subseteq B$ c) Neither set is a subset of the other   d) $B \subseteq A$ e) $A = B$
   f) $A = B$
7. $\dfrac{3}{2}, \dfrac{-3}{2}, \dfrac{-2}{.3}, \dfrac{7943}{2158}, -\dfrac{1}{22}, \dfrac{45}{1}, \dfrac{13}{12}, \dfrac{4}{5}$
9. a) $\dfrac{7}{4}, \dfrac{7}{4}$ b) $\dfrac{4}{7}, \dfrac{14}{5}$ c) $\dfrac{9}{14}, 2$ d) $\dfrac{31}{9}, 7$ e) $12, 12$ f) $0, 0$

Exercise Set 1.6

1. Integers: $7, \frac{6}{3}, -4, 0, \left(\frac{1}{2}\right)^{-1}, \frac{8 \cdot 6}{4 \cdot 2}$

   Rational but not integers: $\frac{1}{3}, \frac{-3}{2}, \frac{8+6}{4+2}$

2. a) Multiplicative inverse   b) Multiplicative identity
   c) Commutative property for addition   d) Distributive property
   e) Equality of rational numbers   f) Commutative property for multiplication
   g) Associative property for addition   h) Additive identity   i) Distributive property
   j) Associative property for multiplication   k) Property of additive inverse

3. a) $-(-a) = a$ b) $(-a)(b) = -(ab)$ c) $-(a + b) = (-a) + (-b) = -a - b$
   d) If $a + b = 0$, then $b = -a$ e) If $ab = 1$, then $b = a^{-1}$ f) $(a^{-1})^{-1} = a$
   g) $(-a)(-b) = ab$ h) $(a^{-1}b^{-1}) = (ba)^{-1}$ i) $a \cdot 0 = 0$ j) If $a + c = b + c$, then $a = b$
   k) If $ac = bc$ and $c \neq 0$, then $a = b$ l) If $ab = 0$, then $a = 0$ or $b = 0$

5. -10 7. -192 9. $\frac{69}{56}$ 11. $\frac{-27}{10}$ 13. $\frac{7\sqrt{2}}{10}$ 15. $\frac{6 + 2\sqrt{3}}{3\sqrt{3}}$ 17. 10 19. $\frac{320}{9}$

21. $\frac{3}{2}$ 23. $\frac{3\pi}{2 + 3\pi}$ 25. 0

27.
$$(-6)\left[1 - \left(\frac{1}{-2}\right)\right] = (-6)\left[1 + \frac{1}{2}\right] \qquad \text{Definition of subtraction}$$
$$= (-6)(1) + (-6)\left(\frac{1}{2}\right) \qquad \text{Distributive property}$$
$$= -6 + -3 \qquad (-a)(b) = -(ab)$$
$$= -9 \qquad -(a + b) = (-a) + (-b)$$

29.
$$\frac{1}{12}\left[\frac{1}{5} - \frac{\frac{2}{3}}{\frac{2}{7}}\right] = \frac{1}{12}\left[\frac{1}{5} - \frac{7}{3}\right] \qquad \frac{a}{b} = ab^{-1}$$
$$= \frac{1}{12}\left[\frac{1}{5} + \frac{-7}{3}\right] \qquad \text{Definition of subtraction}$$
$$= \frac{1}{12}\left[\frac{-32}{15}\right] \qquad \frac{a}{b} + \frac{c}{d} = \frac{ad + bc}{bd}$$
$$= \frac{-32}{180} \qquad \frac{a}{b} \cdot \frac{c}{d} = \frac{ac}{bd}$$
$$= \frac{-8}{45} \qquad \text{Equality of rational numbers}$$

Exercise Set 1.7

1. a) $<$ b) $>$ c) $<$ d) $<$ e) $=$ f) $>$ g) $>$ h) $<$ i) $>$ j) $>$
2. a) F   b) F   c) F   d) T   e) F   f) T   g) T   h) T
3. If $x < 3$, then $x + 2 < 3 + 2$ or $x + 2 < 5$
5. If $x - 7 < -12$, then $x - 7 + 4 < -12 + 4$ or $x - 3 < -8$
7. If $t > 3$, then $12 \cdot t > 12 \cdot 3$ or $12t > 36$
9. If $-1 < t + 2 < 1$, then $-1 - 2 < t + 2 - 2 < 1 - 2$ or $-3 < t < -1$

11. $y > 0$ 13. $2r < 3$ 15. $-\frac{1}{2} \leqslant r + 1 \leqslant \frac{1}{2}$ 17. $r < 1$

19. $1 < x + 2 < 6$
21. $b - a$ is positive   Definition of $a < b$
   $c - b$ is positive   Definition of $b < c$
   $b - a + c - b$ is positive   If $a, b \in P$, then $a + b \in P$
   $b - a + c - b = c - a$ is positive $b - b = 0$
   $a < c$ Definition $a < c$

23. $b - a$ is positive   Definition $a < b$

    c is positive   Given

    $(b - a) \cdot c$ is positive   Order axiom

    $bc - ac$ is positive   Distributive property

    $ac < bc$ Definition of $<$

25. If $a \neq 0$, then either $a > 0$ or $a < 0$. If $a > 0$, then a is positive, whence $a^2 = a \cdot a$ is positive. That is, $a^2 > 0$. If $a < 0$, then $-a$ is positive and $(-a) \cdot (-a) = a^2$ is positive. Again, $a^2 > 0$.

Exercise Set 1.8

1. a) No   b) No   c) Yes

2. a) $c > d \therefore C$ lies to the right of D b) $c < d \therefore C$ lies to the left of D

   c) $c > d \therefore C$ lies to the right of D d) $c < d \therefore C$ lies to the left of D

3. a) 38   b) 38   c) $\sqrt{7}$ e) $a > 0, |38 - \sqrt{7}| = 38 - \sqrt{7}$

   f) $a < 0, |-38 + \sqrt{7}| = 38 - \sqrt{7}$ g) 0   h) $|\pi - 3| = \pi - 3$

4. $\overline{AB} = b - a, |AB| = |b - a|$

   a) $\overline{AB} = \frac{3}{4}, |AB| = \frac{3}{4}$ b) $\overline{AB} = \frac{5}{6}, |AB| = \frac{5}{6}$ c) $\overline{AB} = 1, |AB| = 1$

   d) $\overline{AB} = -2\sqrt{7}, |AB| = 2\sqrt{7}$ e) $\overline{AB} = -7, |AB| = 7$

   f) $\overline{AB} = -2 + \sqrt{3}, |AB| = 2 - \sqrt{3}$

5. Open intervals: c,d,f   Closed intervals: a,b,e,g

6. a) T   c) F, $|-2| + |3| = 2 + 3 \neq |-2 + 3|$ e) T

   g) F, For counterexample see 6.c.

7. a) T   c) F   e) F   g) F

8. a) $\left\{ x: \ -\frac{2}{3} < x < \frac{1}{3} \right\}$ c) $\left\{ x: \ -\sqrt{5} \leqslant x \leqslant -\frac{\sqrt{2}}{2} \right\}$ e) $\left\{ x: \ 1.2 < x < 2.1 \right\}$

9. a) $[1,75]$ c) $(\sqrt{21}, 22)$ e) $\left[-\frac{19}{5}, -2 \right]$

11. a)

      −1 0 1 2 3 4 5 6   b) 0   1   2   3   c) 0 10 20 30 40 50

   d) −40 −20 0   e) 0   1   f) 0   1

Exercise Set 2.1

1. a,b,c,d,f,g   2. a) $3x + (-8) = 0$ b) $x + 12 = 0$ c) $6x + (-2) = 0$

   d) $-2x + 1 = 0$ e) $-8x + 0 = 0$ f) $-\frac{1}{15}x + \frac{5}{4} = 0$ g) $(c - d)x + (p + r) = 0$

3. a) $a = 3, b = -8$ b) $a = 1, b = 12$ c) $a = 6, b = -2$ d) $a = -2, b = 1$

   e) $a = -8, b = 0$ f) $a = -\frac{1}{15}, b = \frac{5}{4}$ g) $a = c - d, b = p + r$

4. a) $7 + 4$ or 11   b) $7 - 12$ or -5 c) $3(7)$ or 21   d) $\left(\frac{-4}{3} \right) 7$ or $\frac{-28}{3}$

   e) 6   f) 6   g) -1 h) -4

5. $\left\{ -\frac{1}{3} \right\}$ 7. $\{-2\}$ 9. $\{-9\}$ 11. $\{\ \}; 0y - 1 \neq 0$, No solution

13. $\left\{ \frac{-224}{225} \right\}$ 15. $\{-12\}$ 17. $\left\{ \frac{5}{14} \right\}$ 19. $\left\{ -\frac{b}{a} \right\}$

Exercise Set 2.2

1. a) $3x + 5$ is greater than 1   b) $3x + 5$ is less than or equal to 1

   c) $-5x - 1$ is greater than or equal to 20   2. a,b,d

3. a) $<$ b) $>$ c) $\geqslant$ d) $>$ e) $\geqslant$ f) $>$ g) $\geqslant$ h) $>$

4. a) 0   b) -11 c) 0   d) -9 e) 7   f) $-3 < x + 1 < 0$ g) $0 < x < \dfrac{5}{2}$ h) $\dfrac{5}{2}$

5. $-3x + 5 > 0$, $-3x + 5 - 5 > 0 - 5$, $-3x > -5$, $\left(-\dfrac{1}{3}\right)(-3x) < \left(-\dfrac{1}{3}\right)(-5)$, $x < \dfrac{5}{3}$.

   Solution set: $x: \left\{ x < \dfrac{5}{3} \right\}$

7. $\{t: \ t > -1\}$ 9. $\left\{ y: \ y > -\dfrac{3}{4} \right\}$ 11. $\left\{ \dfrac{5}{3} \right\}$ 13. $\left\{ x: \ x > \dfrac{2}{7} \right\}$

15. $\{W: \ W < 660\}$ 17. $\{\ \}$ 19. $\left\{ x: \ x < -\dfrac{b}{a} \right\}$

Exercise Set 2.3

1. a) The horizontal line   b) The vertical line
   c) The intersection of the horizontal and vertical lines
   d) 6   e) -8 f) 3   g) -5 h) 0

2. a) $(6,3)$ b) $(6,-5)$ c) $(1,1)$ d) $(-5,-3)$ e) $(6,0)$
   f) $(0,-5)$ g) $(-2,6)$ h) $(-8,0)$

3.

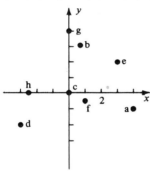

5.

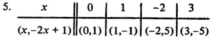

x	0	1	-2	3
$(x,-2x + 1)$	$(0,1)$	$(1,-1)$	$(-2,5)$	$(3,-5)$

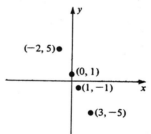

7.

x	-1	$-\dfrac{1}{2}$	0	$\dfrac{1}{2}$	1	$\dfrac{3}{2}$
$(x,4x + 3)$	$(-1,-1)$	$\left(-\dfrac{1}{2},1\right)$	$(0,3)$	$\left(\dfrac{1}{2},5\right)$	$(1,7)$	$\left(\dfrac{3}{2},9\right)$

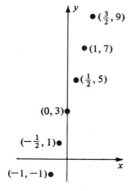

9. For example:

$$(-1,-1), (0,0), \left(\frac{4}{3}, \frac{4}{3}\right), (\pi, \pi)$$

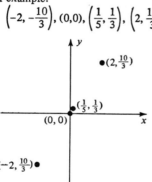

11. For example:

$$\left(-\frac{5}{4}, \frac{5}{2}\right), (0,0), (1,-2), \left(\frac{15}{7}, \frac{-30}{7}\right)$$

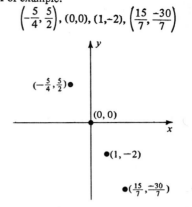

13. For example:

$$\left(-2, -\frac{10}{3}\right), (0,0), \left(\frac{1}{5}, \frac{1}{3}\right), \left(2, \frac{10}{3}\right)$$

15. For example:

$$(-\sqrt{3}, 3.2), (-1, 3.2), (0, 3.2), (1.1, 3.2)$$

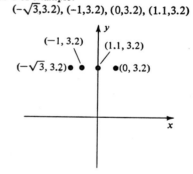

Exercise Set 2.4

1. a) 49   b) $\frac{4}{9}$ c) $\frac{121}{16}$ d) 2.999824   e) 16   f) 1296   g) 1296

   h) 0   i) 5   j) $\frac{5}{22}$ k) a l) c

2. a) 3   b) $\frac{2}{3}$ c) 4   d) 3   e) 8   f) 8   g) 7   h) 0   i) π j) 1.1

   k) 0.8   l) approximately 0.253

3. a) 5   b) 7   c) 6   d) approximately 7.98   e) 49   f) 29   g) $11 - 6\sqrt{2}$

   h) −7

4. a) $x_1 = 2, y_1 = 5$ b) $x_2 = 3, y_2 = -4$

   c) $\sqrt{(x_2 - x_1)^2 + (y_2 - y_1)^2} = \sqrt{(3-2)^2 + (-4-5)^2} = \sqrt{(1)^2 + (-9)^2} = \sqrt{82}$

   d) $x_1 = 3, y_1 = -4$ e) $x_2 = 2, y_2 = 5$

   f) $\sqrt{(x_2 - x_1)^2 + (y_2 - y_1)^2} = \sqrt{(2-3)^2 + (5+4)^2} = \sqrt{(-1)^2 + (9)^2} = \sqrt{82}$

5. $3\sqrt{2}$ 7. $\sqrt{85}$ 9. $\frac{2}{3}\sqrt{10}$ 11. $\sqrt{47}$ 13. a) $\left(\frac{1+4}{2}, \frac{9+6}{2}\right)$ or $\left(\frac{5}{2}, \frac{15}{2}\right)$

b) $\left(\frac{-1+1}{2}, \frac{5-4}{2}\right)$ or $\left(0, \frac{1}{2}\right)$ c) $\left(-\frac{1}{2}, -1\right)$ d) $\left(-\frac{3}{5}, -\frac{13}{12}\right)$

15. Let $A = (0,2)$, $B = (\sqrt{3}, -1)$, $C = (0,-2)$, $|AB| = \sqrt{12}$, $|BC| = 2$, $|AC| = 4$.
Hence, $|AB|^2 + |BC|^2 = 12 + 4 = 16|AC|^2$. Therefore triangle ABC is a right triangle.

17. $|x - y| = \begin{cases} x - y, \text{ if } x \geqslant y \\ y - x, \text{ if } x < y \end{cases}$
If $x \geqslant y$, then $|x - y|^2 = (x - y)^2$, and if $x < y$,
then $|x - y|^2 = (y - x)^2 = [(-1)(x - y)]^2 = (-1)^2(x - y)^2 = (x - y)^2$. Hence, for all cases, $|x - y|^2 = (x - y)^2$.

*19. Let $A = (7,0)$, $B = (4,1)$, $C = 6,7)$.

Area of triangle $ABC = \frac{1}{2}|BC| \cdot |BA| = \frac{1}{2}\sqrt{40} \cdot \sqrt{10} = \frac{1}{2}\sqrt{400} = 10$.

The midpoints are $D = (5,4)$, $E = \left(\frac{13}{2}, \frac{7}{2}\right)$, $F = \left(\frac{11}{2}, \frac{1}{2}\right)$.

Area of triangle $DEF = \frac{1}{2}|DE| \cdot |EF| = \frac{1}{2}\sqrt{\frac{10}{4}}\sqrt{10} = \frac{10}{4}$.

Therefore, the area of $\triangle DEF = \frac{1}{4}$ of the area of $\triangle ABC$.

Exercise Set 2.5

1.

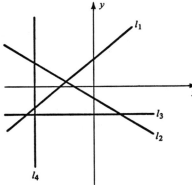

l_1 has positive slope
l_2 has negative slope
l_3 has zero slope
the slope of l_4 is undefined

2. a) $\frac{9-3}{4-2} = \frac{6}{2} = 3$ b) $\frac{3-9}{4-2} = \frac{-6}{2} = -3$ c) $\frac{-8-6}{3+4} = \frac{-14}{7} = -2$ d) $\frac{-5}{9}$ e) $\frac{4}{9}$

f) $\frac{7}{3}$ g) 0 h) $\frac{-7}{26}$

3. a) $\frac{y-5}{x+3} = 1$ or $y - 5 = x + 3$ or $y = x + 8$

b) $\frac{y-5}{x+3} = 4$ or $y - 5 = 4(x + 3)$ or $y = 4x + 17$ c) $y = -2x - 1$ d) $y = 5$

4. a)

x	0	1	-1	4	-4	$\frac{1}{2}$	$-\frac{1}{2}$	$\frac{7}{2}$	$-\frac{7}{2}$
y	6	4	8	-2	14	5	7	-1	13

b)   c)

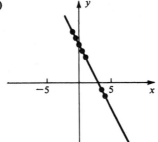

d) $y = -2x + 6$

5. a) $\dfrac{y+2}{x-3} = \dfrac{-3}{5}$

   $5y + 10 = -3x + 9$
   $3x + 5y = -1$

   b) $\dfrac{y-2}{x-3} = \dfrac{-3}{5}$

   $5y - 10 = -3x + 9$
   $3x + 5y = 19$

7. a) $\dfrac{y + \dfrac{2}{3}}{x - \dfrac{1}{2}} = \dfrac{3}{11}$

   $11y + \dfrac{22}{3} = 3x - \dfrac{3}{2}$

   $-3x + 11y = \dfrac{-53}{6}$

   or $\quad 3x - 11y = \dfrac{53}{6}$

   b) $\dfrac{y + \dfrac{3}{2}}{x - 2} = \dfrac{3}{11}$

   $11y + \dfrac{33}{2} = 3x - 6$

   $-3x + 11y = \dfrac{-45}{2}$

   $3x - 11y = \dfrac{45}{2}$

9. a) $3x + y = 10$ b) $x - y = -6$ 11. a) $x + 6y = 0$ b) $x + 6y = 0$

13. a) $y = -18$ b) $x = 4$ 15. a) $x = 3$ b) $y = \sqrt{2}$

17. x-intercept: a) $(1,0)$ b) $(1,0)$ c) $\left(\dfrac{2}{3}, 0\right)$ d) $\left(\dfrac{7}{48}, 0\right)$

    y-intercept: a) $(0,1)$ b) $(0,-1)$ c) $\left(0, \dfrac{1}{3}\right)$ d) $\left(0, -\dfrac{7}{45}\right)$

19. a) l_1: $2x + 2y = 7$

    b) l_2: $2x + 2y = -7$

    a)

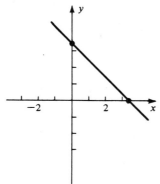

b)

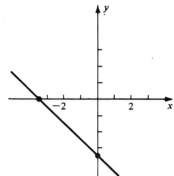

21. a) l_1: $\frac{2}{7}x - 2y = \frac{33}{4}$ a)
                     and

    b) l_2: $-\frac{2}{7}x - 2y = \frac{33}{4}$ b)

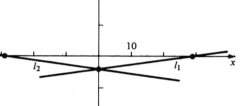

23. 1) $d_3 \leqslant d_1 + d_2$. If $d_3 < d_1 + d_2$, then the three points are the vertices of a triangle. If $d_3 = d_1 + d_2$, then the three points are collinear.

2) $(x_1, y_1), (x_2, y_2), (x_3, y_3) \in \{(x, y): Ax + By = C\}$

3) For example:

$$d_1 = \sqrt{(x_2 - x_1)^2 + (y_2 - y_1)^2} = \sqrt{(x_2 - x_1)^2 + \left(\left[\frac{-A}{B}x_2 + \frac{C}{B}\right] - \left[\frac{-A}{B}x_1 + \frac{C}{B}\right]\right)^2}$$

$$= \sqrt{(x_2 - x_1)^2 + \frac{A^2}{B^2}(x_2 - x_1)^2} = \sqrt{\left(1 + \frac{A^2}{B^2}\right)(x_2 - x_1)^2}$$

$$= \sqrt{\left(1 + \frac{A^2}{B^2}\right)}(|x_2 - x_1|) = \left(\sqrt{1 + \frac{A^2}{B^2}}\right)(x_2 - x_1)$$

Exercise Set 2.6

1. a,b,c   2. a) −5   b) −11   c) −20 or any other number less than −5
   d) −12 or any other number less than −11   e) 2 or any other number greater than −5.
   f) −10.9 or any other number greater than −11.

3. a), b), c), d) See below   e) Yes

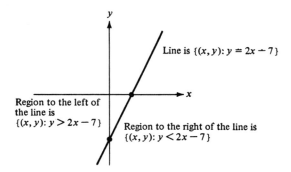

Line is $\{(x, y): y = 2x - 7\}$

Region to the left of the line is $\{(x, y): y > 2x - 7\}$

Region to the right of the line is $\{(x, y): y < 2x - 7\}$

4. The graph of the inequality $3x + 2y \geqslant 4$ includes the graph of the inequality $3x + 2y > 4$, also the graph of the line $3x + 2y = 4$.

5. a,c,f

7. a)

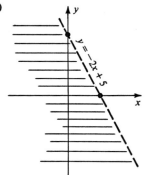

b)

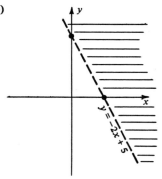

9. a)

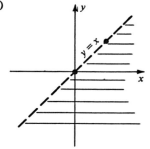

b)

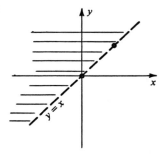

11.

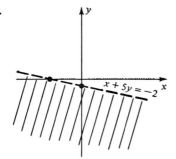

$x + 5y = -2$

13.

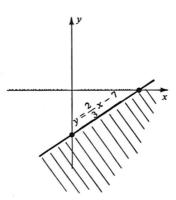

$y = \frac{2}{3}x - 7$

15.

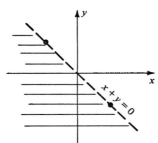

$x + y = 0$

17.

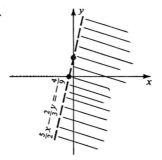

$\frac{5}{2}x - \frac{2}{3}y = -\frac{4}{9}$

19.

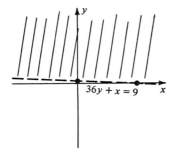

$36y + x = 9$

Exercise Set 2.7

1. a) $A \cup B = \{-1,0,1,2\}, A \cap B = \{0,1\}$
   b) $A \cup B = \{1,2,3,4,5,6,7\}, A \cap B = \{1,2,3,4\}$
   c) $A \cup B = \{x: \ x > 0\} \quad A \cap B = \{x: \ x \text{ is a positive integer}\}$
   d) $A \cup B = \mathcal{R}, A \cap B = \{-1\}$ e) $A \cup B = \mathcal{R}, A \cap B = \{x: \ -5 < x \leqslant 2\}$
   f) $A \cup B = \mathcal{R}, A \cap B = \{ \ \}$ g) $A \cup B = \{x: \ x > -3\}, A \cap B = \{2\}$
   h) $A \cup B = \mathcal{R}, A \cap B = \{ \ \}$

2. For example:
   a) $\{t: \ t < 2\} \cup \{2\}$ b) $\{y: \ 2y < x + 1\} \cup \{y: \ 2y = x + 1\}$
   c) $\{s: \ -s > u + 1\} \cup \{s: \ -s = u + 1\}$ d) $\{x: \ x < -1\} \cup \{x: \ x > 1\}$
   e) $\{1\} \cup \{-1\}$ f) $\{2\} \cup \{-3\}$

3. For example:

a) $\{t: t < 2\} \cap \{t: t > 0\}$ b) $\left\{y: y < \frac{x}{2}\right\} \cap \{y: y > x + 1\}$

c) $\{M: M$ is a multiple of $2\} \cap \{M: M$ is a multiple of $5\}$

d) $\{x: x > 2\} \cap \{x: x < 4\}$ e) $\{x: x > 0\} \cap \{i: i$ is an integer$\}$

f) $\{y: y$ is an integer$\} \cap \{y: y$ is an even number$\}$

4. a) 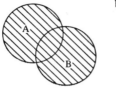   b) 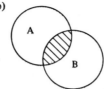   c)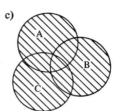

d) 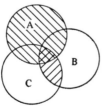   e) 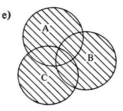   f)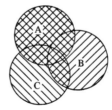

5. $\{1\}$ 7. $\{x: -1 < x\}$ 9. $\{x: -1 \leqslant x \leqslant 1\}, A \cap (B \cup D) \subset (A \cap B) \cup D$

11. $\{-1, 0, 1\}, (A \cap B) \cup (A \cap D) = A \cap (B \cup D)$

13. $B \cup C = \{x: x > -1\}, C \cup D = \{x: -1 \leqslant x < 4\},$

$(B \cup C) \cap (C \cup D) = \{x: -1 < x < 4\}$

15. i)

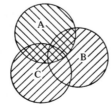

$A \cup (B \cap C)$ is the shaded region

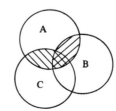

$(A \cup B) \cap (A \cup C)$ is the shaded region

ii)

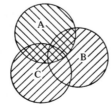

$A \cap (B \cup C)$ is the doubly shaded region

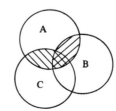

$(A \cap B) \cup (A \cap C)$ is the shaded region

Exercise Set 2.8

1. a) $y = \frac{3}{7}x - 2$ b) $y = -x - 4$ c) $y = \frac{5}{9}x + \frac{2}{9}$ d) $y = \frac{2}{9}x - \frac{2}{3}$

   e) $y = \frac{-118}{95}x + \frac{436}{95}$ f) $y = \frac{4}{\sqrt{2}}x + 0$ or $y = 2\sqrt{2}x + 0$

2. a) $\frac{3}{7}$ b) -1 d) $\frac{5}{9}$ d) $\frac{2}{9}$ e) $\frac{-118}{95}$ f) $2\sqrt{2}$

3. For example: a) $y = -\frac{3}{2}x + 12$ b) $y = 5x + \frac{5}{3}$ c) $-6y = 9x - 10$

4. a) $y = x + 5,$ $m = 1$ $B = 9$ ⎤ Lines are identical

   $y = \frac{\sqrt{2}}{\sqrt{2}}x + \frac{5\sqrt{2}}{\sqrt{2}},$ $m = 1$ $B = 5$ ⎦ System is dependent

   b) $y = \frac{-17}{13}x + \frac{8}{13},$ $m = \frac{-17}{13},$ $B = \frac{8}{13}$ ⎤ Lines are parallel and distinct

   $y = \frac{-34}{26}x + \frac{24}{26},$ $m = \frac{-17}{13}$ $B = \frac{12}{13}$ ⎦ System is inconsistent

   c) $y = \frac{-4}{3}x + 4,$ $m = \frac{-4}{3}$ $B = 4$ ⎤ Lines are not parallel

   $y = \frac{-3}{4}x + \frac{11}{2},$ $m = \frac{-3}{4}$ $B = \frac{11}{2}$ ⎦ System is consistent

   d) Consistent   e) Inconsistent   f) Dependent

5. $\{(x, y) : x = \frac{14}{3}, y = -2\}$ 7. $\{(p,q) : p = \frac{29}{40}, q = \frac{-1}{40}\}$

9. $\{(0,0)\}$ 11. $\{(\frac{21}{4}, \frac{3}{10})\}$ 13. $\{(2, \frac{-9}{2})\}$ 15. Dependent system

17. $\{(\frac{c}{a+b}, \frac{c}{a+b})\}$, provided $a^2 - b^2 \neq 0$

19. $\{(\frac{4}{\sqrt{2}+\sqrt{3}}, \frac{5\sqrt{3}+\sqrt{2}}{3+\sqrt{6}})\}$

21. $x + y = 10$ Solution: $x = 6, y = 4$
    $x = y + 2$

23. $q = d + 3$ Solution: q, the number of quarters = 16
    $25q + 10d = 530$ d, the number of dimes = 13

25. $c + d = 2$ Solution: c, pounds of 60 cent candy = .8
    $.60c + .75d = 1.38$ d, pounds of 75 cent candy = 1.2

27. $x + y = 10,000$ Solution: x, dollars in high risk fund = 4000
    $.07x + .05y = 580$ y, dollars in less speculative fund = 6000

29. $x = y + 135$ Solution: x, airplane speed = 180 m.p.h.
    $\frac{380}{x} = \frac{110}{y}$ y, car speed = 55 m.p.h.

31. Each linear equation represents a line. Hence, the system of three equations is consistent provided that the three lines meet in a single point.

Exercise Set 2.9

1.

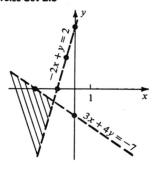

2.

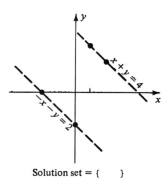

Solution set = {   }

3.

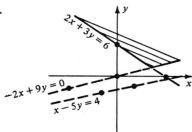

5.

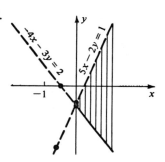

7.

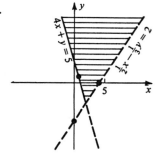

9.

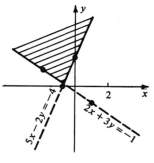

11.

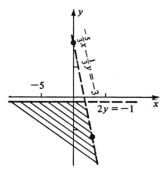

13.

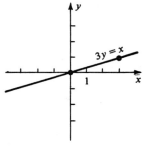

15.

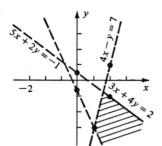

17.

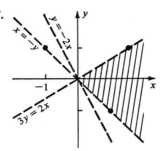

19.

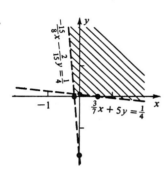

21. Let x = number of blackwalls
   y = number of whitewalls
   $x + y \le 5000$
   $3x \le y$
   $x \ge 0$
   $y \ge 0$

23. Let x = number of blackwalls
   y = number of whitewalls
   $x \ge 500$
   $0 \le y \le 4000$
   $x + y \le 5000$
   $3x \le y$

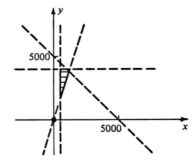

25. Let x = number of ordinary consultants
    y = number of special consultants
    $x \geqslant 1, y \geqslant 1$
    $50x + 100y < 2000$
    $x \geqslant 2y$

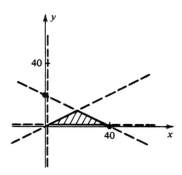

Exercise Set 3.1

1. September, December, February
2. $-30, 3,000,738, 0$
3. $100, 100, 100$
4. a) $D_f = \{h:\ h \text{ is an interstate highway}\}$
   b) $D_f = \{T:\ T \text{ is a professional baseball team}\}$
   c) $D_f = \{r:\ r \text{ is a rational number and } r \neq 0\}$
   d) $D_f = \{x:\ x < 0\}$
5. $0, -1, 0, 39,999, b^2 - 1, (a+b)^2 - 1$
7. $1, 1/2, 2, 1/10, 10, 1/500, 500$
9. $2, 1, 9/4, 7/2, 2 - \frac{1}{2}h, 2 - \frac{1}{2}(x+h)$
11. $16, 10, -26, 24, -6\sqrt{2} + 4$
13. $-3/8, -3/8, -3/8, -3/8, -3/8, -3/8$
15. $4, 3, 3, 16/3, -\sqrt{2} + 2$
17. $D_f = \{x:\ x \neq -8\}$
19. $D_g = \mathcal{R}$
21. $D_f = \mathcal{R}$
23. $i(5) = \$1404, i(7) = \$1965.60, i(31) = \$8704.80$
25. a) Given x, there is not a unique number y assigned to x.
    b) Unless the particular hour is specified, the number of ships passing through a given lock is not uniquely determined.
    c) For each $x > 0$, there are two distinct values of y assigned to it. If $x < 0$, there are no values of y assigned to it.

Exercise Set 3.2

1. $(1, 5), (-1, -1), (8/5, 34/5)$
2. $(1, 4), (-1, 6), (\sqrt{2}, 5 - \sqrt{2})$
3.

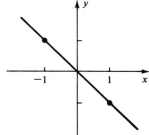

4.

x	1	2	3	4	$\frac{1}{2}$	$\frac{1}{3}$	$\frac{1}{4}$	-1	-2	-3	-4	$-\frac{1}{2}$	$-\frac{1}{3}$	$-\frac{1}{4}$
$g(x)$	1	$\frac{1}{4}$	$\frac{1}{9}$	$\frac{1}{16}$	4	9	16	1	$\frac{1}{4}$	$\frac{1}{9}$	$\frac{1}{16}$	4	9	16

5.

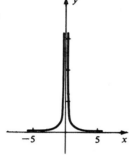

7.

x	-3	-1
$f(x)$	0	4

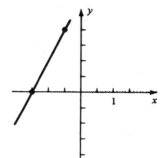

9.

x	-1	3
$G(x)$	0	-1

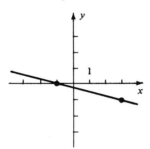

11.

x	0	-2
y	2	-1

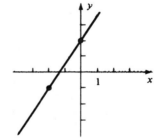

13.

x	-2	1	2	3
$k(x)$	0	3	6	9

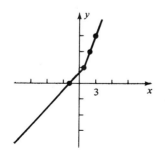

15.

x	-1	$-\frac{1}{2}$	0	$\frac{1}{2}$	1
$h(x)$	3	$\frac{9}{4}$	2	$\frac{9}{4}$	3

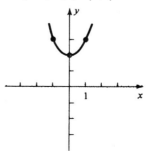

17.

x	0	2
F(x)	−1	4
G(x)	4	$\frac{16}{5}$

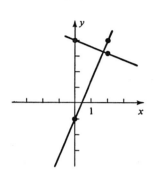

19.

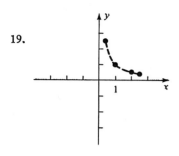

Exercise Set 3.3

1. $\{-2, 1\}$

2.

	2	(x + 2)	(x − 1)	2(x + 2)(x − 1)
x < −2	+	−	−	+
−2 < x < 1	+	+	−	−
1 < x	+	+	+	+

3.

x	−4	−3	−2.5	−2	−1	−0.5	0	0.5	1	1.5	2	3	4
f(x)	20	8	3.5	0	−4	−4.5	−4	−2.5	0	3.5	8	20	36

4.

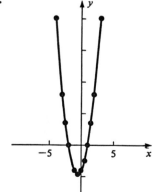

5. a) $\{-2, 1\}$

	-2	$(x+2)$	$(x-1)$	$-2(x+2)(x-1)$
$x < -2$	$-$	$-$	$-$	$-$
$-2 < x < 1$	$-$	$+$	$-$	$+$
$1 < x$	$-$	$+$	$+$	$-$

b)

c)

x	-4	-3	-2.5	-2	-1	-0.5	0	0.5	1	1.5	2	3	4
$f(x)$	-20	-8	-3.5	0	4	4.5	4	2.5	0	-3.5	-8	-20	-36

d)

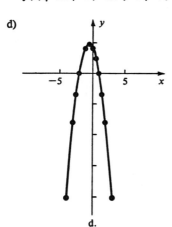

d.

7. $\{2/3, -4\}$ **9.** $\{-4/5, -1\}$ **11.** $\{3/8\}$

13. $\{x: f(x) > 0\} = \{x: x < -4\} \cup \{x: x > 2/3\}$
$\{x: f(x) < 0\} = \{x: -4 < x < 2/3\}$

15. $\{x: f(x) > 0\} = \{x: x < -1\} \cup \{x: x > -4/5\}$
$\{x: f(x) < 0\} = \{x: -1 < x < -4/5\}$

17. $\{x: f(x) > 0\} = \{x: x \neq 3/8\}$
$\{x: f(x) < 0\} = \{\quad\}$

19. Vertex: $(-5/3, -49/9)$ **21.** Vertex: $(-1, 0)$

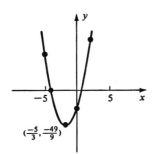

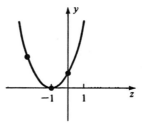

23. Vertex: $(-9/10, -1/20)$ 25. Vertex: $(1, 0)$

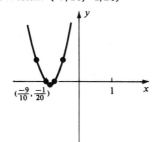

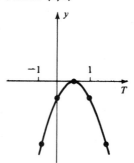

Exercise Set 3.4

1. a) $a = 3, b = 4, c = -5$ b) $a = -2, b = 1, c = 8$ c) $a = 2, b = 1, c = -15$
   d) $a = 1, b = 0, c = -16$ e) $a = -2/3, b = -3/7, c = 21/255$ f) $a = 1, b = -2, c = 1$
2. a) $b = -3, c = 9/4, (-3/2)^2 = 9/4, (x - 3/2)^2$
3. a) $4x^2 + 24x + 36 = 4(x^2 + 6x + 9)$. For $x^2 + 6x + 9$, $b = 6, c = 9$, and $(6/2)^2 = 9$
   b) $4(x^2 + 6x + 9) = 4(x + 3)^2$
4. $b = 4, (b/2)^2 = (4/2)^2 = 4$ $F(x) = x^2 + 4x + 4 - 3 - 4 = (x + 2)^2 - 7$
5. $G(x) = 2(x^2 + 3x + 5/2) = 2(x^2 + 3x + 9/4 + 5/2 - 9/4)$
   $$= 2[(x + 3/2)^2 + 1/4] = 2(x + 3/2)^2 + 1/2$$
7. No   9. Yes, $(t - 2/5)^2$ 11. No   13. Yes, $9(x - 5/3)^2$ 15. No

17. $(x - 3)^2 - 13$ 19. $(r + 1/3)^2 + 5/36$ 21. $(t - 3/14)^2 + \dfrac{641}{980}$

23. $2(x + 5/4)^2 - 41/8$ 25. $9(s - 25/6)^2 + \dfrac{2615}{4}$

Exercise Set 3.5

1. a) $f(x) = (x - 5)^2 - 16$
   b) $f(x) = 0$ if, and only if, $(x - 5)^2 = 16$; that is, $x - 5 = \pm 4$. Hence, $f(x) = 0$ if, and only
      if, $x = 9$ or $x = 1$.   c) $f(x) = (x - 9)(x - 1)$
2. $a = 18, b = -9, c = -20$
$$\left\{\frac{9 + \sqrt{81 + 1440}}{36}, \frac{9 - \sqrt{81 + 1440}}{36}\right\} = \{4/3, -5/6\}$$

3. $18x^2 - 9x - 20 = 18\left(x - \dfrac{4}{3}\right)\left(x + \dfrac{5}{6}\right)$
4. a) $b^2 - 4ac = 784 - 784 = 0$; One solution.
   b) $b^2 - 4ac = 1 - 12 = -11$; No solutions.
   c) $b^2 - 4ac = 1 + 4 = 5$; Two solutions.
5. a) and c) are reducible; b) is irreducible.
   a) $4x^2 - 28x + 49 = 4(x - 7/2)^2$
   c) $x^2 + x - 1 = \left(x - \dfrac{-1 + \sqrt{5}}{2}\right)\left(x - \dfrac{-1 - \sqrt{5}}{2}\right)$
7. $u = \dfrac{5 + \sqrt{13}}{2}$ or $u = \dfrac{5 - \sqrt{13}}{2}$
9. $w = \dfrac{11 + \sqrt{193}}{6}$ or $w = \dfrac{11 - \sqrt{193}}{6}$ 11. 21   13. 0

15. Zeros: 2, 8; $g(x) = \frac{1}{2}(x-2)(x-8)$ 17. Zeros: $-1/2, 5; R(x) = \frac{2}{3}\left(x+\frac{1}{2}\right)(x-5)$

19. Zeros: $\dfrac{-15-\sqrt{15}}{10}, \dfrac{-15+\sqrt{15}}{10}$; $U(V) = 10\left(V-\dfrac{-15-\sqrt{15}}{10}\right)\left(V-\dfrac{-15+\sqrt{15}}{10}\right)$

21. Zeros: $0, \dfrac{1}{\sqrt{7}}$; $f(t) = \sqrt{7}\,t\left(t-\dfrac{1}{\sqrt{7}}\right)$

23. $h(x) = \dfrac{\sqrt{2}}{2}\left(x - \dfrac{\sqrt{5}-\sqrt{\frac{11}{3}}}{\sqrt{2}}\right)\left(x - \dfrac{\sqrt{5}+\sqrt{\frac{11}{3}}}{\sqrt{2}}\right)$ 25. Irreducible

Exercise Set 3.6

1. $-\dfrac{b}{2a} = \dfrac{5}{6}, \dfrac{4ac-b^2}{4a} = \dfrac{23}{12}, \left(\dfrac{5}{6}, \dfrac{23}{12}\right)$

2. $-\dfrac{b}{2a} = 1; \dfrac{4ac-b^2}{4a} = 0, (1, 0)$

3. $\left\{\dfrac{2}{3}, -1\right\}, \left(-\dfrac{1}{6}, -\dfrac{25}{12}\right)$

4.

x	-1	$\frac{2}{3}$	$-\frac{1}{6}$	0	$\frac{1}{2}$	$-\frac{1}{2}$	2	-2
$F(x)$	0	0	$-\frac{25}{12}$	-2	$-\frac{3}{4}$	$-\frac{7}{4}$	12	8

5.

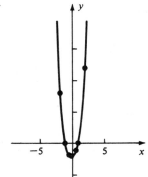

7.

x	1	-1	0	$1/2$	$-1/2$
$Q(x)$	0	0	-1	$-3/4$	$-3/4$

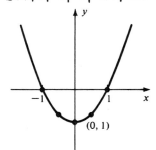

9.

t	-1	3	1	0	2
$G(t)$	0	0	4	3	3

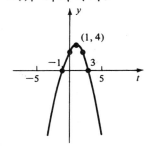

11.

x	$2+\sqrt{3}$	$2-\sqrt{3}$	2	1	3
$f(x)$	0	0	$-3/2$	-1	-1

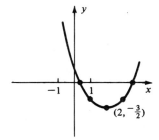

13.

w	1/2	0	1	-1/2	3/2
$P(w)$	0	1	1	4	4

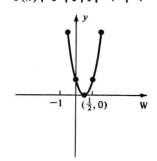

15.

x	$\dfrac{-5 \pm \sqrt{17}}{2}$	-5/2	-1	-3
$G(x)$	0	17/4	2	2

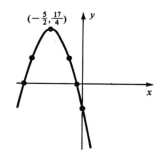

$\left(-\frac{5}{2}, \frac{17}{4}\right)$

17. $\{x:\ f(x) > 0\} = \{x:\ x > 1 \text{ or } x < -1\}$ (see problem 7)
$\{x:\ f(x) < 0\} = \{x:\ -1 < x < 1\}$

19. $\{x:\ f(x) > 0\} = \{x:\ -1 < x < 3\}$ (see problem 9)
$\{x:\ f(x) < 0\} = \{x:\ x < -1 \text{ or } x > 3\}$

21. $D_p = \mathbb{R}, R_p = \{y:\ y \geqslant 3/4\}$　　　**23.** $D_p = \mathbb{R}, R_p = \mathbb{R}$

25. $(1/2, 3/4), x = 1/2$　　　**27.** $(0, -1), x = 0$

Exercise Set 3.7

1. $1, 1, 1/16, 16, 4, 64$

2. $1, 1/16, 81/16, 1/\pi^4, \dfrac{1}{100,000,000},$
$100,000,000$

3. $2, 8, \sqrt{2}, 2\sqrt{2}, 5, 1/5$

4. $3^5 = 243, 3^{-1} = 1/3, 4^6 = 4096, 4^{-6} = \dfrac{1}{4096}, 4^{-6} = \dfrac{1}{4096}, -2^3 = -8$

5. $30^2 = 900, 27^0 = 1, 15^{-1} = 1/15, 1, (4/3)^2 = 16/9, (27/16)^2 = \dfrac{729}{256}.$

7. $-4/5, 25/16, 1, \dfrac{\sqrt[3]{-5}}{\sqrt[3]{4}}, \sqrt[5]{\dfrac{-125}{64}}, -\dfrac{64}{125}$

9. $27, -243, 1/16, 1, 5, 1$　　　**11.** $-3, 16, 1, 1, 9\sqrt{3}, 3\sqrt{3}$

13. $(2/3)^{1/2}, (-5)^{1/5}, 6^{-1/6}, 10^5, 7^{-1/3}, 2^{1/2}$

15. $5^{5/6}, (-8)^{1/3}, 32^{1/3}, 9^{-1}, 162^{1/2}, 9100^{1/2}$

17. $\dfrac{4}{x^2}, \dfrac{a^2 + 2ab + b^2}{a^2 b^2}, x^2, \dfrac{27x^6}{125y^3}, \dfrac{32}{t^5}, \sqrt{2}\, x^{-1}\, y^{-1/2}$

Exercise Set 3.8

1. $1, 40,320, 48, 120, 26, 720$　　　**2.** $3, 10, 10, 1, 6, n$

3. a) $(x + y)(x + y) = x^2 + 2xy + y^2$
$(x^2 + 2xy + y^2)(x + y) = x^3 + 3x^2y + 3xy^2 + y^3$

b) $(x + y)^3 = x^3 + \binom{3}{1}x^2y + \binom{3}{2}xy^2 + y^3$
$= x^3 + 3x^2y + 3xy^2 + y^3$

4. $(2x + 5)^3 = 8x^3 + 60x^2 + 150x + 125$

5. $a^5 + 5a^4b + 10a^3b^2 + 10a^2b^3 + 5ab^4 + b^5$

7. $16a^4 + 32a^3b + 24a^2b^2 + 8ab^3 + b^4$

9. $x^3 + 3x^2h + 3xh^2 + h^3$

11. $\frac{x^4}{16} + 2x^3 + 24x^2 + 128x + 256$

13. $x^6 - 12x^5 + 60x^4 - 160x^3 + 240x^2 - 192x + 64$

15. $81x^4 - 432x^3y + 864x^2y^2 - 768xy^3 + 256y^4$

17. $a^3x^3 + 3a^2bx^2y + 3ab^2xy^2 + b^3y^3$

19. $\frac{t^4}{16} - t^2 + 6 - \frac{16}{t^2} + \frac{16}{t^4}$

21. -224

23. $\binom{n}{n-k} = \frac{n!}{(n-k)!\,[n-(n-k)]!} = \frac{n!}{(n-k)!k!} = \binom{n}{k}$

Exercise Set 3.9

1. a) $(f+g)(x) = \frac{3}{2}x^2 + 4, D_{f+g} = \mathbb{R}$

   $(f-g)(x) = \frac{1}{2}x^2 + 4x - 6, D_{f-g} = \mathbb{R}$

   $(fg)(x) = \frac{1}{2}x^4 - x^3 - \frac{9}{2}x^2 + 12x - 5, D_{fg} = \mathbb{R}$

   $\left(\frac{f}{g}\right)(x) = \frac{2x^2 + 4x - 2}{x^2 - 4x + 10}, D_{\frac{f}{g}} = \mathbb{R}$

   b) $(f+g)(x) = 6x, D_{f+g} = \mathbb{R}$

   $(f-g)(x) = -8, D_{f-g} = \mathbb{R}$

   $(fg)(x) = 9x^2 - 16, D_{fg} = \mathbb{R}$

   $\left(\frac{f}{g}\right)(x) = \frac{3x-4}{3x+4}, D_{\frac{f}{g}} = \{x:\ x \neq -4/3\}$

   c) $(f+g)(x) = \sqrt[3]{x} + 2\sqrt{x} - x + \sqrt{2}, D_{f+g} = \{x:\ x \geqslant 0\}$

   $(f-g)(x) = -\sqrt[3]{x} - x - \sqrt{2}, D_{f-g} = \{x:\ x \geqslant 0\}$

   $(fg)(x) = x^{5/6} + x + \sqrt{2}\,x^{1/2} - x^{4/3} - x^{3/2} - \sqrt{2}\,x, D_{fg} = \{x:\ x \geqslant 0\}$

   $\left(\frac{f}{g}\right)(x) = \frac{\sqrt{x} - x}{\sqrt[3]{x} + \sqrt{x} + \sqrt{2}}, D_{\frac{f}{g}} = \{x:\ x \geqslant 0\}$

   d) $(f+g)(x) = |x| + x + 1, D_{f+g} = \mathbb{R}$

   $(f-g)(x) = |x| - x - 1, D_{f-g} = \mathbb{R}$

   $(fg)(x) = x|x| + |x|, D_{fg} = \mathbb{R}$

   $\left(\frac{f}{g}\right)(x) = \frac{|x|}{x+1}, D_{\frac{f}{g}} = \{x:\ x \neq -1\}$

2. See problem 1 for $g + f = f + g, g - f = -(f - g),$ and $gf = fg.$

   a) $\left(\frac{g}{f}\right)(x) = \frac{x^2 - 4x + 10}{2x^2 + 4x - 2}, D_{\frac{g}{f}} = \{x:\ x \neq -1 \pm \sqrt{2}\}$

   b) $\left(\frac{g}{f}\right)(x) = \frac{3x+4}{3x-4}, D_{\frac{g}{f}} = \{x:\ x \neq 4/3\}$

   c) $\left(\frac{g}{f}\right)(x) = \frac{\sqrt[3]{x} + \sqrt{x} + \sqrt{2}}{\sqrt{x} - x}, D_{\frac{g}{f}} = \{x:\ x \geqslant 0 \text{ and } x \neq 1\}$

   d) $\left(\frac{g}{f}\right)(x) = \frac{x+1}{|x|}, D_{\frac{g}{f}} = \{x:\ x \neq 0\}$

3. a) $f \bigcirc g$ is the function which assigns to each test its letter grade.

   b) $f \bigcirc g$ is the function which assigns to each telephone subscriber in Cleveland the area code 216.

c) $f \circ g$ is the function which assigns to each day of the week that same day.

d) $f \circ g$ is the function which assigns to each musical composition the number of sharps or flats in which it is to be played.

4. a) $R_g = \{x: x \geqslant 0\}, D_f = \Re, R_g \subseteq D_f, (f \circ g)(x) = 2x^2 - 1$

b) $R_g = D_f = \Re, (f \circ g)(x) = (10 - x)^5$

c) $R_g = \{33\}, D_f = \Re, R_g \subseteq D_f, (f \circ g)(x) = 66$

d) $R_g = \{y: y \leqslant -3/4\}, D_f = \{x: x \geqslant 0\}, R_g$ is not a subset of D_f

5. a) $R_f = \Re, D_g = \Re, R_f \subseteq D_g, (g \circ f)(x) = 4x^2 - 4x + 1$

b) $R_f = \Re, D_g = \Re, R_f \subseteq D_g, (g \circ f)(x) = 6 - (x + 4)^5$

c) $R_f = \{x: x \geqslant 0\}, D_g = \Re, R_f \subseteq D_g, (g \circ f)(x) = 33$

d) $R_f = \{x: x \geqslant 0\}, D_g = \Re, R_f \subseteq D_g, (g \circ f)(x) = -x + \sqrt{x} - 1$

7. $(f + g)(x) = 2x^2 + 2x - 2, (f - g)(x) = -2x - 4, (fg)(x) = x^4 + 2x^3 - 2x^2 - 6x - 3,$

$$\left(\frac{f}{g}\right)(x) = \frac{x^2 - 3}{x^2 + 2x + 1}, D_{f+g} = D_{f-g} = D_{fg} = (-10, 10),$$

$D_{\frac{f}{g}} = \{x: -10 < x < 10 \text{ and } x \neq -1\}$

9. $(f + g)(x) = \frac{x^3 + 2x - 1}{(x^2 + 1)(x + 1)}, (f - g)(x) = \frac{-x^3 - 2x^2 + 1}{(x^2 + 1)(x + 1)},$

$(fg)(x) = \frac{x^2 - x}{(x^2 + 1)(x + 1)}, \left(\frac{f}{g}\right)(x) = \frac{x^2 + x}{(x^2 + 1)(x - 1)},$

$D_{f+g} = D_{f-g} = D_{fg} = D_{\frac{f}{g}} = [-4, -2]$

11. $-x^2 - 20$ 13. $\dfrac{1}{x^2 + 1}$ 15. $\dfrac{x^2 - 1}{3}$ 17. x

19. $8 - x^{12}$ 21. $81x^4 - 108x^3 + 54x^2 - 12x + 1$ 23. $x + x^{3/2}$

25. $\dfrac{|x - 1|}{|x - 1| + 1}$ 27. $\sqrt{x^2 + 2xh + h^2 + 4}$ 29. k

31. $(f \circ g)(w) = w$ and $(g \circ f)(w) = w$, no

Exercise Set 3.10

1. a) $f(x) = -8x^3 + 3x^2 + 5x - 15$

b) $f(x) = 7002x^7 + 0x^6 + \dfrac{2}{3}x^5 - 1x^4 + \sqrt{2}x^3 + 0x^2 + 0x + 0$

c) $f(x) = -1x^6 + 9x^5 - 21x^4 + 4x^3 + 0x^2 + 0x + 0$

2. a) 3   b) 7   c) 6

3. a) $a_4 = 2, a_3 = -1, a_2 = 9, a_1 = 1, a_0 = -7/8$

b) $a_4 = 1, a_3 = 0, a_2 = -1, a_1 = 0, a_0 = 1$

c) $a_4 = -13/3, a_3 = \pi, a_2 = \sqrt{3}, a_1 = \sqrt{2}, a_0 = 0$

d) $a_4 = 1/4, a_3 = -2/3, a_2 = 0, a_1 = 0, a_0 = 0$

4. Polynomial functions a, c, and e

Rational functions a, b, c, e, and g

Algebraic functions a, b, c, d, e, f, g, and h

5. a) $f(x) = 3x^{14} - x$

b) $g(x) = x^3 - 3x^2 + 2$

c) $h(x) = 4x - 1$

d) $k(x) = 19$

7. The domain of any polynomial function is $\Re$.

a) Degree 4; $a_4 = 1, a_3 = 0, a_2 = 0, a_1 = 0, a_0 = 15/2$

b) Degree 10; $a_{10} = -1, a_9 = a_8 = a_7 = a_6 = a_5 = a_4 = 0, a_3 = 1, a_2 = 0, a_1 = -\sqrt{2}, a_0 = 2$

c) Degree 5; $h(u) = u^4 + 5u^4 + 10u^3 + 10u^2 + 5u + 1$
$a_5 = 1, a_4 = 5, a_3 = 10, a_2 = 10, a_1 = 5, a_0 = 1$

d) Degree 2; $a_2 = a, a_1 = b, a_0 = c$

e) Degree 1; $a_1 = m, a_0 = b$

f) Degree 0; $a_0 = -\dfrac{12}{\sqrt{19}}$

9. Each is the quotient of two polynomials, but none can be expressed in the form $a_n x^n + \cdots + a_1 x + a_0$.
a) $D_g = \{x: \ x \neq 1\}$ b) $D_g = \{t: \ t \neq \pm 1\}$
c) $D_h = \{t: \ t \neq 0\}$ d) $D_R = \{y: \ y \neq 0\}$

11. None can be expressed as the quotient of two polynomials.

Exercise Set 4.1

1. a) $Q(x) = 4x - 3$ $R(x) = -46$
b) $Q(x) = 3x - 8$ $R(x) = 0$
c) $Q(x) = 12x - 29$ $R(x) = -11$

2. a) $12x^2 - 17x - 40 = (3x - 2)(4x - 3) + (-46)$
b) $12x^2 - 17x - 40 = (4x + 5)(3x - 8) + 0$
c) $12x^2 - 17x - 40 = (x + 1)(12x - 29) + (-11)$

3. a) $Q(x) = 2x^3 + 2x^2 - x + 3$ $R(x) = -5x + 2$
b) $Q(x) = x^3 + 2x^2 - x - 1$ $R(x) = 0$
c) $Q(x) = \dfrac{1}{2}x + \dfrac{7}{8}$ $R(x) = -\dfrac{7}{8}x^3 + \dfrac{9}{4}x^2 - \dfrac{21}{8}x - \dfrac{43}{8}$

4. a) $2x^5 + 4x^4 - x^3 - x - 1 = (x^2 + x - 1)(2x^3 + 2x^2 - x + 3) + (-5x + 2)$
b) $2x^5 + 4x^4 - x^3 - x - 1 = (2x^2 + 1)(x^3 + 2x^2 - x - 1) + 0$
c) $2x^5 + 4x^4 - x^3 - x - 1 = (4x^4 + x^3 - 2x^2 - x + 5)\left(\dfrac{1}{2}x + \dfrac{7}{8}\right)$
$$+ \left(-\dfrac{7}{8}x^3 + \dfrac{9}{4}x^2 - \dfrac{21}{8}x - \dfrac{43}{8}\right)$$

5. a) $R(x) \neq 0$, $3x + 5$ is not a factor
b) $R(x) = 0$, $x - \dfrac{1}{2}$ is a factor
c) $R(x) \neq 0$, $x - 1$ is not a factor
d) $R(x) \neq 0$, $x^2 + x + 2$ is not a factor

7. $Q(r) = 2r^3 + 5r - 8$ $R(r) = 0$

9. $Q(y) = \dfrac{2}{3}y^2 - \dfrac{11}{9}y + \dfrac{62}{27}$ $R(y) = \dfrac{-113}{27}$

11. $Q(x) = \dfrac{1}{3}x^3 + \dfrac{5}{3}x$ $R(x) = -8$

13. $Q(t) = 4t^3 - 2t^2 + 11t - \dfrac{17}{2}$ $R(t) = \dfrac{59}{2}t^2 - \dfrac{39}{2}t + \dfrac{49}{2}$

15. $Q(y) = 4y^2 + 2y + 8$ $R(y) = 11$

17. $D(x)$ is a factor of $P(x)$

19. $D(x)$ is not a factor of $P(x)$

21. $D(x)$ is a factor of $P(x)$

23. $P(x) = \left(2x - \dfrac{1}{3}\right)(3x^2 + 3x - 6) = \left(2x - \dfrac{1}{3}\right)(3x + 6)(x - 1)$

Exercise Set 4.2

1. $\begin{array}{r} 30 + 143 + 117 - \quad 56 \ \underline{|2} \\ 60 + 406 + 1046 \\ \hline 30 + 203 + 523 + \quad 990 \end{array}$ $Q(x) = 30x^2 + 203x + 523, R = 990$

$$\begin{array}{r} 30 + 143 + 117 - 56 \quad \underline{|-7/2} \\ -105 - 133 + 56 \\ \hline 30 + 38 - 16 + 0 \end{array} \quad Q(x) = 30x^2 + 38x - 16, R = 0$$

2. $30x^3 + 143x^2 + 117x - 56 = (x - 2)(30x^2 + 203x + 523) + 990$

$$30x^3 + 143x^2 + 117x - 56 = \left(x + \frac{7}{2}\right)(30x^2 + 38x - 16) + 0$$

3. $x - 2$ is not a factor of $30x^3 + 143x^2 + 117x - 56$

$x + \dfrac{7}{2}$ is a factor of $30x^3 + 143x^2 + 117x - 56$

4. $P(2) = (30)(2)^3 + (143)(2)^2 + (117)(2) - 56 = 990$

$P(-7/2) = (30)(-7/2)^3 + (143)(-7/2)^2 + (117)(-7/2) - 56 = 0$

5. $P(2) = 990 = R, P(-7/2) = 0 = R$

7. a) $P(x) = \left(x + \dfrac{5}{2}\right)(24x^2 - 32x - 70) + (0)$ b) $P(x) = \left(x - \dfrac{5}{2}\right)(24x^2 + 88x + 70) + (0)$

c) $P(x) = \left(x + \dfrac{7}{6}\right)(24x^2 - 150) + (0)$

9. a) $P(z) = (z - 2)(6z^5 + 12z^4 + 24z^3 + 48z^2 + 93z + 186) + (380)$

b) $P(z) = (z + 4)(6z^5 - 24z^4 + 96z^3 - 384z^2 + 1533z - 6132) + (24,536)$

c) $P(z) = (z + 5)(6z^5 - 30z^4 + 150z^3 - 750z^2 + 3747z - 18,735) + (93,683)$

11. a) $P(t) = (t - 1)\left(\dfrac{1}{2}t^3 - \dfrac{3}{2}t - 1\right) + (0)$

b) $P(t) = (t + 4)\left(\dfrac{1}{2}t^3 - \dfrac{5}{2}t^2 + \dfrac{17}{2}t - \dfrac{67}{2}\right) + (135)$

c) $P(t) = (t - 2)\left(\dfrac{1}{2}t^3 + \dfrac{1}{2}t^2 - \dfrac{1}{2}t - \dfrac{1}{2}\right) + (0)$

13. a) $P(y) = (y + 1)(2y^3 + 3y^2 - 3y + 5) + (0)$

b) $P(y) = \left(y + \dfrac{5}{2}\right)(2y^3 + 2) + (0)$

c) $P(y) = \left(y - \dfrac{5}{2}\right)\left(2y^3 + 10y^2 + 25y + \dfrac{129}{2}\right) + \left(\dfrac{665}{4}\right)$

15. a) $P(x) = (x + \sqrt{2})[3x^2 + (1 - 3\sqrt{2})x - \sqrt{2}] + (0)$

b) $P(x) = (x - \sqrt{2})[3x^2 + (1 + 3\sqrt{2})x + \sqrt{2}] + (0)$

c) $P(x) = \left(x + \dfrac{1}{3}\right)(3x^2 - 6) + (0)$

17. $P(-5/2) = 0,\quad P(5/2) = 0,\quad P(-7/6) = 0$

19. $P(2) = 380,\quad P(-4) = 24,536,\quad P(-5) = 93,683$

21. $\begin{array}{r} 5 + 0 - 7 + 2 \quad \underline{|1/2} \\ \dfrac{5}{2} + \dfrac{5}{4} - \dfrac{23}{8} \\ \hline 5 + \dfrac{5}{2} - \dfrac{23}{4} - \dfrac{7}{8} \end{array}$

$$5x^3 - 7x + 2 = \left(x - \frac{1}{2}\right)\left(5x^2 + \frac{5}{2}x - \frac{23}{4}\right) - \frac{7}{8}$$

$$= (2x - 1)\left(\frac{5}{2}x^2 + \frac{5}{4}x - \frac{23}{8}\right) - \frac{7}{8}$$

Exercise Set 4.3

1. $\begin{array}{r} 5 - 14 - 23 - 4 \quad \underline{|2} \\ 10 - 8 - 62 \\ \hline 5 - 4 - 31 \underline{|- 66 = P(2)|} \end{array}$ $P(-1) = 0,\quad P(1/2) = \dfrac{-147}{8},\quad P(1/5) = \dfrac{-228}{25}$

$P(3) = -64,\quad P(-1/5) = 0$

2. $P(2) = 5(2)^3 - 14(2)^2 - 23(2) - 4 = -66$

$P(-1) = 0, \quad P(1/2) = \frac{-147}{8}, \quad P(1/5) = \frac{-228}{25}, \quad P(3) = -64, \quad P(-1/5) = 0$

3. -1 and $-1/5$ are zeros of $P(x) = 5x^3 - 14x^2 - 23x - 4$

4. $5x^3 - 14x^2 - 23x - 4 = 5(x + 1)\left(x + \frac{1}{5}\right)(x - 4)$

5. $(x + 1)$ and $\left(x - \frac{5}{2}\right)$ are factors of $P(x) = 6x^3 - 7x^2 - 18x - 5$

6. $P(x) = (x + 1)(6x^2 - 13x - 5)$ and $P(x) = \left(x - \frac{5}{2}\right)(6x^2 + 8x + 2)$

7. $P(1) = -14, \quad P(-1) = -6, \quad P(3) = -494$

9. $P(0) = 2, \quad P(1/2) = 1/2, \quad P(2/3) = 0$

11. $P(5) = -1228, \quad P(25) = 772, \quad P(-33) = -117{,}432$

13. $P(y) = -3(y - 5)(y - 5)\left(y + \frac{11 + \sqrt{145}}{6}\right)\left(y + \frac{11 - \sqrt{145}}{6}\right)$

15. $P(x) = \left(x - \frac{2}{3}\right)\left(x^2 + \frac{2}{3}x + \frac{4}{9}\right)$

17. $P(y) = \left(y - \frac{1}{2}\right)(4y^2 - 2y + 2)$

19. $P(x) = a\left(x - \frac{1}{2}\right)\left(x + \frac{2}{3}\right)(x + 1) \quad$ Let $a = 6$

$P(x) = 6x^3 + 7x^2 - x - 2$

21. Let $P(x) = x^n - a^n$. Then $P(a) = a^n - a^n = 0$. By the Factor Theorem, $x - a$ is a factor of $P(x)$.

Exercise Set 4.4

1. 4

2. $P(-3) = -15, P(-2) = 3, P(-1) = 3, P(0) = -3, P(1) = -3, P(2) = 15, P(3) = 63.$ P has one zero in each of the intervals $(-3, -2), (-1, 0),$ and $(1, 2)$.

3. Possible rational zeros are 1, 1/2, 1/3, 1/4, 1/6, 1/12, 3, 3/2, 3/4, 5, 5/2, 5/3, 5/4, 5/6, 5/12, 15, 15/2, 15/4 and the negatives of these numbers.

4. $-5/2, -3/2,$ and $1/3$

5. $P(x) = 12\left(x + \frac{5}{2}\right)\left(x + \frac{3}{2}\right)\left(x - \frac{1}{3}\right)$

7. a) 25   b) 1   c) 4

9. Possible rational zeros: $\pm 1, \pm 3, \pm 1/2, \pm 3/2$

Rational zeros: $-3/2$

11. Possible rational zeros: $\pm 1, \pm 3, \pm 1/2, \pm 3/2$

Rational zeros: None

13. Possible rational zeros: $\pm 1, \pm 2, \pm 1/5, \pm 1/7, \pm 1/35, \pm 2/5, \pm 2/7, \pm 2/35$

Rational zeros: $-1/5, 2/7$

15. Possible rational zeros: $\pm 1, \pm 2, \pm 1/3, \pm 2/3$

Rational zeros: None

17. $P(y) = 2\left(y - \frac{1}{2}\right)(y + 3)(y - 1)$

19. $P(x) = (x + 1)(x - 3)(x^2 + 2x + 2)$

21. $Q(0) = -1, Q(3/2) = 361/32.$ Thus, $Q(y)$ has a zero between 0 and 3/2.

23. $P(x)$ can have only $\pm 1, \pm 3$ as rational zeros. None of these numbers are zeros, therefore, $P(x)$ has no rational zeros. $P(-3) = -18, P(-2) = 1, P(0) = 3, P(1) = -2, P(2) = -3, P(3) = 6.$ Therefore, P has one zero in each of the intervals $(-3, -2), (0, 1), (2, 3).$

25. $-1, 1/2$ and 2 are zeros of P.

Exercise Set 4.5

1.

	2	$x + 3$	$x - 1$	$x - 3/2$	$P(x)$
$x < -3$	+	−	−	−	−
$-3 < x < 1$	+	+	−	−	+
$1 < x < 3/2$	+	+	+	−	−
$3/2 < x$	+	+	, +	+	+

2.

	−3	1	3/2	−7/2	−2	−1	0	5/4	2	3
$P(x)$	0	0	0	−45/2	21	20	9	−17/32	5	36

3.

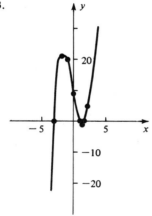

4.   Zeros: −3, −2, −1

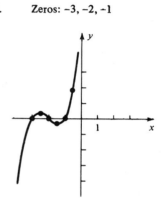

5.   Zeros: −2, 1, 3

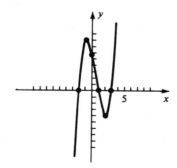

7.   Zeros: None

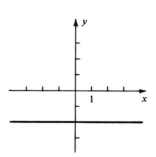

9. Zeros: −1

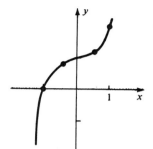

11. Zeros: None

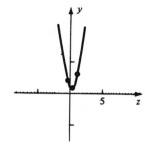

13. Zeros: 1

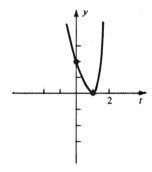

15. Zeros: −3/2

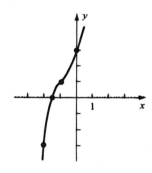

17. Zeros: −1, −2/5, 1/2, 1

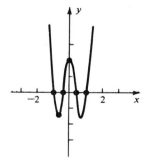

19. Zeros: −√2, −1/2, √2, 3

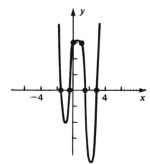

21.   Zeros: −3, 1/2, 1

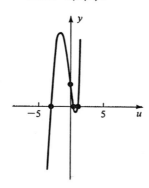

Exercise Set 4.6

1. $\{x: x \neq 2\}$. The line $x = 2$ is a barrier.

2.

	$x - 2$	$R(x)$
$x < 2$	−	−
$x > 2$	+	+

x	−1	0	1	3/2	7/4	9/4	5/2	3	4	5
$R(x)$	−1/3	−1/2	−1	−2	−4	4	2	1	1/2	1/3

3.

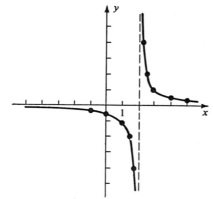

4. $D_R = \{z: z \neq -1, 2\}$. The lines $z = -1$ and $z = 2$ are barriers.

5.

	$z + 1$	$z - 2$	$R(z)$
$z < -1$	−	−	+
$-1 < z < 2$	+	−	−
$2 < z$	+	+	+

x	−2	$-\dfrac{3}{2}$	$-\dfrac{5}{4}$	$-\dfrac{1}{2}$	0	1	$\dfrac{3}{2}$	$\dfrac{5}{2}$	3
$R(x)$	$\dfrac{1}{4}$	$\dfrac{4}{7}$	$\dfrac{16}{13}$	$-\dfrac{4}{5}$	$-\dfrac{1}{2}$	$-\dfrac{1}{2}$	$-\dfrac{4}{5}$	$\dfrac{4}{7}$	$\dfrac{1}{4}$

6. Barrier: $x = -1, x = 2$

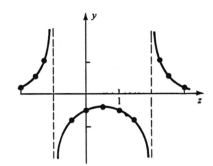

7.   Barrier: $x = 0$

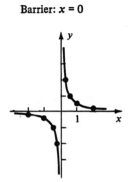

9.   Barrier: $z = 0$

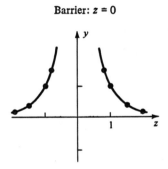

11.   Barrier: $t = -1$

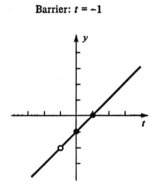

13.   Barrier: $t = \pm 1$

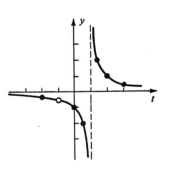

15.   Barrier: $x = -1, x = 5$

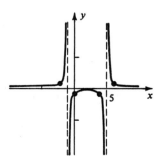

17.   Barrier: $s = 2$ **19.**   Barrier: $w = 0$

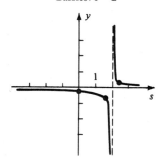

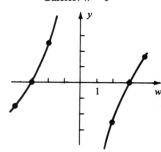

21. $f(x) = \dfrac{x^2 - 1}{x - 1}$ is defined only if $x \neq 1$. For all $x \neq 1$, $f(x) = g(x)$.

Exercise Set 4.7

1. $D_f = \{x: x \geq -1\}$, $(-\infty, -1)$

2.

x	-1	0	1	3	4	8
$f(x)$	0	1	$\sqrt{2}$	2	$\sqrt{5}$	3

3.

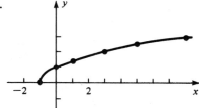

4. Zeros: ± 2

	$x^2 - 4$	$g(x)$
$x < -2$	$+$	$+$
$x > 2$	$+$	$+$

$D_g = \{x: x \geq 2\} \cup \{x: x \leq -2\}$

5.

x	-4	-3	-2	2	3	4
$g(x)$	$\sqrt{12}$	$\sqrt{5}$	0	0	$\sqrt{5}$	$\sqrt{12}$

6.

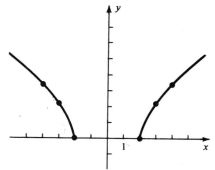

7.   Zeros: $0, D_F = \{x: \; x \ge 0\}$

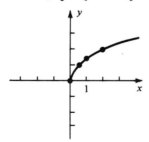

9.   Zeros: $0, D_F = \{x: \; x \le 0\}$

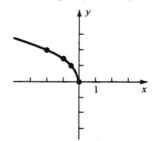

11.   Zeros: $0, D_F = \Re$

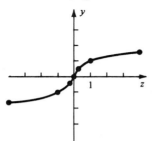

13.   Zeros: $\pm 1, D_F = \{x: \; -1 \le x \le 1\}$

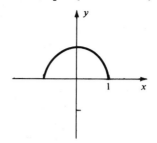

15.   Zeros: $0, D_F = \Re$

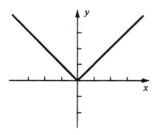

17.   Zeros: $-2, 1, D_F = \{x: \; x \le -2 \text{ or } x \ge 1\}$

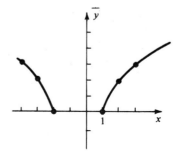

19.   Zeros: $-2, D_F = \Re$

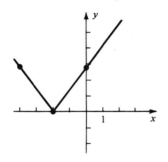

21.   Zeros: $1, D_F = \Re$

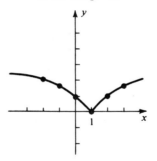

23. Zeros: $\pm 3, D_f = D_g = \{x: -3 \leqslant x \leqslant 3\}$

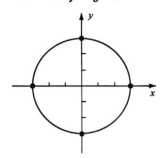

Exercise Set 4.8

1. 5, 2, 2 + 5 or 7, −2 + 5 or 3, $2 + \sqrt{5}, -2 + \sqrt{5}$

2. a) $-8 < x < 8$ b) $x < -8$ or $x > 8$ c) $x = \pm 8$

3. a) $-8 < 2x + 3 < 8$ b) $2x + 3 < -8$ or $2x + 3 > 8$ c) $2x + 3 = 8$ or $2x + 3 = -8$
   $-11 < 2x < 5$ $2x < -11$ or $2x > 5$ $x = 5/2$ or $x = -11/2$
   $-11/2 < x < 5/2$ $x < -11/2$ or $x > 5/2$

4. and 5.

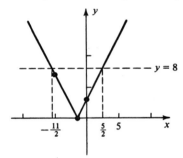

7. a) 0   b) $\frac{33}{20}$ c) $-\frac{17}{20}$ d) $\frac{1}{2}$ e) $-\frac{1}{2}$

9. a) $-2 < y < 8$ b) $y < -2$ or $y > 8$ c) $y = -2$ or $y = 8$
   d) $-5/3 < y < 5/3$ e) $5/3 > y > -5/3$ or $-5/3 < y < 5/3$

11. a) $|x| < 1/2$ b) $|x| > 1/2$ c) $|x| \leqslant 1/2$ d) $|x| < 3/2$ e) $|x| \geqslant 3/2$

13. a) $x = \sqrt{2}$ or $x = -\sqrt{2}$ b) $-\sqrt{2} < x < \sqrt{2}$ c) There are no real numbers which satisfy the inequality.   d) $\{x: x \in \mathcal{R}\}$

15.

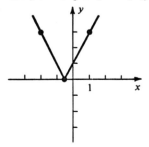

17.

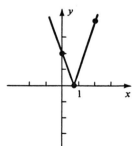

19.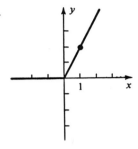

21.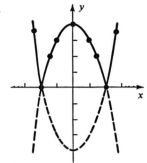

23. a) $-\epsilon < g(t) < \epsilon$ b) $-\epsilon/2 < g(t) < \epsilon/2$ c) $1 - \epsilon < t < 1 + \epsilon$

d) $1 - \dfrac{\epsilon}{2} < t < 1 + \dfrac{\epsilon}{2}$ e) $\dfrac{5}{2} - \dfrac{\epsilon}{2} < t < \dfrac{5}{2} + \dfrac{\epsilon}{2}$ f) $\dfrac{5}{2} - \dfrac{\epsilon}{2} < t < \dfrac{5}{2} + \dfrac{\epsilon}{2}$

Exercise Set 4.9

1. a) 0   b) 1   c) –1   d) 2   e) –4   f) 7   g) –10   h) 60   i) 0   j) 1
2. a) 10   b) 4   c) 0   d) 0.4   e) –1   f) 0   g) 10   h) 11
3. a) 0   b) –1   c) 1   d) –2   e) 2   f) –3

4.

x	$-4 \leqslant x < -2$	$-2 \leqslant x < 0$	$0 \leqslant x < 2$	$2 \leqslant x < 4$
$f(x) = \left[\dfrac{x}{2}\right]$	–2	–1	0	1

5.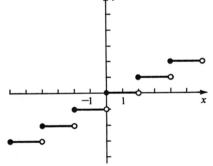

6.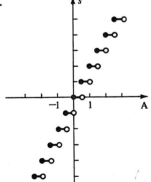

7.

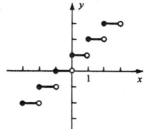

9.

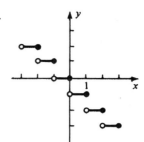

11.

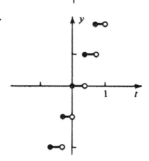

13.

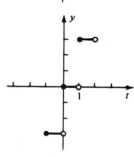

15.

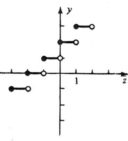

17.

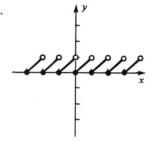

19. a) 24¢   b) 40¢   c) 24¢   d) 24¢   e) 32¢   f) 16¢

21. If $f(x) = x - [x]$, then $f(x + 1) = (x + 1) - [x + 1]$
$$= (x + 1) - ([x] + 1)$$
$$= x - [x]$$
$$= f(x)$$

so $f(x) = f(x + 1)$ for all x.

If f is periodic of period p, then the graph of f repeats itself over each interval of length p (see problem 17).

Exercise Set 5.1

1. a) $8, 27, \dfrac{1}{27}, 32, 25, 64, 3\sqrt{3}, \dfrac{27}{8}$

b) $\dfrac{1}{8}, -27, -\dfrac{1}{27}, \dfrac{1}{2}, 1, \dfrac{1}{64}$, undefined, $\dfrac{8}{27}$

c) $2, \dfrac{1}{81}, 81, \dfrac{1}{2}, \sqrt[3]{4}, \ \sqrt[3]{\dfrac{2}{3}}, 8, \dfrac{2}{3}$

2. a)

x	-3	-2	-1	$-\frac{1}{2}$	0	$\frac{1}{2}$	1	2	3
3^x	$\frac{1}{27}$	$\frac{1}{9}$	$\frac{1}{3}$	$\frac{1}{\sqrt{3}}$	1	$\sqrt{3}$	3	9	27

b)

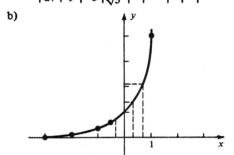

c) $3^{\frac{1}{3}}$ is approximately 1.5, $3^{\frac{\sqrt{2}}{2}}$ is approximately 2.1, $3^{-\frac{1}{3}}$ is approximately 0.7

3. a) $D_E = \mathcal{R}$,   b) $R_E = \{y: y > 0\}$

5. a)

t	-2	-1	$-\frac{1}{2}$	0	$\frac{1}{2}$	1	2
10^t	$\frac{1}{100}$	$\frac{1}{10}$	$\frac{1}{\sqrt{10}}$	1	$\sqrt{10}$	10	100

b)

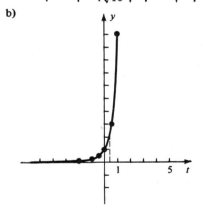

c) No   d) If $t = -3$, $J(t) = .001 < .005$ e) If $t = 4$, $J(t) = 10{,}000 > 5000$

f) $\sqrt[3]{10}$ is approximately 2.2

7. $D_E = \mathcal{R}, R_E = \{y: y > 0\}$

9. $3 < 3^{\sqrt{2}} < 9, \frac{1}{9} < 3^{-\sqrt{2}} < \frac{1}{3}, 27 < 3^{\sqrt{10}} < 81, \frac{1}{27} < 3^{-\sqrt{5}} < \frac{1}{9}, 27 < 3^{\pi} < 81,$

$\frac{1}{81} < 3^{-\pi} < \frac{1}{27}$

11. a)

x	-4	-2	-1	0	1	2	3	4
$(\sqrt{2})^x$	$\frac{1}{4}$	$\frac{1}{2}$	$\frac{1}{\sqrt{2}}$	1	$\sqrt{2}$	2	$2\sqrt{2}$	4

b)

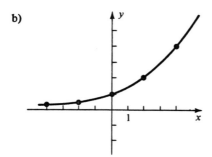

13. a) $\frac{1}{2}$ b) No c) 0 d) $-\frac{1}{2}$ e) No f) No

15. If $a = 0$, then $E(x) = a^x$ would be the constant function $E(x) = 0$ $(x \neq 0)$. This function has already been studied. If $a < 0$, then $E(x) = a^x$ would be defined only for certain values of x, and in fact would be quite difficult to study and of very limited use in applications.

Exercise Set 5.2

1. a) 25% b) 25¢ c) $1.25 d) $1.56 e) $2.44

2. a) 2.37, 2.49, 2.52 b) As n increases, $\left(1 + \frac{1}{n}\right)^n$ better approximates e

3. a)

x	-2	-1	$-\frac{1}{2}$	0	$\frac{1}{2}$	1	2
e^x	.1353	.3679	.6065	1	1.649	2.718	7.389

b)

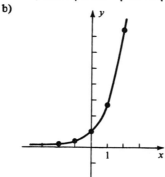

4. a) $D_E = \Re$ b) $R_E = \{y : y > 0\}$

5. a)

x	-2	-1	$-\frac{1}{2}$	0	$\frac{1}{2}$	1	2
e^{-x}	7.389	2.718	1.649	1	.6065	.3679	.1353

b)

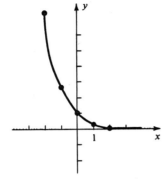

7. a)

x	-2	-1	$-\dfrac{1}{2}$	0	$\dfrac{1}{2}$	1
e^{x+1}	.3679	1	1.649	2.718	4.482	7.389

b)

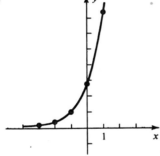

9. a)

t	± 1	$\pm\dfrac{1}{2}$	$\pm\dfrac{1}{4}$	0
e^{t^2}	2.718	1.284	1.06	1

b)

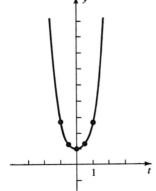

c) $D_E = \Re$ d) $R_E = \{y: y \geqslant 1\}$
11. a) 2.717   b) $e = 2.7183$ (accurate to four places)

Exercise Set 5.3

1. a) 2 b) 1 c) 0 d) −1 e) $\frac{1}{2}$ f) 15 g) $-\frac{1}{3}$ h) x i) $m+n$

2. a)

x	0.1	0.5	1	1.5	2.0	3.0	4.0
$L(x) = \ln x$	−2.3	−0.7	0	0.4	0.7	1.1	1.4

b)

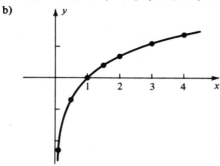

3. a) $D_L = \{x : x > 0\}$ b) $R_L = \mathcal{R}$ c) $D_L = R_E$ and $R_L = D_E$

4. a) $\ln 10 = \ln (5 \cdot 2) = \ln 5 + \ln 2 = 2.3025$

b) $\ln 4 = \ln 2^2 = 2\ln 2 = 1.3862$

c) $\ln \frac{1}{2} = \ln 1 - \ln 2 = -0.6931$

d) $\ln 2.5 = \ln \frac{5}{2} = \ln 5 - \ln 2 = 0.9163$

e) $\ln 25 = \ln 5^2 = 2\ln 5 = 3.2188$

f) $\ln 0.4 = \ln \frac{2^2}{10} = -0.9163$

g) $\ln \frac{1}{5} = -\ln 5 = -1.6094$

h) $\ln 8 = \ln 2^3 = 3\ln 2 = 2.0793$

i) $\ln \frac{16}{25} = \ln 2^4 - \ln 5^2 = -0.4464$

j) $\ln \sqrt{2} = \ln 2^{\frac{1}{2}} = \frac{1}{2}\ln 2 = 0.3466$

k) $\ln \sqrt{5} = \ln 5^{\frac{1}{2}} = \frac{1}{2}\ln 5 = 0.8047$

l) $\ln \sqrt{10} = \ln 10^{\frac{1}{2}} = \frac{1}{2}\ln 10 = 1.1513$

5. a) 3.15 b) 2.00 c) −1.04 d) .40 e) 1.5 f) 100 g) u h) t

7. a) $x^2 = 6x - 9$ or $x^2 - 6x + 9 = 0$ or $(x - 3)^2 = 0, x = 3$

b) $e^{x+5} = 5$ or $\ln 5 = x + 5, x = -5 + \ln 5$

c) $e^{-x} = \frac{1}{3}$ or $\ln \frac{1}{3} = -x, x = \ln 3$

d) $x^2 = 3x - \frac{3}{4}$ or $x^2 - 3x + \frac{3}{4} = 0, x = \frac{3 \pm \sqrt{6}}{2}$

9. $2\sqrt[3]{12} = \ln 2 + \frac{1}{3} (\ln 4 + \ln 3) = \frac{u}{2} + \frac{u}{3} + \frac{v}{3} = \frac{5u}{6} + \frac{v}{3}$

11. a) $\ln\dfrac{3}{2}$ b) $\ln 15$

13. a)

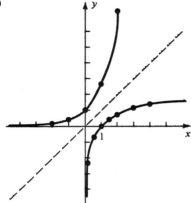

b) If (a, b) is on the graph of $y = e^x$, then (b, a) is on the graph of $y = \ln x$ and conversely (symmetry about line $y = x$).

Exercise Set 5.4

1. $\dfrac{9}{8}, \dfrac{13}{8}, \dfrac{21}{8}, \dfrac{33}{8}$ 2. $\dfrac{1}{16}, \dfrac{1}{16}, \dfrac{1}{16}, \dfrac{1}{16}$

3. a) 3.446 b) 3.478 c) 3.494
4. a) 2.118 b) 2.308 c) 2.422
5. a) 1.3937 b) −0.1525
6. a) 2.16   b) 3.74
7. 1.1568   9. 0.8196   11. 1.2669
13. 5.33   15. 0.821   17. 3.7851
19. 0.5657

21. $\sqrt[3]{\dfrac{11}{15}}$ is approximately equal to $\dfrac{83}{95}$, using interpolation on the interval $\left[\dfrac{8}{27}, 1\right]$.

$\sqrt[3]{\dfrac{1}{39}}$ is approximately equal to $\dfrac{37}{123}$, using interpolation on the interval $[27, 64]$ in order to compute $\sqrt[3]{39}$.

Exercise Set 5.5

1. a) 3.431 b) 2.192   2. a) 0.508 b) 3.810
3. $(1.32)(2.6) = 3.432, (2.89)(0.76) = 2.1964, \dfrac{1.32}{2.6} = .5077, \dfrac{2.89}{0.76} = 3.803$
4. a) 1.097 b) 1.554
5. $(1.097)^3 = 1.3197, (1.554)^3 = 3.753$
7. 3.623   9. 0.196   11. 2.283
13. 0.216   15. 1.041   17. 2.279
19. a) $3.31

Exercise Set 6.1

1. a) $2, \dfrac{4}{3}, -1$ b) $-6, -4, 3$ c) $5, \dfrac{13}{3}, 2$

2. a) 2, Yes, $\frac{4}{3}$, Yes, –1, Yes

   b) –6, Yes, –4, Yes, 3, Yes

   c) 5, Yes, $\frac{13}{3}$, Yes, 2, No

3. For example: a) $x_0 = 1.97, x_0 = 1.34, x_0 = -.98$
   b) $x_0 = 2.01, x_0 = 1.32, x_0 = -.99$
   c) $x_0 = 2.02, x_0 = 1.34, x_0 = -.98$

4. a)

x	1.5	1.6	1.7	1.71	1.73	1.732	1.734	1.74	1.75	1.8	1.9
$f(x)$	2.25	2.56	2.89	2.9241	2.9929	2.9998	3.0068	3.0276	3.0625	3.24	3.61

   b) 3   c) 3

5. a)

x	.5	$\frac{\sqrt{2}}{2}$	$\frac{4}{5}$	$\frac{\sqrt{3}}{2}$	$\frac{7}{8}$	$\frac{\sqrt[3]{6}}{2}$	$\frac{\pi}{3}$	$\frac{4}{3}$	$\frac{7}{5}$	$\sqrt{2}$	1.5
$f(x)$	2	0	2	0	2	0	0	2	2	0	2

   b) This function does not have a limit as x approaches 1.

7. 2   9. $\sqrt{3}$ 11. k 13. –2   15. 5   17. 0

19. $\delta = \dfrac{1}{2000}$ 21. $\delta = \dfrac{1}{1000}$ 23. $\delta = \dfrac{1}{20}$

Exercise Set 6.2

1. a)

x	1.5	1.9	1.99	1.999	2.001	2.01	2.1	2.5
$f(x)$	5.5	5.9	5.99	5.999	6.003	6.03	6.3	7.5

   b) 6, 6, 6, 6
   c)

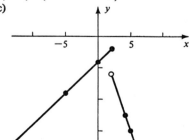

2. a)

x	1.5	1.9	1.99	1.999	2.001	2.01	2.1	2.5
$f(x)$	–2.5	–2.1	–2.01	–2.001	–6.003	–6.03	–6.3	–7.5

   b) –2, –6, does not exist, –2
   c)

3. a)

x	-2	-1	-.5	0	0.5	1	2
$f(x)$	-1	-1	-1	0	1	1	1

b) -1, 1, does not exist, 0

c)

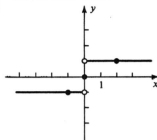

4. a)

x	-2	-1	-.5	-.1	-.01	.01	.1	.5	1	2
$f(x)$	10	4	1.75	.31	.0301	-.0299	-.29	-1.25	-2	-2

b) 0, 0, 0, 0

5. -14, -14, -14   7. 0, 0, 0   9. -4, -4, -4

11.

x	0	.5	.8	.9	.99	1	1.01	1.1	1.2	1.5	2
$g(x)$	0	.25	.64	.81	.9801	1	.99	.9	.8	.5	0

$$\lim_{x \to 1^-} g(x) = 1 \qquad \lim_{x \to 1^+} g(x) = 1$$

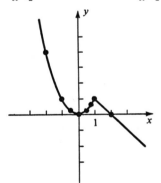

13.

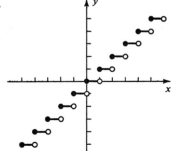

15.

z	-2	-1	$-\frac{1}{2}$	$-\frac{1}{8}$	0	$\frac{1}{8}$	$\frac{1}{2}$	1	2
$f(z)$	$\frac{1}{5}$	$\frac{1}{2}$	$\frac{4}{5}$	$\frac{64}{65}$	1	$\frac{64}{65}$	$\frac{4}{5}$	$\frac{1}{2}$	$\frac{1}{5}$

$$\lim_{z\to 0^+} f(z) = 1 \qquad\qquad \lim_{z\to 0^-} f(z) = 1$$

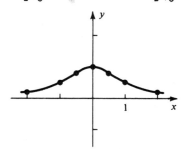

17. $\lim_{t\to 1^-} f(t) = -1$, $\lim_{t\to 1^+} f(t) = -1$ Note: D_f does not contain 1

Exercise Set 6.3

1. $\lim_{x\to 2} 6f(x) = 6 \lim_{x\to 2} f(x) = 14$, $\lim_{x\to 2} [f(x) + g(x)] = \lim_{x\to 2} f(x) + \lim_{x\to 2} g(x) = -\frac{5}{3}$,

$\lim_{x\to 2} [f(x) - g(x)] = \lim_{x\to 2} f(x) - \lim_{x\to 2} g(x) = \frac{19}{3}$,

$\lim_{x\to 2} f(x) \cdot g(x) = \lim_{x\to 2} f(x) \lim_{x\to 2} g(x) = -\frac{28}{3}$

$\lim_{x\to 2} \frac{f(x)}{g(x)} = \frac{\lim_{x\to 2} f(x)}{\lim_{x\to 2} g(x)} = -\frac{7}{12}$, $\lim_{x\to 2} [-2f(x) + 3g(x)] = (-2)\left(\frac{7}{3}\right) + (3)(-4) = -\frac{50}{3}$

2. $\lim_{x\to 2} f(x) = 3(2)^2 + (2) - 5 = 9$, $\lim_{x\to 2} g(x) = 2(2) + 3 = 7$,

$\lim_{x\to 2} \frac{f(x)}{g(x)} = \frac{9}{7}$, $\lim_{x\to 2} f(x) \cdot g(x) = 63$,

$\lim_{x\to 2} [f(x)]^5 = 59,049$, $\lim_{x\to 2} \frac{g(x)}{f(x)} = \frac{7}{9}$

3. $-12, -12, -126, 3018$

4. a) All polynomials are continuous. $P(x)$ is continuous.

b) $\lim_{x\to 1} g(x) = \frac{1}{2}, g(1) = \frac{1}{2}, g(x)$ is continuous at 1.

c) $\lim_{x\to 1} h(x) = \lim_{x\to 1} \frac{(x + 1)(x - 1)}{x - 1} = 2, h(1)$ is undefined, $h(x)$ is not continuous at 1.

d) $\lim_{x\to 1} [x]$ does not exist, $f(x)$ is not continuous at 1.

e) $\lim_{x\to 1^-} j(x) = 2$, $\lim_{x\to 1^+} j(x) = 2$, $\lim_{x\to 1} j(x) = 2$, $j(1) = 2$, $j(x)$ is continuous at 1.

f) $\lim_{x\to 1^-} j(x) = 1$, $\lim_{x\to 1^+} j(x) = 2$, $\lim_{x\to 1} j(x)$ does not exist, $j(x)$ is not continuous at 1.

5. -5 7. $\dfrac{-27}{4}$ 9. $\dfrac{14}{13}$ 11. 2   13. 3   15. $\dfrac{1}{7a^4 + 4a^2 - 3}, a^2 \neq \dfrac{3}{7}$

17. 5

19. $f(x)$ is not continuous at $x = 0$

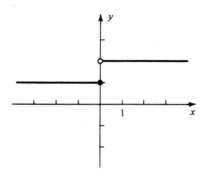

21. $g(x)$ is continuous at $x = 1$

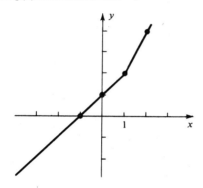

23. $h(x)$ is continuous at $x = 1$

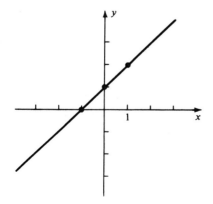

Exercise Set 6.4

1. $\lim_{x \to a} [f(x)]^{25} = (-1)^{25} = -1$ $\lim_{x \to a} [g(x)]^{\frac{1}{2}} = 4^{\frac{1}{2}} = 2$

    $\lim_{x \to a} [f(x)]^{\frac{1}{3}} = (-1)^{\frac{1}{3}} = -1$ $\lim_{x \to a} [g(x)]^{\frac{3}{4}} = 4^{\frac{3}{4}} = (\sqrt[4]{4})^3 = 2\sqrt{2}$

2. $\lim_{x \to 4} f(x) = 20$ $\lim_{x \to 4} [f(x)]^{\frac{1}{2}} = \sqrt{20} = \sqrt{4 \cdot 5} = 2\sqrt{5}$

    $\lim_{x \to -2} f(x) = 8$ $\lim_{x \to -2} [f(x)]^{-\frac{1}{3}} = (8)^{-\frac{1}{3}} = \frac{1}{\sqrt[3]{8}} = \frac{1}{2}$

3. a) $g(u) = e^w, u(x) = 2x, g(u(x)) = e^{2x}$
   b) $g(u) = \ln u, u(x) = x^2 + x - 1, g(u(x)) = \ln (x^2 + x - 1)$
   c) $g(u) = |u|, u(x) = 4 - x^2$ d) $g(u) = u^{-7}, u(x) = x^3 + 1$

4. a) 1   b) 0   c) 5   d) $\dfrac{8^7}{3^7}$

5. −27   7. $\sqrt{\dfrac{5}{6}}$ 9. $\lim_{x \to a} (x^2 + ax + a^2) = 3a^2$ 11. e^2 13. $\ln 2$

15. $\dfrac{\ln 6}{16}$ 17. $2e\sqrt{2}$ 19. 0   21. 0   23. $4x^3$

Exercise Set 6.5

1. a)

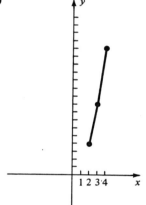

b)

Δx	.5	.3	.1	.01	−.01	−.1	−.3	−.5
$f(x_0 + \Delta x)$	12.25	10.89	9.61	9.0601	8.9401	8.41	7.29	6.25
$\dfrac{f(x_0 + \Delta x) - f(x_0)}{\Delta x}$	6.50	6.3	6.1	6.01	5.99	5.9	5.7	5.5

c) 6   d) $6, \dfrac{y - 9}{x - 3} = 6$ or $y = 6x - 9$

2. a)

t = 2   t = 2.5   t = 3   t = 3.5   t = 4

```
0    4    8    12   16   20   24   28   32   36   40
```

b)

Δt	.5	.3	.1	.01	−.01	−.1	−.3	−.5
$s(t_0 + \Delta t)$	24.5	21.78	19.22	18.1202	17.8802	16.82	14.58	12.5
$\dfrac{s(t_0 + \Delta t) - s(t_0)}{\Delta t}$	13.0	12.6	12.2	12.02	11.98	11.8	11.4	11.0

c) 12   **d)** 12

3. a)

Δt	.5	.3	.1	.01	−.01	−.1	−.3	−.5
$v(t_0 + \Delta t)$	987.75	989.11	990.39	990.0399	991.0599	991.59	992.71	993.75
$\dfrac{v(t_0 + \Delta t) - v(t_0)}{\Delta t}$	−6.5	−6.3	−6.1	−6.01	−5.99	−5.9	−5.7	−5.5

5. a)

t = 2   t = 2.5   t = 3   t = 3.5   t = 4

```
0    2    4    6    8    10   12
```

b)

Δt	.5	.3	.1	.01	−.01	−.1	−.3	−.5
$s(t_0 + \Delta t)$	8.75	7.59	6.51	6.0501	5.9501	5.51	4.59	3.75
$\dfrac{s(t_0 + \Delta t) - s(t_0)}{\Delta t}$	5.5	5.3	5.1	5.01	4.99	4.9	4.7	4.5

c) $\displaystyle \lim_{\Delta t \to 0} \dfrac{s(t_0 + \Delta t) - s(t_0)}{\Delta t} = 5$ **d)** Velocity is 5

7. $\dfrac{3(x + \Delta x) + 1 - 3x + 1}{\Delta x} = \dfrac{3\Delta x}{\Delta x} = 3$ **9.** −6   **11.** 0   **13.** $2x + \Delta x$

15. $8x + 4\Delta x - 1$ **17.** $\dfrac{\sqrt{x + \Delta x} - \sqrt{x}}{\Delta x}$ **19.** $\dfrac{e^{x + \Delta x} - e^x}{\Delta x} = e^x \dfrac{(e^{\Delta x} - 1)}{\Delta x}$

21. $\dfrac{\ln(x + \Delta x) - \ln x}{\Delta x} = \dfrac{1}{\Delta x} \ln \dfrac{x + \Delta x}{x}$

23. $\dfrac{1}{\Delta x} \left(\dfrac{1}{2x + 2\Delta x} - \dfrac{1}{2x} \right) = \dfrac{-1}{2(x^2 + x\Delta x)}$

25. $m = 2x$ thus, m evaluated at the point $(1,2)$ has the value $2(1) = 2$

Exercise Set 6.6

1. a) $\displaystyle \lim_{\Delta x \to 0} \dfrac{(4 + \Delta x - 5) - (4 - 5)}{\Delta x} = \lim_{x \to 0} \dfrac{\Delta x}{\Delta x} = 1$

b) $\displaystyle \lim_{\Delta x \to 0} \dfrac{(x + \Delta x - 5) - (x - 5)}{\Delta x} = \lim_{x \to 0} \dfrac{\Delta x}{\Delta x} = 1$

2. a) 1   **b)** 1   **c)** 1

3. a) 24   **b)** $6x$

4. a) $6z$ **b)** $6x$ **c)** $6u$

5. $\dfrac{y - 48}{x - 4} = 24$ or $y = 24x - 48$

7. $g'(u) = -6$ 9. $\dfrac{dy}{dx} = -x$

11. $\dfrac{dw}{dz} = 14z - 2$ 13. $f'(x) = 9x^2 + 14x$ 15. $\dfrac{du}{dx} = 3x^2 - 1$

17. $y = 4x + 7$ 19. $y = 4$ 21. $y = -3x + 4$ 23. $\dfrac{dy}{dx} = 6x + 1,\ 6x + 1 = 0$ if $x = -\dfrac{1}{6}$

Exercise Set 6.7

1. a) $2x$ b) $12x^{11}$ c) 1 d) $-2x^{-3}$ e) $-3x^{-4}$

2. a) $10x$ b) $\dfrac{12x^{11}}{5}$ c) -1 d) $-2\sqrt{2}\ x^{-3}$ e) $-x^{-4}$

3. a) $10x - 1$ b) $-2\sqrt{2}\ x^{-3} - x^{-4}$
4. a) $14x^{13}$ b) $10x^9$ c) $9x^8$

5. a) $-3x^{-4}$ b) $\dfrac{11}{(2x + 1)^2}$ c) $-5x^{-2} + 2x^{-3}$

7. $3x^2 + 10x - 6$ 9. $6u^5 - 5u^4 + 4u^3 - 3u^2 + 2u - 1$
11. $(z + z^{-1})(-2z^{-3} + 2z) + (1 - z^{-2})(z^{-2} + z^2)$
13. $(-t^2 + 5t - 8)(2t - 5) + (-2t + 5)(t^2 - 5t + 8) = -4t^3 + 30t^2 - 82t + 80$
15. $\dfrac{x}{4}$ 17. $\dfrac{-v^2 - 4v + 2v^2}{v^4} = v^{-2} - 4v^{-3}$ 19. 1 21. $\left(-\dfrac{4}{3}, -\dfrac{133}{3}\right)$

Exercise Set 6.8

1. a) $\dfrac{1}{2}u^{-\frac{1}{2}}$ b) $-\dfrac{1}{2}u^{-\frac{3}{2}}$ c) $\dfrac{3}{2}u^{\frac{1}{2}}$ d) $\dfrac{1}{3}u^{-\frac{2}{3}}$

2. a) $g(u) = u^{\frac{1}{2}},\ u(x) = x^2 - 4$

   b) $g(u) = -8u^{-\frac{1}{2}},\ u(x) = 3x + 4$

   c) $g(u) = u^{\frac{1}{3}},\ u(x) = x^2 - x + 10$

3. a) $g'(u) = \dfrac{1}{2}u^{-\frac{1}{2}},\ u'(x) = 2x$

   $h'(x) = \dfrac{1}{2}(x^2 - 4)^{-\frac{1}{2}}(2x) = x(x^2 - 4)^{-\frac{1}{2}}$

   b) $h'(x) = 12(3x + 4)^{-\frac{3}{2}}$

   c) $h'(x) = \dfrac{1}{3}(2x - 1)(x^2 - x + 10)^{-\frac{2}{3}}$

4. a) $\lim\limits_{\Delta x \to 0} \dfrac{\sqrt{x + \Delta x + 1} - \sqrt{x + 1}}{\Delta x} = \lim\limits_{\Delta x \to 0} \dfrac{\Delta x}{\Delta x(\sqrt{x + \Delta x + 1} + \sqrt{x + 1})} = \dfrac{1}{2\sqrt{x + 2}}$

   b) $\dfrac{d(\sqrt{x + 1})}{dx} = \dfrac{d(x + 1)^{\frac{1}{2}}}{dx} = \dfrac{1}{2}(x + 1)^{-\frac{1}{2}} = \dfrac{1}{2\sqrt{x + 1}}$

5. a) $g'(x) = 3(x^2 - 1)^2(2x) = 6x(x^2 - 1)^2$
   b) $g'(x) = 6x(x^2 - 1)^2$

7. $g'(x) = 320x(5x^2 + 1)^{31}$ 9. $y' = \dfrac{6}{5}z^2(z^3 - 6)^{-\frac{3}{5}}$

11. $s' = -24(3t - 12)^{-9}$ 13. $-2(3x^2 + 2x - 1)(x^3 + x^2 - x)^{-3}$
15. $4x^3(7x + 1)^{-3} - 21(7x + 1)^{-4}(x^4 - 3)$

17. $y' = -\frac{4}{5}(x^2 - 8x + 4)(x + 3)^{-\frac{9}{5}} + (2x - 8)(x + 3)^{-\frac{4}{5}}$

19. $\frac{1}{3}(w - 1)^{-\frac{2}{3}}(2w^2 + 3)^{\frac{1}{2}}(4 - w)^{\frac{2}{9}} + (w - 1)^{\frac{1}{3}}(2w)(2w^2 + 3)^{-\frac{1}{2}}(4 - w)^{\frac{2}{9}}$

$\qquad -\frac{2}{9}(w - 1)^{\frac{1}{3}}(2w^2 + 3)^{\frac{1}{2}}(4 - w)^{-\frac{7}{9}}$

21. $\frac{1}{2}(2x^3 - 11x^2)(x^2 - 11x - 3)^{-\frac{1}{2}} + 2x(x^2 - 11x - 3)^{\frac{1}{2}}$

23. $h'(t) = 5t^2(t^3 - 4)^{\frac{2}{3}}$
$\qquad h'(t) = 0$ if $t = 0$ or $t = \sqrt[3]{4}$

Exercise Set 6.9

1. a) e^u b) $2e^u$ c) $-e^{-u}$ d) $\frac{1}{2}e^{\frac{u}{2}}$

2. a) $u(x) = x^2 + x + 1,$ $g(u) = e^u$
   b) $u(x) = \pi x,$ $g(u) = 2e^u$
   c) $u(x) = -(x + \sqrt{2})$ $g(u) = e^u$
   d) $u(x) = \frac{x^2}{2}$ $g(u) = e^u$

3. a) $(2x + 1)e^{x^2 + x + 1}$ b) $2\pi e^{\pi x}$
   c) $-e^{-(z + \sqrt{2})}$ d) $\frac{2t}{2}e^{\frac{t^2}{2}} = te^{\frac{t^2}{2}}$

4. a) $\frac{dy}{dx} = xe^x + e^x = (x + 1)e^x$ b) $y = xe^{-x}, \frac{dy}{dx} = -xe^{-x} + e^{-x} = e^{-x}(1 - x)$
   c) $\frac{dy}{dx} = \frac{xe^x - e^x}{x^2} = \frac{e^x(x - 1)}{x^2}$ d) $\frac{dy}{dx} = 19(e^x + x)^{18}(e^x + 1)$

5. $3e^{3x}$ 7. e^{x-1} 9. $3t^2e^{t^3}$ 11. $\frac{e^{\sqrt{t}}}{2\sqrt{t}}$

13. $\frac{1}{3}e^{y^{-\frac{2}{3}}}$ 15. $2\sqrt{2}xe^{x^2-3}$ 17. $y' = 2e^x(1 + e^x)$

19. $\frac{dy}{dz} = e^{2z} - e^{-2z}$ 21. $\frac{1}{3}(2x - 5)e^{-x}(x^2 - 5x + 4)^{-\frac{2}{3}} - e^{-x}(x^2 - 5x + 4)^{\frac{1}{3}}$

23. $y' = xe^x + e^x = e^x(x + 1)$
$\qquad m = 2e$ at the point $(1, e)$
$\qquad \frac{y - e}{x - 1} = 2e$ or $y = 2ex - e$

Exercise Set 6.10

1. a) $\frac{1}{u}$ b) $\frac{d}{du}[\ln 2u] = \frac{d}{du}[\ln 2 + \ln u] = \frac{1}{u}$
   c) $\frac{2}{u}$ d) $\frac{2\ln u}{u}$

2. a) $u(x) = x^2, g(u) = \ln u$
   b) $u(x) = \frac{x + 1}{x - 1}, g(u) = \ln u$
   c) $u(x) = x^3, g(u) = 2\ln u$
   d) $u(x) = \frac{1}{x}, g(u) = \ln u$

3. a) $\dfrac{d}{dx}[\ln x^2] = \underbrace{\dfrac{1}{u}}\cdot 2x = \dfrac{1}{x^2}\cdot 2x = \dfrac{2}{x}$

b) $\dfrac{d}{dx}\left[\ln\dfrac{x+1}{x-1}\right] = \underbrace{\dfrac{1}{u}}\cdot\dfrac{-2}{(x-1)^2} = \dfrac{x-1}{x+1}\cdot\dfrac{-2}{(x-1)^2} = \dfrac{-2}{x^2-1}$

c) $\dfrac{d}{dx}[2\ln x^3] = \underbrace{\dfrac{2}{u}}\cdot 3x^2 = \dfrac{2}{x^3}\cdot 3x^2 = \dfrac{6}{x}$

d) $\dfrac{d}{dx}\left[\ln\dfrac{1}{x}\right] = \underbrace{\dfrac{1}{u}}\cdot\left(\dfrac{-1}{x^2}\right) = x\left(-\dfrac{1}{x^2}\right) = -\dfrac{1}{x}$

4. a) $\dfrac{dy}{dx} = x\cdot\dfrac{1}{x}+1\cdot\ln x = 1+\ln x$

b) $\dfrac{dy}{dx} = \dfrac{x\cdot\dfrac{1}{x}-\ln x}{x^2} = \dfrac{1-\ln x}{x^2}$

c) $\dfrac{dy}{dx} = \dfrac{1}{x}+\dfrac{2}{x} = \dfrac{3}{x}$

d) $\dfrac{dy}{dx} = (\ln x)\left(\dfrac{2}{x}\right)+\left(\dfrac{1}{x}\right)(\ln x^2) = \dfrac{2\ln x}{x}+\dfrac{2\ln x}{x} = \dfrac{4\ln x}{x}$

5. $\dfrac{1}{1+x}$ 7. $\dfrac{2x}{1+x^2}$ 9. $\dfrac{2t-6}{t^2-6t+5}$ 11. $\dfrac{-4e^x}{3-4e^x}$

13. $\ln xe^x = \ln x + \ln e^x = \ln x + x, f'(x) = \dfrac{1}{x}+1$

15. $\ln\dfrac{u}{u+3} = \ln u - \ln(u+3), h'(u) = \dfrac{1}{u}-\dfrac{1}{u+3} = \dfrac{3}{u(u+3)}$

17. $\ln\dfrac{1}{\sqrt[3]{x}} = \ln x^{-\frac{1}{3}} = -\dfrac{1}{3}\ln x, F'(x) = \dfrac{-1}{3x}$

19. $y' = x(1+2\ln x)$ 21. $\dfrac{(1+e^{-x})\dfrac{2}{2x+1}+e^{-x}(\ln 2x+1)}{(1+e^{-x})^2}$

23. $y' = \dfrac{\dfrac{2x\ln(x^3+1)}{x^2+1}-\dfrac{3x^2\ln(x^2+1)}{x^3+1}}{[\ln(x^3+1)]^2}$

25. $y' = 1+\ln x, m = 2; \dfrac{y-e}{x-e} = 2, y = 2x-e$

Exercise Set 6.11

1. $f^{(1)}(x) = 6x-7, f^{(2)}(x) = 6, f^{(3)}(x) = 0$

2. $g'(x) = xe^x + e^x, g''(x) = xe^x + 2e^x$

3. $\dfrac{dy}{dx} = \dfrac{5}{2}(5x+12)^{-\frac{1}{2}}, \dfrac{d^2y}{dx^2} = \dfrac{-25}{4}(5x+12)^{-\frac{3}{2}}, \dfrac{d^3y}{dx^3} = \dfrac{375}{8}(5x+12)^{-\frac{5}{2}}$

4. $f(2) = 9; f'(x) = 3x^2+\dfrac{1}{2}, f'(2) = \dfrac{25}{2}; f''(x) = 6x, f''(2) = 12$

5. $g(0) = 1; g'(x) = 2x\,e^{x^2}, g'(0) = 0; g''(x) = 4x^2\,e^{x^2}+2e^{x^2}, g''(0) = 2$

7. $\dfrac{d^2y}{dx^2} = 36x-11$ 9. $\dfrac{d^2y}{dx^2} = 4e^x$ 11. $\dfrac{4}{(x-1)^3}$

13. $\dfrac{d^2y}{dx^2} = e^{-x}$ 15. $\dfrac{d^2y}{dx^2} = e^{-x}(x^2-4x+2)$

17. $f^{(1)}(x) = 28x^3 - 15x^2 + 16x + 1$
    $f^{(2)}(x) = 84x^2 - 30x + 16$
    $f^{(3)}(x) = 168x - 30$

19. $f^{(1)}(x) = 4(3x - 4)^{\frac{1}{3}}$

    $f^{(2)}(x) = 4(3x - 4)^{-\frac{2}{3}}$

    $f^{(3)}(x) = -8(3x - 4)^{-\frac{5}{3}}$

21. $f^{(1)}(0) = 0 \qquad f^{(2)}(0) = 12 \qquad f^{(3)}(0) = -24$

23. $f^{(1)}(-1) = 4(-7)^{\frac{1}{3}} \qquad f^{(2)}(-1) = 4(-7)^{-\frac{2}{3}} \qquad f^{(3)}(-1) = -8(-7)^{-\frac{5}{3}}$

Exercise Set 7.1

1. a) $f'(x) = 4x - 1$, $f'\left(-\frac{1}{2}\right) = -3$, $f'\left(\frac{1}{2}\right) = 1$

   b) $-3, 1$ c) $y = -3x - \frac{7}{2}$, $y = x - \frac{7}{2}$

2.

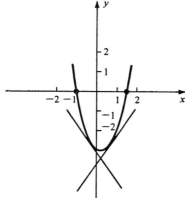

3. Slope is positive on $\left(-\infty, \frac{1}{2}\right)$, negative on $\left(\frac{1}{2}, \infty\right)$, and zero at $x = \frac{1}{2}$.

4.

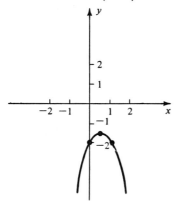

Graph is rising on $\left(-\infty, \frac{1}{2}\right)$

Graph is falling on $\left(\frac{1}{2}, \infty\right)$

5. $-1, 7, 39$ 7. $3, 3, 3$

9. $\dfrac{23}{484}, \dfrac{207}{196}, \dfrac{23}{328,329}$

11. $-2, 0, \dfrac{40}{3\sqrt[3]{99}}$ 13. $0, 1, 2e$

15. Slope negative on $(-\infty, \infty)$

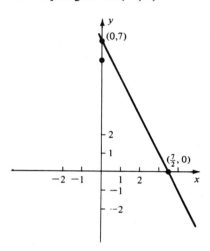

17. Slope positive on $\left(-\dfrac{3}{4}, \infty\right)$

Slope negative on $\left(-\infty, -\dfrac{3}{4}\right)$

Slope zero at $x = -\dfrac{3}{4}$

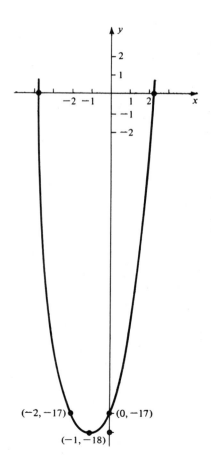

19. Slope positive on $(0, \infty)$
    Slope negative on $(-\infty, 0)$
    Slope zero at $x = 0$

21. Slope positive on $(0, \infty)$
    Slope negative on $(-\infty, 0)$
    Slope never equals 0

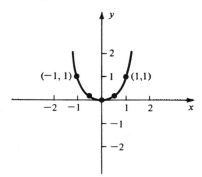

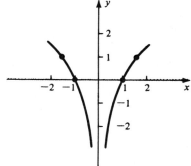

23. $y = \dfrac{3\sqrt{14}}{28}x + \dfrac{13\sqrt{14}}{28}$

25. $y = -6e^2 x - 11e^2$

27. $y = 3x - 2$

29. $y = 1$

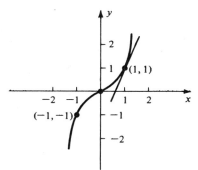

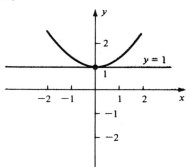

Exercise Set 7.2

1. f is increasing on $(-\infty, \infty)$

2. f is decreasing on $\left(\dfrac{5}{3}, \infty\right)$

3. a) $f'(x) = 4x^3 - 12x^2 - 16x = 4x(x - 4)(x + 1)$

b)	x	$x-4$	$x+1$	$4x(x-4)(x+1)$
$x < -1$	−	−	−	−
$-1 < x < 0$	−	−	+	+
$0 < x < 4$	+	−	+	−
$4 < x$	+	+	+	+

   c) f increasing on $(-1, 0)$ and $(4, \infty)$
   d) f decreasing on $(-\infty, -1)$ and $(0, 4)$

4. $f'(x) < 0$ for all x, so f is decreasing function

5. Increasing on $\left(-\infty, \frac{7}{8}\right)$, decreasing on $\left(\frac{7}{8}, \infty\right)$

7. Increasing on $(-\infty, -1)$, $(1, \infty)$, decreasing on $(-1, 1)$

9. Increasing on $(-\infty, \infty)$

11. Increasing on $(-3 - \sqrt{5}, -3 + \sqrt{5})$, decreasing on $(-\infty, -3 - \sqrt{5})$, $(-3 + \sqrt{5}, \infty)$

13. Increasing on $(-\infty, \infty)$

15. Increasing on $(-\infty, -1)$, $(0, \infty)$, decreasing on $(-1, 0)$

17. Increasing on $(0, \infty)$, decreasing on $(-\infty, 0)$

19. Increasing on $(-\infty, \infty)$

21. Increasing on $\left(-\frac{8}{9}, \infty\right)$, decreasing on $\left(-\frac{5}{2}, -\frac{8}{9}\right)$

23. Decreasing on $(-\infty, 0)$, $(0, \infty)$

Exercise Set 7.3

1. Relative minima at x_2, x_5, and x_7.
   Relative maxima at x_3 and x_6.

2. a) $g'(x) = 0$ for $x = 0, -1, \frac{9}{2}$. Thus, relative extrema might occur at $x = 0, -1, \frac{9}{2}$.

   b) $G'(x) > 0$ for all x, so G has no relative extrema.

3. a) $f'(x) < 0$ on $(-\infty, -2)$, $f'(x) > 0$ on $(-2, \infty)$, and $f'(-2) = 0$. By Theorem 7.6b, f has a relative minimum at $x = -2$. $f(-2) = -16$.

   b) $f'(x) > 0$ on $(-\infty, 2)$, $f'(x) < 0$ on $(2, \infty)$. By Theorem 7.6a, f has a relative maximum at 2. $f(2) = -8$.

4. a) $f'(x) = 12x^2 - 6x - 6 = 6(2x + 1)(x - 1)$.

   b) $Z = \left\{-\frac{1}{2}, 1\right\}$

   c) $f'(x) > 0$ on $\left(-\infty, -\frac{1}{2}\right)$, $f'(x) < 0$ on $\left(-\frac{1}{2}, 1\right)$, and $f'(x) > 0$ on $(1, \infty)$. Therefore, condition a holds at $x_0 = -\frac{1}{2}$ and condition b holds at $x_0 = 1$.

   d) Relative maximum is $f\left(-\frac{1}{2}\right) = \frac{27}{4}$, relative minimum is $f(1) = 0$.

5. Decreasing on $(-\infty, 2)$, increasing on $(2, \infty)$. Relative minimum: $h(2) = -31$.

7. If $a > 0$, f is increasing on $\left(-\frac{b}{2a}, \infty\right)$ and decreasing on $\left(-\infty, -\frac{b}{2a}\right)$. Relative minimum: $f\left(-\frac{b}{2a}\right) = \frac{4ac - b^2}{4a}$. If $a < 0$, f is increasing on $\left(-\infty, -\frac{b}{2a}\right)$ and decreasing on $\left(-\frac{b}{2a}, \infty\right)$. Relative maximum: $f\left(-\frac{b}{2a}\right) = \frac{4ac - b^2}{4a}$.

9. Increasing on $(-\infty, \infty)$. No relative extrema.

11. Increasing on $\left(-\frac{4}{3}, -\frac{50}{63}\right)$, decreasing on $\left(-\frac{50}{63}, \infty\right)$. Relative maximum: $G\left(-\frac{50}{63}\right)$.

13. Decreasing on $\left(-\infty, \frac{5}{4}\right)$, increasing on $\left(\frac{5}{4}, 7\right)$, decreasing on $(7, \infty)$. Relative minimum: $P\left(\frac{5}{4}\right)$; relative maximum: $P(7)$.

15. Decreasing on $(-\infty, -4)$, increasing on $(-4, 1)$, decreasing on $(1, \infty)$. Relative minimum: $G(-4) = -\frac{1}{4}$; relative maximum: $G(1) = 1$.

17. Increasing on $\left(-\infty, -\frac{5}{2}\right)$, decreasing on $\left(-\frac{5}{2}, 3\right)$, increasing on $(3, \infty)$. Relative maximum: $G\left(-\frac{5}{2}\right)$, relative minimum: $G(3)$.

19. Decreasing on $(-\infty, -1)$, increasing on $\left(-1, \frac{29}{21}\right)$, decreasing on $\left(\frac{29}{21}, \frac{7}{3}\right)$, increasing on $\left(\frac{7}{3}, \infty\right)$. Relative minima: $g(-1), g\left(\frac{7}{3}\right)$; relative maximum: $g\left(\frac{29}{21}\right)$.

21. Decreasing on $(-\infty, -2)$, increasing on $(-2, \infty)$. Relative minimum: $h(-2) = 1$.

23. If $A \geqslant 0$, there are no relative extrema; if $A < 0$ there are two relative extrema. It is impossible for f to have exactly one relative extremum.

Exercise Set 7.4

1. a) $f'(x) = 2x + 4$, $f''(x) = 2$, $f'(-2) = 0$, $f''(-2) > 0$.
   By Theorem 2.7, $f(-2)$ is a relative minimum.
   b) $f'(x) = -2x + 4$, $f''(x) = -2$, $f'(2) = 0$, $f''(2) < 0$.
   By Theorem 7.7, $f(2)$ is a relative maximum.

2. a) $f'(x) = 12x^2 - 6x - 6 = 6(2x + 1)(x - 1)$
   b) $f''(x) = 24x - 6$
   c) $Z = \left\{-\frac{1}{2}, 1\right\}$
   d) $f''\left(-\frac{1}{2}\right) = -18$, $f''(1) = 18$
   e) $f\left(-\frac{1}{2}\right)$ is a relative maximum; $f(1)$ is a relative minimum.
   f) $f\left(-\frac{1}{2}\right) = \frac{27}{4}$; $f(1) = 0$

3. a) $g'(x) = 3x^2 - 6x - 9, g''(x) = 6x - 6$
   b) $Z = \{3, -1\}$
   c) $g''(-1) = -12, g''(3) = 12$. Relative maximum at $x = -1$, relative minimum at $x = 3$.
   d) $g(-1) = 7, g(3) = -25$.

4. $E'(x) = 2e^{2x} - 2, E''(x) = 4e^{2x}, E'(0) = 0, E''(0) = 4, E(0) = 1$ is a relative minimum.

5. Relative minimum at $x = 2$.

7. Relative minimum at $x = \frac{5}{4}$, relative maximum at $x = 7$.

9. Relative minimum at $z = -\frac{1}{3}$, relative maximum at $z = \frac{1}{3}$

11. Relative minimum at $x = e^{-1}$

13. Relative maximum at $x = \frac{3}{2}$

15. Relative minimum at $x = 0, f(0) = -9$

17. Relative minima at $x = \frac{-1 - \sqrt{17}}{8}$ and $x = 1$, relative maximum at $x = \frac{-1 + \sqrt{17}}{8}$

19. Relative maximum at $x = -2, h(-2) = 4e^{-2}$, relative minimum at $x = 0, h(0) = 0$

21. Relative minima at $x = 2$ and $x = -2, f(-2) = f(2) = -12\sqrt{6}$, relative maximum at $x = 0$, $f(0) = -16\sqrt{2}$

23. Relative minimum at $x = 0, g(0) = 0$

Exercise Set 7.5

1. Concave downward on (x_1, x_3), (x_5, x_7), and (x_8, x_9); concave upward on (x_3, x_5) and (x_7, x_8). Points of inflection at x_3, x_5, x_7, and x_8.

2. $f''(x) < 0$ for $x < \frac{1}{3}$; $f''(x) > 0$ for $x > \frac{1}{3}$. The graph of f is concave downward on $\left(-\infty, \frac{1}{3}\right)$ and concave upward on $\left(\frac{1}{3}, \infty\right)$.

3. $g''(x) = 0$ for $x = 0, -1, 3$

4. a) $f'(x) = 3x^2 - 4x + 5, f''(x) = 6x - 4, f'''(x) = 6$

   b) Point of inflection at $x = \frac{2}{3}$

5. Concave upward on $(-\infty, \infty)$, no points of inflection

7. Concave downward on $(-\infty, -2)$, concave upward on $(-2, \infty)$; point of inflection at $x = -2$

9. Concave upward on $(0, \infty)$, no points of inflection

11. Concave upward on $(-\infty, \infty)$, no points of inflection

13. Concave upward on $\left(-\infty, \frac{1}{4}\right)$, concave downward on $\left(\frac{1}{4}, \infty\right)$; no points of inflection since the function is undefined at $t = \frac{1}{4}$.

15. Concave downward on $(0, \infty)$, no points of inflection

17. Concave upward on $(-\infty, 0) \cup (2, \infty)$, concave downward on $(0, 2)$; points of inflection at $x = 0, 2$

19. Concave upward on $(-\infty, \ 2 - \sqrt{2}) \cup (2 + \sqrt{2}, \ \infty)$, concave downward on $(2 - \sqrt{2}, \ 2 + \sqrt{2})$; points of inflection at $x = 2 \pm \sqrt{2}$.

Exercise Set 7.6

1. a) $\frac{2}{5}, \frac{1}{5}, \frac{1}{1000}, \frac{1}{4,000,000}$; $\lim\limits_{x \to \infty} \frac{2}{x} = 0$

   b) $-\frac{2}{5}, -\frac{1}{5}, -\frac{1}{1000}, -\frac{1}{4,000,000}$; $\lim\limits_{x \to -\infty} \frac{2}{x} = 0$

2. a) $2, 32, 1024, (1024)^{10}$; $\lim\limits_{x \to \infty} 2^x = \infty$

   b) $\frac{1}{2}, \frac{1}{32}, \frac{1}{1024}, \frac{1}{(1024)^{10}}$; $\lim\limits_{x \to -\infty} 2^x = 0$

3. a) $3, \frac{3}{2}, \frac{101}{99}, \frac{5,000,000}{4,999,998}$; $\lim\limits_{x \to \infty} \frac{x + 1}{x - 1} = 1$

   b) $\frac{1}{3}, \frac{2}{3}, \frac{99}{101}, \frac{4,999,998}{5,000,000}$; $\lim\limits_{x \to -\infty} \frac{x + 1}{x - 1} = 1$

4. a) $5; 21; 201; 50,001$; $\lim\limits_{x \to 1^+} \frac{x + 1}{x - 1} = \infty$

   b) $-3, -19, -1999, -3,333,332$; $\lim\limits_{x \to 1^-} \frac{x + 1}{x - 1} = -\infty$

5. a) 0 b) ∞ c) $\frac{4}{3}$

7. $-\frac{3}{2}, -\frac{3}{2}$ 9. $0, 0$ 11. $-\infty, \infty$ 13. $\frac{\sqrt{2}}{7}, -\frac{\sqrt{2}}{7}$ 15. $\infty, -\infty$

17. $\infty, -\infty$ 19. ∞, ∞ 21. $\infty, -\infty$ 23. $\infty, -\infty$ 25. $1, -\infty$

Exercise Set 7.7

1. a)

x	-1	$-\frac{1}{2}$	$-\frac{1}{4}$	0	$\frac{1}{2}$	1
$f(x)$	$-\frac{9}{4}$	-1	$-\frac{9}{8}$	$-\frac{5}{4}$	0	$\frac{23}{4}$

b) and c)

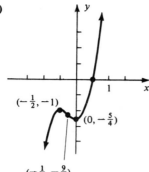

$(-\frac{1}{2},-1)$

$(0,-\frac{5}{4})$

$(-\frac{1}{4},-\frac{9}{8})$

2. a) $f'(x) = 3x^2 - 2x - 8$, $f''(x) = 6x - 2$, $f'''(x) = 6$

b) $Z = \{-1 - \sqrt{3}, -1 + \sqrt{3}, 3\}$

c) f is increasing on $\left(-\infty, -\frac{4}{3}\right)$, $(2, \infty)$; f is decreasing on $\left(-\frac{4}{3}, 2\right)$

d) $f\left(-\frac{4}{3}\right) = \frac{338}{27}$ is a relative maximum, $f(2) = -6$ is a relative minimum

e) Concave downward on $\left(-\infty, \frac{1}{3}\right)$, concave upward on $\left(\frac{1}{3}, \infty\right)$

f) $\left(\frac{1}{3}, \frac{88}{27}\right)$

g) $\lim\limits_{x \to \infty} f(x) = \infty$, $\lim\limits_{x \to -\infty} f(x) = -\infty$

h)

x	$-1 - \sqrt{3}$	$-\frac{4}{3}$	0	$\frac{1}{3}$	$-1 + \sqrt{3}$	2	3	4
$f(x)$	0	$\frac{338}{27}$	6	$\frac{88}{27}$	0	-6	0	22

i)

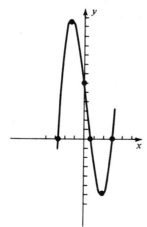

3.

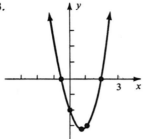

5.

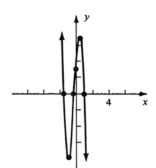

7.

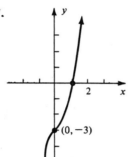

9.

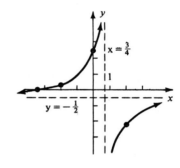

11.

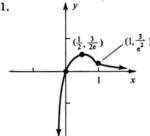

13.

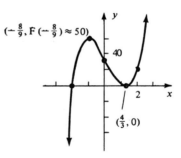

15.

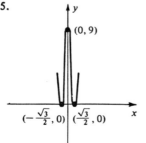

17.

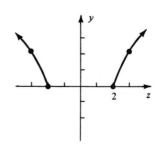

19.

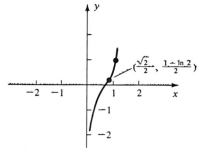

21.

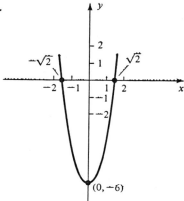

Exercise Set 7.8

1. a)

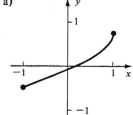

b)

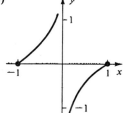

c)

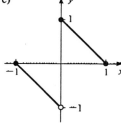

2. a) Yes, yes
   b) Extrema of g over $[-2, 5]$ must occur at $x = -2, -1, 2, 3,$ or 5.

3. a) $Z = \left\{\frac{1}{4}\right\}$ b) $f\left(\frac{1}{4}\right) = \frac{19}{4}$ c) $f(-1) = 11$, $f(1) = 7$ d) $f(-1) = 11$

   e) $f\left(\frac{1}{4}\right) = \frac{19}{4}$ f) There is none g) $f\left(\frac{1}{4}\right) = \frac{19}{4}$

4. a) $Z = \left\{-2, \frac{1}{2}, 1\right\}$ b) $f(-2) = -54$, $f\left(\frac{1}{2}\right) = \frac{11}{16}$, $f(1) = 0$
   c) $f(-3) = 16$, $f(3) = 196$ d) 196 e) -54
   f) There is none g) -54

5. Maximum on $[-1, 1]$ is $19 = f(1)$; minimum on $[-1, 1]$ is $\frac{79}{16} = f\left(-\frac{7}{8}\right)$. No maximum
   on $(-1, 1)$; minimum on $(-1, 1)$ is $\frac{79}{16}$.

7. Maximum on $[0, 3]$ is $10 = f(3)$; minimum on $[0, 3]$ is $1 = f(0)$. No maximum or minimum on $(-\infty, \infty)$.

9. Maximum on $(-\infty, 0)$ is $-32 = f(-2)$; no minimum on $(-\infty, 0)$. No maximum on $(0, \infty)$;
   minimum on $(0, \infty)$ is $32 = f(2)$.

11. Maximum on $[-10, 10]$ is $28 = f(10)$; minimum on $[-10, 10]$ is $-52 = f(-10)$. No maximum or minimum on $(-10, 10)$.

13. Maximum on $(-\infty, -2)$ is $-8 = f(-4)$; no minimum on $(-\infty, -2)$. No maximum on
    $(-2, \infty)$; minimum on $(-2, \infty)$ is $0 = f(0)$.

15. Maximum on $[0, 1]$ is 1024; minimum on $[0, 1]$ is 243. Maximum on $[-4, 1]$ is 1024 minimum on $[-4, 1]$ is -1.
17. No maximum or minimum on $(-1, 1)$ or on $(-\infty, \infty)$.
19. Maximum on $[-10, 10]$ is -1; minimum on $[-10, 10]$ is $10 - e^{10}$. Maximum on $(-\infty, \infty)$ is -1; no minimum on $(-\infty, \infty)$.
21. Maximum on $[-1, 1]$ is $\sqrt{3}$; minimum on $[-1, 1]$ is $-\sqrt{3}$. Maximum on $[-2, 2]$ is 2; minimum on $[-2, 2]$ is -2.

Exercise Set 7.9

1. $x = \dfrac{21 - \sqrt{129}}{6}$ which is approximately 1.6; $V(1.6) = 75.2$ cu. in.
2. $10\sqrt{3} \times 200\sqrt{3}$ 3. $t = 12$; $u(12) = 5000 + e^{-8}$
4. \$9.82 per month
5. $x = \dfrac{122 - \sqrt{5308}}{6}$ which is approximately 12; $V(12) = 10{,}080$ cu. in.
7. $6'' \times 9''$ 9. $x = y = 22$ 11. $5\sqrt{5}$ feet
13. x is approximately 27 feet; y is approximately 30 feet.
15. $r = \sqrt[3]{\dfrac{192}{4\pi}}$, $h = 2\sqrt[3]{\dfrac{192}{4\pi}}$
17. $\dfrac{1}{36\pi}$ feet per second   19. 7

Exercise Set 8.1

1. a) $a_1 + a_2 + a_3 + a_4 + a_5$ b) $a_2 + a_3 + a_4 + a_5$ c) $c_1 + c_2 + c_3 + c_4$
   d) $(a_1 + 2b_1) + (a_2 + 2b_2) + (a_3 + 2b_3) + (a_4 + 2b_4) + (a_5 + 2b_5) + (a_6 + 2b_6)$
2. a) $1 + 2 + 3 + 4 + 5 + 6 + 7 = 28$
   b) $(2 + 3) + (4 + 3) + (6 + 3) + (8 + 3) = 32$
   c) $-1 + 1 - 1 + 1 - 1 + 1 = 0$
   d) $\dfrac{1}{2} + \dfrac{1}{4} + \dfrac{1}{6} = \dfrac{11}{12}$
3. a) $0 + 1 + 2 + 3 = 6$
   b) $\dfrac{1}{2} + \dfrac{1}{3} + \dfrac{1}{4} + \dfrac{1}{5} = \dfrac{77}{60}$
   c) $(0 + 1 + 2 + 3) + \left(\dfrac{1}{2} + \dfrac{1}{3} + \dfrac{1}{4} + \dfrac{1}{5}\right) = 6 + \dfrac{77}{60} = \dfrac{437}{60}$
   d) $\dfrac{1}{2} + \dfrac{4}{3} + \dfrac{9}{4} + \dfrac{16}{5} = \dfrac{437}{60}$
4. a) 325   b) 9455
5. a) 204   b) 408   c) 408
7. $\dfrac{437}{60}$ 9. 58   11. 15   13. 228   15. $\dfrac{63}{32}$ 17. 7140   19. -360
21. $(x_1 - x_0) + (x_2 - x_1) + (x_3 - x_2) + \cdots + (x_{n-1} - x_{n-2}) + (x_n - x_{n-1})$
23. $M_1 \Delta x_1 + M_2 \Delta x_2 + M_3 \Delta x_3 + \cdots + M_{n-1} \Delta x_{n-1} + M_n \Delta x_n$
25. $(x_1 - x_0)f(z_1) + (x_2 - x_1)f(z_2) + (x_3 - x_2)f(z_3) + \cdots + (x_{n-1} - x_{n-2})f(z_{n-1}) + (x_n - x_{n-1})f(z_n)$
27. $\dfrac{1}{n} + \dfrac{1}{n} + \dfrac{1}{n} + \cdots + \dfrac{1}{n} + \dfrac{1}{n}$
29. a) 255   b) 1024   c) $-\dfrac{12}{13}$ d) $c_1 - c_0$

Exercise Set 8.2

1. a) $\dfrac{1}{1^2}, \dfrac{1}{2^2}, \dfrac{1}{3^2}, \dfrac{1}{4^2}, \dfrac{1}{5^2}, \dfrac{1}{6^2}, \dfrac{1}{7^2}$ or $1, \dfrac{1}{4}, \dfrac{1}{9}, \dfrac{1}{16}, \dfrac{1}{25}, \dfrac{1}{36}, \dfrac{1}{49}$

   b) $\dfrac{1-1}{1+1}, \dfrac{2-1}{2+1}, \dfrac{3-1}{3+1}, \dfrac{4-1}{4+1}, \dfrac{5-1}{5+1}, \dfrac{6-1}{6+1}, \dfrac{7-1}{7+1}$ or $0, \dfrac{1}{3}, \dfrac{1}{2}, \dfrac{3}{5}, \dfrac{2}{3}, \dfrac{5}{7}, \dfrac{3}{4}$

   c) $e^1, e^2, e^3, e^4, e^5, e^6, e^7$

2. a) $b_1 = 10, b_{10} = \dfrac{1}{10^2}, b_{100} = \dfrac{1}{10^5}, b_{1000} = \dfrac{1}{10^8}$

   b) $f(x) = \dfrac{10}{x^3}$ c) 0   d) 0

3. a) 0   b) 1   c) ∞

5. $-\infty$ 7. $-\dfrac{1}{2}$ 9. $\dfrac{2}{3}$ 11. $\dfrac{1}{2}$ 13. $\dfrac{2}{3}$ 15. -2

17. $b_1 = \dfrac{1}{2}, b_2 = \dfrac{1}{4}, b_5 = \dfrac{1}{32}, b_{100} = \dfrac{1}{2^{100}}, b_{5000} = \dfrac{1}{2^{5000}}, \displaystyle\lim_{n\to\infty} b_n = 0$

19. $b_1 = 1, b_2 = -\dfrac{1}{4}, b_5 = \dfrac{1}{25}, b_{100} = -\dfrac{1}{10{,}000}, b_{5000} = -\dfrac{1}{25{,}000{,}000}, \displaystyle\lim_{n\to\infty} b_n = 0$

21. $b_1 = -1{,}999{,}995, b_2 = -999{,}995, b_5 = -399{,}995, b_{100} = -19{,}995, b_{5000} = -395,$
    $\displaystyle\lim_{n\to\infty} b_n = 5$

23. $b_1 = \dfrac{2}{3}, b_2 = \dfrac{1}{12}, b_5 = \dfrac{50}{57}, b_{100} = \dfrac{20{,}285}{970{,}482}, b_{5000} = \dfrac{50{,}014{,}985}{124{,}925{,}024{,}982}, \displaystyle\lim_{n\to\infty} b_n = 0$

25. $b_1 = \dfrac{135}{4}, b_2 = \dfrac{43}{8}, b_5 = \dfrac{443}{100}, b_{100} = 75 + \dfrac{1}{20} - \dfrac{17}{10{,}000} + \dfrac{45}{1{,}000{,}000},$

    $b_{5000} = 3750 + \dfrac{1}{1000} - \dfrac{17}{25{,}000{,}000} + \dfrac{45}{125{,}000{,}000{,}000}, \displaystyle\lim_{n\to\infty} b_n = \infty$

Exercise Set 8.3

1. a) $\underline{A_5} = 6$

   b) $\overline{A_5} = 11$

   c)

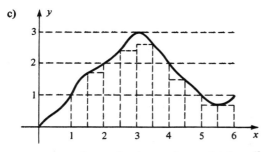

$\underline{A_{10}} = \frac{1}{2}(1 + \frac{5}{3} + 2 + \frac{5}{2} + \frac{8}{3} + 2 + \frac{3}{2} + 1 + \frac{2}{3} + \frac{2}{3}) = \frac{47}{6}$

   d) $\overline{A_{10}} = \dfrac{1}{2}\left(\dfrac{5}{3} + 2 + \dfrac{5}{2} + 3 + 3 + \dfrac{8}{3} + 2 + \dfrac{3}{2} + 1 + 1\right) = \dfrac{61}{6}$

   e) $\underline{A_5} \leqslant \underline{A_{10}} \leqslant \overline{A_{10}} \leqslant \overline{A_5}$ f) estimated area is 9 square units

2.

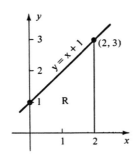

a) $\dfrac{1}{2}, \dfrac{1}{10}, \dfrac{2}{n}$

b) $I_k = \left[\dfrac{k-1}{2}, \dfrac{k}{2}\right]$ for $k = 1,2,3,4$

$I_k = \left[\dfrac{k-1}{10}, \dfrac{k}{10}\right]$ for $k = 1,2,\dots,20$

$I_k = \left[\dfrac{2(k-1)}{n}, \dfrac{2k}{n}\right]$ for $k = 1,2,\dots,n$

c) height $(\underline{r_k}) = \dfrac{k+1}{2}$ for $k = 1,2,3,4$

height $(\underline{r_k}) = \dfrac{k+9}{10}$ for $k = 1,2,\dots,20$

height $(\underline{r_k}) = \dfrac{2k-2+n}{n}$ for $k = 1,2,\dots,n$

d) Area $(\underline{r_k}) = \dfrac{1}{2}\left(\dfrac{k+1}{2}\right)$ for $k = 1,2,3,4$

Area $(\underline{r_k}) = \dfrac{1}{10}\left(\dfrac{k+9}{10}\right)$ for $k = 1,2,\dots,20$

Area $(\underline{r_k}) = \dfrac{2}{n}\left(\dfrac{2k-2+n}{n}\right)$ for $k = 1,2,\dots,n$

e) height $(\overline{r_k}) = \dfrac{k+2}{2}$, Area $(\overline{r_k}) = \dfrac{1}{2}\left(\dfrac{k+2}{2}\right)$ for $k = 1,2,3,4$

height $(\overline{r_k}) = \dfrac{k+10}{10}$, Area $(\overline{r_k}) = \dfrac{1}{10}\left(\dfrac{k+10}{10}\right)$ for $k = 1,2,\dots,20$

height $(\overline{r_k}) = \dfrac{2k+n}{n}$, Area $(\overline{r_k}) = \dfrac{2}{n}\left(\dfrac{2k+n}{n}\right)$ for $k = 1,2,\dots,n$

f) $\underline{A_4} = \dfrac{7}{2}, \underline{A_{20}} = \dfrac{39}{10}, \underline{A_n} = \dfrac{2(2n^2-n)}{n^2}$

g) $\overline{A_4} = \dfrac{9}{2}, \overline{A_{20}} = \dfrac{41}{10}, \overline{A_n} = \dfrac{2(2n^2+n)}{n^2}$

h) $\lim\limits_{n\to\infty} \underline{A_n} = 4$, $\lim\limits_{n\to\infty} \overline{A_n} = 4$ i) Area $(R) = 4$

3.

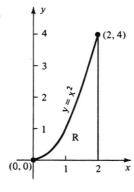

a) $\dfrac{1}{2}, \dfrac{1}{10}, \dfrac{2}{n}$

b) $I_k = \left[\dfrac{k-1}{2}, \dfrac{k}{2}\right]$ for $k = 1,2,3,4$

$I_k = \left[\dfrac{k-1}{10}, \dfrac{k}{10}\right]$ for $k = 1,2,\dots,20$

$I_k = \left[\dfrac{2(k-1)}{n}, \dfrac{2k}{n}\right]$ for $k = 1,2,\dots,n$

c) height $(\underline{r_k}) = \left(\dfrac{k-1}{2}\right)^2$ for $k = 1,2,3,4$

height $(\underline{r_k}) = \left(\dfrac{k-1}{10}\right)^2$ for $k = 1,2,\dots,20$

height $(\underline{r_k}) = \left(\dfrac{2(k-1)}{n}\right)^2$ for $k = 1,2,\dots,n$

d) Area $(\underline{r_k}) = \dfrac{1}{2}\left(\dfrac{k-1}{2}\right)^2$ for $k = 1,2,3,4$

Area $(\underline{r_k}) = \dfrac{1}{10}\left(\dfrac{k-1}{10}\right)^2$ for $k = 1,2,\ldots,20$

Area $(\underline{r_k}) = \dfrac{2}{n}\left(\dfrac{2(k-1)}{n}\right)^2$ for $k = 1,2,\ldots,n$

e) height $(\overline{r_k}) = \left(\dfrac{k}{2}\right)^2$, Area $(\overline{r_k}) = \dfrac{1}{2}\left(\dfrac{k}{2}\right)^2$ for $k = 1,2,3,4$

height $(\overline{r_k}) = \left(\dfrac{k}{10}\right)^2$, Area $(\overline{r_k}) = \dfrac{1}{10}\left(\dfrac{k}{10}\right)^2$ for $k = 1,2,\ldots,20$

height $(\overline{r_k}) = \left(\dfrac{2k}{n}\right)^2$, Area $(\overline{r_k}) = \dfrac{2}{n}\left(\dfrac{2k}{n}\right)^2$ for $k = 1,2,\ldots,n$

f) $\underline{A_4} = \dfrac{7}{4}, \underline{A_{20}} = \dfrac{247}{100}, \underline{A_n} = \dfrac{8}{n^3}\left(\dfrac{n(n+1)(2n+1)}{6} - 2\dfrac{n(n+1)}{6} + n(1)\right)$

g) $\overline{A_4} = \dfrac{15}{4}, \overline{A_{20}} = \dfrac{287}{100}, \overline{A_n} = \dfrac{8}{n^3}\left(\dfrac{n(n+1)(2n+1)}{6}\right)$

h) $\lim\limits_{n\to\infty} \underline{A_n} = \dfrac{8}{3}, \lim\limits_{n\to\infty} \overline{A_n} = \dfrac{8}{3}$ i) Area $(R) = \dfrac{8}{3}$

5. 3   7. 21   9. $\dfrac{11}{3}$ 11. $\dfrac{5}{6}$

Exercise Set 8.4

1. a) $x_0 = 0, x_1 = \dfrac{1}{2}, x_2 = \dfrac{7}{8}, x_3 = \dfrac{4}{3}, x_4 = 2$

   $\Delta x_1 = \dfrac{1}{2}, \Delta x_2 = \dfrac{3}{8}, \Delta x_3 = \dfrac{11}{24}, \Delta x_4 = \dfrac{2}{3}, \|P\| = \dfrac{2}{3}$

   b) $x_0 = 0, x_1 = 0.1, x_2 = 0.4, x_3 = 1.4, x_4 = 1.9, x_5 = 2.0$

   $\Delta x_1 = 0.1, \Delta x_2 = 0.3, \Delta x_3 = 1.0, \Delta x_4 = 0.5, \Delta x_5 = 0.1, \|P\| = 1.0$

   c) for example, $\{0, 0.3, 0.6, 0.9, 1.2, 1.5, 1.8, 2.0\}$

2. a) $x_0 = 1, x_1 = 2, x_2 = 3, x_3 = 4, x_4 = 5, x_5 = 6, \Delta x_k = 1$ for $k = 1,\ldots,5$

3. a) $x_0 = 1, x_1 = \dfrac{5}{4}, x_2 = \dfrac{4}{3}, x_3 = \dfrac{3}{2}, x_4 = 2$

   b) $\Delta x_1 = \dfrac{1}{4}, \Delta x_2 = \dfrac{1}{12}, \Delta x_3 = \dfrac{1}{6}, \Delta x_4 = \dfrac{1}{2}$

   c) $\underline{S} = \dfrac{449}{720}\left(z_1 = \dfrac{5}{4}, z_2 = \dfrac{4}{3}, z_3 = \dfrac{3}{2}, z_4 = 2\right)$

   d) $\overline{S} = \dfrac{31}{40}\left(z_1 = 1, z_2 = \dfrac{5}{4}, z_3 = \dfrac{4}{3}, z_4 = \dfrac{3}{2}\right)$

   e) $S_1 = \dfrac{22912}{33201}\left(z_1 = \dfrac{9}{8}, z_2 = \dfrac{31}{24}, z_3 = \dfrac{17}{12}, z_4 = \dfrac{7}{4}\right)$

   g) $\underline{S} < S_1 < \overline{S}$ (S_2 depends upon the choice in part f)

   h) Area (R) is approximately .693 (actual value is ln 2)

4. b) $\displaystyle\int_1^2 \dfrac{1}{x}\, dx = \ln 2$

5. $\underline{S} = .6456, \overline{S} = .7456$ 7. $\underline{S} = .6687, \overline{S} = .7187$

9. $\overline{S} = 3.07, \overline{S} = 5.06$

11. $\overline{S} = 1.44$ for the choices $z_1 = 0, z_2 = .1, z_3 = .3, z_4 = .8$

13. $S = 3.57$ for the choices $z_1 = 1.5, z_2 = 2, z_3 = 2.5, z_4 = 3, z_5 = 3.5, z_6 = 4$

15. $S = 1.033$ for the choices $z_1 = 0, z_2 = \frac{1}{9}, z_3 = \frac{1}{4}, z_4 = \frac{3}{4}, z_5 = 1, z_6 = \frac{10}{9}$

17. $\int_0^1 e^x dx = 1.718$ 19. $\int_1^4 \ln x \, dx = 2.545$ 21. $\int_0^{3/2} \sqrt{x} \, dx = 1.22$

23. a) $\frac{9}{2}$ b) 2   c) $\frac{27}{2}$ 25. $\int_0^2 (x^2 + x - 4) dx = \frac{2}{3}$

Exercise Set 8.5

1. a) $\frac{8}{3}$ b) -1 c) $\frac{4}{3}$ d) 2   e) $-\frac{7}{3}$ f) $\frac{9}{5}$ g) $\frac{2\sqrt{2}}{3}$ h) $\frac{1}{3}$

2. a) $\frac{23}{4}$ b) 8   c) $\frac{23}{2} \leqslant \int_0^2 (x^2 - 3x + 8) dx \leqslant 16$

3. $\frac{5}{4}$ 5. $\frac{13}{12}$ 7. $e - 5$ 9. $-\frac{15}{8}$ 11. $\frac{34}{3}$ 13. $\frac{4}{5} \leqslant \int_1^5 \frac{1}{x} dx \leqslant 4$

15. $-\frac{21}{4} \leqslant \int_{-2}^{-1} (x^2 + x - 5) dx \leqslant -3$ 17. $-28 \leqslant \int_{-2}^2 (x^3 + 1) dx \leqslant 36$

19. 78   21. 5

Exercise Set 8.6

1. a) $6x - 5$ b) $3x^2 - 5x + 2$ c) $3x^2 - 5x + C$

2. a) $\frac{1}{2\sqrt{x}}$ b) $\sqrt{x}$ c) $\sqrt{x} + C$

3. a) ke^{kx} b) $\int 8e^x dx = 8e^x + C, \int e^{-x} dx = -e^{-x} + C, \int e^{2x} dx = \frac{1}{2}e^{2x} + C,$

$\int -3e^{2x} dx = -\frac{3}{2}e^{2x} + C$

4. a) $\frac{x^4}{4} + C$ b) $\frac{x^5}{5} + C$ c) $\frac{3}{4}x^{\frac{4}{3}} + C$ d) $-x^{-1} + C$ e) $\frac{2}{3}x^{\frac{3}{2}} + C$

   f) $\frac{t^6}{6} + C$ g) $\frac{5}{6}y^{\frac{6}{5}} + C$ h) $-\frac{1}{3}z^{-3} + C$

5. a) 30   b) 1   c) $1 - e^{-1}$ d) $\frac{3}{2}(e^{-2} - e^2)$ e) 0   f) $\frac{5}{14}$ g) $\frac{91}{48}$ h) $\frac{21}{64}$

7. $\frac{x^4}{12} + C$ 9. $\frac{2}{15}x^{\frac{3}{2}} + C$ 11. $-6x^{-1} + C$ 13. $\frac{2}{9}x^3 - \frac{1}{3}x^2 + \frac{3}{2}x + C$

15. $-2x + C$ 17. $\frac{3}{7}\ln x + C$ 19. $\frac{64}{5}$ 21. 2   23. $3(e^4 - 1)$

25. 3   27. $-\frac{259}{12}$ 29. $\frac{1}{2}(e^{-1} - e^{-2})$ 31. $\frac{13}{2}$

33. a)

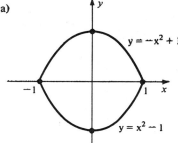

$y = -x^2 + 1$

-1 1 x

$y = x^2 - 1$

b) $\dfrac{8}{3}$

35. $x^3 e^x$ 37. $\dfrac{2x}{x^2 + 1}$ 39. $\dfrac{v^4 - 4}{v^4 + 4}$ 41. $w^4(\ln w - \ln 2) + \dfrac{1}{5w}(w^5 - 243)$

43. $\dfrac{46}{3}$

Exercise Set 8.7

1. a) $du = 3\,dx$ b) $du = 2x\,dx$ c) $du = \dfrac{dx}{x + 1}$

2. a) $\dfrac{1}{6}(3x - 2)^6 + C$ b) $\dfrac{2}{3}(3x - 2)^{\frac{3}{2}} + C$ c) $\ln(3x - 2) + C$

   d) $\dfrac{1}{18}(3x - 2)^6 + C$ e) $\dfrac{2}{9}(3x - 2)^{\frac{3}{2}} + C$ f) $\dfrac{1}{3}\ln(3x - 2) + C$

3. a) $-\dfrac{1}{2}(x^2 + 5)^{-2} + C$ b) $\dfrac{2}{3}(x^2 + 5)^{\frac{3}{2}} + C$ c) $\ln(x^2 + 5) + C$

   d) $-\dfrac{3}{4}(x^2 + 5)^{-2} + C$ e) $-\dfrac{1}{6}(x^2 + 5)^{\frac{3}{2}} + C$ f) $\dfrac{1}{2}\ln(x^2 + 5) + C$

4. a) $\dfrac{1}{2}(\ln(x + 1))^2 + C$ b) $\dfrac{2}{3}(\ln(x + 1))^{\frac{3}{2}} + C$ c) $\ln(\ln(x + 1)) + C$

5. a) $-\dfrac{21}{2}$ b) $-\dfrac{21}{2}$ c) 14 d) $\dfrac{1}{2}\ln 6$

7. $2\sqrt{x^2 + 1} + C$ 9. $\dfrac{3}{40}(x^5 + 1)^8 + C$ 11. $\dfrac{3}{16}(2x^2 + 3)^{\frac{4}{3}} + C$

13. $\dfrac{1}{2}e^{x^2+3} + C$ 15. $\dfrac{1}{2}\ln(2 + e^{2x}) + C$ 17. $2\sqrt{2x^3 + x^2 + 5} + C$

19. $\dfrac{4}{3}(\sqrt{2} - 4)$ 21. $\dfrac{1}{2}$ 23. $2(e^9 - e)$ 25. $2 - \sqrt{3}$ 27. $\dfrac{14\sqrt{7}}{3}$

29. $\dfrac{1}{6}\ln\dfrac{5}{3}$

Exercise Set 8.8

1. a) $du = dx$ b) $v = -e^{-x}$ c) $-xe^{-x}$ d) $e^{-x} + C$ e) $-xe^{-x} - e^{-x} + C$

2. a) $du = \dfrac{1}{x}dx$ b) $v = \dfrac{x^2}{2}$ c) $\dfrac{x^2}{2}\ln 3x$ d) $\dfrac{x^2}{4} + C$ e) $\dfrac{x^2}{2}\ln 3x - \dfrac{x^2}{4} + C$

3. a) $du = dx$ b) $v = \dfrac{2}{3}(x + 4)^{\frac{3}{2}}$ c) $\dfrac{2}{3}x(x + 4)^{\frac{3}{2}}$ d) $\dfrac{4}{15}(x + 4)^{\frac{5}{2}} + C$ e) $\dfrac{506}{15}$

4. a) $du = \dfrac{2\ln t}{t} dt$ b) $v = t$ c) $t\ln^2 t$ d) $2(t\ln t - t) + C$

 e) $t\ln^2 t - 2t\ln t + 2t + C$

5. $\dfrac{1}{3}xe^{3x} - \dfrac{1}{9}e^{3x} + C$ 7. $\dfrac{1}{4}x^4\ln x - \dfrac{1}{16}x^4 + C$ 9. $\dfrac{1}{2}x^2 e^{2x} - \dfrac{1}{2}xe^{2x} + \dfrac{1}{4}e^{2x} + C$

11. $\dfrac{2}{7}x^{\frac{7}{2}}\ln x - \dfrac{4}{49}x^{\frac{7}{2}} + C$ 13. $\dfrac{2}{3}x(x+1)^{\frac{3}{2}} - \dfrac{4}{15}(x+1)^{\frac{5}{2}} + C$

15. $12\ln 8 - \dfrac{135}{16}$ 17. $-\dfrac{1}{18}\ln 3 + \dfrac{2}{9}$ 19. $\dfrac{40}{9}e^{-39} - \dfrac{1}{9}$

Exercise Set 8.9

1. a) $-e^{-x^2} + C$ b) $-e^{-t^2} + 1$ c) 1   d) 1   e) Yes, to the number 1

2. a) $-\dfrac{1}{x} + C$ b) $-\dfrac{1}{t} + 1$ c) 1   d) 1   e) Yes, to the number 1

3. a) $\ln(x^2 + 1) + C$ b) $\ln(t^2 + 1)$ c) ∞ d) ∞ e) No

4. a) $-(e^x + 1)^{-1} + C$ b) $-(e^t + 1)^{-1} + \dfrac{1}{2}$ c) $\dfrac{1}{2}$ d) $-\dfrac{1}{2} + (e^t + 1)^{-1}$

 e) $\dfrac{1}{3}e^{-6}$ 7. $\dfrac{1}{5}$ 9. 2   11. Diverges   13. 1   15. Diverges

17. Diverges   19. $5e^{-1}$ 21. 1

Exercise Set 8.10

1. a)

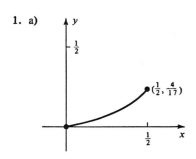

 b) $\displaystyle\int_0^{\frac{1}{2}} \dfrac{2x}{x^2 + 4}\, dx$ c) $\ln\left(\dfrac{17}{16}\right)$

2. a)

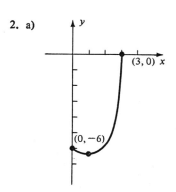

 b) $\displaystyle\int_0^3 -(x^2 - x - 6)\, dx$ c) $\dfrac{27}{2}$

3. a)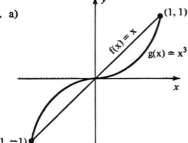

b) $\int_0^1 (x - x^3)\,dx = \dfrac{1}{4}$

c) $\int_{-1}^0 (x^3 - x)\,dx = \dfrac{1}{4}$

d) $\dfrac{1}{2}$

4. a) $(0,0)$ and $(2,-2)$

c) $\int_0^2 ((x - x^2) - (-x))\,dx$

d) $\dfrac{4}{3}$

b)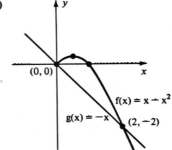

5. $\dfrac{1075}{6}$ feet

7.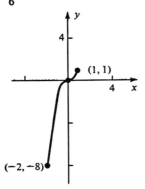

$\dfrac{17}{4}$

9.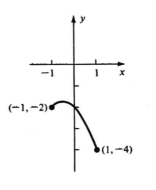

$\dfrac{14}{3}$

11.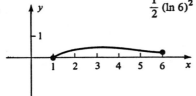

$\dfrac{1}{2} (\ln 6)^2$

13.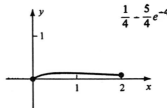

$\dfrac{1}{4} - \dfrac{5}{4} e^{-4}$

15. $(-1, 0)$ and $(1,0), \frac{8}{3}$ 17. $(0,0)$ and $(1,1), \frac{1}{12}$

19. $\left(\frac{3-\sqrt{5}}{2}, \frac{3+\sqrt{5}}{2}\right)$ and $\left(\frac{3+\sqrt{5}}{2}, \frac{3-\sqrt{5}}{2}\right)$, $\ln\left(\frac{7}{2} - \frac{3\sqrt{5}}{2}\right)$

21. $42\frac{2}{3}\%$, $(150)^{\frac{2}{3}}$ seconds   23. $\frac{e+1}{2} + \frac{1}{e}$ 25. $2(\sqrt{30} - 1)$

27. 96 minutes   29. $2\sqrt{10}$

Exercise Set 8.11

1. a) $x_0 = 0, x_1 = \frac{1}{2}, x_2 = 1, x_3 = \frac{3}{2}, x_4 = 2, \frac{b-a}{2n} = \frac{1}{4}$

   b) $f(x_0) = 0, f(x_1) = \frac{1}{4}, f(x_2) = 1, f(x_3) = \frac{9}{4}, f(x_4) = 4$

   c) $\frac{11}{4}$ d) $\frac{8}{3}$

2. a) .69   b) .69   c) $\ln 2 = .6931$

3. a) $x_0 = 0, x_1 = 0.2, x_2 = 0.4, x_3 = 0.6, x_4 = 0.8, \frac{b-a}{2n} = 0.1$

   b) $f(x_0) = 1, f(x_1) = \sqrt{.96}, f(x_2) = \sqrt{.84}, f(x_3) = 0.8, f(x_4) = 0.6$
   c) 0.7
   d) $x_0 = 0, x_1 = 0.1, x_2 = 0.2, x_3 = 0.3, x_4 = 0.4, x_5 = 0.5, x_6 = 0.6,$
   e) $f(x_0) = 1, f(x_1) = \sqrt{.99}, f(x_2) = \sqrt{.96}, f(x_3) = \sqrt{.91}, f(x_4) = \sqrt{.84},$
     $f(x_5) = \sqrt{.75}, f(x_6) = 0.8, f(x_7) = \sqrt{.51}, f(x_8) = 0.6$
   f) .703   g) .603

5. 45, 45   7. 2.04, 2   9. $-1.7, -1.472$ 11. approximately .785
13. approximately 0.140

Exercise Set 8.12

1. a) $y = \frac{2}{3}x^{\frac{3}{2}} + C$ b) $y = \frac{2}{3}x^{\frac{3}{2}} + \frac{1}{3}$

2. a) $y = \frac{K}{2-x}$ b) $y = \frac{2}{2-x}$

3. a) $y' + \left(-\frac{1}{x}\right)y = x^2\sqrt{1-x^2}$ b) $\frac{1}{x}$ c) $\frac{1}{x}y' - \frac{1}{x^2}y = x\sqrt{1-x^2}$

   d) $y = -\frac{1}{3}x(1-x^2)^{\frac{3}{2}} + Cx$ e) $y = -\frac{1}{3}x(1-x^2)^{\frac{3}{2}} - x$

5. $y = \frac{2}{3\ln x + C}$ 7. $y = -1 + Ce^{x^3}$ 9. $y = -\frac{4}{3} + Ce^{4x^3}$

11. $y = -\frac{1}{2} + Ce^{\frac{2}{3}x^3}, y = -\frac{1}{2} + \frac{21}{2}e^{\frac{2}{3}x^3}$

13. $y = -\frac{1}{3x^4} - \frac{1}{4x^5} + \frac{C}{x}, y = -\frac{1}{3x^4} - \frac{1}{4x^5} + \frac{1}{12x}$

15. $\frac{dP}{dt} = kP$ so $P = Ce^{kt}$ where $C = 3200$ and $k = 0.059$. The population in 1958 is $P = 3200\,e^{(0.059)(8)} = 5120$ pheasants.

17. $\frac{dA}{dt} = kA$ so $A = Ce^{kt}$ where $C = 100$ and $k = -0.056$.   (i) 49%   (ii) 41.1 minutes

Exercise Set 9.1

1. a) $2x_1 - 3x_2 + 7x_3 - x_4 = 3$

   b) $\frac{4}{3}x_1 - \frac{1}{3}x_2 + \frac{7}{5}x_3 + \frac{31}{2}x_4 = 0$

   c) $\sqrt{2}x_1 - 64x_2 - \frac{1}{2}x_3 + (\sqrt{2} + 5)x_4 = -\frac{1}{2}$

2. a) $a_{11} = 1, a_{34} = 1, a_{23} = \frac{2}{3}, a_{32} = 0, a_{21} = -\frac{1}{3}$

   b) $b_1 = -4, b_2 = \sqrt{7}, b_3 = \frac{3}{8}$

   c) $t_1 - t_2 + 2t_3 + \sqrt{5}\ t_4 = -4$

   $-\frac{1}{3}t_1 + \frac{1}{4}t_2 + \frac{2}{3}t_3 - t_4 = \sqrt{7}$

   $t_1 \qquad\qquad\qquad + t_4 = \frac{3}{8}$

3. a) No   b) No   c) No   d) No   e) No
4. $x_1 = 16, x_2 = 1, x_3 = 2$
5. $t_1 = -13, t_2 = 2, t_3 = 2$
7. $l_1 = -\frac{65}{2}, l_2 = \frac{1}{2}, l_3 = 7$
9. $y_1 = -\frac{1}{12}, y_2 = \frac{5}{6}, y_3 = \frac{1}{2}$
11. $x_1 = \frac{409}{20}, x_2 = \frac{23}{8}, x_3 = 8$
13. $w = \frac{31}{6}, x = -\frac{11}{30}, y = \frac{2}{3}, z = -\frac{2}{5}$
15. $(x_3 - 2, 5 - x_3, x_3), (-2, 5, 0)$
17. $\left(\frac{4}{3}t_3 + 10, -2 - \frac{1}{3}t_3, t_3\right), (14, -3, 3)$
19. $(1 + 2m_4, -2, m_4, m_4), (3, -2, 1, 1)$
21. $\left(\frac{3}{2}t_3 + 2t_5 + \frac{17}{2}, -\frac{1}{2}t_3 - t_5 - \frac{5}{2}, t_3, t_5, t_5\right), \left(\frac{27}{2}, -\frac{9}{2}, 2, 1, 1\right)$

Exercise Set 9.2

1. a) $6x_1 - 4x_2 + 5x_3 - \frac{6}{17}x_4 = 82$

   $-9x_1 + 6x_2 - \frac{15}{2}x_3 + \frac{9}{17}x_4 = -123$

   $x_1 - \frac{2}{3}x_2 + \frac{5}{6}x_3 - \frac{1}{17}x_4 = \frac{41}{3}$

   $-\frac{15}{2}x_1 + 5x_2 - \frac{25}{4}x_3 + \frac{15}{34}x_4 = -\frac{205}{2}$

   $102x_1 - 68x_2 + 85x_3 - 6x_4 = 1394$

   b) Yes   c) $0 = 0$ d) No

2. a) $L_3 : \frac{23}{2}w + \frac{17}{3}x - y - 5z = -4$

   b) No   c) No   d) Yes

   e) $L_4 : 44x + 45y - \frac{215}{2}z = 19$

   f) Yes

3. a) $x + 2y - z = 9$
       $- 5y + 3z = -21$
    $-3x + 5y + 8z = -4$

  c) $x + 2y - z = 9$
       $y - \dfrac{3}{5}z = \dfrac{21}{5}$
      $11y + 5z = 23$

  e) $x = 1, y = 3, z = -2$

5. $x = 2, y = 1, z = -2$

9. $x = 0, y = \dfrac{7}{4}, z = \dfrac{3}{4}$

13. $u = 4, v = 17, w = -5$

17. $x = 0, y = -2, z = -3$

21. $\left(\dfrac{2}{17} + \dfrac{1}{17}x_3, -\dfrac{11}{17} - \dfrac{20}{17}x_3, x_3\right)$

23. $\left(\dfrac{1}{20}x_4 + \dfrac{48}{5}, \dfrac{14}{5} - \dfrac{3}{5}x_4, \dfrac{1}{4}x_4 - 3, x_4\right)$

25. $(5 - 4x_2 - 8x_3, x_2, x_3, -2)$

27. 300 in apartments, 150 in own home, 200 renting home

29. 68 abrasive discs, 32 polishers, 108 cutting tools

31. 10 shares of Coco, 30 shares of Echo, 20 shares of Whirlwind

b) $x + 2y - z = 9$
   $- 5y + 3z = -21$
    $11y + 5z = 23$

d) $x + 2y - z = 9$
   $y - \dfrac{3}{5}z = \dfrac{21}{5}$
    $\dfrac{58}{5}z = -\dfrac{116}{5}$

f) $x = 1, y = 3, z = -2$

7. $x_1 = \dfrac{1}{2}, x_2 = -1, x_3 = 2$

11. $t_1 = 1 - \dfrac{1}{3}t_3, t_2 = 2t_3 - 3, t_3 \in \mathcal{R}$

15. This system is inconsistent.

19. $w = 1, x = 1, y = 1, z = 1$

Exercise Set 9.3

1. a) $a_{22} = 2, a_{31} = 1, a_{34} = 3, a_{24} = 0$
  b) 3×4
  c) $3x_1 \qquad + x_3 = 5$
    $2x_1 + 2x_2 - 3x_3 = 0$
    $x_1 - 4x_2 + 2x_3 = 3$

2. a) $\begin{bmatrix} 1 & -4 & 2 & 3 & 0 \\ -3 & \frac{1}{2} & 4 & 1 & 6 \\ 2 & 0 & 2 & 9 & 8 \end{bmatrix}$

  b) $\begin{bmatrix} 4 & 2 & 6 \\ 8 & -3 & 0 \\ 2 & 1 & 3 \\ 7 & 6 & 4 \\ 8 & 3 & -4 \end{bmatrix}$

  c) $\begin{bmatrix} 8 & 1 & -3 & -3 \\ -2 & 2 & 0 & 4 \\ -5 & 4 & 4 & 1 \\ 6 & 4 & 3 & 0 \end{bmatrix}$

  d) $[3, 1, 7]$

  e) $\begin{bmatrix} 4 \\ 4 \\ 1 \\ -2 \end{bmatrix}$

3. a) $x = 3, y = 0$

  b) $x = 9, y = 2$

4. $\begin{bmatrix} 3 & 0 & -1 \\ 2 & 4 & 3 \\ 1 & -8 & 0 \end{bmatrix}, \begin{bmatrix} 3 & 0 & -1 & 4 \\ 2 & 4 & 3 & 6 \\ 1 & -8 & 0 & 0 \end{bmatrix}$

5. a) $\begin{bmatrix} 1 & -1 & 3 & 8 \\ 2 & 4 & -1 & -5 \\ -3 & 3 & 3 & 12 \end{bmatrix}$

  b) $\begin{bmatrix} 1 & -1 & 3 & 8 \\ 0 & 6 & -7 & -21 \\ -3 & 3 & 3 & 12 \end{bmatrix}$

c) $\begin{bmatrix} 1 & -1 & 3 & 8 \\ 0 & 6 & -7 & -21 \\ 0 & 0 & 12 & 36 \end{bmatrix}$
d) Yes

7. $\begin{bmatrix} 3 & -6 & 19 \\ -1 & -1 & 35 \\ -2 & 5 & -8 \end{bmatrix}$, $\begin{bmatrix} 3 & -6 & 19 & 144 \\ -1 & -1 & 35 & 2 \\ -2 & 5 & -8 & -39 \end{bmatrix}$

9. $\begin{bmatrix} 3 & -2 & 1 & -5 \\ -\frac{1}{2} & -3 & 1 & 0 \\ 0 & 14 & 0 & -3 \end{bmatrix}$, $\begin{bmatrix} 3 & -2 & 1 & -5 & 0 \\ -\frac{1}{2} & -3 & 1 & 0 & 8 \\ 0 & 14 & 0 & -3 & 1 \end{bmatrix}$

11. $\begin{bmatrix} \frac{1}{2} & \frac{1}{3} & -1 & 7 \\ 0 & -2 & 6 & -48 \end{bmatrix}$

13. $\begin{bmatrix} 1 & -1 & 1 & -1 \\ 0 & 4 & -3 & -8 \\ 0 & 0 & 0 & 0 \end{bmatrix}$

15. $\begin{bmatrix} 1 & 0 & -1 & 1 \\ 0 & 1 & 0 & 1 \\ 0 & 0 & 2 & -3 \\ 0 & 0 & 0 & -6 \end{bmatrix}$

17. $\begin{bmatrix} -1 & 0 & 2 \\ 0 & 1 & -\frac{9}{2} \\ 0 & 0 & \frac{75}{8} \\ 0 & 0 & \frac{11}{6} \end{bmatrix}$

19. $x = -\frac{1}{5}, y = \frac{2}{5}, z = 1$

21. $x_1 = 0, x_2 = 12, x_3 = -8$

23. The system is inconsistent.

25. $x_1 = \sqrt{2}, x_2 = 3, x_3 = -1$.

Exercise Set 9.4

1. $A + B = \begin{bmatrix} -1 & -1 \\ -1 & 10 \end{bmatrix}$, $AB = \begin{bmatrix} -10 & -6 \\ 31 & 24 \end{bmatrix}$

2. $A + B = \begin{bmatrix} -1 & -1 \\ -1 & 10 \end{bmatrix}$, $AB = \begin{bmatrix} -6 & 3 \\ -22 & 20 \end{bmatrix}$

3. $A - B = \begin{bmatrix} 5 & -1 \\ -9 & -2 \end{bmatrix}$, $B - A = \begin{bmatrix} -5 & 1 \\ 9 & 2 \end{bmatrix}$

4. a) 6 b) 3 c) −11 d) 4 e) $c_{22} = -29, c_{23} = 24, c_{31} = 2,$
$c_{32} = -14, c_{33} = 8$

5. a) 2×4 b) 4×3 c) 2×3 d) $\begin{bmatrix} -2 & 10 & -15 \\ 14 & 2 & 31 \end{bmatrix}$

7. $A + B = \begin{bmatrix} \frac{3}{4} & \frac{1}{2} \\ -\frac{3}{2} & -6 \end{bmatrix}$, $A - B = \begin{bmatrix} -\frac{1}{4} & -\frac{7}{6} \\ \frac{9}{2} & 18 \end{bmatrix}$, $B - A = \begin{bmatrix} \frac{1}{4} & \frac{7}{6} \\ -\frac{9}{2} & -18 \end{bmatrix}$

9. $A + B = [-11 \quad 10 \quad 7 \quad 49], A - B = [13 \quad -4 \quad 11 \quad -19],$
$B - A = [-13 \quad 4 \quad -11 \quad 19]$

11. $A + B = \begin{bmatrix} -4 & 5 \\ 3 & 1 \\ 2 & \frac{17}{21} \end{bmatrix}$, $A - B = \begin{bmatrix} 22 & -1 \\ -11 & 0 \\ -2 & -\frac{11}{21} \end{bmatrix}$, $B - A = \begin{bmatrix} -22 & 1 \\ 11 & 0 \\ 2 & \frac{11}{21} \end{bmatrix}$

13.
$$A + B = \begin{bmatrix} 1 & 3 & 5 \\ 4 & 1 & 11 \\ -3 & 10 & -4 \end{bmatrix}, \quad A - B = \begin{bmatrix} 5 & -1 & -1 \\ 4 & -1 & 7 \\ 3 & 0 & -4 \end{bmatrix}, \quad B - A = \begin{bmatrix} -5 & 1 & 1 \\ -4 & 1 & -7 \\ -3 & 0 & 4 \end{bmatrix}$$

15. $1 \times 4, 4 \times 1, 1 \times 1, AB = [-4]$

17. $2 \times 2, 2 \times 2, 2 \times 2, AB = \begin{bmatrix} -1 & -2 \\ -11 & -22 \end{bmatrix}$

19. $2 \times 3, 3 \times 2, 2 \times 2, AB = \begin{bmatrix} 17 & -10 \\ -8 & 33 \end{bmatrix}$

21.
$$A + A = \begin{bmatrix} 6 & 8 \\ -2 & 4 \end{bmatrix}, \quad A + B = \begin{bmatrix} 4 & 4 \\ -1 & 3 \end{bmatrix}$$
$$A - B = \begin{bmatrix} 2 & 4 \\ -1 & 1 \end{bmatrix}, \quad AA = \begin{bmatrix} 5 & 20 \\ -5 & 0 \end{bmatrix}$$
$$AB = \begin{bmatrix} 3 & 4 \\ -1 & 2 \end{bmatrix}, \quad BA = \begin{bmatrix} 3 & 4 \\ -1 & 2 \end{bmatrix}$$

23.
$$A + A = \begin{bmatrix} 0 & 2 \\ -2 & 0 \end{bmatrix}, \quad A + B = \begin{bmatrix} 5 & -3 \\ 2 & 7 \end{bmatrix}$$
$$A - B = \begin{bmatrix} -5 & 5 \\ -4 & -7 \end{bmatrix}, \quad AA = \begin{bmatrix} -1 & 0 \\ 0 & -1 \end{bmatrix}$$
$$AB = \begin{bmatrix} 3 & 7 \\ -5 & 4 \end{bmatrix}, \quad BA = \begin{bmatrix} 4 & 5 \\ -7 & 3 \end{bmatrix}$$

25.
$$A + A = \begin{bmatrix} -6 & 4 \\ 10 & 2 \end{bmatrix}, \quad A + B = \begin{bmatrix} -\dfrac{40}{13} & \dfrac{28}{13} \\ \dfrac{70}{13} & \dfrac{16}{13} \end{bmatrix}$$
$$A - B = \begin{bmatrix} -\dfrac{38}{13} & \dfrac{24}{13} \\ \dfrac{60}{13} & \dfrac{10}{13} \end{bmatrix}, \quad AA = \begin{bmatrix} 19 & -4 \\ -10 & 11 \end{bmatrix}$$
$$AB = \begin{bmatrix} 1 & 0 \\ 0 & 1 \end{bmatrix}, \quad BA = \begin{bmatrix} 1 & 0 \\ 0 & 1 \end{bmatrix}$$

27.
$$A + A = \begin{bmatrix} 12 & -30 & 16 \\ -2 & -2 & 78 \\ 2\sqrt{2} & 2\sqrt{3} & -2\sqrt{5} \end{bmatrix}, \quad A + B = A, \quad A - B = A,$$
$$AA = \begin{bmatrix} 51 + 8\sqrt{2} & -75 + 8\sqrt{3} & -537 - 8\sqrt{5} \\ -5 + 39\sqrt{2} & 16 + 39\sqrt{3} & -47 - 39\sqrt{5} \\ 6\sqrt{2} - \sqrt{3} - \sqrt{10} & -15\sqrt{2} - \sqrt{3} - \sqrt{15} & 8\sqrt{2} + 39\sqrt{3} + 5 \end{bmatrix}$$
$$AB = BA = \begin{bmatrix} 0 & 0 & 0 \\ 0 & 0 & 0 \\ 0 & 0 & 0 \end{bmatrix}$$

29.
$$A + A = \begin{bmatrix} 4 & 2 & -4 \\ 0 & 2 & 2 \\ 6 & 0 & -4 \end{bmatrix}, \quad A + B = \begin{bmatrix} \dfrac{8}{5} & \dfrac{7}{5} & -\dfrac{7}{5} \\ \dfrac{3}{5} & \dfrac{7}{5} & \dfrac{3}{5} \\ \dfrac{12}{5} & \dfrac{3}{5} & -\dfrac{8}{5} \end{bmatrix}$$

$$A - B = \begin{bmatrix} \dfrac{12}{5} & \dfrac{3}{5} & -\dfrac{13}{5} \\ -\dfrac{3}{5} & \dfrac{3}{5} & \dfrac{7}{5} \\ \dfrac{18}{5} & -\dfrac{3}{5} & \dfrac{12}{5} \end{bmatrix}, \quad AA = \begin{bmatrix} -2 & 3 & 1 \\ 3 & 1 & -1 \\ 0 & 3 & -2 \end{bmatrix}$$

$$AB = BA = \begin{bmatrix} 1 & 0 & 0 \\ 0 & 1 & 0 \\ 0 & 0 & 1 \end{bmatrix}.$$

31.
$$A + A = \begin{bmatrix} 4 & 4 & -4 \\ 0 & 6 & -2 \\ 0 & 0 & 2 \end{bmatrix}, \quad A + B = \begin{bmatrix} 6 & -1 & 1 \\ 0 & 6 & 7 \\ 0 & 0 & -8 \end{bmatrix}$$

$$A - B = \begin{bmatrix} -2 & 5 & -5 \\ 0 & 0 & -9 \\ 0 & 0 & 10 \end{bmatrix}, \quad AA = \begin{bmatrix} 4 & 10 & -8 \\ 0 & 9 & -4 \\ 0 & 0 & 1 \end{bmatrix}$$

$$AB = \begin{bmatrix} 8 & 0 & 40 \\ 0 & 9 & 33 \\ 0 & 0 & -9 \end{bmatrix}, \quad BA = \begin{bmatrix} 8 & -1 & -2 \\ 0 & 9 & 5 \\ 0 & 0 & -9 \end{bmatrix}.$$

Exercise Set 9.5

1. a) Use commutativity of addition in $\mathcal{R}$.
   b) Use associativity of addition in $\mathcal{R}$.
   c) $0 = \begin{bmatrix} 0 & 0 \\ 0 & 0 \end{bmatrix}$ d) Just add.   e) $-A = \begin{bmatrix} -a_{11} & -a_{12} \\ -a_{21} & -a_{22} \end{bmatrix}$
   f) Just add.

2. a) If $A = \begin{bmatrix} 1 & 3 \\ -2 & 2 \end{bmatrix}$ and $B = \begin{bmatrix} 2 & 4 \\ 4 & -6 \end{bmatrix}$, then $AB = \begin{bmatrix} 14 & -14 \\ 4 & -20 \end{bmatrix}$ while $BA = \begin{bmatrix} -6 & 14 \\ 16 & 0 \end{bmatrix}$.
   b) Use necessary properties of $\mathcal{R}$.   c) $I = \begin{bmatrix} 1 & 0 \\ 0 & 1 \end{bmatrix}$
   d) Both A and C are nonsingular.

3. a) $\begin{bmatrix} 3 & 4 \\ -1 & 2 \end{bmatrix}$ b) $AX = \begin{bmatrix} 3x_1 + 4x_2 \\ -x_1 + 2x_2 \end{bmatrix}$ and $B = \begin{bmatrix} -20 \\ 5 \end{bmatrix}$. Thus $AX = B$
   if, and only if, $3x_1 + 4x_2 = -20$ and $-x_1 + 2x_2 = 5$.
   c) $\begin{bmatrix} 3 & 4 \\ -1 & 2 \end{bmatrix} \begin{bmatrix} \dfrac{1}{5} & -\dfrac{2}{5} \\ \dfrac{1}{10} & \dfrac{3}{10} \end{bmatrix} = \begin{bmatrix} 1 & 0 \\ 0 & 1 \end{bmatrix}$ d) $A^{-1} B = \begin{bmatrix} -6 \\ \dfrac{1}{2} \end{bmatrix}$
   e) $x_1 = -6, x_2 = -\dfrac{1}{2}$ f) $x_1 = -6, x_2 = -\dfrac{1}{2}$

In problems 5–12, simply multiply A by C and verify that $AC = I$.

13. $\begin{bmatrix} x \\ y \end{bmatrix} = \begin{bmatrix} -\dfrac{3}{2} \\ -5 \end{bmatrix}$ 15. $\begin{bmatrix} x_1 \\ x_2 \end{bmatrix} = \begin{bmatrix} \dfrac{24}{7} \\ \dfrac{1}{14} \end{bmatrix}$ 17. $\begin{bmatrix} t_1 \\ t_2 \end{bmatrix} = \begin{bmatrix} \dfrac{5}{2} \\ \dfrac{5}{2} \end{bmatrix}$

19. $\begin{bmatrix} x_1 \\ x_2 \\ x_3 \end{bmatrix} = \begin{bmatrix} 4 \\ 17 \\ -18 \end{bmatrix}$ 21. $\begin{bmatrix} x \\ y \\ z \end{bmatrix} = \begin{bmatrix} \frac{14}{13} \\ \frac{14}{13} \\ -\frac{10}{13} \end{bmatrix}$

*23. If $AB = I$ and $AC = I$, then $AB = AC$, so $A^{-1}(AB) = A^{-1}(AC)$. Thus, $(AA^{-1})B = (AA^{-1})C$, or $IB = IC$, so that $B = C$.

Exercise Set 9.6

1. a) 14   b) 0   c) 16   d) $4 - \frac{1}{2}\sqrt{2}$

2. a) $|A| = -54, |B| = 12$

   b) $AB = \begin{bmatrix} 11 & 25 \\ -15 & -93 \end{bmatrix}$, $|AB| = -648$

   c) $|AB| = |A||B|$

3. a) $\begin{bmatrix} 8 & 0 \\ -2 & -9 \end{bmatrix}$ b) $a_{13} = 5$ c) $A_{13} = -72$

   d) $(-1)^{1+3} A_{13} = -72$ e) $(-1)^{1+1} A_{11} = 18, (-1)^{1+2} A_{12} = -12$

4. a) 54   b) 12   c) -360 d) -294

5. a) $(-1)^{1+2} A_{12} = -12, (-1)^{2+2} A_{22} = 13, (-1)^{3+2} A_{32} = 34$

   b) $|A| = (-1)(-12) + 0(13) - 9(34) = -294$

   c) $|A| = -294$

7. $|A| = -80, |B| = 1, AB = \begin{bmatrix} -6 & 14 \\ 7 & -3 \end{bmatrix}$, $|AB| = |A||B| = -80$

9. $|A| = 0, |B| = 183, AB = \begin{bmatrix} -18 & 13 \\ -18 & 13 \end{bmatrix}$, $|AB| = |A||B| = 0$

11. $\frac{113}{8}$ 13. -6 15. 173   17. -30 19. 0   21. -15

23. 16   25. $\begin{vmatrix} 1 & -3 \\ 2 & 4 \end{vmatrix} = 10$ and $\begin{vmatrix} 2 & 4 \\ 1 & -3 \end{vmatrix} = -10$.   27. They are equal.

Exercise Set 9.7

1. a) The rows were interchanged.

   b) Row 1 was multiplied by -1.

   c) The second row was replaced by the sum of itself and the first row.

   d) The columns were interchanged.

   e) The first column was replaced by the sum of itself and the second column.

   f) A factor of -2 was removed from each term of the second row.

2. a) Row 1 is -3 times row 3.

   b) Column 2 is -2 times column 3.

   c) Column 1 is $-\frac{3}{2}$ times column 2.

3. a) -3 b) -3 c) $\frac{1}{6}$

4. $|A| = 31 \begin{vmatrix} -20 & 63 \\ 0 & 107 \end{vmatrix} = 31(-20)(107) = -66,340.$

5. a) $|A| = 16 \begin{vmatrix} 1 & 0 & 3 & 9 \\ 0 & 1 & 1 & 5 \\ 0 & 0 & 2 & 5 \\ 0 & 0 & 0 & -11 \end{vmatrix}$ b) $|A| = 16(-22) = -352$

7. –2   9. –12   11. 23   13. –2   15. 45   17. –76

19. –121   21. –18   23. 108

25. $\begin{vmatrix} 1 & x & y \\ 1 & x_1 & y_1 \\ 1 & x_2 & y_2 \end{vmatrix} = x_1y_2 - x_2y_1 - x(y_2 - y_1) + y(x_2 - x_1).$

Thus $\begin{vmatrix} 1 & x & y \\ 1 & x_1 & y_1 \\ 1 & x_2 & y_2 \end{vmatrix} = 0$ if, and only if, $\dfrac{y - y_1}{x - x_1} = \dfrac{y_2 - y_1}{x_2 - x_1}.$

Exercise Set 9.8

1. a) $\begin{bmatrix} 2 & 3 & 1 \\ 4 & 0 & -1 \\ -1 & 6 & -4 \end{bmatrix}$

b) $\begin{bmatrix} \frac{5}{3} & \frac{21}{6} & \frac{6}{7} \\ \frac{2}{5} & -7 & \frac{5}{3} \\ -\frac{1}{8} & \frac{3}{4} & \frac{2}{7} \end{bmatrix}$

c) $\begin{bmatrix} a & d & g \\ b & e & h \\ c & f & i \end{bmatrix}$

2. a) $C_{11} = -10,\ C_{12} = -1,\ C_{13} = -6$
$C_{21} = \ \ 4,\ C_{22} = -7,\ C_{23} = -5$
$C_{31} = -13,\ C_{32} = -5,\ C_{33} = \ \ 7$

b) $\begin{bmatrix} -10 & 4 & -13 \\ -1 & -7 & -5 \\ -6 & -5 & 7 \end{bmatrix}$

3. a) $|A| = -37$ b) $A^{-1} = \begin{bmatrix} \frac{10}{37} & -\frac{4}{37} & \frac{13}{37} \\ \frac{1}{37} & \frac{7}{37} & \frac{5}{37} \\ \frac{6}{37} & \frac{5}{37} & \frac{7}{37} \end{bmatrix}$

4. $\begin{bmatrix} x \\ y \\ z \end{bmatrix} = \begin{bmatrix} \frac{10}{37} & -\frac{4}{37} & \frac{13}{37} \\ \frac{1}{37} & \frac{7}{37} & \frac{5}{37} \\ \frac{6}{37} & \frac{5}{37} & -\frac{7}{37} \end{bmatrix} \begin{bmatrix} -2 \\ 7 \\ -2 \end{bmatrix} = \begin{bmatrix} -2 \\ 1 \\ 1 \end{bmatrix}$

5. $x = -2,\ y = 1,\ z = 1.$

7. $\begin{bmatrix} -\frac{5}{2} & 6 \\ -3 & 7 \end{bmatrix}$

9. $\begin{bmatrix} -\frac{2}{5} & \frac{1}{5} & \frac{2}{5} \\ 1 & 0 & -\frac{1}{2} \\ \frac{6}{5} & \frac{2}{5} & -\frac{7}{10} \end{bmatrix}$

11. $\begin{bmatrix} \frac{44}{107} & -\frac{3}{214} & \frac{33}{214} & -\frac{27}{107} \\ \frac{17}{107} & \frac{11}{214} & -\frac{7}{107} & \frac{8}{107} \\ \frac{-20}{107} & \frac{25}{107} & \frac{15}{214} & \frac{22}{107} \\ \frac{4}{107} & -\frac{5}{107} & \frac{3}{214} & \frac{17}{107} \end{bmatrix}$

13. $\begin{bmatrix} r \\ s \end{bmatrix} = \begin{bmatrix} 16 \\ 18 \end{bmatrix}$

15. $\begin{bmatrix} x_1 \\ x_2 \\ x_3 \end{bmatrix} = \begin{bmatrix} \dfrac{36}{5} \\ -9 \\ -\dfrac{43}{5} \end{bmatrix}$ 17. $\begin{bmatrix} y_1 \\ y_2 \\ y_3 \\ y_4 \end{bmatrix} = \begin{bmatrix} \dfrac{312}{107} \\ \dfrac{140}{107} \\ \dfrac{278}{107} \\ \dfrac{458}{107} \end{bmatrix}$

INDEX